# Biogeochemistry of Amino Acids

# BIOGEOCHEMISTRY OF AMINO ACIDS

---

*Papers presented at a conference at Airlie House, Warrenton, Virginia, October 29 to November 1, 1978*

Edited by

P. E. HARE

T. C. HOERING

K. KING, JR.

**JOHN WILEY & SONS**

**New York • Chichester • Brisbane • Toronto**

***Library of Congress Cataloging in Publication Data:***

International Conference on the Biogeochemistry of Amino Acids, Airlie House, 1978.
Biogeochemistry of amino acids

Sponsored by the Carnegie Institution of Washington and the National Science Foundation.
Includes indexes.
1. Amino acids—Congresses. 2. Biogeochemistry—Congresses. I. Hare, Peter Edgar, 1933– II. Hoering, C. III. King, Kenneth. IV. Carnegie Institution of Washington. V. United States. National Science Foundation. VI. Title.

QH344.I56 1978 551.9 79-25824
ISBN 0-471-05493-3

Printed in the United States of America

10 9 8 7 6 5 4 3 2 1

Carnegie Institution of Washington Conference

# ADVANCES IN BIOGEOCHEMISTRY OF AMINO ACIDS

## List of Participants

Abelson, Philip H.
*Carnegie Institution of Washington*

Akiyama, Masahiko
*Hokkaido University, Japan*

Bada, Jeffrey L.
*Scripps Institution of Oceanography*

Belknap, Daniel F.
*University of Delaware*

Berger, R.
*University of California*

Blake, Weston, Jr.
*Geological Survey of Canada*

Blunt, David J.
*U.S. Geological Survey*

Bonner, William
*Stanford University, California*

Broecker, Wallace S.
*Lamont-Doherty Geological Observatory*

Carey, Julia
*Virginia Polytechnical Institute and State University*

Crawford, R. J.
*University of Alberta*

Cronin, John R.
*Arizona State University*

Cunningham, Robert
*University of Texas, Dallas*

Demarais, David J.
*NASA Ames Research Center*

DeNiro, Michael J.
*California Institute of Technology*

Dungworth, Graham
*BP Research Centre*

Ebert, James D.
*Carnegie Institution of Washington*

Eglinton, Geoffrey
*University of Bristol, England*

Engel, Michael H.
*University of Arizona*

Estep, Marilyn
*Geophysical Laboratory, Carnegie Institution of Washington*

Eyre, David R.
*Children's Hospital Medical Center, Harvard University*

Firsching, F. H.
*Southern Illinois University*

Gerow, Bert A.
*Stanford University, California*

Gil-Av, E.
*Weizmann Institute of Science Israel*

Halstead, L. B.
*University of Reading, England*

Harada, Kaoru
*University of Tsukuba, Japan*

Hare, P. Edgar
*Geophysical Laboratory, Carnegie Institution of Washington*

Hassan, Afifa
*U.S. Geological Survey*

Hauschka, Peter V.
*Children's Hospital Medical Center, Harvard University*

Hedges, John I.
*University of Washington*

Henrichs, Susan
*Wood's Hole Oceanographic Institution*

Hoering, Thomas C.
*Geophysical Laboratory, Carnegie Institution of Washington*

Hood, Leroy
*California Institute of Technology*

Hoopes, E. A.
*Scripps Institution of Oceanography*

Jope, E. M. and
Jope, Margaret
*Queen's University, No. Ireland*

Katz, Barry J.
*University of Miami*

Keenan, Everly
*University of Delaware*

Kennedy, George
*U.S. Geological Survey*

King, Kenneth, Jr.
*Lamont-Doherty Geological Observatory*

Kriausakul, Nivat
*University of Texas, Dallas*

Kvenvolden, Keith A.
*U.S. Geological Survey*

Lajoie, Kenneth R.
*U.S. Geological Survey*

Lee, Cindy
*Wood's Hole Oceanographic Institution*

Liddicoat, Joseph
*Lamont-Doherty Geological Observatory*

Lowenstam, Heinz A.
*California Institute of Technology*

Lowenstein, Jerold M.
*University of California*

Lugenbeal, Edward N.
*Andrews University*

Man, Eugene H.
*University of Miami*

Masters, Patricia M.
*Scripps Institute of Oceanography*

Michaelis, Walter
*University of Hamburg, Federal Republic of Germany*

Miller, Gifford H.
*University of Colorado*

Mitterer, Richard M.
*University of Texas*

Nagy, Bartholomew
*University of Arizona*

Nelson, Alan R.
*University of Colorado*

Peterson, Etta
*NASA Ames Research Center*

Piez, Karl A.
*National Institute of Dental Research*

Rapp, John B.
*U.S. Geological Survey*

Ritland, Richard
*Andrews University*

Rutter, N. W.
*University of Alberta*

Schroeder, Roy
*U.S. Geological Survey*

Sellers, George A.
*U.S. Geological Survey*

Shou, M.-Y.
*Scripps Institution of Oceanography*

Shimoyama, Akira
*University of Maryland*

Smith, Grant Gill
*Utah State University*

Taylor, R. E.
*University of California*

Towe, Kenneth M.
*Smithsonian Institution*

Tuross, Noreen
*University of Idaho*

Von Endt, David W.
*Smithsonian Institution*

Wallace, C. T. A.
*Oxford University, England*

Wehmiller, John F.
*University of Delaware*

Weiner, Steven
*Weizmann Institute of Science, Israel*

Wyckoff, Ralph W. G.
*University of Arizona*

Yoder, H. S., Jr.
*Geophysical Laboratory, Carnegie Institution of Washington*

Zumberge, John E.
*University of Arizona*

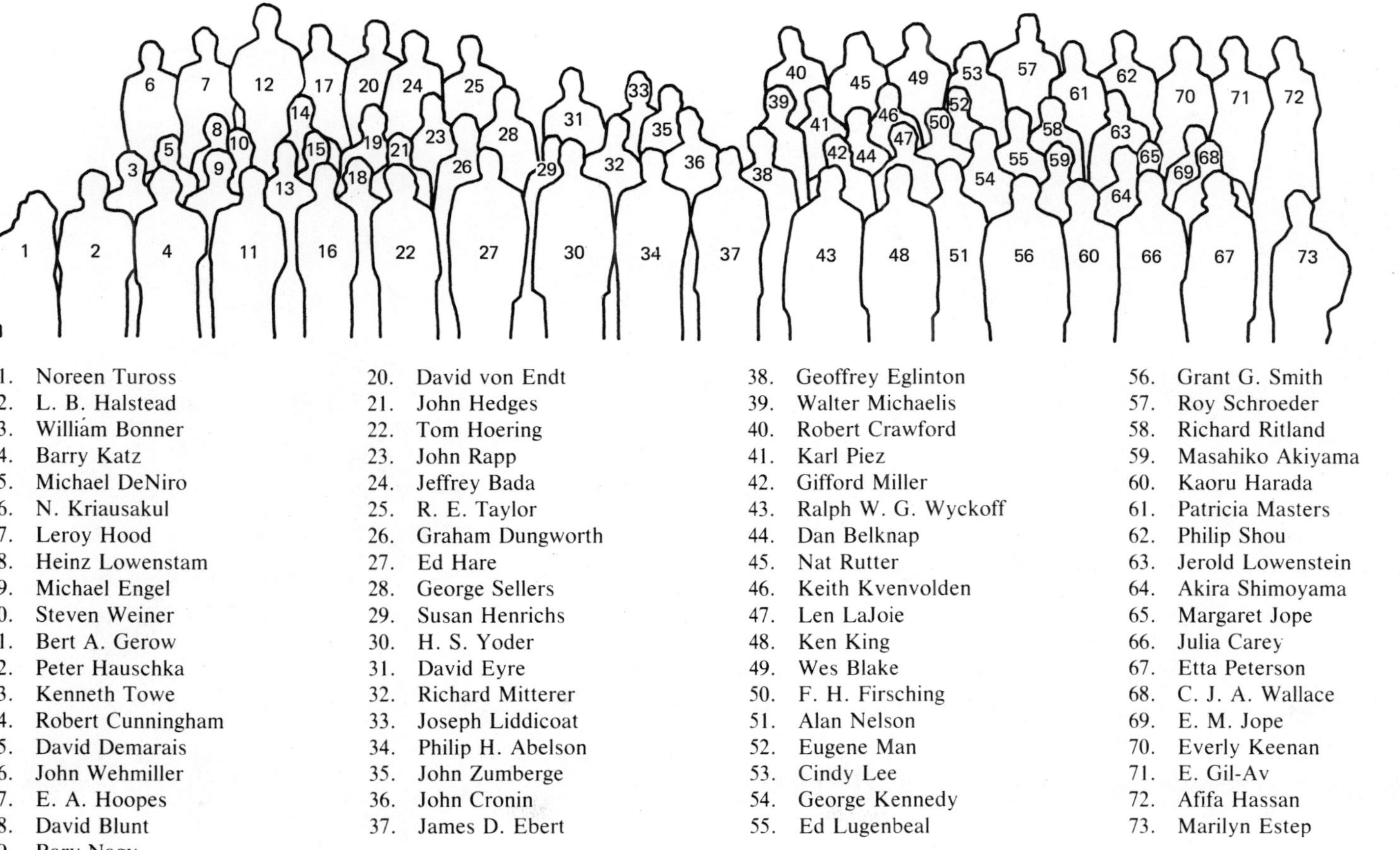

1. Noreen Tuross
2. L. B. Halstead
3. William Bonner
4. Barry Katz
5. Michael DeNiro
6. N. Kriausakul
7. Leroy Hood
8. Heinz Lowenstam
9. Michael Engel
10. Steven Weiner
11. Bert A. Gerow
12. Peter Hauschka
13. Kenneth Towe
14. Robert Cunningham
15. David Demarais
16. John Wehmiller
17. E. A. Hoopes
18. David Blunt
19. Bary Nagy
20. David von Endt
21. John Hedges
22. Tom Hoering
23. John Rapp
24. Jeffrey Bada
25. R. E. Taylor
26. Graham Dungworth
27. Ed Hare
28. George Sellers
29. Susan Henrichs
30. H. S. Yoder
31. David Eyre
32. Richard Mitterer
33. Joseph Liddicoat
34. Philip H. Abelson
35. John Zumberge
36. John Cronin
37. James D. Ebert
38. Geoffrey Eglinton
39. Walter Michaelis
40. Robert Crawford
41. Karl Piez
42. Gifford Miller
43. Ralph W. G. Wyckoff
44. Dan Belknap
45. Nat Rutter
46. Keith Kvenvolden
47. Len LaJoie
48. Ken King
49. Wes Blake
50. F. H. Firsching
51. Alan Nelson
52. Eugene Man
53. Cindy Lee
54. George Kennedy
55. Ed Lugenbeal
56. Grant G. Smith
57. Roy Schroeder
58. Richard Ritland
59. Masahiko Akiyama
60. Kaoru Harada
61. Patricia Masters
62. Philip Shou
63. Jerold Lowenstein
64. Akira Shimoyama
65. Margaret Jope
66. Julia Carey
67. Etta Peterson
68. C. J. A. Wallace
69. E. M. Jope
70. Everly Keenan
71. E. Gil-Av
72. Afifa Hassan
73. Marilyn Estep

# PREFACE

In the twenty-five years since amino acids were discovered in fossils, there has been steadily increasing growth in research on the distribution and transformation of amino acids and proteins in the geological environment. So many scientific disciplines are involved and the results are published in such a variety of journals that communication has been difficult.

To facilitate the exchange of ideas and information, a three-day conference on Advances in the Biogeochemistry of Amino Acids was held October 29 to November 1, 1978, at Arlie House, Warrenton, Virginia, under the sponsorship of the Carnegie Institution of Washington and the National Science Foundation. For the first time, most of the active participants in this area of research gathered together at one place to discuss concepts, techniques, and results. This book is a result of the conference and gives an overview of the wide range of studies that have been carried out. It is likely that few of the participants were totally aware of the quantity and quality of the research.

The papers presented here are representative of the major areas in which amino acid biogeochemistry is being developed. Some problems have come to a focus, but many are in a preliminary stage and are controversial and unresolved. As a result, the papers are not necessarily in agreement and in some cases present conflicting viewpoints. These differences and the lively debate they stimulated at the conference attest to the vitality of the field. Most important, the papers point to new areas for future research.

Amino-acid biogeochemistry is an interdisciplinary science. Methods of biochemistry, combined with principles from physical and organic chemistry, are applied to problems in geology, archeology, paleontology, geochemistry, and other fields. Because no one scientist can be expert in all of these, an interchange of ideas is essential. This goal of the conference was achieved and there was communication across the traditional boundaries of science. Organic chemists discovered mutual areas of interest with stratigraphers, and enzymologists compared notes with archeologists.

The organization of this volume parallels the conference program. It is divided into four major areas. The first presents data on the occurrence of proteins and amino acids in geological environments, discusses the relationships between ultrastructural integrity of fossils and the degree of organic preservation, and attacks the problem of organic contamination. The second deals with the all-important topic of methods and techniques. The papers in this section show how a variety of available analytical tools can be applied to the detection and characterization of organic molecules at low concentration levels. The third contains papers describing how theoretical and experimental methods of physical and organic chemistry can be used as a guide for studies of naturally occurring materials and for the interpretation of their results. The fourth, and probably most significant, section presents a number of applications of amino acid biogeochemistry. The most heated arguments and most vigorous dissents of the conference happened during the oral presentation of this group of papers. Obviously this area is in a state of rapid development.

The conference took place during the year of the twenty-fifth anniversary of Philip H. Abelson's discovery of amino acids in fossils. It also marked the occasion of his retirement as President of the Carnegie Institution of Washington. Everyone doing research on the biogeochemistry of amino acids acknowledges and appreciates Dr. Abelson's pioneering research and vision. The vigor and excitement of this area of science reflects his efforts.

The organizers of the conference and editors of this volume thank Sheila McGough and A. David Singer of the Carnegie Institution of Washington for their assistance in the operation of the meeting and the preparation of its proceedings.

P. EDGAR HARE
THOMAS C. HOERING
KENNETH KING, JR.

*Washington D.C.*
*March, 1980*

# Introductory Remarks

H. S. Yoder, Jr.
*Geophysical Laboratory, Carnegie Institution of Washington*

It is a pleasure to extend a warm welcome to the participants of the First International Conference on the Biogeochemistry of Amino Acids. The Carnegie Institution has fostered many innovative and imaginative programs, and the field of biogeochemistry has proven to be one of the most successful. Many of you have no doubt wondered why a geophysical laboratory would support such studies. The initiative was taken by Dr. Philip H. Abelson some 25 years ago, and an incredible number of achievements have been recorded by a small group of people focusing on biogeochemically related topics. You have all received a copy of their collective bibliography. Needless to say, the Carnegie Institution of Washington supports excellence in research across field boundaries or wherever a bright idea emerges—and will continue to do so irrespective of the site of origin. The relationship of biogeochemistry is closer to other projects, or areas of focus, at the Geophysical Laboratory than one might first imagine. I believe the organic geochemistry group has contributed significantly to many areas of petrology, the principal field of research at the Geophysical Laboratory.

I would like to suggest three areas where organic geochemistry might make substantial contributions in the future. Oil formation is believed by some investigators to be at a maximum in the temperature range 66°–150°C at pressures in the order of 1–2 kilobars. You may disagree with these ranges of temperature and pressure, but they are the ranges in which the great lead and zinc ore deposits of the world also form. Organic material is present in those ore deposits, and one would like to know if the kerogen or bitumens, or both, played an active or passive role in the ore formation. Organic geochemistry will probably play a significant part in the ore deposits program now evolving at the Geophysical Laboratory.

The groundwork laid for the stable isotopes of hydrogen, carbon, oxygen, and sulfur in organic compounds is equally applicable to the hydrous, carbonaceous, and sulfide minerals of the igneous and metamorphic rocks. These isotopes help in defining the source of volatile constituents in the earth, and I believe they will eventually provide geobarometric and geothermometric calibrations of the conditions of origin of their host rocks.

A third opportunity that lies primarily in organic geochemistry, biomineralization, concerns practically all the disciplines that you represent here today. Consider a 0.1 micron thick organic substrate that produces 1 micron of perfect, optically oriented single crystals of calcite, aragonite, or even strontianite, barite, or hematite in a shell. We need to understand the processes at the organic membrane. How does the structure of the organic substrate influence the crystallography, structure, and composition of the inorganic crystals and vice versa? What is the memory device whereby the organic and inorganic layers alternate repetitively? The trace elements concentrated in those crystals are very specific, and we need to understand the biologically controlled partitioning mechanisms that were operative even in the Precambrian. Do they follow Henry's Law or is the process more complex? The applications of such knowledge to problems in bone, teeth, gall and kidney stone growth, not to neglect sea bottom ore deposits, are obvious to you all.

These are only a few of the many exciting opportunities in organic geochemistry. Each of you probably has his or her own list of future projects, but I hope sufficient numbers of you will be attracted to those problems relating organic compounds to ore deposits so that we can meet here again in November 1980, to discuss the results.

# CONTENTS

# I

# DISTRIBUTION AND CHARACTERIZATION OF PROTEINS, PEPTIDES, AND AMINO ACIDS

---

Proteins are major constituents of all living organisms, and their molecular remains are present in fossils and sediments. Amino acids have also been detected in meteorites. These organic molecules are unstable and are not in thermodynamic equilibrium with their environment, and they are subject to bacterial metabolism. Thus, some fundamental questions arise. Why are there any amino acids or peptides remaining after geological lengths of time and in what state are they preserved? What conditions are necessary for such long-term preservation? The papers in this section bear on these questions. The chemical and physical habitats of amino acids in geological situations are described, and their molecular preservation is characterized.

It is logical to attempt a correlation of morphological fossils with chemical fossils. The specimens with the best physical preservation are apt to have a representative collection of indigenous organic matter. Well-preserved bone is common in the geological record, and collagen in live bone has been studied carefully. Several of the papers examine the potential of fossil collagen as a paleobiochemical indicator. Two sets of views emerge. One is that proteinaceous material with appreciable amounts of biological organization can persist for geological lengths of time. The other view is that because of the chemical instability of proteins and the "openness" of fossils to their environment, molecular organization rapidly disappears. It may be possible to decide between these two possibilities by some of the theoretical and experimental procedures given in the third section of this book.

Several papers combined microscopic examination of fossils with chemical studies. They concluded that while materials may have their detailed microstructure preserved, their chemical structure has been greatly altered. It will be interesting to learn how this process takes place.

There are many problems in sample preparation and interpretation of results. Organic matter in a fossil can be attacked by microorganisms, contaminated by younger organic matter, and leached by ground waters. A number of papers consider this problem and several criteria are proposed for determining the integrity of fossils.

In the past, amino acid geochemistry has been concerned mainly with the analysis of individual amino acids. Several contributors to this volume have now considered the possibility of characterizing fossil protein macromolecules and determining their amino acid sequence. Experimental techniques for these measurements exist and need only to be adapted to biogeochemical research.

Amino acids in meteorites continue to be a puzzle. It now seems likely that these molecules observed in recent experiments are not the modern contamination that plagued early experimenters, but their origin and mechanism of preservation are not yet known.

# Bioinorganic Constituents of Hard Parts

**HEINZ A. LOWENSTAM**
*Division of Geological and Planetary Sciences*
*California Institute of Technology*

*ABSTRACT*

Thirty-one kinds of material have been identified among bioinorganic constituents of extant life. Of these, 75% are crystalline and 25% are in the colloidal state. $Ca^{2+}$, $H_2O$, and OH-bearing compounds are dominant. Carbonates are the most widely distributed, and opal is the next most common constituent in the phyla spectrum of extant life. Bioinorganic constituents at anatomical sites range from numerous separate bodies to single aggregates of high structural complexity and mineralogy. Changes in bioinorganic materials take place during in situ maturation and in the ontogeny of some organisms. Extracellular mineral precipitation started more than $2 \times 10^9$ years ago, whereas intracellular biomineralization began only about $6 \times 10^8$ years B.P. Some evolutionary changes in Phanerozoic biomineralization products are seen in calcareous hard parts. Phosphatic materials outranked carbonates initially. Thereafter, carbonates become dominant. Some lineages show an evolutionary trend of aragonite substitution for calcite or the addition of aragonite to calcite.

## INTRODUCTION

Biomineralization is widespread in present-day life. It occurs intra- and extracellularly. The biomineralization products range from colloidal through crypto-crystalline to crystalline materials, which form either simple entities or aggregates and can attain sizable dimensions in mineralized skeletons of animals. Organisms of the cellular grade have usually a single biomineralization site, whereas those of the tissue grade may have multiple sites where this process takes place. Also, in tissue grade organisms, the bioinorganic materials at a given body site are either monomineralic or polymineralic, and this is true for species with multiple monomineralization sites.

Cellular biomineralization products are intimately associated with organic matrices which occur on the inside or as a surface veneer of these materials and in crystalline minerals may be found at both sites (Gregoire 1967; Mutvei 1969; Jope 1965; Watabe 1965; Erben and Watabe 1974).

Based on data on the amino acid compositions and in particular on the amino acid sequence of one of the glycoprotein fractions, the organic matrices of carbonate skeletons have been strongly implicated to constitute a mineral specific templet (Hare 1963, 1969; Weiner and Hood 1975). Identifications of the molecular structures of organic matrices, again based on carbonate skeletons, strengthen the concept of their template function (Weiner and Traub 1979).

A study of the organic matrices associated with noncarbonate, bioinorganic materials should show whether this is generally the case. Extended to the fossil records of the Phanerozoic, already initiated for carbonate skeletons (Weiner et al., 1976, 1979), the study of their organic and bioinorganic fractions holds promise to trace pathways of biochemical evolution.

With this in mind, the following presentation is designed to provide background information on: (1) the diversity of the bioinorganic materials in hard parts of the extant biomass; (2) the spectral range of structural complexity in mineral hard parts; (3) in-situ maturation modes of bioinorganic materials; (4) ontogeny, related changes in biomineralization products; and (5) aspects of the evolutionary history of biomineralization.

## DIVERSITY OF BIOMINERALIZATION PRODUCT

A total of 31 different biomineralization products have been thus far identified by means of X-ray diffraction, electron diffraction, and infrared spectroscopy (Lowenstam 1974, and in preparation). Figure 1 presents a list of 30 of their bioinorganic constituents and shows their distribution in the phyla of the five kingdoms in which species have been found to synthesize them. Octacalcium phosphate $[Ca_8H_2(PO_4)_6 \cdot 5H_2O]$ is omitted because it appears to be only a transient phase in bone mineralization (Degens 1976). The figure shows a nearly three-fold increase in the number

| KINGDOM | MONERA | PROTOCTISTA | | | | | | | | | | | | | | | FUNGI | | ANIMALIA | | | | | | | | | | | PLANTAE | |
|---|---|---|---|---|---|---|---|---|---|---|---|---|---|---|---|---|---|---|---|---|---|---|---|---|---|---|---|---|---|---|---|
| PHYLUM | | DINOFLAGELLATA | HAPTOPHYTA | BACILLARIOPHYTA | PHAEOPHYTA | RHODOPHYTA | CHLOROPHYTA | ZYGNEMATOPHYTA | RHIZOPODEA | SIPHONOPHYTA | CHAROPHYTA | HELIOZOATA | RADIOLARIATA | FORAMINIFERA | MIXOMYCOTA | CILIOPHORA | BASIDIOMYCOTA | DEUTEROMYCOTA | PORIFERA | COELENTERATA | PLATYHELMINTHES | ECTOPROCTA | BRACHIOPODA | ANNELIDA | MOLLUSCA | ARTHROPODA | SIPUNCULA | ECHINODERMATA | CHORDATA | BRYOPHYTA | TRACHAEPHYTA |
| CARBONATES: | | | | | | | | | | | | | | | | | | | | | | | | | | | | | | | |
| CALCITE | + | + | + | | | + | | | | + | + | | | + | + | | | | + | + | + | + | + | + | + | + | + | + | + | + | + |
| ARAGONITE | + | | ? | | + | + | + | | | + | | | | + | | | | | + | + | | + | | + | + | + | + | | + | | + |
| VATERITE | | | | | | + | | | | | | | | | | | | | | | | | | | + | + | | | + | | + |
| MONOHYDROCALCITE | + | | | | | | | | | | | | | | | | | | | | | | | | + | | | | + | | |
| AMORPH. HYDR. CARB. | | | | | | | | | | | | | | | + | | | | | | + | | | | + | + | | | + | | |
| PHOSPHATES: | | | | | | | | | | | | | | | | | | | | | | | | | | | | | | | |
| DAHLLITE | + | | | | | | | | | | | | | | | + | | + | | | | | | | | | | | + | | ? |
| FRANCOLITE | | | | | | | | | | | | | | | | | | | | | | | + | | + | | | | + | | |
| $Ca_3Mg_3(PO_4)_4$ | | | | | | | | | | | | | | | | | | | | | | | | + | | | | | | | |
| BRUSHITE | | | | | | | | | | | | | | | | | | | | | | | | | + | | | | | | |
| AMORPH. DAHLLITE PRECURSER | | | | | | | | | | | | | | | | | | | | | | + | | | + | + | | | + | | |
| AMORPH. BRUSHITE PRECURSER | | | | | | | | | | | | | | | | | | | | | | | | | + | | | | | | |
| AMORPH. WHITLOCKITE PRECURSER | | | | | | | | | | | | | | | | | | | | | + | | | + | + | | | | | | |
| AMORPH. HYDR. FERRIC PHOSPHATE | | | | | | | | | | | | | | | | | | | | | | | | + | + | | | + | | | |
| HALIDES: | | | | | | | | | | | | | | | | | | | | | | | | | | | | | | | |
| FLUORITE | | | | | | | | | | | | | | | | | | | | | | | | | + | + | | + | | | |
| AMORPH. FLUORITE PRECURSER | | | | | | | | | | | | | | | | | | | | | | | | | + | | | | + | | |
| OXALATES: | | | | | | | | | | | | | | | | | | | | | | | | | | | | | | | |
| WHEWELLITE | | | | | | | ? | + | | + | | | | | | | + | ? | | | | | | | | + | | | | + | + |
| WEDDELITE | | | | | | | | | | | | | | | | | + | ? | | | | | | | + | | | + | + | | |
| SULFATES: | | | | | | | | | | | | | | | | | | | | | | | | | | | | | | | |
| GYPSUM | | | | | | | | | | | | | | | | | | | | + | | | | | | | | | | ? | + |
| CELESTITE | | | | | | | | | | | | | + | | | | | | | | | | | | | | | | | | |
| BARITE | | | | | | | | | + | | + | | | | | | | | | | | | | | | | | | | | |
| SILICA: | | | | | | | | | | | | | | | | | | | | | | | | | | | | | | | |
| OPAL | | | | + | | | | | ? | | | + | + | + | | | | | + | | | | | + | + | + | | + | | | + |
| Fe-OXIDES: | | | | | | | | | | | | | | | | | | | | | | | | | | | | | | | |
| MAGNETITE | + | | | | | | | | | | | | | | | | | | | | | | | | + | + | | | + | | |
| MAGHEMITE | ? | | | | | | | | | | | | | | | | | | | | | | | | | | | | | | |
| GOETHITE | | | | | | | | | | | | | | | | | | | | | | | | | + | | | | | | |
| LEPIDOCROCITE | | | | | | | | | | | | | | | | | | | + | | | | | | + | | | | | | |
| FERRIHYDRITE | + | | | | | | | | | | | | | | | | + | + | | | | | | + | + | | | | + | + | + |
| AMORPH. FERRIHYDRATES | | | | | | | | | | | | | | + | | | | | | | | | | + | | | | | + | | |
| Mn-OXIDES: | | | | | | | | | | | | | | | | | | | | | | | | | | | | | | | |
| "TODOROKITE" | + | | | | | | | | | | | | | | | | | | | | | | | | | | | | | | |
| Fe-SULFIDES: | | | | | | | | | | | | | | | | | | | | | | | | | | | | | | | |
| PYRITE | + | | | | | | | | | | | | | | | | | | | | | | | | | | | | | | |
| HYDROTROILITE | + | | | | | | | | | | | | | | | | | | | | | | | | | | | | | | |

**Figure 1.** Diversity and phyla distribution of biomineralization products in extant organisms.

of biomineralization products compared to those recognized 15 years ago (Lowenstam 1963). The additions register a noticeable increase in the diversity of phosphates and iron oxides and extend the range of biomineralization products to iron sulfides and manganese oxide minerals, earlier suggested but only recently confirmed to constitute biologic precipitates (Hallberg 1965; Ehrlich 1972; Krumbein and Altmann 1973; Nealson 1978, and personal communication).

With the data now at hand, it is possible for the first time to discern some dominant traits in biomineralization products, namely, that two thirds are calcium minerals and that a similar percentage is characterized by having $H_2O$ or OH molecules. A new aspect is that almost one fourth of the biomineralization products consists of colloidal materials. These form, as in the previously known case, opal ($SiO_2 \cdot nH_2O$) stable hard part constituents.

Considering the mineral diversity at the kingdom level, 25 of the biomineralization products are synthesized by animals, 11 by the protoctists, 8 by monarans, 7 by vascular plants, and 4 by fungi. Animals are thus the most versatile organisms in the synthesis of bioinorganic materials. However, one should keep in mind that their biomineralization products have been far more extensively investigated than those of species from the other kingdoms. This is particularly true of the monarans and fungi.

As to the phyla distribution, the data shown in Figure 1 further elaborate the long-established fact that carbonate minerals are by far the most common bioinorganic constituents in the spectrum of extant life. The data also show that the vaterite phase, monohydrocalcite and amorphous calcium carbonate cannot be considered biological oddities anymore inasmuch as they are found in species of many genera in the phyla shown in the figure. Opal emerges as the second most widely utilized biomineralization product as a hard part constituent. It is noteworthy that magnetite ($FeO \cdot Fe_2O_3$), first discovered as a biomineralization product in chiton teeth (Lowenstam 1962), is also synthesized by magnetotactic bacteria (Blakemore 1975; Frankel et al., 1979), honeybees (Gould et al., 1978), and possibly by homing pigeons (Walcott et al., 1979).

## Spectral Range of Structural Complexity in Mineralized Hard Parts

Investigations of the bioinorganic constituents, organic matrices, and microarchitecture have been extended in recent decades in animals, from their mineralized skeletons to other, less conspicuous biomineralization sites, and to inconspicuous occurrences in a wide range of other biologic systems. The information now at hand reveals a considerable diversity in building plans of the mineral-bearing structures at homologous and anatomically unrelated body sites of extant life. The various building plans, when considered irrespective of their biologic sources, can be grouped into a sequence of stages of increasing complexity in constructional grade. Stage 1 is characterized by the development of single crystals or single colloidal bodies in great profusion at their depositional site. Decalcification of the crystals from the otocysts of some mammals have revealed intracrystalline matrices and possibly an organic nucleus (Lim 1973, Fig. 2, *a*, *b*, *c*). Stage 2 is seen in aggregates of smaller crystals growing in crystalographic alignment upon a larger crystal. No intercrystalline matrices are in evidence. This kind of aggregation occurs among the calcitic statoconia of the spiny dogfish, *Squalus acantheus*,

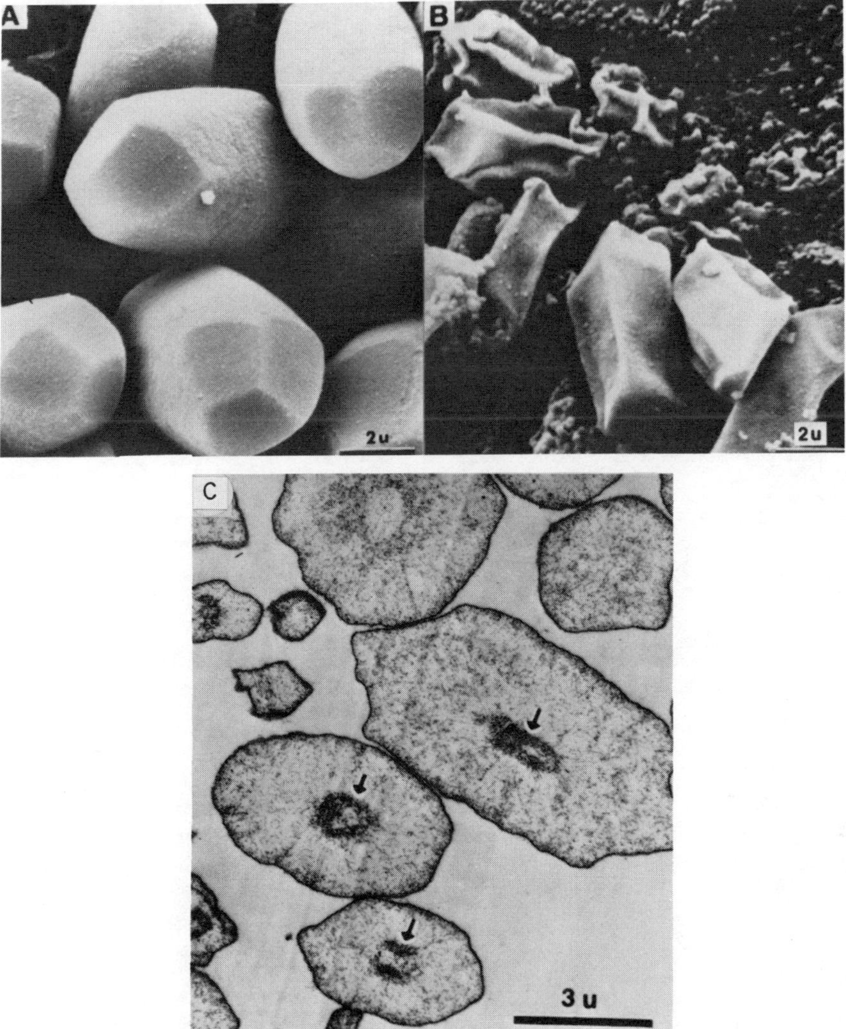


**Figure 2.** Otoconia from the squirrel monkey. A. Crystal habit of the calcitic otoconia. B. Collapsed crystals, stored in 70% ethanol after osmic acid fixation. C. Intracrystalline matrix of the crystals, as seen after decalcification. From Lim, 1973.

which consist mostly of single crystals representing stage 1 (Lowenstam and Fitch, in preparation). At stage 3, the numerous bodies that are developed at an anatomical site consist now exclusively of mineral aggregates displaying only minor variations in their basic building plans. A wide range of vertebrate otoconia, described by Carlström (1963), can be assigned to this structural grade. Stage 4 is characterized by a large polycrystalline structure associated with numerous minute polycrystalline bodies. This stage is illustrated by the development of a discrete otolith in association with numerous isolated otoconia, (Carlström 1963) in the labyrinths of primitive bony fishes. Stages 5 and 6 are characterized by

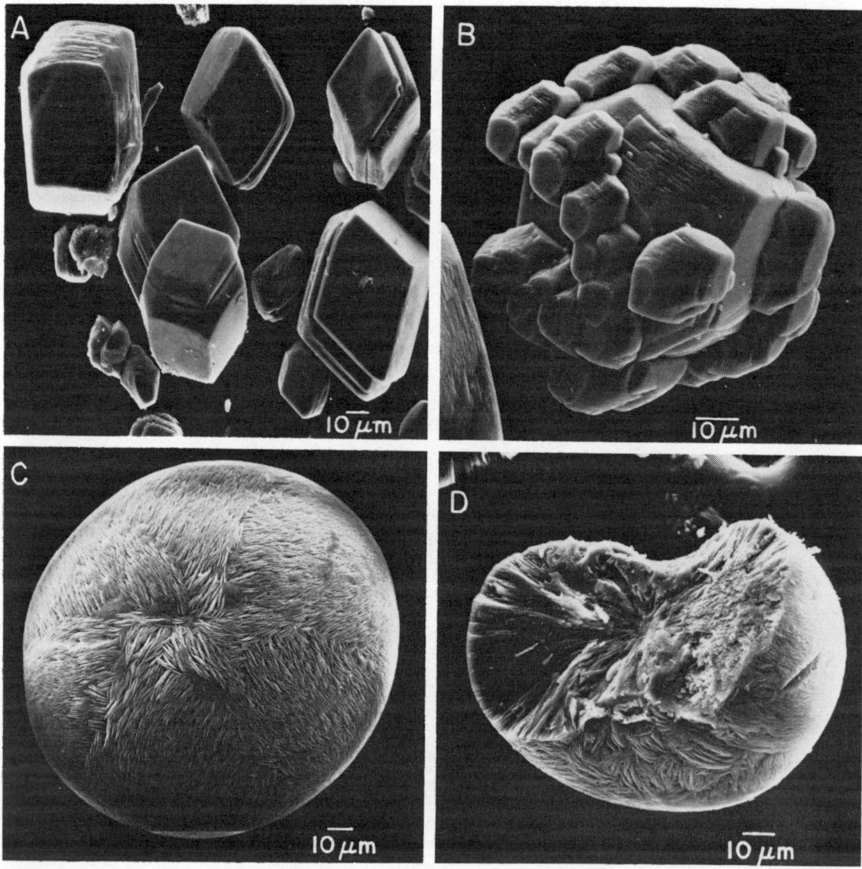


**Figure 3.** A. An otoconium of the shark *Squalus acantheus* consisting of a single calcite crystal. B. Polycrystalline calcitic otoconium from the same species. C. Otoconium of the shark *Somniosus pacificus*, consisting of an polycrystalline aggregate of vaterite. D. Crystal fabric of an otoconium from the same species, as seen on the inside, after crushing.

the integration of all bioinorganic constituents that are elaborated at a given body site into a single structural entity. At stage 5, there is little or no textural and mineralogic differentiation of its component parts. The compound skeletons of the echinoids are an illustration of this building plan. Stage 6 is marked by structural differentiation into several microarchitectural components with distinct textural designs which may but do not have to be related to mineral differentiation. At stage 5, the bioinorganic constituents of the structure are permeated by mesodermal cells or as in stage 6 are sheathed by organic matrices. Their structural relations have been extensively investigated in molluscan shells (Gregoire 1967; Mutvei 1969). Thin sections of some untreated samples indicate that the mineral aggregates of structures referred to stages 3 and 4 also have organic matrices. More data are needed to determine whether intracrystalline matrices occur in these polycrystalline aggregates, since aragonitic constituents of some molluscan shells, assigned to stage 6, have been reported to contain them (Watabe 1965).

The mineralized constituents in the labyrinth of the elasmobranches encompass stages 1 through 3 (Fig. 3, *a, b, c, d*). Those of primitive members in the bony fishes correspond to stage 4 and in the more advanced teleosts, to stage 5. In the Mollusca the cuticular spicules of the Aculifera represent stage 3; the perinotum and girdle spicules and the mineralized plates of Placophora represent stage 4; and the shells of the Conchifera correspond to stage 6. These data suggest that the evolution of mineralized hard parts proceeded in the direction of increase in structural complexity, outlined here. With better information on pathways within various lineages, a more complex evolutionary pattern is likely to emerge.

## IN SITU MATURATION OF BIOINORGANIC MATERIALS

Most data on biomineralization products are based on the study of growth increments of hard parts, where the mineralization process has come to completion. There are, however, a few studies where in situ mineralization has been traced from its initiation to the completion of this process. Among them are examples showing that the mature minerals are conversion products of a precursor. In species of two gastropod genera the mature shells are mineralized by the crystalline carbonate phase aragonite. However, the seed crystals on the organic matrix of newly elaborated and regenerated shell margins are composed of the most metastable carbonate phase vaterite (Kessel 1933; Levetzow 1932).

The iron oxide minerals in chiton teeth illustrate the most radical change of in situ maturation of bioinorganic materials noted to date (Towe and

Lowenstam 1967; Kirschvink and Lowenstam 1979). Mineralization of the major lateral teeth is initiated by the deposition of a paracrystalline, hydrated ferric oxide compound: $5Fe_2O_3 \cdot 9H_2O$, which has been structurally defined by Towe and Bradley (1967) and named ferrihydrite by Chukhrov et al. (1974). This mineral, which also forms the micelle of the iron storage protein ferritin, is subsequently converted in the chiton teeth to magnetite ($FeO \cdot F_2O_3$). Here the transformation is initiated within the ferrihydrite deposits, is followed by mass conversion in one tooth row, and continues thereafter on the magnetite surfaces of subsequent tooth rows until iron mineralization of the teeth is completed. The immediate source of the magnetite precursor, ferrihydrite, is a crowded, highly concentrated array of ferritin granules in the proximal parts of the superior epithelial cells of the radular sac. These cells are firmly attached to the tooth denticles until biomineralization is completed. The same cells perform two distinct biochemical functions: (1) they supply the iron for the magnetite precursor, and (2) they are involved in the transformation of the ferrihydrite to magnetite. This involves reduction of one third of the ferric iron to the ferrous state, removal of oxygen, dehydration, and a change from hematite-like hexagonal close packing to a cubic-inverse spinal structure. It has been suggested that one, or possibly several enzymes may mediate this phase transformation (Nesson, unpublished thesis) but further study is needed to clarify this biochemical process.

Mineralization by ferrihydrite extends over two to three consecutive tooth rows. Conversion to magnetite is initiated at the inside of the ferrihydrite deposits in the next tooth row, and mass transformation of the remainder of the precursor takes place in the tooth row anterior to it. It is important to recall that the cells involved in the formation of ferrihydrite are the ones that mediate its transformation to magnetite.

Hence these chiton teeth provide a unique opportunity to study the chemistry and molecular structure of the organic matrices from two radically different minerals formed in succession at a single biomineralization site. The data should show whether the organic matrix undergoes structural reorganization when conversion from ferrihydrite to magnetite takes place. This would be expected in accordance with the template function assigned to the organic matrix by current theory (Hare 1963; Weiner and Hood 1975).

## ONTOGENY-RELATED CHANGES IN BIOMINERALIZATION PRODUCTS

Many groups of organisms initiate mineralized hard part development at the embryonic stage. Most mineralogic determinations of these structures have been performed on samples of mature adults. Hence information on the biomineralization products of larval and early ontogenetic growth stages is limited.

The following examples illustrate the various modes and ontogenetic ranges of this phenomenon. In the bivalve species *Crassostrea virginica* the larval shell is mineralized by aragonite. In the postlarval shell the mantle-secreted layers contain calcite, and the myostracal layer formed by the fibres of the adductor and auxilliary muscles is mineralized by aragonite. This is also true of the ligament (Stenzel 1962, 1963, 1964).

The otoconia in the two chambers of the labyrinth found in a number of species within three superorders of sharks illustrate a different mode of mineralogic change during ontogeny (Lowenstam and Fitch 1978). This is related to the fact that the number of otoconia increases exponentially from the embryonic to the adult stage.

Where ontogenetic changes in otoconia mineralogy occurs, the initiation and termination of these changes can be traced, since newly formed minerals are confined to newly formed otoconia. In the blue shark (*Prionace glauca*) the otoconia of the embryos are mineralized by amorphous hydrous calcium phosphate (ACP), whereas those formed at the juvenile and adult stages are composed of aragonite (Lowenstam and Fitch 1978). Therefore, the otoconia consist in the embryos entirely of ACP, in the juveniles of ACP plus aragonite, and in the adults of aragonite with a trace of ACP. This ACP fraction can be detected only in SEM micrographs by its distinct botryoidal surface geometry. In some other species of the same superorder, where the otoconia of the embryos are similarly composed of ACP and postembryologically of aragonite, the ACP fraction clearly seen in juveniles cannot be located in the mature adult otoconia. In those cases, it appears that at that stage of growth the endolymph fluids of the labyrinth resorb the ACP fraction of the embryonic otoconia.

An entirely different mode of mineralogic change occurs in the holothurian species *Molpadia intermedia* s.l. (Lowenstam and Rossman 1975). The mesodermal deposits in very early postlarval individuals consist of calcitic spicules, as in other species of this class. During subsequent growth, most spicules outside the oral and caudal region acquire a reddish color and shortly thereafter disaggregate into reddish to purplish granules in the deep-seated part of the connective tissue. However, in the outer part of the dermis of the body region resorption reduces the calcite spicules to mere remnants. The dermal granules, as seen in SEM micrographs of up to 100,000 times magnification, consist of multi-layered, roughly spherical subunits of 0.1 to 0.2 μm in diameter, which are made up of 80 to 110 Å sized ovoidal sub-subunits. Throughout, the granules are composed of amorphous, hydrous ferric phosphate plus opal. Organic material is an integral part of the granules. With increase in age, calcite spicule formation ceases, whereas granule formation increases by further accretion of the initially developed granules and the formation of new ones, principally in the dermis of the main body.

## ASPECTS OF THE EVOLUTIONARY HISTORY OF BIOMINERALIZATION

Recent discoveries of fossils in Precambrian rocks have extended the documentation of life to about $3.2 \times 10^9$ years B.P. (Schopf 1969). Up to late Vendian time, that is, close to $6 \times 10^8$ years B.P., the fossils are all soft-bodied organisms. This has led to the startling realization that mineralized hard part formation, so widespread in present-day life, is a surprisingly late development in the biochemical evolution of life. However, this development was preceded by extracellularly mediated mineral precipitation by at least $2 \times 10^9$ years. This is indicated by the $S^{34}$ values of many sedimentary pyrites, which show isotopic fractionation values similar to those produced by extant sulfate reducing bacteria (Thode et al., 1951; Kemp and Thode 1968) and also by bacteria, which are morphologically indistinguishable from extant strains that precipitate iron oxide minerals (Barghoorn 1965; Cloud 1965).

The inception of biochemically mediated intracellular processes that led to mineralized hard part formation enhanced enormously the preservation of life records as fossils. It is generally recognized that the rapid spread of this innovation to a wide range of organisms and its continuing proliferation in the biomass accounts for the wealth of biologic records of the Phanerozoic phase of the history of life in contrast to the comparatively meager records of its preceding history.

The abundance of mineralized hard parts in Phanerozoic deposits enabled paleontologists to define the phylogenies of those groups of life which acquired mineralized hard parts. For some groups it provided data to discern details of their phylogeny. Synthesis of phylogenetic trends has provided data on tempos and modes in the evolution of Phanerozoic life (Simpson 1944; Raup and Stanley 1978). In fact, this encompasses only those facets of evolution which can be gleaned from trends in morphologic changes of skeletal designs and skeletal microarchitecture. Few attempts have been made to explore evolutionary information from the bioinorganic constituents and organic matrices of mineralized hard parts. Such information would provide insight into the underlying causes of biochemical evolution as documented by fossils. The fact that skeletal mineralogy and chemistry of extant species are expressions of the biochemistry indicates the need for such studies of the fossil records (Weiner et al., 1976).

In the average state of preservation, the mineral content of fossils has undergone diagenetic changes that range from structural alteration to substitution by minerals unrelated to those originally present. This accounts mainly for the limited interest in extending the study of evolution in fossils to their bioinorganic constituents. However, determined efforts to locate carbonate hard parts with original mineral constituents for the study of their trace elements, oxygen and carbon isotopic compositions, and the chemistry of their organic matrices has succeeded in locating unaltered aragonitic hard parts as old as the early Carboniferous (Hallem and O'Hara 1962). These investigations provided data about the types of sedimentary rocks and the geologic setting necessary to preserve this metastable carbonate phase in the fossil record. Similar strategies are likely to succeed and accelerate progress in detecting and tracing evolutionary changes when extended to other biomineralization products such as the phosphate minerals. At present only some gross trends are discernible in the evolution of calcareous hard parts of Phanerozoic life.

The progression of substituting carbonate minerals for phosphates in many groups of organisms is the most fundamental evolutionary trend in hard part development so far in evidence. The observation that phosphate hard parts are found in a greater diversity of higher taxa in Cambrian deposits than in younger Phanerozoic rocks has been known for some time (Müller 1963; Rhoades and Bloxom 1971). A recent assay of the oldest calcareous hard parts among the taxa represented shows that in the late Vendian deposits, two thirds and in the Lower Tommotian stage of the Lower Cambrian, about one half are phosphatic (Lowenstam and Margulis, in press). This demonstrates that phosphates were the most common biomineralization product of calcareous hard parts in the initial phase of biomineralization processes of life.

Replacement of phosphates by carbonates in hard parts can be clearly seen in lineages such as the Ostracoda, where this substitution occurred at the turn of the Cambrian to the Ordovician (Müller 1963). In a general way, this trend is also discernible at the phylum level. The coelenterata are an example. Phosphate is the biomineralized product of the Conulariina, the first of this group to acquire mineralized hard parts. All later evolving groups, had, and still have, only carbonate skeletons.

With reference to the history of the carbonate secreting biomass, there are data to show that calcite was the initial biomineralization product and that aragonite was added at a later stage in the evolution of several higher taxa (Lowenstam 1973). The anthozoa exemplify this trend, for the Paleozoic Rugosa had calcitic skeletons, and the Scleractinia which came into existence in Triassic time have aragonite hard parts (Lowenstam 1954, 1963; Stehli 1956). There is some evidence to support this thesis among the bryozoa and possibly in the Nautiloidea (Bøggild 1930).

The phosphatic content from fossils in their average state of preservation has been shown to consist of fluorapatite. Since this is clearly a diagenetic alteration product, it is not clear whether the phosphate minerals originally present were also subject to evolutionary changes. These data indicate a growth trend in the evolution of calcareous skeletons from phosphates to carbonates and a tendency to replace calcite by arangonite within the carbonates.

## ACKNOWLEDGMENTS

P. E. Hare and S. Weiner were a stimulus to my studies on biomineralization processes, by raising questions and providing background information on the biogeochemistry of organic matrices in mineralized hard parts, while participating in my seminar and in the years thereafter. J. E. Fitch got me interested and participated in the studies of fish otolith and otoconia which contributed to the formulation of the mineral aggregation trend in hard part development presented here. D. J. Lim provided the SEM micrographs of the mammalian otoconia. Research was supported by N.S.F. grants EAR76-03725 and DES75-08659.

## REFERENCES

Barghoorn, E. S. and S. A. Tyler, Microorganisms from the Gunflint chert, *Science,* **147,** 563–577, 1965.

Blakemore, R., Magnetotactic bacteria, *Science,* **190,** 377–379, 1975.

Bøggild, O. B., The shell structure of the mollusks, *K. Danske Vidensk. Selk. Skr.,* **2,** 232–325, 1930.

Carlström, D., A crystallographic study of vertebrate otoliths, *Bio. Bull.,* **125,** 441–463, 1963.

Chukrov, F. V., et al., Ferrihydrite, *Int. Geol. Rev.,* **16,** 1131–1143, 1974.

Cloud, P. E., Significance of the Gunflint (Precambrian) microflora, *Science,* **148,** 27–34, 1965.

Degens, E. T., Molecular mechanisms on carbonate, phosphate, and silica deposition in the living cell, *Fortschr. Chem. Forsch.,* **64,** 1–112, 1976.

Ehrlich, H. L., Iron oxidation and reduction-microbial, in *The Encyclopedia of Geochemistry and the Environmental Sciences,* R. W. Fairbridge, ed., Van Nostrand-Reinholt, New York, pp. 610–612, 1972.

Erben, H. K., and N. Watabe, Crystal formation and growth in bivalve nacre, *Nature,* **248,** 128–130, 1974.

Frankel, R. B., R. Blakemore, and R. S. Wolfe, Magnetite in freshwater magnetic bacteria, *Science* (in press).

Gould, J. L., J. L. Kirschvink, and K. S. Deffeyes, Bees have magnetic remanence, *Science,* **201,** 1026–1028, 1978.

Grégoire, C., Sur la structures des matrices organiques des coquilles de mollusques, *Bio. Rev.*, **42,** 653–688, 1967.

Hallberg, R., Notes on biosynthesis of pyrite, *Stockholm Contr. Geol.*, **13,** 35–37, 1965.

Hallem, A., and J. M. O'Hara, Aragonitic fossils in the Lower Carboniferous of Scotland, *Nature*, **195,** 273–274, 1962.

Hare, P. E., Amino acids in the proteins from aragonite and calcite in the shells of *Mytilus californianus*, *Science*, **139,** 216–217, 1963.

Hare, P. E., Geochemistry of proteins, peptides and amino acids, in *Organic Geochemistry*, G. Eglinton and M. T. J. Murphy, eds., Springer-Verlag, New York, pp. 438–463, 1969.

Jope, H. M., Composition of brachiopod shells, *Treatise on Invertebrate Paleontology*, Pt. H. Brachiopoda, **1,** R. C. Moore, ed., pp. H.156–H.164, 1965.

Kemp, A. L., and H. G. Thode, The mechanism of bacterial reduction of sulfate from isotopic fractionation studies, *Geochim. Cosmochim. Acta*, **32,** 71–91, 1968.

Kessel, E., Über die Schale von *Viviparus viviparus* L. und *Viviparus fasciatus* Müll., *Zeitschr. Morphol. Ökol. der. Tiere*, **27,** 9–198, 1933.

Kirschvink, J. L., and H. A. Lowenstam, Mineralization and magnetization of Chiton teeth: paleomagnetic, sedimentologic and biologic implications, *Earth Planet. Sci. Lett.* (in press).

Krumbein, W. E., and H. J. Altmann, A new method for the detection and enumeration of manganese oxidizing and reducing bacteria, *Helgoländer Wiss. Meeresunters.*, **25,** 347–356, 1973.

Levetzow, K. G., Die Structur einiger Schneckenschalen und ihre Entstehung durch typisches und atypisches Wachstum, *Jena Ztschr. Naturwiss.*, **68,** 41–108, 1932.

Lim, D. J., Formation and fate of the otoconia, *Ann. Otol., Rhinol. Laring.*, **82,** 23–36, 1973.

Lowenstam, H. A., Systematics, paleoecologic and evolutionary aspects of skeletal building materials, *Mus. Comp. Zool., Harvard Coll. Bull.*, **112,** 287–317, 1954.

Lowenstam, H. A., Magnetite in denticle capping in recent chitons (Polyplacophora), *Geol. Soc. Amer. Bull.*, **73,** 435–438, 1962.

Lowenstam, H. A., Biologic problems relating to the composition and diagenesis of sediments, in *The Earth Sciences: Problems and Progress in Current Research*, T. W. Donnelly, ed., Univ. of Chicago Press, Chicago, pp. 137–195, 1963.

Lowenstam, H. A., Evolutionary changes in hard part mineralogy of some invertebrates, *Geol. Soc. Amer. Annu. Meet., 1973 Abstr.*, 719, 1973.

Lowenstam, H. A., Impact of life on chemical and physical processes, *The Sea 5.* Marine Chemistry, E. D. Goldberg, ed., Wiley, New York, pp. 715–796, 1974.

Lowenstam, H. A. and G. R. Rossman, Amorphous, hydrous, ferric phosphatic dermal granules in *Molpadia* (Holothuroidea): physical and chemical characterization and ecologic implications of the bioinorganic fraction, *Chem. Geol.*, **15,** 15–51, 1975.

Lowenstam, H. A., and J. E. Fitch, Morphology, mineral diversity and ontogenetic mineralogic changes in extant elasmobranch otoconia, *Geol. Soc. Amer. Abstr.* **10,** 114, 1978.

Lowenstam, H. A., and L. Margulis, Calcium regulations and the appearance of calcareous skeletons in the fossil record, *Proc. 3rd International Symposium on the Mechanism of Biomineralization*, Tokyo, (in press).

Müller, K. J., Phosphatische Gehäuse bei Ostracoden aus dem oberen Kambrium und die

Bedeutung von Apatit als Schalenbildner altpaläozoischer Metazoa, *Naturwis.*, **51,** 36–37, 1964.

Mutvei, H., On the micro- and ultrastructure of the conchiolin in the nacreous layer of some recent and fossil molluscs. *Stockholm Contr. Geol.*, **20,** 1–17, 1969.

Nealson, K. H., The isolation and characterization of marine bacteria which catalyze manganese oxidation, in *Environmental Biogeochemistry and Geomicrobiology*, **3,** W. E. Krumbein, ed., Ann Arbor Science, Ann Arbor, pp. 847–858, 1978.

Nesson, M. H., Studies on radula tooth mineralization in the Polyplacophora, Ph.D. thesis, Cal. Tech., Biol. Div., 1968.

Raup, D. M., and S. M. Stanley, *Principles of Paleontology*, W. H. Freeman & Co., San Francisco, 1978.

Rhodes, F. H. T., and T. W. Bloxom, Phosphatic organisms in the Paleozoic and their evolutionary significance, *Proc. North Amer. Paleont. Conv.* Pt. K., 1489–1513, 1971.

Schopf, J. W., Recent advances in Precambrian paleobiology, *Grana Palynolog.*, **9,** 147–168, 1969.

Simpson, G. G., Tempo and mode in evolution, *Columbia Biological Series*, **15,** Columbia Univ. Press, New York, 1944.

Stehli, F. G., Shell mineralogy in Paleozoic invertebrates, *Science*, **123,** 1031, 1956.

Stenzel, H. B., Aragonite in the resilium of oysters, *Science*, **136,** 1121–1122, 1962.

Stenzel, H. B., Aragonite and calcite as constituents of adult oyster shells, *Science*, **142,** 232–233, 1963.

Stenzel, H. B., Oysters: composition of the larval shell, *Science*, **145,** 155–156, 1964.

Thode, H. G., H. Kleerekoper, and D. McElcheran, Isotopic fractionation in the bacterial reduction of sulfate, *Research*, **4,** 581, 1951.

Towe, K. M., and W. F. Bradley, Mineralogical constitution of colloidal "hydrous ferric oxides," *J. Colloid Interface Sci.*, **24,** 382–392, 1967.

Towe, K. M., and H. A. Lowenstam, Ultrastructure and development of iron mineralization in the radular teeth of *Cryptochiton stelleri* (Mollusca), *J. Ultrastr. Res.*, **17,** 1–13, 1967.

Walcott, C., J. L. Gould, and J. L. Kirschvink, Pigeons have magnets, *Science* (in press).

Watabe, N., Studies on shell formation, 11 crystal-matrix relationships in the inner layers of mollusk shells, *J. Ultrastruct. Res.*, **12,** 351–370, 1965.

Weiner, S., and L. Hood, Soluble protein of the organic matrix of mollusk shells: a potential template for shell formation, *Science*, **190,** 987–989, 1975.

Weiner, S., H. A. Lowenstam and L. Hood, Characterization of 80-million-year-old mollusk shell proteins, *Proc. Nat. Acad. Sci. U.S.A.*, **73,** 2541–2545, 1976.

Weiner, S., H. A. Lowenstam, B. Taborek, and L. Hood, Fossil mollusk shell organic matrix components preserved for 80 million years, *Paleobiology*, in press.

Weiner, S., and H. A. Lowenstam, Well preserved fossil mollusk shells: characterization of mild diagenesis, This volume.

Weiner, S. and W. Traub, X-ray diffraction of the insoluble organic matrix of mollusk shells: Framework for crystal growth by epitaxi?, in preparation.

# Collagen in Fossil Bones

**Ralph W. G. Wyckoff**
*Department of Physics, University of Arizona*

Because collagen is the most plentiful and one of the most stable organic components of vertebrate hard tissues, it is necessarily of primary concern to those interested in fossil bones and teeth. My concern with it began about 20 years ago, one can say almost by accident. For years at the National Institutes of Health I had been involved with people at the Dental Institute in seeking to apply the electron microscope to dental structures. When I moved to the University of Arizona, I found that the Arizona desert abounds in mammalian fossils, and it seemed natural to begin examining them in the light of some of the mineralogical interests of my first job—at the Carnegie Institution's Geophysical Laboratory in 1919, almost 60 years ago.

Having at hand, in my new laboratory, equipment for x-ray spectroscopy and diffraction as well as an electron microscope, the first question I faced was this: to what extent were our fossils remineralized artifacts rather than original bones and teeth? With this in mind I investigated sectioned specimens with optical and electron microscopes and finally with an early electron probe made in our laboratory. It was found that though recrystallization and replacements with foreign minerals were not uncommon, the apatite in well-preserved specimens had lost none of its initial fine structure. When extraneous minerals were present, it was noted that they might on occasion faithfully replace organic as well as inorganic details. For example, we found ordered fibrous masses having the configuration of muscle in a well-preserved Eocene fish; the electron probe made it clear that these were siliceous and ferrous replacements which faithfully reproduced the periodic striations of a single muscle fiber. Early in our work we were made aware that mineral replacements could preserve the appearance of even more minute organic structures. In one

Present address: Duval Corporation, 4715 E. Fort Lowell Road, Tucson, Arizona 85712.

series the electron micrographs demonstrated what were certainly fossilized bacteria within a tubule of an Oligocene brontotherium tooth. Though there are numerous references in the older literature to such bacteria, it soon became clear that electron microscopic magnifications were required to establish their existence.

Among the first fossils we examined were Pleistocene specimens from the LaBrea tar pits in Los Angeles; it came as a surprise to me that under the electron microscope the fibers in some of them manifested the 640 Å striations of native collagen. This immediately raised the question of whether these also were replacement artifacts or were indeed the original collagen preserved intact over a geologically significant length of time. In view of the then-accepted chemical instability of proteins, the former seemed the more likely; nevertheless, with chromatography a definite choice between these alternatives could be made, and this possibility determined the direction of much of our further work.

Abelson, Hare, and co-workers at the Geophysical Laboratory had already shown the presence of amino acids in fossil shells; we needed only to make a similar application to bones, with the difference that our attention was directed toward the insoluble material remaining after weak HCl or EDTA had freed a bone or tooth of its apatite. Several LaBrea fossils subjected to this treatment left residues that showed striated collagen-like fibers under the electron microscope and were converted into a mixture of free amino acids when hydrolyzed with stong HCl. Liquid chromatography proved these amino acids to be those of collagen in the same proportions as in fresh material.

We next sought to determine if these LaBrea fossils were exceptional in their content of insoluble protein. Many specimens collected around Tucson and elsewhere in Arizona, though less well preserved than those from the tar pits, did contain protein. The local specimens were, however, of no more than Pliocene age, and it seemed important to find out if intact protein could be recovered from still older fossils. At the time we were worried by the possibility that what we found might be a recent contamination, and to minimize this we decided to restrict ourselves to single bones so large that surface layers could be removed before the material to be analyzed was selected.

This method led us to concentrate on dinosaur fossils, which were certainly large enough and could be obtained in practically unlimited quantities. Analyses were carried out on more than a hundred Cretaceous and Jurassic reptilian bones and teeth drawn from a wide range of North American localities. Some protein was obtained from most, but only a few contained appreciable amounts of the hydroxyproline characteristic of collagen or yielded ratios of their other amino acids typical of this protein. Instead they commonly had relatively large amounts of alanine, valine, and glutamic acid and reduced quantity of glycine. We have now come to consider such a protein to be a product of putrefaction, presumably of the freshly dead animal. Most specimens yielded less than 100 μg/g of protein and contained considerable quantities of iron soluble in strong HCl. The need to eliminate this iron from the large samples, because of the limited sensitivity of ninhydrin, led us to turn to gas chromatography. We found it required only about 1/50 as large a sample and for the study of collagen it has the great advantage of responding as well to the critically diagnostic hydroxyproline as to other amino acids. Though the development of high pressure capillary columns and replacement of ninhydrin by the more sensitive Fluoram has increased the sensitivity of liquid chromatography, we have continued to rely on gas chromatography for our work with collagen.

Work with dinosaur remains demonstrated that enough protein for analysis could often be recovered from bones and teeth as old as the Jurassic. At the time we were unable to obtain older vertebrate material suitable for analysis, so we turned to invertebrate shells for samples that would tell if insoluble protein was to be found in still older fossils. At this point Dr. Akiyama came to spend a year with us. He may refer to some of the results obtained then, but for my purpose it is sufficient to say that Ordovician and Devonian shells have furnished protein.

Our accumulated experience has strongly suggested that collagen is more likely to be found in fossils preserved in alkaline than in acidic rocks. Some years ago a note in *Nature* from Poland indicated the presence of much collagen in fossils from the Gobi desert, and we have found hydroxyproline in a small sample collected there 75 years ago by Roy Chapman Andrews. These results were so promising that I tried without success to obtain from East European sources additional material from the Asian deserts. Turning to the Sahara as the next most likely desert region, I should note that we are publishing with Professor Lehman and Dr. Taquet of the Museum National d'Histoire Naturelle in Paris analyses of collagen-like protein from specimens they collected from the southern and western borders of the Sahara. Some of these fossils are from Cretaceous dinosaurs and some are from far older Devonian arthrodire bone and armor. All were within calcitic matrices. Thus it would appear that though most initial collagen may be lost with the passage of time, what remains after hundreds of millions of years may still have the amino acid composition of collagen.

Having shown that protein, some of it collagen, can be recovered from individual fossils of extreme antiquity, we next turned to learning as much as possible about its molecular structure. The first step in such a program must always be solubilization with the smallest possible molecular alteration. Modern collagens of adult animals are rendered soluble by gelatinization, which is a partial hydrolysis that splits the molecules into a collection of linear peptides. This is effected by a variety of procedures, some of which we have successfully applied to our fossil proteins. We have compared the fresh and the fossil gelatins obtained by heating with water at temperatures above 100°C, with dilute HCl at lower temperatures, with acetic acid and acetates, and most recently with dilute formic acid at temperatures as low as 60°C. These gelatins have been roughly fractionated by passage through Sephadex or polyacrylamide gels or graded Pellicon filters. The fractions thus produced have then been completely hydrolyzed with 6 N HCl and their amino acid contents determined. Qualitatively different results have been obtained from the collagen-like and the non-collagenous fossil proteins.

Collagens, both fossil and fresh, are split by any one of the treatments mentioned above into peptides of various lengths, the mean length being greater the gentler the treatment. Except for gelatins from obviously degraded fossils or those made by drastic heating, all fractions have yielded amino acids in ratios close to that of collagen itself. The lightest bed fraction has often seemed to be an exception, but that is because a considerable amount of aspartic acid and some glycine have been freed during solubilization and collected along with the smallest peptides in this terminal fraction. Clearly, the collagen molecule breaks apart most readily at these residues. The amount of liberated aspartic acid has depended on the severity of the solubilizing treatment and apparently on the quality of the fossil; where its preservation has been obviously poor, at least 50% of the total aspartic acid of the fossil protein has been freed. In contrast to what takes place with fresh collagen, none of the less drastic treatments we have used has solubilized all the fossil collagen of a specimen; the percentage has varied with the specimen and may well be a measure of its preservation.

The different results with the non-collagenous dinosaur fossil proteins have often indicated that a smaller percentage of the total protein in a specimen is solubilized by a given treatment; invariably the peptides produced have been shorter. Thus up to 80% of the fossil collagens we have studied have been rendered soluble by being heated in 5% formic acid at 75°C for 18 hours, but not more than 40% of the mosasaur and only about 10% of our dinosaur proteins have responded to the same treatment. Similarly, whereas most of the peptides obtained by heating a fossil collagen in 5% formic acid at 75°C have been so large that they collect in the void and intermediate fractions of a gel filtration, the reverse has been true for the non-collagenous dinosaur proteins. Their smaller peptides are nevertheless big enough to contain all the amino acids. A larger percentage of these dinosaur proteins can be solubilized by using stronger acids or higher temperatures. Thus we have at least doubled the yield of peptides without substantially changing their amino acid composition by heating (1) in water for three days at 110°C; (2) for 24 hours at 110°C in 0.03 *N* HCl; (3) for four hours at 90°C in 6 *N* HCl.

There is one aspect of this general question to which adequate attention has not been given. It concerns the fate of most of the protein that undoubtedly was present in the living bone or tooth. It could have been gradually solubilized and washed away; or it could have become so thoroughly denatured that it no longer is hydrolyzed by strong HCl but remains in the fossil as a kind of kerogen. A choice between these alternatives could be made by burning the insoluble residue with oxygen and measuring the evolved $CO_2$.

Much time has been spent trying to determine the nature of the non-collagenous fossil proteins. As part of this effort we analyzed the residues remaining when fresh collagen is heated, moist or dry, at temperatures above 100°C. It is significant that the nonsoluble residues thus obtained have not resembled in their amino acid ratios the non-collagenous fossil proteins; and as a consequence, we have come to believe that such proteins have their origin in putrid carcasses.

Our long-term goal in studying fossil collagen has been to see if it has undergone molecular changes during the course of evolution. It can be stated that analytical results thus far give no reason to believe that the ratios of the amino acids of this protein have perceptibly altered during evolution. There remains the possibility that the sequence of these acids has not always been the same, and I had started to concern myself with this rather difficult question. Our present detailed knowledge of the collagen molecule is a consequence of the fact that immature collagen can be extracted with $CaCl_2$ or split with CNBr to yield soluble peptides of uniform length. This probably cannot be done with fossil collagen. I do not know of other ways to obtain uniform peptides or how much useful information can result from sequencing the less uniform fractions we now can produce. Nevertheless we had begun to try.

All the usual amino acids except those containing sulfur have been part of the dinosaur proteins we analyzed, but their chromatograms also contained an important peak in the position of β-alanine as well as several other large unknown peaks, notably those between aspartic and glutamic acids and before lysine. Without the means of isolating the substances causing them and lacking a mass spectrometer, we have not been able to identify them. They may be products of the gradual denaturation of their parent protein; if not, it is conceivable that older proteins contained amino acids which are no longer among their essential ingredients.

Were we able to continue work with fossil proteins, we would certainly be seeking the source of these unknown peaks. We would also be trying to obtain in larger amounts the very ancient collagens in Professor Lehman's and Dr. Taquet's Saharan specimens. We had likewise turned again to well-preserved Pleistocene and mummified materials in hopes of isolating from them proteins more likely than collagen to reflect species differences and evolutionary changes.

We have also been trying to make D-L determinations of amino acids from some of our older fossils, thus far with rather indifferent results. We failed using the technique in which hydrolyzed amino acids are converted to their volatile diastereomers and chromatographed with an ordinary glass column coated by an inactive stationary phase. Since controls were satisfactorily separated, we assumed that this was due to the low efficiency of the conversion reaction when our very small amounts of amino acids were involved. We are continuing our efforts using an active stationary phase coating a capillary column. There are two things we hope to learn from this. One is an incontrovertible demonstration that the protein being studied is fossil and not a contaminant of recent date. We also hope to gain information about the comparative ease of racemization with time of an amino acid in the free state and as part of an intact protein.

I cannot conclude without expressing my deep impression of the expansion in knowledge of the organic remains of earlier life over the past 20 years. This expansion has scarcely begun, and an inability to participate longer in it is one of the few things that make me wish I were younger.

The work described above has been carried out with the support of series of grants from the National Institute of Dental Research and the National Science Foundation.

# Ancient Bone and Plant Proteins: The Molecular State of Preservation

**E. M. JOPE**
*Queen's University, Belfast, Northern Ireland*

## INTRODUCTION

Living bone consists of about 23% collagen fibrils disposed irregularly through the phosphatic calcified (hydroxyapatite) matrix. It is now well known that this collagen can survive as a supramolecular (microfibril) and therefore molecular structural entity at least since the Pleistocene (over 12,000 years ago), as Wyckoff and his co-workers so elegantly showed over a decade ago (Doberenz and Wyckoff 1967; Piez and Miller 1974). Amino acid compositions also now show that, with careful selection, bone and antler with well-preserved collagens can be found at least from Pleistocene and later contexts (Table 1; Wyckoff 1972; Jope 1976).

Collagen is a highly ordered assemblage of polypeptide molecular units each 1055 amino acids long (the α-chain), which are the products of precise genetically coded biosynthesis (E. J. Miller 1976; Bornstein 1974). Insofar as these collagen α-chains are preserved in detail as molecular structures in ancient bone, antler, or teeth, they are a potential source of paleogenetic information direct from the remote past—almost the only *direct* source of such surviving evidence we have—compared with the retrospective extrapolation derivable from comparative studies on living species. It is thus vital that all such possibilities with ancient collagens should be fully explored.

This ancient collagen is also the source of the carbon used for $^{14}C$ and $^{13}C$ studies on bone or antler material (the latter often giving anomalous results, Burleigh 1976). For precision dating, ideally we should know the credentials of each collagen carbon atom in the ancient protein material as an authentic product of the original animal. Knowledge of the molecular state of preservation of the collagen can help to ensure this authenticity of carbon source.

The primary ancient protein is also the source of the amino acids so freely used with meaningful intent in racemization dating studies on ancient bone and analogous material such as shells or foraminifera. These amino acids are inevitably the multistage catabolic products from the original protein. If such dating studies are to be of real value, we must understand any constraints imposed on racemization probability when amino acids are locked in protein supramolecular structures or in the various stages of their catabolic peptides. No such work has yet been reported on the fundamental principles underlying observed racemization rates. These many-staged degradation processes must, however, give complex racemization/time relations, as is evident in data reported in other papers in this volume.

Bone collagen studies are at a disadvantage in that most fundamental collagen molecular research is done on skin or tendon collagen. Too little is known of the microstructural and mechanical relations between the collagen and inorganic components in bone or antler (A. Miller 1976b, pp. 67f).

## THE MOLECULAR AND SUPRAMOLECULAR STRUCTURE OF COLLAGENS

Collagens are the main organic structural material of bone, antler, teeth, tendon, skin, and connective and many other specialized tissues. They are complex but fully ordered protein assemblages, organized at four molecular and supramolecular levels: (1) the *monomer*, the α-chain, 1055 amino acids long; (2) the *trimer*, coiled-coil assemblage of three parallel α-chains supercoiled (tropocollagen) tube-like, 300 nm long and 1.4 nm wide; (3) the *microfibril*, structured in staggered repeating units of five trimer coiled-coils; (4) the *fibril*, lateral bundles of microfibrils, of varying length and thickness.

1. The *monomer* α-chain of 1055 amino acids (Piez and Trus 1978, for revision from 1052 of Hulmes et al., 1973) has a central 1014 residue of tight, strictly helical conformation that is made up of triplets -Gly-X-Y- and requires glycine (the smallest amino acid) at every third position, and about 300 proline residues to turn the tight helix (Fietzek and Kühn 1975), about half being hydroxylated postbiosynthesis. At the ends are nonhelical tails of 16 and 25 residues, from which hydroxylysine residues can give covalent interchain linking. (For specimen length of atomic model showing H-bonding, Pauling 1959, pp. 499 ff).

2. The *trimer* molecule, Tropocollagen, consists of three of these monomer α-chains (lefthanded helices) assembled side by side, with a one-residue (0.286 nm) stagger (Parry 1977) and wound together (right-handed); this coiled-coil triple helix is then supercoiled (left-handed) to give a close-packed rod-like unit 300 nm long and 1.4 nm across.

   The trimer tropocollagens of bone, antler, skin, and tendon are identical in constitution (Type I), made up of two α-chains of one constitution (α-1(1)) and one of another (α-2), i.e. $(\alpha\text{-}1(1)_2\alpha\text{-}2$ (E. J. Miller 1976; Piez and Trus 1978, pp. 420 f). In cartilage collagen all three α-chains are α-1(2).

3. The *microfibril* is an assemblage of repeating groups of five trimer units 300 nm long, staggered lengthwise by 66.8 nm (Piez and Miller 1974), each trimer with its ends spaced 36.8 nm from the next.

   The microfibril is a rope-like structure, with the α-helix, the triple-helix, and the supercoil turned alternately left handed and right handed, thus resisting uncoiling (Piez and Trus 1977, p. 704).

4. The *fibril* is made up of bundles of aligned microfibrils, which can be assembled in various ways, adding either lengthwise or laterally. Fibrils are usually long (20,000 nm) and of very variable thickness, in tendon, 10–300 nm, according to species and depending on water content (Piez and Miller 1974, pp. 135 ff; Miller 1976b, pp. 59–60).

Up to the microfibril stage the constitution and structure of all Type I collagens seem identical; it is only at the next stage of supramolecular assembly, to fibrils, that differentiation according to varied tissue function—bone, tendon, skin, and so on—becomes apparent. This stage is versatile, as fibrils can be assembled by adding microfibrils either lengthwise or laterally; it is not yet known, however, how this differentiation according to tissue function is biologically regulated (Piez and Miller 1974, pp. 135–136).

The completed collagen fibril thus shows close-packed trimer molecular units (though supercoiled with a hollow center) built up into a more open but still precisely ordered microfibril structure, giving space for water and for enzyme access and perhaps also resilience for the 30° oscillation or rotation noted in $^{13}C$ nuclear magnetic resonance observations (Torchia and VanderHart 1976; Piez and Trus 1978, p. 431).

The collagens, especially in conjunction with ancillary substances (e.g., apatite), provide a versatile biological material with very varied biological and mechanical properties, for both calcified and the most varied non-calcified tissues, such as skin, tendon, and connective tissue, as well as cartilage (Type II, $(\alpha\text{-}1(2)_3)$ and other specialized tissues (E. J. Miller

1976). This complex, internally compensated and hydrogen-bonded supramolecular structure has considerable stability and resistance to degradation.

## THE MOLECULAR STATE OF SURVIVAL OF COLLAGENS IN ANCIENT BONE AND ANTLER

The 66.8-nm periodicity seen in electronmicrographs of Pleistocene bone (Doberenz and Wyckoff 1967) shows survival of a high degree of supramolecular order in fibrils assembled out of microfibrils and therefore at molecular level (trimer and monomer). As this periodicity is revealed by heavy metal staining of polar groups in the amino acid residues of the protein chain after decalcification (thereby locating these in detail (Piez and Miller 1974; A. Miller 1976b, p. 60), it is a direct observation on the surviving organic component of Pleistocene bone, providing potentially a resolution to amino acid level (A. Miller 1976b, p. 60), and not mere preservation of a palimpsest in a calcified matrix.

Amino acid analyses (Table 1; Wyckoff 1969; Jope 1976) show that it is possible to find ancient bone and antler with even the more unstable amino acids (e.g., methionine and hydroxyproline) well preserved since the Pleistocene. Tyrosine is not well preserved, but it is a special case because it is confined to the nonhelical tails of type I collagen (Hulmes et al., 1973; Fietzek and Kühn 1975), and may indicate that these tails are more susceptible to degradation with time then the helical part. Armstrong and Halstead (1978) have used enzymatic splitting of fossil collagens, and the resulting peptide analyses indicate good molecular preservation; the differences they note from recent collagens could be due to changes in the nonhelical tails and in interchain crosslinking.

Gel electrophoresis also shows preservation of these collagen α-chains to full length in ancient material, though the final check by end group analysis is difficult, as the end groups in collagens are conjugated.

To recover paleogenetic information from ancient collagens the α-1(1) and α-2 chains first have to be separated, for which the collagen must be taken into solution. This can be done in a nonpolar solvent such as anhydrous formic acid (Jope and Jope 1979) after decalcification with EDTA in ethanol (Scott and Kyffyn 1978). Interchain covalent cross-links (e.g., hydroxylysyl aldehyde links) inevitably restrict the yield of monomer α-chains, but there can still be sufficient material available. The palaeogenetic information must then be recovered by breaking the α-chains into CNBr peptides with cyanogen bromide at the methionine positions (Miller, Martin, Piez, and Powers 1967; E. J. Miller 1976, pp. 167 ff) and

**Table 1. Amino Acid Compositions[a] of Collagens from Recent and Ancient Bone and Antler (parts per 1000)**

Collagen compositions

| | Red Deer—Modern | | Megaceros c.10,000 BC | Ox 3100 BC |
|---|---|---|---|---|
| | Skull | Antler | Antler | Tibia |
| Gly | 316 | 329 | 340 | 331 |
| Pro | 141 | 130 | 101 | 128 |
| OHPro | 110 | 100 | 77 | 88 |
| Ala | 107 | 104 | 138 | 118 |
| Val | 22 | 21 | 25 | 23 |
| Leu | 27 | 27 | 26 | 25 |
| Ileu | 12 | 12 | 11 | 11 |
| Ser | 32 | 31 | 35 | 26 |
| Thr | 18 | 18 | 17 | 13 |
| Asp | 49 | 49 | 47 | 45 |
| Glu | 74 | 72 | 84 | 80 |
| Met | 6.4 | 6.8 | 6.0 | 3.0 |
| Phe | 13 | 13 | 18 | 13 |
| Tyr | 3.8 | 3.5 | — | 2.0 |
| Lys | 25 | 24 | 22 | 33 |
| OHLys | 10 | 7 | 3 | 5 |
| Arg | 47 | 49 | 49 | 49 |
| His | 4.8 | 5.3 | 10 | 5 |

[a] Data from Jope and Jope, 1979.

ultimately by full amino acid sequencing of selected sections of these CNBr peptides. Enzymic splitting is also informative (e.g., Armstrong and Halstead 1978). Work is now in progress on antler and bone of the extinct giant deer Megaceros and on excavated bovid and ovicaprid material.

## PALAEOGENETIC INFORMATION PRESERVED IN THE COLLAGEN MOLECULE

Ancient collagens are products of coded biosynthesis, and if the molecular structure is well preserved, should show in their amino acid sequences some evidence of the evolution of that protein system and perhaps also of the species involved. But the collagen system, though very polymorphic, is in some ways conservative, with many restrictions on its amino acid configuration imposed by functional requirements: for example, the -Gly-X-Y- triplets (with about 230 Pro + Hyp) throughout the helical length of 1014 residues, disposition of polar and hydrophobic groups, H-bonding, and spatial conformation. Nevertheless there is still scope for viable genetic variance in the remaining 300 or so positions. Total amino acid sequence data are not yet known for inter-species comparisons (the only complete α-chain sequence so far available is a composite of calf and rat skin α-1(1)s (Piez and Trus 1978). However, the possibilities can be tested by extrapolation from the fuller data for one large CNBr peptide: for the sequence range 404–477 in the α-1(1) chain there are eight differences between chick and rat and two between rat and bovid (E. J. Miller 1976, p. 179). Extrapolating for the full 1014 helical length this implies the probability of some thirty mutational variations between the rat and bovid collagen α-1(1) chain, and as the α-2 chain is genetically distinct (E. J. Miller 1976; cf. Sundar Raj et al., 1977; Sykes and Solomon 1978), it would also contribute its separate quota. Even subspecies variance may occasionally be observable in collagen α-chain sequences; recent evidence on calf skin collagens seems to reveal a subspecies mutation involving methionine, thus giving a variant pattern of CNBr peptides (K. A. Piez and D. Volpin, personal communication).

It remains to be shown where in the pattern of species development these 30 or so mutations have occurred. We are assembling the data for *Cervus giganteus* (Megaceros) and other Pleistocene and living deer, and we hope that ancient collagen data may help to throw light also on early ovicaprid development and domestication.

Other paleogenetic information might be derivable from blood group bodies (polysaccharides) preserved in ancient bone, especially if decalcification is done with EDTA in ethanol (Scott and Kyffyn 1978); promising results have been obtained on mummy material (Harrison 1969; Ray 1975), and serological techniques may indeed prove usefully sensitive in extracting information from ancient proteins such as collagens.

## ANTLER

Antler is a variant of bony structure. Little biochemical work has been done on it. It has a high protein content, about twice as much as bone (Kim et al., 1975), and with good state of preservation (Table 1) it should prove a useful material for paleo-biochemical and paleo-biophysical studies.

In animals that shed their antlers annually, antler regrowth (about three months) is one of the most rapid processes of calcified tissue generation, comparable only with some osteosarcomas. Biosynthesis of the collagen is therefore rapid. Antler collagen seems genetically the same as the collagen of the skull it surmounts, with closely comparable amino acid composition (Table 1; Jope and Jope 1979). However, the multiple processes of antler collagen deposition have not been studied (see Gay and Miller 1978, and personal communication, for bone). The lysine and OHLys data on antler do not suggest incomplete lysyl interchain cross-linking.

Ancient antler tends to give anomalous data in $^{14}C$ and $^{13}C$ studies (Burleigh 1976); the reasons for this still need clarifying.

## RACEMIZATION STUDIES

When L-amino acids are locked in the close-packed triple helix of the collagen trimer, limitations of space, polarities, and H-bonding imply that some constraints will be imposed on the probability of their epimerization. This is especially so for polar amino acids lysine, arginine, glutamic, and aspartic, as positioning of their charged groups is crucial for the cohesion and stability of the supramolecular microfibril and fibril structure (Fietzek and Kühn 1975, pp. 147 ff; Trus and Piez 1976, figs. 5 and 15; Parry 1977). On the other hand, triple-helix assembly is sensitive to nonpolar group dispositions (Parry 1977).

Examination of the structural models of the collagen triple helix (Dickerson and Geis 1969, p. 42) shows that some aspartic residues are less constricted than others in more proline-hydroxyproline-rich regions of the helix, but even these become constricted when the triple-helix is supercoiled to form the tropocollagen unit. Epimerization, if at all possible with aspartic residues in such positions (and their side-chain -COOH is an H-bond former), would tend to raise the free energy of the whole molecule, and so reduce stability and lead even to disruption of the molecular structure. The same would apply in varying degree to other asymmetric residues.

There are, however, some elements of freedom in the collagen molecule, such as the azimuthal oscillation or rotation around the helix axis, observed in the $^{13}C$ nuclear magnetic resonance, which suggests that intermolecular forces do not depend strongly on azimuthal angle and hence do not arise from a unique set of interactions between side chains (Torchia and VanderHart 1976). The cis/trans relations of the proline rings also remain to be defined.

Epimerization of amino acid residues that are initially components of a protein system (and thus the only amino acids ascribable with certainty to the life span of the original organism) must be seen as a multistage process, and its kinetics treated accordingly. This will give a composite curve with several process rates superposed, as is shown by much of the data reported below (King and Neville 1977). Some amino acids (e.g., isoleucine or hydroxyproline) have two asymmetric centers, introducing further complexities in the curve form.

Shell proteins do not show evidence of such highly ordered supramolecular structure as in collagens (H. M. Jope 1979), and these constraints on epimerization are probably less operative; but they should not be forgotten, and shell systems do give composite epimerization curves (King and Neville 1977, using isoleucine from foraminifera).

Epimerization data may turn out to be of great value in constructing time scales for the distant past; but they should not be used indiscriminately, and very little of the necessary fundamental research on its kinetics has been carried out.

## ANCIENT PLANT PROTEINS

Something must be said of the proteins preserved in ancient plant material, which have been given little attention. One reason is probably that destructive and tanning enzymes are very quickly brought into play when a moist plant tissue is damaged, and the resulting brown mess is difficult to process.

Recent studies have shown that storage proteins can survive for some 4000 years in maize grains preserved under the very dry conditions on sites in Arizona (Derbyshire et al., 1977) and from the lowest levels in Bat Cave in New Mexico dated by Libby to the 4th millennium B.C. (Libby 1955, pp. 111–112; Boulter and Jope 1979).

These ancient maize proteins show features closely analogous to the storage proteins of modern maizes. Interrelations still have to be worked out in detail, but such data should help to clarify the progress of selective breeding in corn and other crop plants. Preservation under dry conditions seems essential.

Neither the 1000-year-old maize grains from northeastern Arizona nor those much older from Bat Cave would germinate, apparently because of damage to the embryos. Nonviability may have resulted from loss of labile enzyme proteins, even though some of the embryo proteins had survived. The earliest viable seeds so far recorded are 600 year old seeds from Santa Rose Tastil in Argentina (Lerman and Cigliano 1971), which shows that the whole genetic apparatus of DNA could survive in working order. This opens up yet another possible field of research in DNA sequencing of ancient plant material.

## CONCLUSION

Quite apart from bone, antler, and teeth, there is unexploited excavated material that might provide paleogenetic information through protein and perhaps DNA molecular structures. The keratins of wool and horn are troublesome because they have undergone so many postbiosynthetic changes. Yet the fully preserved human and animal cadavers from the frozen tombs of Pazyryk in Siberia (Rudenko 1967) have never been comprehensively investigated biologically and biochemically. There is in such material a whole field of paleobiochemistry to be developed.

## *ACKNOWLEDGMENTS*

I am most grateful to the following for discussions and help in various ways: Professor D. Boulter, Professor D. T. Elmore, Professor L. Gotte, Dr. P. P. Fietzek, Dr. S. Gay, Dr. H. M. Jope, Dr. H. King, Professor K. Kühn, Dr. A. Miller, Dr. E. J. Miller, Dr., K. A. Piez, Professor J. E. Scott, Dr. R. Timpl, Dr. D. Volpin, Dr. C. Wallace, Professor R. W. G. Wyckoff, and Dr. G. T. Young.

## *REFERENCES*

Armstrong, W. J. and L. B. Halstead, Personal Communication, 1978.

Bear, R. S., J. B. Adams, and J. W. Poulton, Disclosure by Fourier Methods of a Long-range Pattern of Non-polar Residues in the α1(1) Sequence of Collagen, *J. Mol. Biol.*, **118,** 123–126, 1978.

Bornstein, P. The structure and assembly of procollagen, *J. Supramol. Structure,* **2,** 108–120, 1974.

Boulter, D., and E. M. Jope, Proteins in maize grains from Bat Cave, New Mexico, In press, 1979.

Burleigh, R. Archaeological evidence for short term natural $^{14}C$ variations, *Proc. Ninth Int. Radiocarbon Conf.,* 1976.

Derbishire, E., N. Harris, D. Boulter, and E. M. Jope, The extraction, composition, and intra-cellular distribution of protein in early maize grains from an archaeological site in N.E. Arizona, *New Phytologist,* **78,** 499–504, 1977.

Dickerson, R. E., and I. Geis, *The Structure and Action of Proteins* W. A. Benjamin, Menlo Park, CA, 1969.

Doberenz, A. R., and R. W. G. Wyckoff, Fine structure in fossil collagen, *Proc. Nat. Acad. Sci. U.S.A.,* **57,** 539–541, 1967.

Fietzek, P. P., and K. Kuhn, Information contained in the amino acid sequence of the α-1(1) chain of collagen and its consequences upon the formation of the triple helix of fibrils and cross links, *Mol. Cell. Biochem.,* **8,** 141–157, 1975.

Fietzek, P. P., and K. Kuhn, The primary structure of collagen, *Int. Rev. Conn. Tissue Res.,* **7,** 1–60, 1976.

Gay, S., and E. J. Miller, *Collagen in the Physiology and Pathology of Connective Tissue,* Fischer, Stuttgart, 1978.

Hulmes, D., A. Miller, D. A. D. Parry, K. A. Piez, and J. Woodhead-Galloway, Analysis of the primary structure of collagen for the origin of molecular packing, *J. Mol. Biol.,* **79,** 137–148, 1973.

Jope, E. M., The evolution of plants and animals under domestication: the contribution of studies at molecular level, *Phil. Trans. Roy. Soc. Lond.,* **275,** 99–116, 1976.

Jope, E. M., and H. M. Jope, The selection and preparation of protein material for radiocarbon dating, *Proc. Ninth Int. Radiocarbon Conf.,* 1976.

Jope, H. M., The proteins of brachiopod shell IV: shell protein from fossil inarticulates, *Comp. Biochem. Physiol.,* **30,** 225–232, 1969.

Jope, H. M., The proteins of brachiopod shell VI: C-terminal end groups and SDS-gel electrophoresis: Molecular constitution and structure of the protein, *Comp. Biochem. Physiol.,* **63b,** 163–173, 1979.

Kim, Y. E., S. K. Lee, U. C. Yoon, and J. S. Kim, Biochemical studies on Antler, *Korean Biochem. J.,* **8,** 89–107, 1975.

King, K., and C. Neville, Isoleucine epimerization for dating marine sediments: importance of analyzing monospecific foraminiferal samples, *Science,* **195,** 1333–1335, 1977.

Lerman, J. C., and E. M. Cigliano, New carbon-14 evidence for six hundred year old Canna compacta see, *Nature,* **232,** 568–569, 1971.

Libby, W. F., *Radiocarbon Dating,* 2nd ed., Univ. of Chicago Press, Chicago, 1955.

MacNeish, R. S., The origins of New World civilization, *Sci. Amer.,* **211,** 29–37, 1964.

Mangelsdorf, P. C., *Corn: its Origin, Evolution and Improvement,* Harvard Univ. Press, Cambridge, 1974.

Miller, A., Chapter 3 in *Biochemistry of Collagen,* G. N. Ramachandran and E. Reddi, eds., Plenum Press, New York, 1976a.

Miller, A, The structure of collagen in *Structure-Function Relationships of Proteins,* R. Markham and R. W. Horne, eds., Elsevier, New York, 1976b.

Miller, A., and D. A. D. Parry, Structure and packing of microfibrils in collagen, *J. Mol. Biol.,* **75,** 441–447, 1973.

Miller, E. J., Biochemical characteristics and biological significance of the genetically distinct collagens, *Mol. Cell. Biochem.,* **13,** 165–192, 1976.

Miller, E. J., G. R. Martin, K. A. Piez, and M. J. Powers, Characterization of chickbone collagen and compositional changes associated with maturation, *J. Biol Chem.,* **242,** 5481–5489, 1967.

Parry, D. A. D., Axial arrangement of α-chains in the Collagen Molecule: polar and apolar interactions, *Polymer,* **18,** 1091, 1977.

Pauling, L., *The Nature of the Chemical Bond and the Structure of Molecules and Crystals,* 3rd ed., Cornell Univ. Press, Ithaca, NY, 1960.

Piez, K. A., and A. Miller, The structure of collagen fibrils, *J. Supramol. Structure* 2, 121–137, 1974.

Piez, K. A., and D. A. Torchia, Possible contribution of ionic clustering to molecular packing of collagen, *Nature,* **258,** 87, 1975.

Piez, K. A., and B. L. Trus, Sequence regularities and packing of collagen molecules, *J. Mol. Biol.,* **122,** 419–432, 1978.

Piez, K. A., and B. L. Trus, Microfibrillar structure and packing of collagen: hydrophobic interactions, *J. Mol. Biol.,* **110,** 701–704, 1977.

Ranterberg, J., and K. Kuhn, The renaturation behaviours of modified collagen molecules, *Hoppe Zeyler's Z. Physiol. Chim.,* **349,** 611–622, 1968.

Rudenko, S. I., *The Frozen Tombs of Siberia, The Pazyryk Burials of Iron Age Horsemen,* translated by M. W. Thompson, Univ. of California, Berkeley, 1970.

Scott, J. E., and T. W. Kyffin, Demineralization in organic solvents by alkylammonium salts of EDTA, *Biochem. J.,* **169,** 697–701, 1978.

Sundar Raj, C. V., R. L. Church, L. A. Klobutcher, and P. H. Tuddle, Genetics of connective tissue protein: assignment of a type I procollagen to chromosome 17 by analysis of cell hybrids and microcell hybrids, *Proc. Nat. Acad. Sci. U.S.A.,* **74,** 4444–4448, 1977.

Sykes, B., and E. Solomon, Assignment of a Type I collagen structural gene to human chromosome 7, *Nature,* **272,** 548–549, 1978.

Torchia, D. A., and D. L. VanderHart, $^{13}C$ Magnetic resonance evidence for anisotropic molecular motion in collagen fibrils, *J. Mol. Biol.,* **104,** 315–322, 1976.

Trus, B. L., and K. A. Piez, Molecular packing of collagen: three-dimensional analysis of electrostatic interactions, *J. Mol. Biol.,* **108,** 705–732, 1976.

Wyckoff, R. W. G., *The Biochemistry of Animal Fossils,* Scientechimca, Bristol, 1972.

Van der Rest, M., W. G. Cole, and F. H. Glorieux, Human collagen fingerprints produced by Clostridiopeptidase A, Digestion and High Pressure Liquid Chromatography, *Biochem. J.,* **161,** 527–534, 1977.

# The Survival of Protein in Bone

**R. E. M. HEDGES**
*Research Laboratory for Archaeology and the History of Art, Oxford, U.K.*

**CARMICHAEL J. A. WALLACE**
*Department of Zoology, Laboratory of Molecular Biophysics, Oxford, U.K.*

This work was undertaken to study the possibility of the survival of genetic information in archaeological bone (Hedges and Wallace 1979). Information at the molecular level was sought (Bodmer and Cavalli-Sforza 1971), and it was therefore natural to consider the survival of protein in buried bone (Garlick 1969). While collagen is the most abundant protein, it is not necessarily the most genetically informative. Our work has not been exclusively concentrated on collagen, but we believe that a study of its fate during burial will also help us understand the situation for other proteins, glycoproteins, and so on.

We took mainly human skeletal compact bone from a Medieval site in Oxford, which had been buried for 700 years in saturated but not anaerobic conditions. The bone was ground (in one case, for the estimation of surviving albumin, at 77°K) and decalcified in EDTA (de Jong et al., 1974). Experimental conditions are outlined below. A detailed description is given in reference 1. Extractions under various conditions of pH were compared, and molecular weight distributions and amino acid compositions were determined for the extracted proteinaceous material. Controls were also run on modern (unburied) human bone. We found that the relative proportion of soluble to insoluble protein was increased in the archaeological samples and that the amino acid compositions had been modified.

Since collagen accounts for such a high proportion of protein in bone, it is difficult to isolate other possible proteins. We have made some attempts to identify other proteins immunochemically (Clausen 1969), in all cases with negative results. Modern controls (from bone) were not run for the immunochemical work, except for the estimation of albumin, where we could establish that the level of albumin surviving in the Medieval bone was less than 1% of its modern level.

We give below an account of the experimental methods, the results obtained, and a discussion of these results in terms of the suggested changes taking place in bone protein during burial.

## EXPERIMENTAL WORK AND RESULTS

Bone samples of 1–5 g were ground and decalcified in 10% EDTA buffered to pH 4, 6, 7, 8, and 10 at room temperature for two days. Some insoluble residue collected, and after dialysis and freeze-drying, the soluble portion and the precipitated nondiffusible residue were subjected to gel filtration.

Table 1 shows a comparison of total soluble and insoluble residue fractions from modern and archaeological bone, at pH 7. Note that the long-buried samples have an increased soluble to insoluble ratio.

Figure 1 shows an elution profile from gel filtration of the soluble fraction from extraction at pH 4. At this pH the proportion of low molecular weight extract is at its lowest, but the curve is similar for extracts at other pH's. The eluted material was analyzed for amino acid composition for each of the five molecular weight ranges. The N-terminal amino acids present were identified after dansylation. However, the complete range

**Table 1. Percentage of Soluble and Insoluble Fractions Extracted from Modern and Archaeological Bone**

| | % Insoluble | % Soluble |
|---|---|---|
| Modern bone | | |
| Sample a | 17 | 1 (wet) |
| Sample b | 23 | 0.7 (wet) |
| Sample c | 58 | 5.5 (dry) |
| Mediaeval bone | | |
| Sample 1 | 13 | 1.5 (dry) |
| Sample 2 | 9 | 1.3 (dry) |
| Sample 3 | 40 | 8.5 (dry) |

of amino acids was found to be present in each fraction, indicating the presence of a complex mixture of polypeptides.

Table 2 shows the amino acids compositions found for the various molecular weight fractions, including the insoluble residue. The same analysis was applied to modern bone extract. A comparison is made with the known composition of collagen (Colla in Table 2).

The insoluble material was further treated with CNBr to cleave the polypeptide chains at methionine residues. The solubilized material was then treated as before, and found to be resolvable into two peaks of molecular weight range 30,000–70,000 and two peaks with molecular weight of about 3000.

A number of immunological tests were made on the soluble fraction, in the hope that intact antigens might be recognized within the mass of chemically varied proteinaceous material of the extract. The results are set out in Table 3.

**Table 2A. Amino Acid Analysis of Ancient Bone**

| Amino acid | Ippt | Isol | LIa | Colla | IIppt | IIsol | IIIppt | IIIsol | Colla | insol |
|---|---|---|---|---|---|---|---|---|---|---|
| Aspartic acid | 5.1 | 5.4 | 5.6 | 4.7 | 17.0 | 6.8 | 19.8 | 8.4 | 4.7 | 5.8 |
| Threonine | 2.3 | 2.9 | 2.4 | 1.8 | 3.3 | 3.0 | 3.8 | 2.8 | 1.8 | 2.2 |
| Serine | 3.7 | 3.8 | 4.0 | 3.6 | 4.5 | 4.3 | 5.0 | 5.9 | 3.6 | 3.8 |
| Glutamic acid | 8.7 | 9.2 | 9.1 | 7.2 | 21.3 | 10.3 | 20.6 | 12.7 | 7.2 | 8.9 |
| Proline | 12.7 | 13.3 | 10.9 | 12.3 | 9.1 | 12.6 | n/c | 8.8 | 12.4 | 14.4 |
| Glycine | 37.2 | 33.8 | 39.1 | 31.9 | 18.1 | 31.2 | 18.0 | 27.6 | 31.9 | 32.7 |
| Alanine | 13.6 | 14.4 | 14.8 | 11.3 | 8.4 | 12.5 | 9.5 | 11.1 | 11.3 | 13.4 |
| Valine | 2.5 | 2.5 | 2.4 | 2.4 | 3.4 | 3.1 | 2.5 | 2.6 | 2.4 | 2.5 |
| Half Cystine | n/c | n/c | n/c | | | | | | | |
| Methionine | n/c | 0.5 | 0.5 | 0.5 | 1.0 | 0.6 | 0.5 | 0.3 | 0.5 | 0.9 |
| Isoleucine | 1.2 | 1.0 | 1.0 | 1.3 | 1.1 | 1.2 | 2.2 | 1.3 | 1.3 | 1.5 |
| Leucine | 3.0 | 3.0 | 2.7 | 2.6 | 4.3 | 3.5 | 5.0 | 4.2 | 2.6 | 3.5 |
| Tyrosine | 0.3 | 0.1 | 0.3 | 0.5 | 0.8 | 0.8 | 1.9 | 2.0 | 0.5 | 0.5 |
| Phenylalanine | 1.6 | 1.5 | 1.5 | 1.4 | 1.3 | 1.6 | 1.8 | 2.6 | 1.4 | 1.4 |
| Histidine | 0.5 | 1.0 | 0.6 | 0.6 | 1.1 | 1.2 | 4.4 | 3.0 | 0.6 | 1.2 |
| Lysine | 2.8 | 3.1 | 2.2 | 2.8 | 1.7 | 2.5 | 1.2 | 2.1 | 2.8 | 2.1 |
| Arginine | 4.8 | 4.7 | 5.3 | 4.7 | 3.5 | 4.9 | 3.7 | 1.2 | 4.7 | 4.8 |
| Molecular weight | >50,000 | | 10,000 to 50,000 | | 3000 to 10,000 | | <3000 | | | |
| % of bone mass | 0.5 | 0.05 | 0.3 | | 0.05 | 0.1 | 1.5 | 1.0 | | |

**Table 2B. Amino Acid Analysis of Modern Bone**

| Amino acid | Ialsol | Colla | IIppt | IIsol | IIIppt | IIIsol | Colla | Insol |
|---|---|---|---|---|---|---|---|---|
| Aspartic acid | 5.2 | 4.7 | 11.6 | 5.9 | 9.1 | 5.6 | 4.7 | 8.3 |
| Threonine | 2.2 | 1.8 | 2.2 | 3.0 | 2.9 | 2.6 | 1.8 | 3.8 |
| Serine | 4.1 | 3.6 | 3.6 | 3.6 | 6.1 | 4.6 | 3.6 | 5.4 |
| Glutamic acid | 10.1 | 7.2 | 13.5 | 8.1 | 13.1 | 10.0 | 7.2 | 10.6 |
| Proline | 12.8 | 12.3 | 14.2 | 11.0 | 10.7 | 10.3 | 12.4 | 10.3 |
| Glycine | 37.3 | 31.9 | 25.9 | 39.4 | 27.5 | 32.1 | 31.9 | 23.3 |
| Alanine | 11.5 | 11.3 | 10.4 | 12.2 | 9.9 | 12.5 | 11.3 | 12.1 |
| Valine | 2.6 | 2.4 | 2.6 | 2.6 | 2.8 | 2.9 | 2.4 | 4.8 |
| Half Cystine | n/c | | | | | | | |
| Methionine | 0.6 | 0.5 | n/c | 0.3 | n/c | 0.6 | 0.5 | n/c |
| Isoleucine | 0.8 | 1.3 | 0.9 | 1.1 | 1.3 | 1.3 | 1.3 | 2.2 |
| Leucine | 2.5 | 2.6 | 4.3 | 2.8 | 3.8 | 4.5 | 2.6 | 6.6 |
| Tyrosine | 0.1 | 0.5 | 1.2 | 0.2 | 1.3 | 0.7 | 0.5 | 1.1 |
| Phenylalanine | 1.4 | 1.4 | 1.6 | 1.4 | 2.2 | 2.6 | 1.4 | 2.3 |
| Histidine | 0.7 | 0.6 | 1.7 | 1.8 | 2.5 | 1.6 | 0.6 | 2.4 |
| Lysine | 2.9 | 2.8 | 1.6 | 2.9 | 1.8 | 1.8 | 2.8 | 2.6 |
| Arginine | 5.2 | 4.7 | 4.6 | 3.8 | 4.9 | 6.4 | 4.7 | 4.2 |
| % of bone mass | 0.3 | | 0.05 | 0.5 | 2.75 | 1.5 | | |

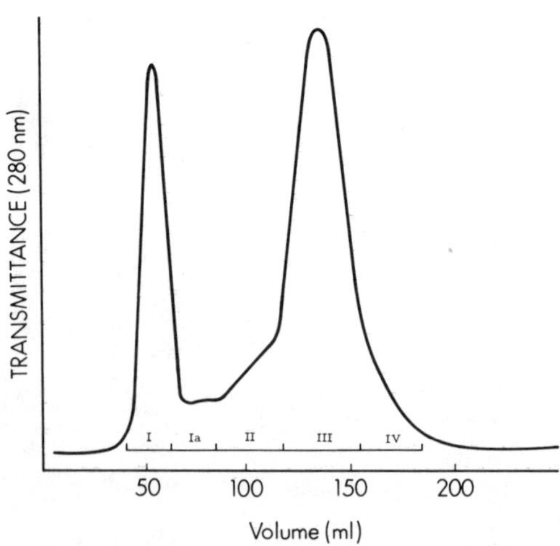


**Figure 1.** Elution profile from gel-filtration of bone extract on Sephadex G75 in 7% HCOOH.

**Table 3.**

| Reaction to | Result |
|---|---|
| (a) Whole human serum anti-serum | Negative |
| (b) A highly specific hemoglobin A anti-serum | Negative |
| (c) The most frequent components of the human H-LA complex | Negative |
| (d) Human globulin and albumin anti-serum | Negative |
| (e) As above, but for modern bone | Positive |
| (f) Human albumin on selected molecular weight fraction | <1% modern bone |
| (g) As above, but on modern bone | |

## DISCUSSION

1. There are important chemical differences between the protein extracted from modern bone and the protein from long-buried bone. The buried bone has a greater proportion of soluble, nonspecific polypeptides in relation to the total organic content. The composition of the soluble fraction for the buried bone closely resembles collagen in amino acid composition on the whole, although some specific fractions (especially the precipitate formed on dialysis, which proved to have a molecular weight below 10,000) were found to be quite different. These same fractions in modern bone are much more collagen-like. On the other hand, the insoluble component in modern bone is *less* collagen-like than for the ancient bone.
2. The loss of other proteins from bone appears to be more complete than the loss of collagen. This is perhaps not surprising, given the structural stability, insolubility, and intimate relationship with the inorganic matrix possessed by collagen, but it does indicate that the study of proteins other than collagen in surviving bone will require extremely sensitive techniques.

These observations suggest that much of the soluble extract in ancient bone is derived from collagen as it degrades on burial. This accounts for the low molecular weight fractions. It is obvious (from Tables 2A and B) that the soluble extract does not reflect the original soluble material in bone. We conclude that at least half of the soluble portion originates in the bone, on the basis of the amino acid composition and the relative change in soluble and insoluble quantities of protein. Possibly the rest of the proteinaceous extract is a result of bacterial processing, with the amount of soluble material in the bone at a steady state where leaching of bacterial and degraded protein is balanced by the rate at which collagen is made soluble. Likewise, the remaining insoluble protein, although shown to be much more collagen-like in its amino acid composition, does not resemble the insoluble protein found in recent bone, presumably because recent bone contains proteins other than collagen which have been lost at a faster rate.

## ENVIRONMENTAL INFLUENCES AND AMINO ACID RACEMIZATION RATES

The organic residue in the bone we studied is a complex mixture of degradative and intrusive material. It is hard to believe that a measurement of the degree of amino acid racemization on the whole mixture would have any significance. While undoubtedly the extent of racemization increases with time, this may reflect the possibility that racemization is dependent on the collagen's becoming partially degraded, a process which itself depends on many factors other than time. It is first desirable to determine if the simple time/temperature relationship assumed for racemization applies to any one well-characterized fraction of the organic residue. The extrapolation from laboratory experiments on collagen at elevated temperatures to the situation for archaeological bone can be made only if it is established that the same chemical process is taking place. The sensitivity of laboratory racemization rates to temperature and of collagen breakdown and extraction to pH suggest that the complex residue in buried bone may well contain much information on its burial environment; for example, knowledge of the bacterial ecology could be archaeologically valuable. However, much more detailed study of the degradative mechanisms will be required before that is possible.

## *REFERENCES*

Bodmer, W. F. and L. L. Cavalli-Sforza, *The Genetics of Human Populations,* Freeman, San Francisco, 1979.

Clausen, J., *Immunochemical Techniques for Identification and Estimation of Macromolecules,* Interscience, Holland, 1969.

De Jong, E. V., P. Westbroek, and J. F. Westbroek, Preservation of antigenic properties of macromolecules over 70Myr, *Nature,* **252,** 63–64, 1974.

Garlick, J. D., in *Science and Archaeology,* D. Brothwell and E. Higgs, eds., Thames and Hudson, London, 437–446, 1969.

Hedges, R. E. M. and C. J. A. Wallace, The survival of biochemical information in archaeological bone, *J. Arch. Sci.,* **5,** 337–386, 1978.

# Immunospecificity of Fossil Collagens

**JEROLD M. LOWENSTEIN**
*University of California Department of Medicine, San Francisco, California*

## *ABSTRACT*

The evolutionary relationships of living species can be evaluated either by morphological comparisons or by molecular comparisons of proteins such as cytochrome c or albumin. The relationships of fossil species to one another and to living species have been estimated only from their morphology. A molecular method for comparing fossils, based on their residual proteins, might provide valuable taxonomic data. To this end, a sensitive radioimmunoassay for collagen has been developed that distinguishes between the collagens of humans, common and pygmy chimpanzees, rhesus monkeys, dogs, cattle, rats, mice, and guinea pigs. Human fossils as old as 1.9 million years have been tested and found to have immunologically reactive collagen with species-specific reactions.

Comparisons of the proteins of living species by amino acid sequencing, immunological methods, or electrophoresis, have provided data independent of morphology for estimating genetic distance between species and evolutionary divergence times (Ayala 1976). The advantage of the molecular over the morphological evidence is that it is quantitative and replicable, whereas morphology can be interpreted differently by different observers. In some cases, the molecular findings conflict with established ideas. For example, protein similarities suggest that the human line diverged from that of the African apes 4 to 5 million years ago (Sarich and Cronin, in Goodman 1976), but it is the prevailing paleontological opinion that the human line began with the Miocene hominoid *Ramapithecus* 15 to 30 million years ago (Simons 1977).

The interpretation of fossils and their place in evolution is based almost entirely on morphological considerations. It would add another dimension to the taxonomy of fossil species if one could make meaningful comparisons of their residual proteins. Numerous studies on fossils have shown the persistence of polypeptides and amino acids for millions of years (Wyckoff 1972). The analytical methods used, however—principally quantitative amino acid determination—have not provided species-specific information.

Determining the species specificity of fossil proteins presents problems. Bone, tooth, and shell proteins change in both quantity and quality with the passage of time. A fresh bone contains 20% protein by weight, but bones a million years old contain only 0.1% or less (Wyckoff 1972). Those protein molecules that remain may have suffered bond breakage, fragmentation, racemization of amino acids, or transformation of one type of amino acid to another. It would therefore require a very sensitive probe to detect species-specific sequences in fossil proteins.

In principal, a radioimmunoassay for collagen might be such a probe. Radioimmunoassay (RIA) combines the sensitivity of radioactive measurements with the specificity of immunological reactions (Odell 1971), making it possible to measure nanogram ($10^{-9}$ g) or picogram ($10^{-12}$ g) amounts of protein in blood or tissues without separating them chemically from the many other factors present. Collagen is the main protein of bones and teeth and is so durable that its presence has been detected by amino acid analysis and electron microscopy in fossils millions of years old (Wyckoff 1972). It is one of the largest known proteins, a triple helix with about 1000 amino acids in each helix and short nonhelical terminal peptides (telopeptides): total molecular weight 300,000. Collagen is species specific (Balazs, 1970; Fietzek, 1976; Furthmayr, 1976; Pikkarainen, 1968), even though there is a great deal of homology in all species from sponges to man.

To carry out a radioimmunoassay, three things are needed: (1) a specific antibody to the target protein (collagen, in this case), (2) an antigen-antibody reaction, with separation of the free and bound fractions, and (3) a radioactive label on either antigen or the antibody (Odell 1971). The ratio of bound to free radioactivity, measured with a scintillation counter, is precisely related to the amount of target protein present.

Species-specific antibodies to collagen can be produced in rabbits, rats, mice, guinea pigs, sheep, and chickens (Adelmann, 1973; Furthmayr, 1976; Kirrange, 1968; Nowack, 1976; Rothbard, 1965). Rabbits tend to make antibodies to the nonhelical terminal peptides, rats to the helical structures (Furthmayr 1976). Antibodies can be made to fractions of the collagen molecule obtained by cyanogen bromide cleavage of the methionine residues. Such antibodies, specific for the N-terminal telopeptides, have been used to distinguish Type I from Type III human collagen (Nowack 1976). Collagen altered by heating or chemical treatment retains some of its species-specific reaction with antibody (Adelmann, 1973; Furthmayr, 1976; Kirrane, 1968). Heating accelerates changes in proteins that occur with aging and fossilization (Abelson, 1956; Hare, 1974, 1980). If heat-denatured collagen retains its immunological species specificity, then fossilized collagen might be expected to do so too.

An RIA for fossil collagen has proved difficult to develop because most fossil collagen (in fact, most collagen) is insoluble at physiological pH 7.4, necessary for antigen-antibody reactions. Adelmann and co-workers (1973) reported an RIA for neutral, salt-soluble collagen, in which collagen was tagged with $^{125}$I and the free and bound fractions separated by double antibody precipitation. In the present investigator's hands, this RIA did not prove applicable to fossils.

During the past two years, I have developed a solid-phase RIA that can detect extremely small amounts of soluble or insoluble collagen and distinguish between the collagens of even closely related species like human and chimpanzee. Solvent pH does not affect the assay. Small polyvinyl microtiter plates are employed. The procedure follows—first, for insoluble bone collagen (Fig. 1):

Bone fragments are ground to a fine powder and decalcified with EDTA 0.2 *M*, 10 volumes for 1–2 weeks, then the insoluble fraction is suspended in phosphate buffered saline (PBS) pH 7.4. Two-hundredths (0.02) ml of this suspension (containing about 10$\mu$g collagen when recent bone is used) is pipetted into the cups of the microtiter plate. Rabbit anti-collagen antibody is added. After 24 hr, the cups are filled three times with 1% bovine serum in PBS and the fluid removed by suction, leaving the insoluble bone collagen in the bottom of the cups. $^{125}$I-labeled goat antibody to rabbit gamma globulin ($^{125}$I GARGG) is added, and after 24 hr washed

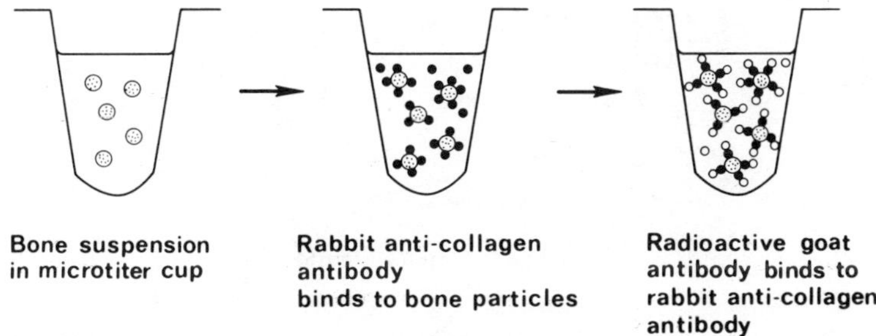


**Figure 1.** Collagen radioimmunoassay for decalcified bone.

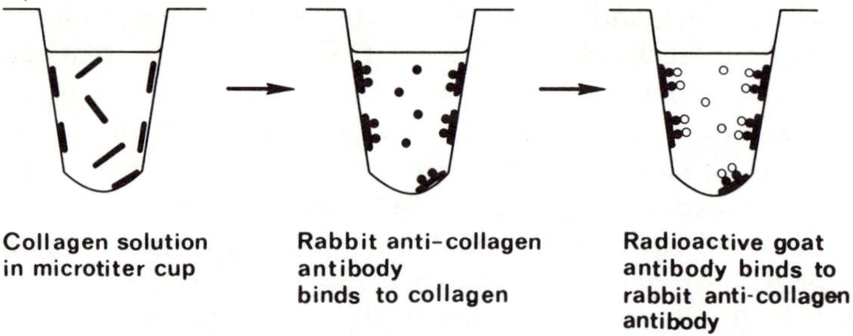


**Figure 2.** Collagen radioimmunoassay for soluble collagen. Collagen binds firmly to the polyvinyl cup, so that the radioactivity in the cup increases with the amount of collagen present.

three times with bovine serum, as above. In this procedure, the rabbit antibody binds to the bone collagen, and the $^{125}$I GARGG binds to the rabbit antibody. The radioactivity in the cups, which are cut from the plate and counted in a scintillation counter, increases with the amount of collagen present. For soluble collagen, the method is identical except for the first step (Fig. 2). Two-hundredths (0.02) ml collagen solution (the usual solvent is acetic acid 0.5 *M*) is pipetted into the microtiter cups and allowed to remain for one hour. During this time, a fraction of the collagen is firmly bound to the polyvinyl cup. The solution is removed by suction, the cups washed twice with bovine serum, and the rabbit and goat antisera added seriatim as above. At the end, the radioactivity in each cup reflects

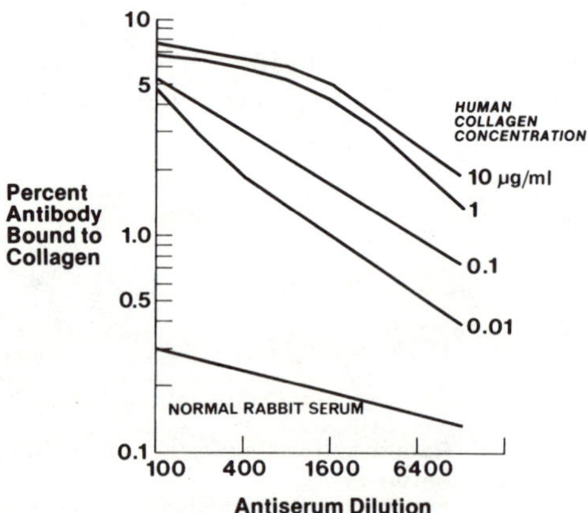


**Figure 3.** Percentage of the radioactive antibody bound to collagen with various human collagen concentrations and serial antiserum dilutions.

the amount of collagen present. The relation of $^{125}I$ binding to collagen and antiserum concentration is shown in Fig. 3. Binding versus collagen concentration for a single dilution of rabbit antiserum (1:1600) is shown in Figure 4. The minimum detectable amount of collagen at that dilution is 5 nanograms.

For these studies, skin collagen of various species was purified by extraction with acetic acid and repeated precipitation with salt (Michaeli 1968). Rabbits were immunized with 5 mg collagen in acetic acid 0.05 *M*, emulsified with Freund's complete adjuvant, injected subcutaneously. After three weeks, a second 5-mg injection was given intraperitoneally without adjuvant. Three weeks later the rabbits were bled. That the antibodies obtained were directed against collagen and not against protein contaminants was confirmed in several ways: by demonstration of specific lines of reaction to gelatin on Ouchterlony plates; by reactions with collagen highly purified by column chromatography; and by inhibition of binding with highly purified cyanogen bromide fractions. High-titer antibodies to human albumin had very low reactions to the human collagen used in this study.

By using equal amounts of acid soluble collagen of various species, good species specificity of the antigen-antibody reactions could be demonstrated. For instance, as shown in Figure 5, anti-rat antibody binds best to rat collagen, anti-guinea pig to guinea pig. The collagens of very closely related primate species, common and pygmy chimpanzees, can be distinguished (Fig. 6), not by the antibody to common chimp, which binds equally to both collagens in the experiment shown, but by the antibody

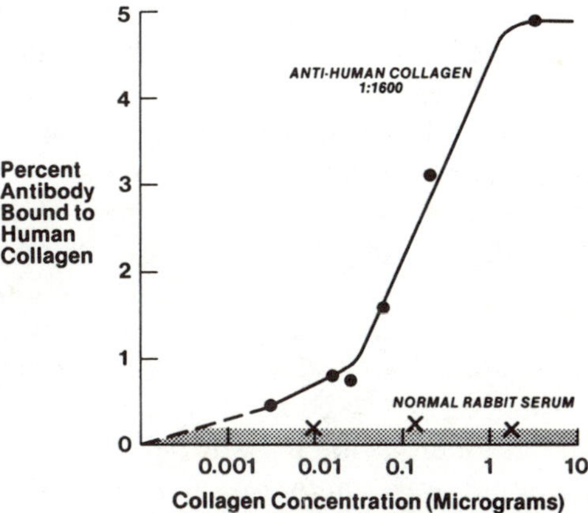


**Figure 4.** For a single antiserum dilution, antibody binding increases rapidly with collagen concentration.

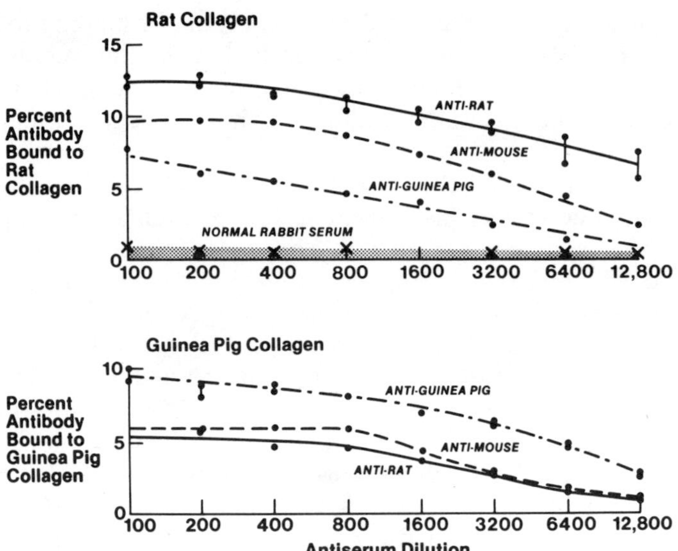


**Figure 5.** Anti-rat serum binds best to rat collagen; anti–guinea pig to guinea pig collagen.

to pygmy chimp, which evidently contains a fraction directed against determinants that differentiate the two species. From these curves, one can calculate the "immunological distance" (ID) between two collagen species, which is ID $= 100 \log_{10}\Delta$ where $\Delta$ is the dilution difference between points of equal binding. Table 1 summarizes the relative binding affinities of anti-collagen antibodies to collagen of nine species. As would be expected, the strongest cross reactions occur within orders—primates to primates, and rodents to rodents. Table 2 summarizes the immunological distances between the various collagen species.

To apply this technique to fossils, it is necessary to separate the effects of aging and fossilization from those of speciation. Collagen, like other proteins, changes with time in a variety of ways: the chains dissociate from one another and break into smaller fragments, amino acids racemize and are converted into other amino acids. Many of these transformations result in loss of some immunological activity. As a preliminary simulation of these changes, collagen dissolved in acetic acid was heated for 24 hours at 100°C and aliquots taken at intervals for RIA. Of the several species tested, different amounts of immunological reactivity were lost, but all retained some species-specific activity at 24 hours (Fig. 7). Hence the method should be appropriate for fossils.

**Table 1. Relative Binding of Anti-Collagen Antisera to Different Species of Collagen**

(In each case, binding of antiserum to the homologous collagen is taken as 1.0)

| Antiserum against | Collagen Human (3) | P.chim (1) | C.chim (1) | Rhesus (1) | Dog (1) | Calf (unk) | Rat (20) | Mouse (20) | GP (20) |
|---|---|---|---|---|---|---|---|---|---|
| Human (4) | 1.0 | 0.76 | 0.84 | 0.51 | 0.26 | 0.49 | 0.28 | 0.31 | 0.32 |
| Pygmy chimp (4) | 0.64 | 1.0 | 0.59 | 0.33 | 0.25 | 0.26 | 0.38 | 0.16 | 0.34 |
| Common chimp (4) | 0.88 | 0.80 | 1.0 | 0.42 | 0.23 | 0.29 | 0.28 | 0.39 | 0.27 |
| Rhesus monkey (3) | 0.42 | 0.26 | 0.13 | 1.0 | 0.21 | 0.29 | 0.39 | 0.24 | 0.26 |
| Dog (4) | 0.40 | 0.24 | 0.24 | 0.21 | 1.0 | 0.32 | 0.12 | 0.63 | 0.40 |
| Calf (4) | 0.77 | 0.50 | 0.78 | 0.28 | 0.11 | 1.0 | 0.37 | 0.10 | 0.17 |
| Rat (1) | 0.58 | 0.30 | 0.65 | 0.19 | 0.35 | 0.54 | 1.0 | 0.90 | 0.74 |
| Mouse (1) | 0.48 | 0.22 | 0.51 | 0.12 | 0.26 | 0.27 | 0.97 | 1.0 | 0.90 |
| Guinea pig (1) | 0.64 | 0.47 | 0.82 | 0.32 | 0.15 | 0.27 | 0.40 | 0.41 | 1.0 |

The numbers in parentheses are the number of individuals of each species from which collagen was obtained; or, in the column on the left, the number of antisera pooled for this experiment. The lack of reciprocity in some cases (e.g., anti-human binds less well to rat collagen than anti-rat to human collagen) may result from the fact that some antisera (like anti-human) are clearly less discriminatory than others (like anti-rat). Antisera from different rabbits show considerable variability in their pattern of cross-reactions, so that a single antiserum (as for rat, mouse, GP) is likely to be less "representative" for that species than multiple pooled antisera.

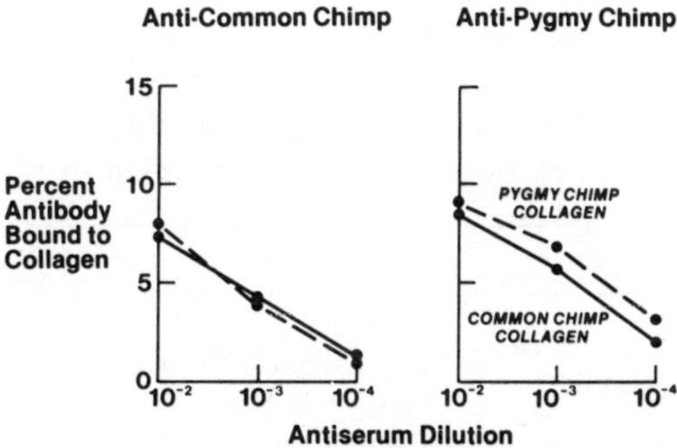


**Figure 6.** Antiserum to common chimpanzee collagen does not distinguish between the collagens of the two chimpanzee species, but antibody to pygmy chimpanzee collagen does. This suggests that the rabbits are responding to an antigenic determinant on pygmy chimpanzee collagen that is not shared by the common species.

**Table 2. Immunological Distances Between Species**

(ID = 100 $\log_{10}$ dilution of antiserum between points of equal binding)

| | P.chim | C.chim | Rhesus | Dog | Calf | Rat | Mouse | GP |
|---|---|---|---|---|---|---|---|---|
| Human | 50 | 45 | 120 | 210 | 120 | 200 | 170 | 150 |
| Pygmy chimp | | 48 | 115 | 175 | 130 | 190 | 210 | 125 |
| Common chimp | | | 130 | 175 | 125 | 205 | 150 | 115 |
| Rhesus | | | | 170 | 190 | 210 | 215 | 120 |
| Dog | | | | | 190 | 235 | 140 | 100 |
| Calf | | | | | | 170 | 230 | 205 |
| Rat | | | | | | | 40 | 90 |
| Mouse | | | | | | | | 85 |

These numbers are each the average of two ID's—for instance, binding of anti-human to human and rat collagen gives one ID, binding of anti-rat to human and rat collagen, another. Ideally, these should be the same, yielding perfect reciprocity. In this system, as discussed in the note for Table 1, there is considerable nonreciprocity, probably due to the small number of individuals and/or antisera sampled. It is assumed that the average shown above is closer to the "true" immunological distance than either of the individual ID's. These correspond very roughly to ID's obtained from other proteins: humans and chimpanzees are close, as are rats, mice, and guinea pigs. There are clear discrepancies also: humans are much farther from cattle than from rhesus monkeys, though these ID's are the same. Perhaps larger numbers of collagens and antisera will adjust these anomalies. (GP on top line is guinea pig.)

Collagen RIA was done on a series of human fossils ranging in age from 1000 to 1.9 million years. Approximately 20 mg of each fossil bone was ground to a fine powder and decalcified for two weeks with 5 ml EDTA 0.2 *M*. Three fractions were obtained for RIA: (1) the EDTA solution, (2) the acetic acid 0.5 *M* extract of the decalcified bone powder, and (3) the insoluble decalcified bone powder, suspended in phosphate-buffered

| Species | Source | Age (approx.) |
|---|---|---|
| Homo sapiens | Hungary | 1,000 years |
| Homo sapiens | Cro-Magnon France | 20,000 |
| Neanderthal | Qafzeh Israel | 50,000 |
| Homo erectus | Briache France | 250,000 |
| Homo erectus | Vertesszöllos Hungary | 500,000 |
| Australopithecus robustus | Omo Ethiopia | 1,900,000 |

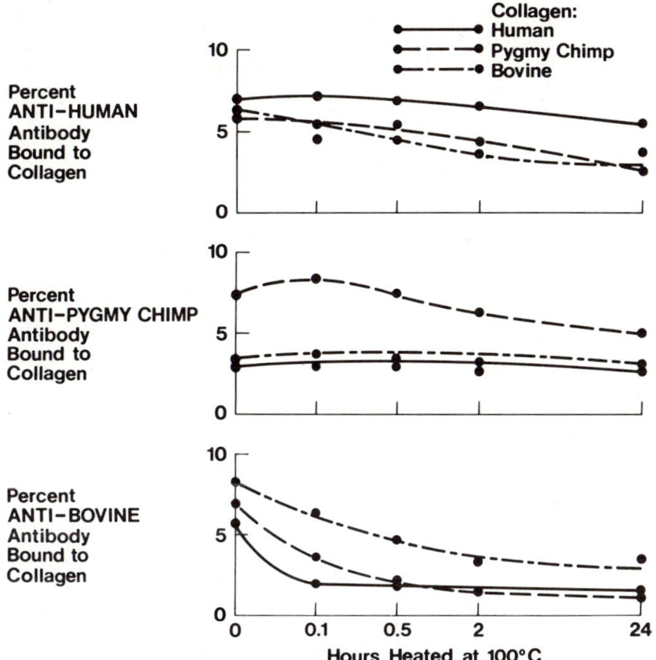


**Figure 7.** Heating collagen in acetic acid solution at 100°C for 24 hours reduced its immunological reactivity but did not abolish the species specificity.

saline (PBS). All three fractions were tested for their ability to bind a 1:1000 dilution of anti-human collagen and anti–common chimpanzee collagen antisera. Purified human collagen was used as the "zero time" control, bone ash containing no organic material was the "zero collagen" control. Total collagen reactivity of each specimen was the sum of antibody binding to all three fractions, shown in Figure 8. The more recent specimens react more strongly with antibody to human collagen; the older specimens react more strongly with antibody to common chimpanzee collagen.

These preliminary experiments demonstrate that immunologically reactive collagen or collagen fragments persist in human fossils for at least two million years and that species-specific binding to antibody is detectable. Collagen radioimmunoassay may become a useful method for investigating genetic relationships among fossil species.

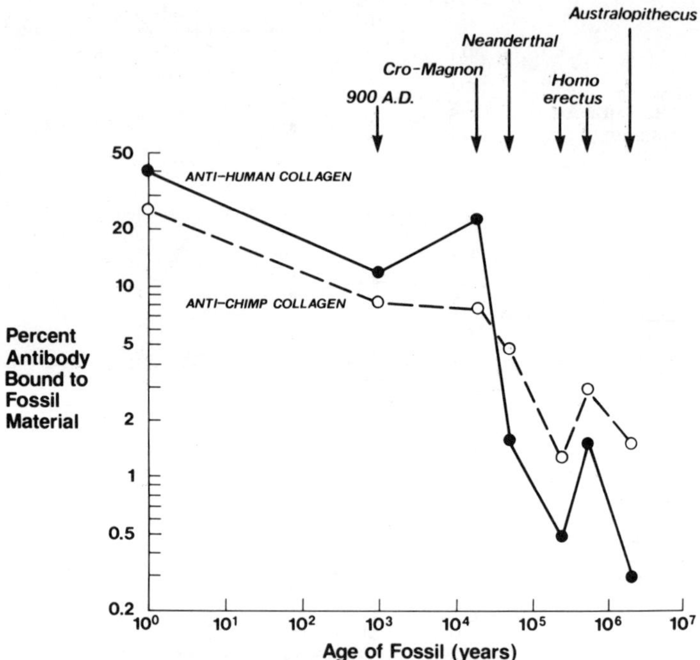


**Figure 8.** Binding of anti-human and anti–common chimpanzee antisera to purified human collagen and to human and hominid fossils going back 1.9 million years. Immunological activity decreases with time and changes from being "like human" in the more recent fossils to being more "like chimpanzee" in the older ones.

## *ACKNOWLEDGMENTS*

This research was supported in part by grants from the L. S. B. Leakey Foundation and the Research Committee of the University of California, San Francisco. I thank Adrienne L. Zihlman for primate material and for valuable suggestions; Dov Michaeli, Robert Siegel, and Steffen Gay for purified collagens, antibodies, and guidance; Sherwood L. Washburn, John E. Cronin, and Vincent Sarich for helpful discussions; Philip Grant and the Yerkes Regional Primate Center for primate skins; and P. E. Hare, F. Clark Howell, Imre Lengyel, Yves Coppens, Farish Jenkins, and Bernard Vandermeersch for fossil material. William Lew and Heinz and Gary Scheuenstuhl provided invaluable technical assistance.

## *REFERENCES*

Abelson, P. H., Paleobiochemistry, *Sci. Amer.*, **195**(1), 83–92, 1956.

Adelmann, B. C., and J. A. Kirrane, The structural basis of cell-mediated immunological reactions of collagen; the species-specificity of the cutaneous delayed hypersensitivity reactions, *Immunology*, **25**, 123–130, 1973.

Adelmann, B. C., G. J. Gentner, and K. Hopper, A sensitive radioimmunoassay for collagen, *J. Immunol. Methods*, **3**, 319–334, 1973.

Ayala, F. J., ed., *Molecular Evolution*, Sinauer Assoc., Sunderland, MA, 1976.

Balazs, E. A., ed., *Chemistry and Molecular Biology of the Intercellular Matrix*, Vol. 1, Academic Press, 1970.

Carmichael, D. J. et al., A partial biochemical characterization of fossilized bone from Makapansgat, Swartkrans, and Queen Charlotte Islands, *Comp. Biochem. Physiol. (B)*, **51**, 257–262, 1975.

Epstein, E. H., R. D. Scott, E. J. Miller, and K. A. Piez, Isolation and characterization of the peptides derived from soluble human and baboon skin collagen after cyanogen bromide cleavage, *J. Biol. Chem.*, **246**, 1718–1724.

Fietzek, P. P. and K. Kuhn, The primary structure of collagen, *Int. Rev. Connective Tissue Res.*, **7**, 61–101, 1976.

Furthmayr, H., and R. Timpl, Immunochemistry of collagens and procollagens, *Int. Rev. Connective Tissue Res.*, **7**, 61–101, 1976.

Goodman, M., ed., *Molecular Anthropology*, Plenum, New York, 1976.

Hare, P. E., G. H. Miller, and N. C. Tuross, Simulation of natural hydrolysis of proteins in fossils, *Carnegie Inst. Wash. Year Book*, **74**, 609–612, 1974.

Hare, P. E., Organic geochemistry of bone and its relation to the survival of bone in the natural environment, in *Taphonomy and Vertebrate Paleontology*, Behrensmeyer, A. K., and Hill, A., Eds., Univ. of Chicago Press, 1980.

Kirrane, J. A. and W. B. Robertson, The antigenicity of native and tyrosylated neutral-salt-soluble rat collagen, *Immunology*, **14**, 139–149, 1968.

Michaeli, D., H. Kamenecka, E. Benjamini, et al., Immunochemical studies on collagen—I. The binding and precipitation of guinea-pig skin collagen and gelatin with rabbit anti-collagen, *Immunochemistry*, **5**, 433–440, 1968.

Nowack, H., S. Gay, G. Wick, et al., Preparation and use in immunohistology of antibodies specific for Type I and Type III collagen and procollagen, *J. Immunol. Methods*, **12**, 117–124, 1976.

Odell, W. D., and H. Daughaday, eds., *Principles of Competitive Protein-binding Assays*, J. B. Lippincott, Philadelphia, 1971.

Pikkarainen, J., The molecular structures of vertebrate skin collagens. A comparative study, *Acta Physiol. Scand. Supp.* **309**, 1968.

Rothbard, S., and R. F. Watson, Immunological relations among various animal collagens, *J. Exp. Med.*, **122**, 441–454, 1965.

Simons, E. L., Ramapithecus, *Sci. Amer.*, **236**(5), 28–35, 1977.

Wyckoff, R. W. G., *The Biochemistry of Animal Fossils*, Williams and Wilkins, Baltimore, 1972.

# Collagen in Fossil Bones

**NOREEN TUROSS**
*Carnegie Institution of Washington, Washington, D.C.*

**DAVID R. EYRE**
**MARŸKE E. HOLTROP**
**MELVIN J. GLIMCHER**
*Children's Hospital Medical Center, Boston, Massachusetts*

**P. E. HARE**
*Carnegie Institution of Washington, Washington, D.C.*

## INTRODUCTION

The study of protein degradation in bone is largely the study of collagen, which makes up 90%–95% of the total protein content in bone. In electron microscopy studies, fossil bone samples have shown 640 Å striations, characteristic of collagen (Wyckoff and Doberenz 1965). Collagen-like amino acid patterns have been seen in the acid hydrolysates of the bone and dentin of many fossil mammals (Ho 1965). In recent years major emphasis has been given to the study of racemization rates of amino acids in fossil bones (Bada 1972; Williams and Smith 1977).

The present study assesses the molecular integrity of the collagen molecule in several fossil bones. Significant degradation was found in all cases when extracted material from fossil bones was compared to that extracted from modern bone. Intact α chains and Type I collagen cyanogen bromide peptides were seen in two of the samples. The study suggests that in fossil bones there is complex diagenesis of the collagen molecule in which the stabilizing and destabilizing forces are not fully understood.

## MATERIALS AND METHODS

Normal human compact bone and bovine compact bone were used as controls throughout the study. Five fossil samples of varying ages were chosen (Table 1).

**Table 1. Ages and Nitrogen Contents in Bones**

| Bone Sample | Estimated Age (yr B.P.) | % Nitrogen |
|---|---|---|
| Modern (bovine) | – | 5.2 |
| Baffin Island whale | >53,000 | 4.8 |
| Delaware whale | <500 | 3.8 |
| La Brea smilodon | 12,650 ± 160 | 5.0 |
| New Zealand moa | >500 | 3.0 |
| Egyptian mummy | 2,850 ± 53 | 3.6 |

## Bone Preparation and Demineralization

The bones were cut into small pieces and demineralized in 0.5 *M* EDTA, pH 7.5 at 4°C for several days. For pepsin extractions, whole bone chunks were powdered in a Spex freezer/mill for 3 minutes. The bone was then demineralized as described above. The samples were washed thoroughly with water and freeze-dried.

## Nitrogen and Amino Acid Analysis

Small chunks (2–20 mg) of whole bone were used in a carbon/hydrogen/nitrogen analyzer (F & M model 185). Cyclohexanone 2-4 dinitrophenylhydrazone was used as a standard.

For the amino acid analysis, whole bone and pepsin solubilized material was hydrolyzed in 6 *N* HC1 under nitrogen at 157°C for 20 minutes. A column, 0.2 × 25 cm, filled with DC-4A cation exchange resin was eluted with citrate buffers of pH 2.94, 3.25, 4.00, 5.50 and a borate buffer of pH 11.4. Both ninhydrin and o-phthaldehyde were used in the detection of the amino acids (Hare 1977).

## Extraction with Guanidine Hydrochloride

Chunks of demineralized bone were extracted by stirring in 4 *M* guanidine hydrochloride (l ml/ 4 gram of dry tissue) for 24 hours at 25°C (Glimcher and Katz 1964). The supernatant was dialyzed against water and freeze-dried.

## Digestion with Pepsin

Demineralized bone powder was suspended in 3% acetic acid (l ml/ 2 g dry tissue) and pepsin added ( 1 g enzyme/ 10 g dry tissue). Digestion proceeded at 4°C for 24 hours (Miller 1972). The solution was raised to pH 8 and centrifuged at 4°C for one hour. The supernatant was dialyzed against water and freeze-dried.

### Cleavage with Cyanogen Bromide

Demineralized bone chunks were solubilized with cyanogen bromide in 70% formic acid for 20 hours at 25°C (Eyre and Glimcher 1973). The solution was diluted tenfold with water and freeze-dried.

### Slab Electrophoresis in Sodium Dodecyl Sulphate Polyacrylamide Gels

The procedure described by Neville (1971) was used with a lower gel buffer of pH 9.18. Ten percent (w/v) acrylamide gels were used. The gels were stained with Coomassie Brilliant Blue in water-isopropanol-glacial acetic acid (13: 5: 2) for 24 hours at 4°C. Destaining was done in 7% acetic acid at 25°C.

### Electronmicroscopy

Fossil specimens were fixed in 2.5% glutaraldehyde in 0.1 *M* cacodylate (1 vol) mixed with 1% $OsO_4$ in 0.1 *M* cacodylate (2 vols) for 3 hours. They were then decalcified in 7.5% EDTA, 2.5% glutaraldehyde, 0.1 *M* cacodylate, pH 7.5, for 1 week, and embedded in Spurr's embedding medium. Thin sections (silver) were stained with uranyl acetate and lead citrate on the grid.

The amino acid analyses of both the whole bone and pepsin extracted material are consistent with a collagen amino acid pattern. Glycine was present as approximately one third of all residues, and hydroxyproline and hydroxylysine were present in all samples.

### Guanidine Hydrochloride Solubilized Collagen

Guanidine hydrochloride solubilized varying amounts of material from the fossil samples (Table 2). Electrophoresis in SDS polyacrylamide gels showed that much of the solubilized material was degraded to a wide range of molecular weight fragments (Fig. 1). The Baffin Island and Delaware whale extracts showed faint bands corresponding to the α chains of the collagen molecule seen in the modern bone extract. No discrete molecular weight units were seen in the Egyptian mummy, the La Brea smilodon, or the New Zealand moa.

**Table 2. Percent of the Demineralized Matrix Solubilized in 4 *M* Guanidine Hydrochloride in 24 Hours at 25°C.**

| Bone Sample | % Solubilized |
|---|---|
| Modern (human) | 5 |
| Baffin Island whale | 9 |
| Delaware whale | 11 |
| La Brea smilodon | 13 |
| New Zealand moa | 16 |
| Egyptian mummy | 74 |

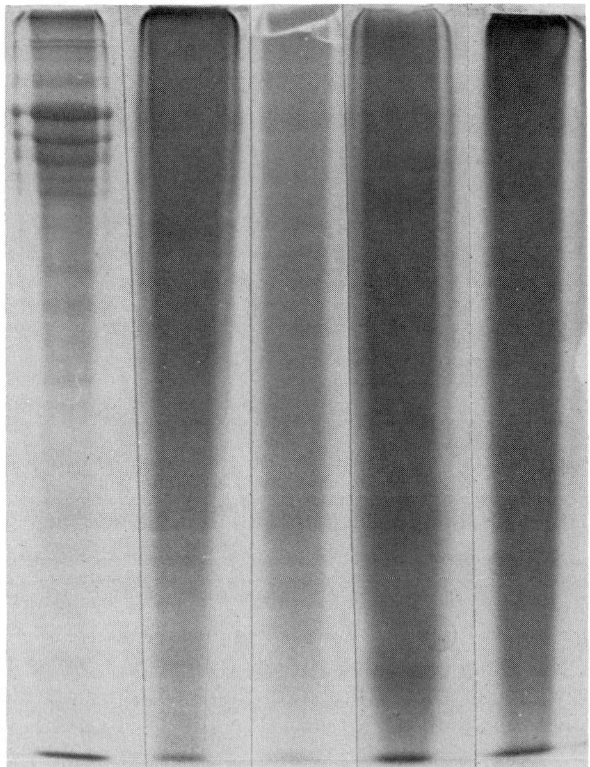

**Figure 1.** Slab electrophoresis in an SDS polyacrylamide gel of guanidine hydrochloride solubilized collagen. L to R, modern human, New Zealand moa, La Brea smilodon, Baffin Island whale, Egyptian mummy.

## Pepsin-Solubilized Collagen

On SDS polyacrylamide gels, protein bands corresponding to α chains could be seen in the Baffin Island and Delaware bone digests. The bands were greatly diminished in intensity as compared to the bands from the modern bone digest. No banding could be seen in the extracts of the Egyptian mummy, the La Brea smilodon, or the New Zealand moa.

## Cyanogen Bromide Peptides

Several of the fossils were resistant to digestion by cyanogen bromide, and samples were not fully solubilized. Peptides from the modern human bone showed the expected Type I collagen pattern when fractionated by electrophoresis on SDS polyacrylamide gels (Fig. 2). The Baffin Island

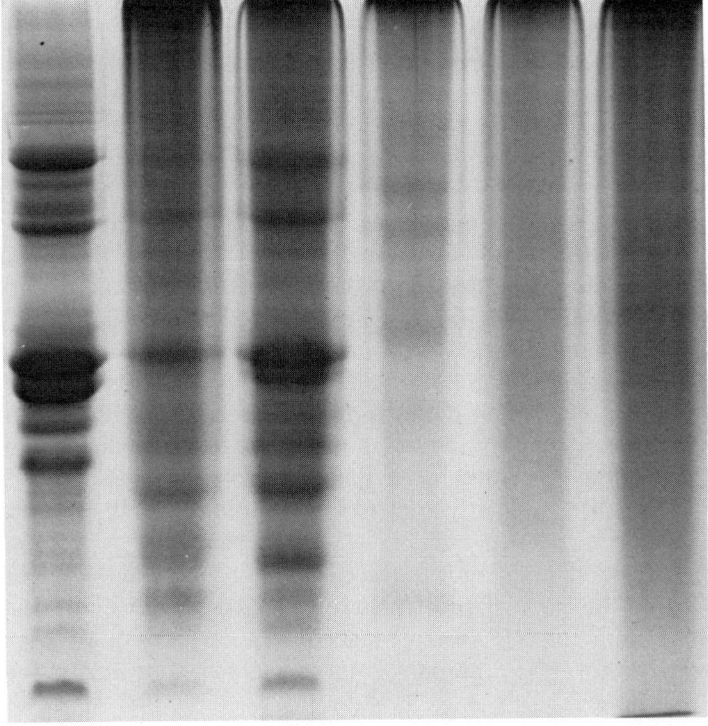

**Figure 2.** Slab electrophoresis in an SDS polyacrylamide gel of cyanogen bromide digested peptides. L to R, modern human, Baffin Island whale, Delaware whale, New Zealand moa, La Brea smilodon, Egyptain mummy. One half the amount of modern bone peptides were loaded on the gel, as compared to the five fossil samples.

and Delaware whale digests showed peptide banding, some of which corresponds to Type I collagen CNBr peptides. Additional non-Type I collagen peptides were seen in the Baffin Island and Delaware whale extracts. Different but also non-Type I collagen peptides were seen in the New Zealand moa extract. No peptide banding could be seen when CNBr extracts of the LaBrea smilodon or the Egyptain mummy were fractionated on polyacrylamide gels. The dark staining at the top of each of the fossil extract gels suggests that a large amount of protein remained undigested.

## DISCUSSION

The present work suggests that much of the collagen in fossil bones has become severely degraded with time. It is now clear that the degradation will not necessarily be reflected in an amino acid analysis or by electronmicroscopy of the fibrils. Nor can degradation be assumed to correspond to the age of the fossil. The oldest sample, the Baffin Island whale, had the best preserved collagenous material. This could well be due to the low temperatures at which this bone has been stored for most of its history.

The wide rage of molecular weight material seen in the guanidine hydrochloride extracts suggests that extensive peptide cleavage has occurred along the collagen molecules in the fossil bone samples. The molecular fragments must have remained *in situ* in the fibrils, however, to account for the excellent preservation seen by electronmicroscopy. Presumably, glutaraldehyde fixation before decalcification prevented the molecular fragments from falling apart.

The gels of the guanidine hydrochloride and cyanogen bromide extracts show a much darker stain down the full length of the gel, compared to the gels of the pepsin extracted material. This could be due to the pepsin further degrading the already partially hydrolyzed collagen to smaller molecular weight peptides which would be lost on dialysis (Tuross and Hare 1978).

The fossil bones' resistance to the CNBr digestion and the dark staining at the top of the gels of the CNBr digest suggest an alteration in the methionine molecule. It seems reasonable that methionine could have been oxidized to either the sulfoxide or the sulphone in the history of these fossil bones.

The existence of the collagen degradation products in fossil bones through such long periods suggests that the lysyl and hydroxylysyl derived crosslinks in the collagen molecule may play a role in stabilizing the breakdown products. It is also possible that additional crosslinking may occur in the geologic setting.

## RESULTS

### Electronmicroscopy

Figures 3 to 6 show that the collagen fibrils are well preserved in the fossil bones. The typical 64 nm periodicity is evident in many of the fibrils. In contrast to modern bone, however, all specimens show some degradation of collagen fibrils, with amorphous regions and blurred fibrils that lack periodicity.

### Nitrogen and Amino Acid Analysis

The high percentage of nitrogen (Table 1) confirms that a significant amount of proteinaceous material remains in the fossil bones.

The amino acid analyses of both the whole bone and pepsin extracted material are consistent with a collagen amino acid pattern. Glycine was present as about one third of all residues, and hydroxyproline and hydroxylysine were present in all samples.

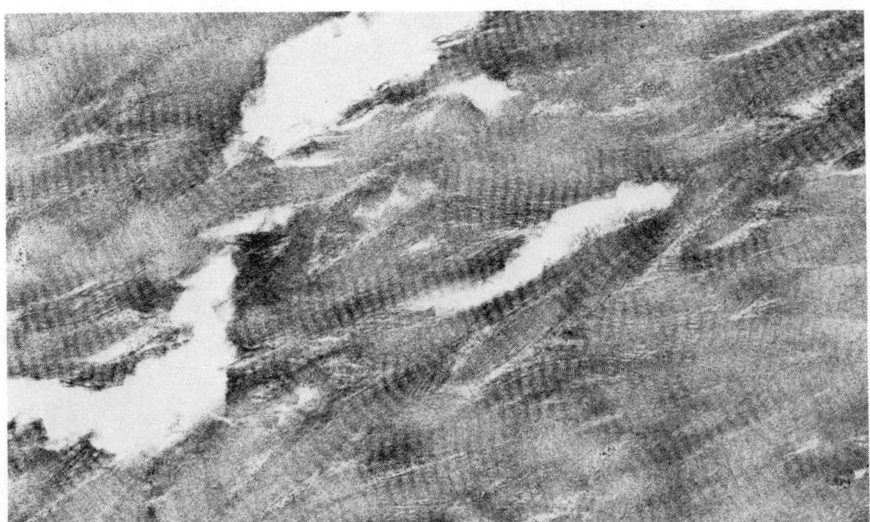

**Figure 3.** Electronmicrograph of Baffin Island whale bone. ×28,200.

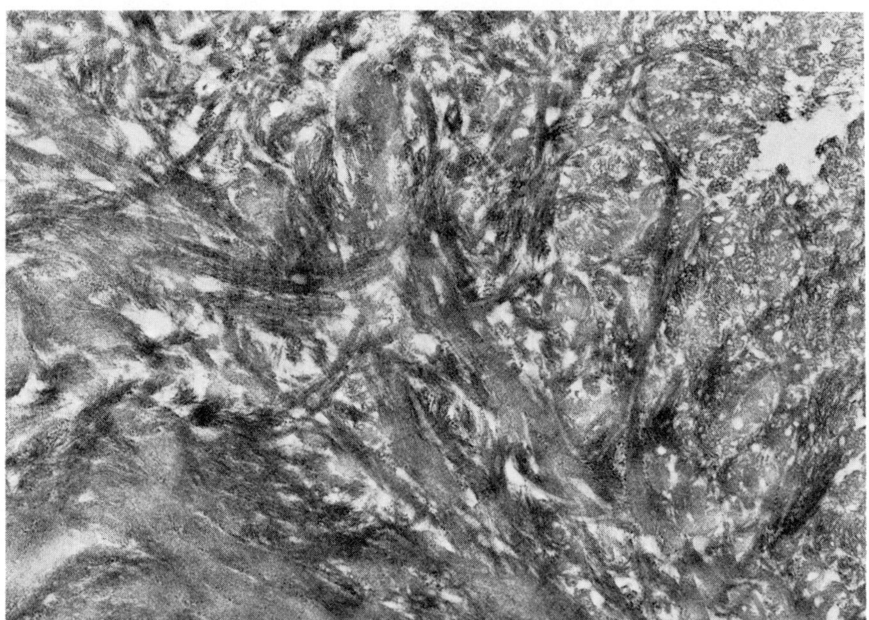

**Figure 4.** Electronmicrograph of La Brea smilodon bone. ×23,500.

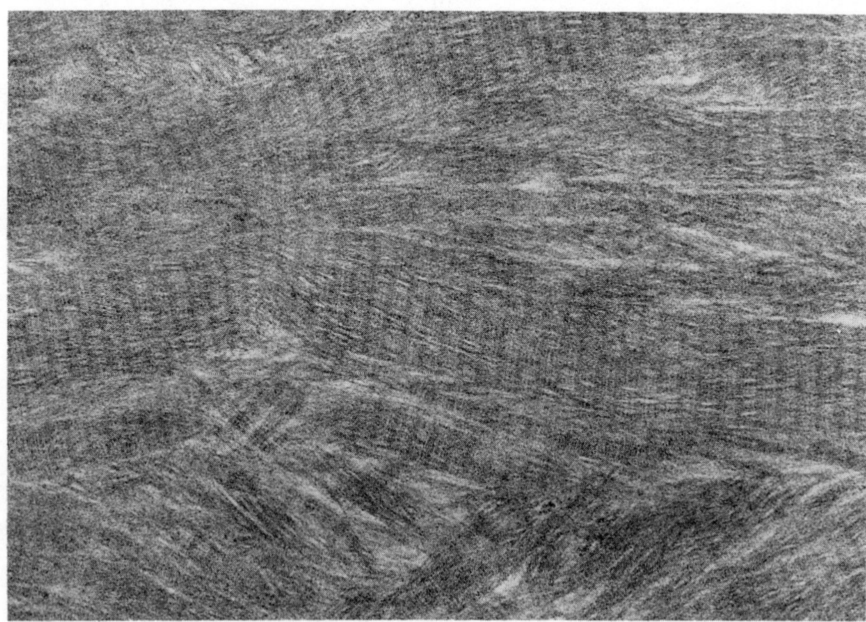

**Figure 5.** Electronmicrograph of New Zealand moa bone. ×48,250.

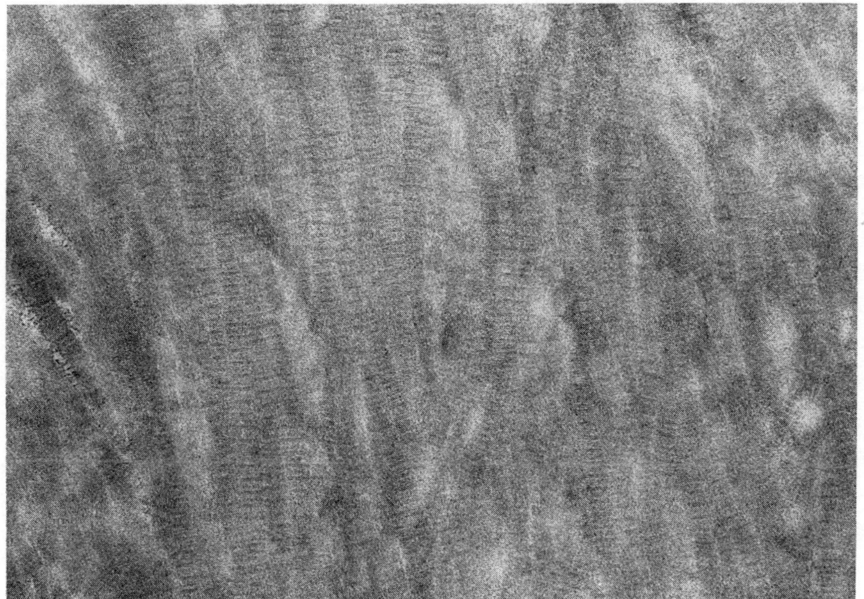

**Figure 6.** Electronmicrograph of Egyptian mummy bone. ×30,750.

## Guanidine Hydrochloride Solubilized Collagen

Guanidine hydrochloride solubilized varying amounts of material from the fossil samples (Table 2). Electrophoresis in SDS polyacrylamide gels showed that much of the solubilized material was degraded to a wide range of molecular weight fragments (Fig. 1). The Baffin Island and Delaware whale extracts showed faintbands corresponding to the α chains of the collagen molecule seen in the modern bone extract. No discrete molecular weight units were seen in the Egyptian mummy, the La Brea smilodon or the New Zealand moa.

## Pepsin-Solubilized Collagen

On SDS polyacrylamide gels, protein bands corresponding to α chains could be seen in the Baffin Island and Delaware bone digests. The bands were greatly diminished in intensity compared to the bands from the modern bone digest. No banding could be seen in the extracts of the Egyptian mummy, the La Brea smilodon, or the New Zealand moa.

## Cyanogen Bromide Peptides

Several of the fossils were resistant to digestion by cyanogen bromide, and samples were not fully solubilized. Peptides from the modern human bone showed the expected Type I collagen pattern when fractionated by electrophoresis on SDS polyacrylamide gels (Fig. 2). The Baffin Island and Delaware whale digests showed peptide banding, some of which corresponds to Type I collagen CNBr peptides. Additional non-Type I collagen peptides were seen in the Baffin Island and Delaware whale extracts. Different, but also non-Type I collagen peptides were seen in the New Zealand moa extract. No peptide banding could be seen when CNBr extracts of the La Brea smilodon or the Egyptain mummy were fractionated on polyacrylamide gels. The dark staining at the top of each of the fossil extract gels suggests that a large amount of protein remained undigested.

The study clearly shows that biochemical investigations, such as isolating and sequencing individual peptides from fossil bone samples, will be a formidible task. The fossil bones used in this study were a priori good candidates for possessing intact collagen. Most fossil bones a thousand years old or more will not be likely to yield much information by conventional means in the way of intact discrete peptide units.

## *ACKNOWLEDGMENTS*

We thank Ms. Cynthia Richardson for electronmicroscopy and Ms. Linda Lewi for gel electrophoresis. Supported in part by a grant to MJG from the Macy Foundation.

## *REFERENCES*

Bada, J. L., The dating of fossil bones using racemization of isoleucine, *Earth Planet. Sci Lett.*, **15**, 223–231, 1972.

Eyre, David R., and Melvin J. Glimcher, Isolation of cross-linked peptides from collagen of chicken bone, *Biochemical J.*, **135**, 393–403, 1973.

Glimcher, Melvin J., and Elton P. Katz, The organization of collagen in bone:The role of noncovalent bonds in the relative insolubility of bone collagen, *J. Ultrastructure Research*, **12**, 705–729, 1965.

Hare, P. E., Subnanomole-range amino acid analysis, in *Methods in Enzymology*, Vol. XLVII,3, C. H. W. Hirs and S. H. Timasheff, eds., Academic Press, New York, pp. 3–18, 1977.

Ho, Tong-Yun, Amino acid composition of bone and tooth proteins in late Pleistocene mammals, *Proc. Nat. Acad. Sci. U.S.A.*, **54**, 26–31, 1965.

Miller, Edward J., Structural studies on cartilage collagen employing limited cleavage and solubilization with pepsin, *Biochemistry,* **11,** 4903–4909, 1972.

Neville, David M., Molecular weight determination of protein-dodecyl sulphate complexes by gel electrophoresis in a discontinuous buffer system, *J. Biol. Chem.*, **246,** 6328–6334, 1971.

Tuross, Noreen, and P. E. Hare, Collagen in fossil bone, *Carnegie, Inst. Wash. Year Book,* **77,** 891, 1978.

Williams, K. M., and G. G. Smith, A critical evaluation of the application of amino acid racemization to geochronology and geothermometry, *Origin of Life,* **8,** 91–144, 1977.

Wyckoff, R. W. G., and A. R. Doberenz, Electron microscopy of Rancho La Brea bone, *Proc. Nat. Acad Sci. U.S.A.,* **53,** 230–233, 1965.

# Preserved Organic Ultrastructure: An Unreliable Indicator for Paleozoic Amino Acid Biogeochemistry

KENNETH M. TOWE
*Department of Paleobiology, Smithsonian Institution, Washington, D.C.*

## INTRODUCTION

The physical preservation of invertebrate and vertebrate organic materials in ancient sediments has been used as both a criterion for sample selection in amino acid biogeochemical studies and as *ex post facto* evidence for the chemical integrity of such samples. The purpose of this paper is to introduce a note of caution regarding the reliability of such evidence. To this end, two types of fossil organic materials will be described. One of these is a calcified matrix and the other is an unmineralized organic skeleton. Both are from Paleozoic rocks more than 350 million years old.

## THE TRILOBITE CUTICLE

A calcified organic matrix occurs within the cuticular exoskeleton of the common Devonian trilobite arthropod *Phacops rana*. Numerous specimens of trilobites of this genus are available from rocks of Devonian age near Buffalo, New York. Most of them appear to be well-preserved because the delicate microstructural features of the carapace are unaltered and the mineralogy of microcrystalline calcite having a preferred orientation shows no evidence of substantial recrystallization. The organic matrix of this species was examined as part of a larger study of this long extinct group of arthropods (Teigler and Towe 1975).

Fragments of the exoskeleton of *Phacops rana* were freed from the enclosing sedimentary rock. They were placed in vessels containing EDTA solutions of pH>7 and left undisturbed until the calcareous ma-

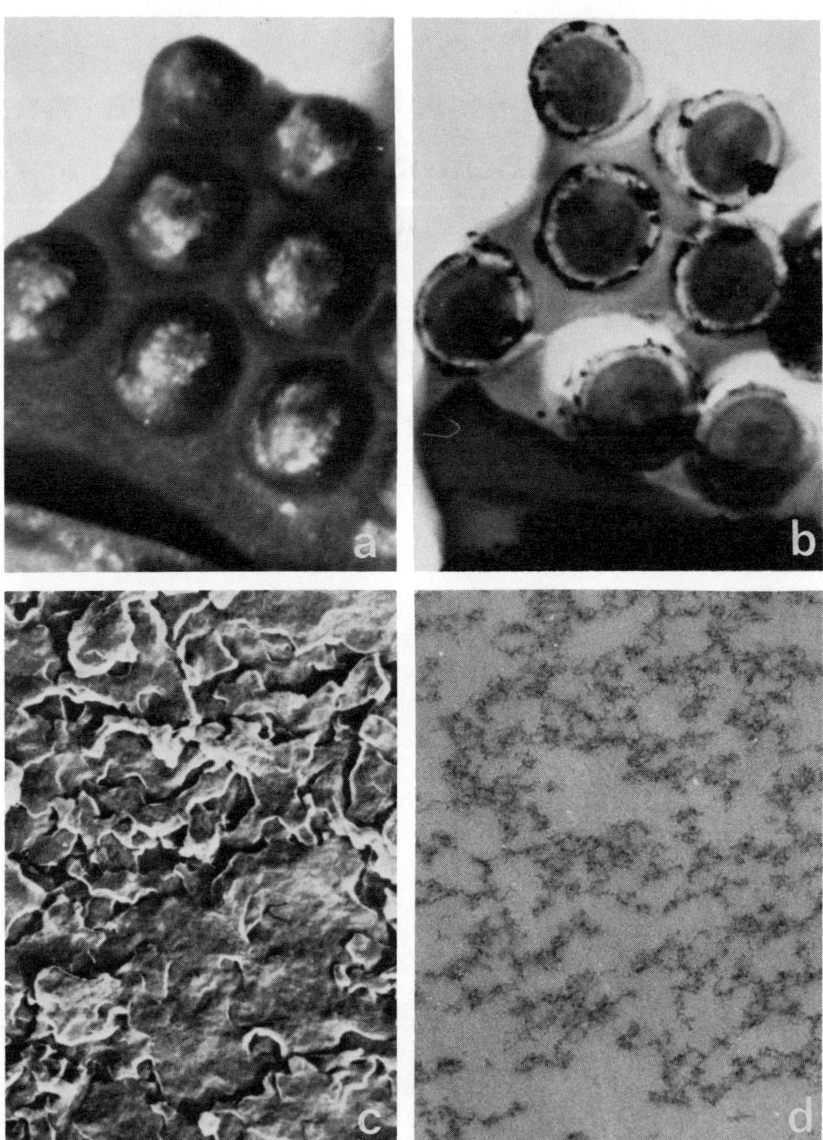


**Figure 1.** A. Fragment of calcified exoskeleton of the trilobite *Phacops rana* including several calcified eye lenses each about 0.5 mm. in diameter. B. Same fragment after decalcification in EDTA. Dark blotches around the eye lenses are pyrite; everything else is the organic matrix. C. Etched surface of the trilobite cuticle showing the irregular boundaries between calcitic elements. D. Decalcified exoskeletal matrix, embedded and sectioned. Magnification of C, D is about 8000×. (after Teigler and Towe 1975).

terial was dissolved. The 'before-and-after' results of such treatment are shown in Figure l,A and B. This fragment of exoskeletal material included several of the calcified eye lenses (Towe 1973). The decalcified organic matrix (Fig. 1B) retains the shape of the original fragment (Fig. 1A) to a high degree. This matrix is brownish colored and very delicate, subject to breakage with any but the slightest movement in the EDTA solution. In spite of this exceptional fragility, small pieces of intact matrix were embedded for ultrathin- section preparation. Viewed in the transmission electron microscope, unfixed and unstained, the sectioned matrix shows a reticulate meshwork (Fig. 1C) that is consistent with an organic material isolated from the calcified cuticle, as shown in Figure 1D. The size of the openings in the matrix and the size of the calcitic elements in the mineralized cuticle from which it was derived are in reasonable agreement with one another. The organic matrix of these trilobites thus appears to be very well preserved physically.

Using appropriate precautions to avoid contamination (Hare 1965), similar fragments of *Phacops rana* exoskeleton were decalcified and acid hydrolyzed for amino acid analysis. The analyses were performed by P. E. Hare at Carnegie Institution's Geophysical Laboratory, Washington, D. C. In spite of the physical preservation that one can see in the matrix, there were no meaningful peptide-bound amino acids detectable at a sensitivity of $10^{-9}$ moles. Because the living arthropod cuticle is composed of chitin and protein, a careful search for the amino-sugar glucosamine was made, but it was not found. The organic matrix of these trilobites is completely degraded biochemically, having been altered to an undefined carbonaceous material. The physical preservation of the organic matrix provided no clues to its chemical degradation, offering instead a misleading criterion for the biogeochemical work we attempted.

## THE GRAPTOLITE PERIDERM

The second type of fossil organic material to be described is the skeletal periderm of another extinct group of animals, the colonial graptolites. Unlike the trilobites and many of the other animals commonly preserved as fossils, the graptolite skeleton was completely unmineralized in life, and there was no protective calcite or other inorganic phase to inhibit postmortem diagenetic alteration. Although most commonly found preserved as flattened films on rock surfaces, three-dimensional preservation has been reported from some limestones, especially those of northern Europe. Early work with the light microscope (Kozlowski 1949) revealed that microstructural preservation of such material was often good. A wide variety of similar graptolite fragments were isolated for study. Compared to the trilobite organic matrices described above, the graptolite material is robust and resistant to either mechanical or chemical treatment. Ultrathin sections of epoxy-embedded fragments were examined, unfixed and unstained, in the transmission electron microscope. The complete results of these studies have been published in a series of papers on graptolite ultrastructure (Towe and Urbanek 1972, 1974; Urbanek and Towe 1974, 1975).

The physical preservation of the graptolite organic skeleton is astonishing. Although the identification of the organic phase is not known for certain, its physical characteristics are closely similar to those of collagens. Figure 2 illustrates this point with a direct comparison at the same magnification of the sectioned graptolite periderm with a section taken from rabbit eye scleral tissue. The alternating layers of fibrous plies are remarkably alike in both materials in spite of the fact that the graptolite is from Ordovician rocks that are 450 million years old. Even a banding can be seen with a periodicity that is very similar in both samples.

Another group of graptolites, taxonomically distinct from the first, also shows exquisite physical preservation (Figure 3). Here a collection of fibrils is preserved embedded in a dense, homogeneous organic matrix that lacks any fine structure of its own. The fibrous material shows incredible detail in the transmission electron microscope. A stellate-septate ultrastructure is seen in cross section, and an obliquely banded pattern appears in longitudinal sections. The vertical repeat of the banding is about 70 nm. Towe and Urbanek (1974) have discussed the modern affinities of this material and, again, a collagen-like substance is likely, although a fibrous material unknown to date in modern organisms is also a possibility.

Because of the exceptional physical preservation encountered, it was thought that amino acid analysis might be useful, especially in helping to identify the fibrous component of these skeletons. Independent amino acid analyses on 6 *M* HC1 hydrolyzates of the graptolites were made by P. E. Hare at the Geophysical Laboratory and by D. Von Endt at the Smithsonian Institution. Both analysts observed that the acid hydrolyses seemed to have little effect on the skeletal fragments. And, as for the trilobite material, only marginal amounts of peptide-bound amino acids were detected. The results provided no data of any value to the problem of protein identification. For example, there was no evidence whatever for the usual collagen markers, hydroxyproline or hydroxylysine. Proline and glycine were present in very small amounts, but the possibility of contamination could not be ruled out.

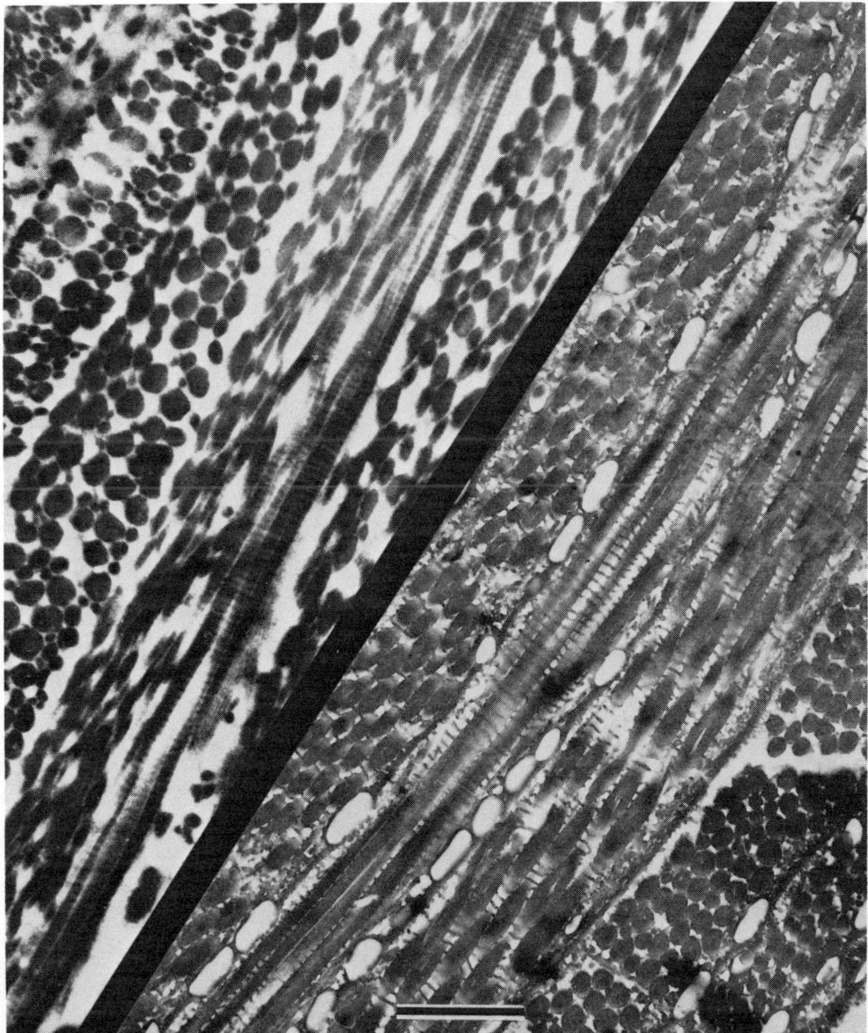

**Figure 2.** A comparison at the same magnification of collagen fibrils in the scleral tissue of the rabbit (courtesy of M. A. Jakus) with that of the Paleozoic graptolite peridermal tissue, lower right. Scale bar is 1μm. (after Towe and Urbanek 1972).

The results show that the excellent physical preservation is clearly independent of any chemically intact protein preservation. The trilobite organic matrix and the graptolite peridermal skeleton are preserved primarily as carbonized fossils. The exact composition of this carbonaceous material is unknown as is the mechanism for its astonishing physical pres-

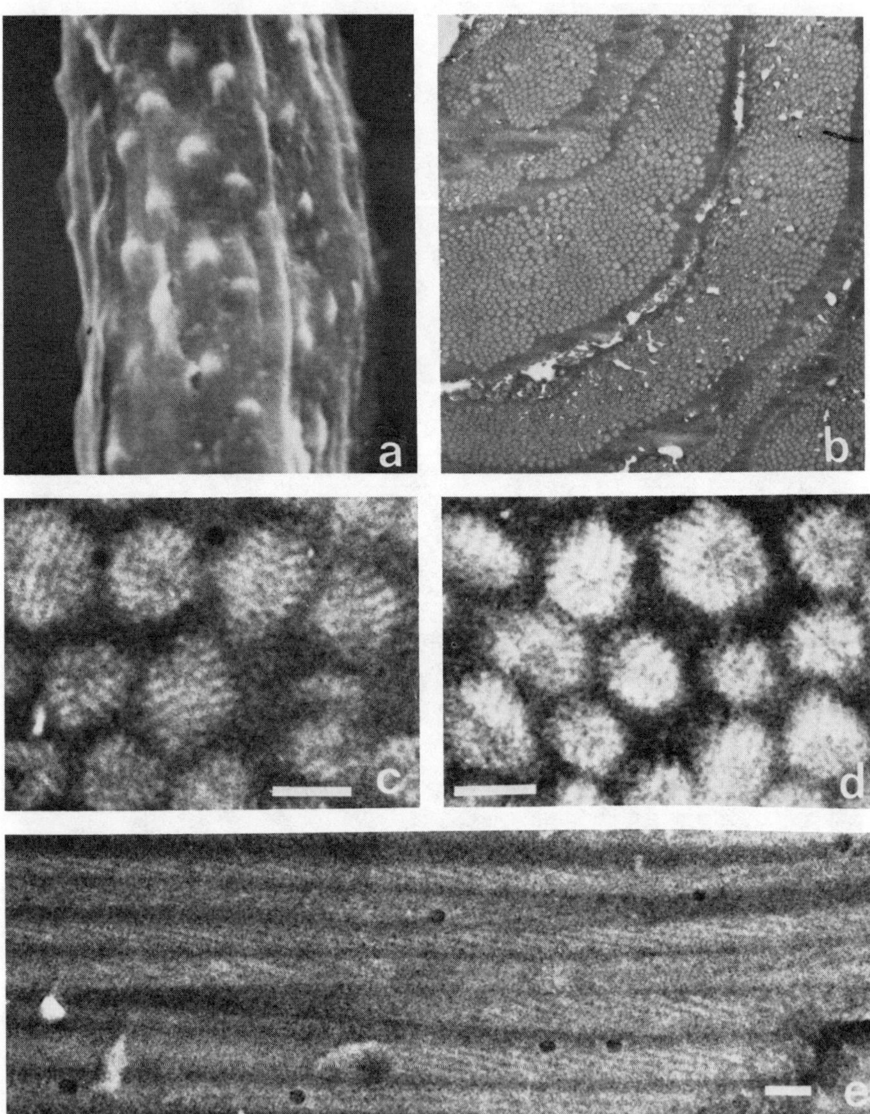


**Figure 3.** A. SEM photo of graptolite external surface. 1800×. B. Cross section through the periderm showing the fibrils embedded in a homogeneous matrix. 9000×. C,D. Higher magnification of the fibrils that have a stellate-septate ultrastructure. E. Longitudinal section shows an oblique periodicity with a vertical repeat of about 70 nm. (Scale bars are 0.1 μm).

ervation. In the case of the graptolites this is especially troublesome because of the absence of a protective mineral phase. Perhaps a post-mortem 'self-fixation' process was operative.

## DISCUSSION

There are many steps in transforming an unaltered protein or glycoprotein into a carbonaceous fossil. Among the first of these steps is the hydrolysis of peptide linkages. But because the breakage of the peptide bond requires water, it is clear that to short-circuit this alteration the preservation of proteinaceous material over extended geologic time will be more likely under anhydrous conditions. It is obvious then that although both of the fossil materials described above were inaccessible to bacterial action, they had ready access to water. For the unmineralized graptolite skeletons there is no mechanism to prevent or even inhibit access to water. The trilobite exoskeleton, on the other hand, is a calcified structure, and it might be expected that the mineral phase could protect the organic phase from water somewhat better. The electron microscopic data help explain why this is not the case. The reticulate meshwork of the organic matrix (Figure 1C) shows good agreement with the mineral structure (Figure 1D) so that it appears primarily as an *inter*crystalline matrix. As such it would have been subject to capillary forces attracting ground waters sufficient to result in extensive leaching and amino acid hydrolysis. Thus regardless of what mechanism allows for the ultrastructural preservation of these organic materials, their present nonproteinaceous composition should come as no surprise. Indeed, if intact protein had been found it would have been more difficult to explain.

From these very basic considerations it would appear that, other things being equal, the best hope for the biochemical preservation of proteinaceous material should be found not just among the calcified matrices but among the *intra*crystalline calcified matrices. Intracrystalline matrices have been found in some modern organisms where intact proteins can still be isolated from calcareous shells even after extensive oxidation of the shells with peroxides or hypochlorites. These intracrystalline materials, which lack the structural coherence of the intercrystalline matrices, appear to have been trapped within single crystals of the carbonate minerals during the growth of the shell and thus protected from the action of either the strong oxidants or of water (Towe and Thompson 1972).

Initially, the idea that a single crystal of solid calcite or aragonite could trap a significant amount of organic matter seems suspect. Yet data from the chemical literature (Nickl and Henisch 1969; Henisch 1970) on the

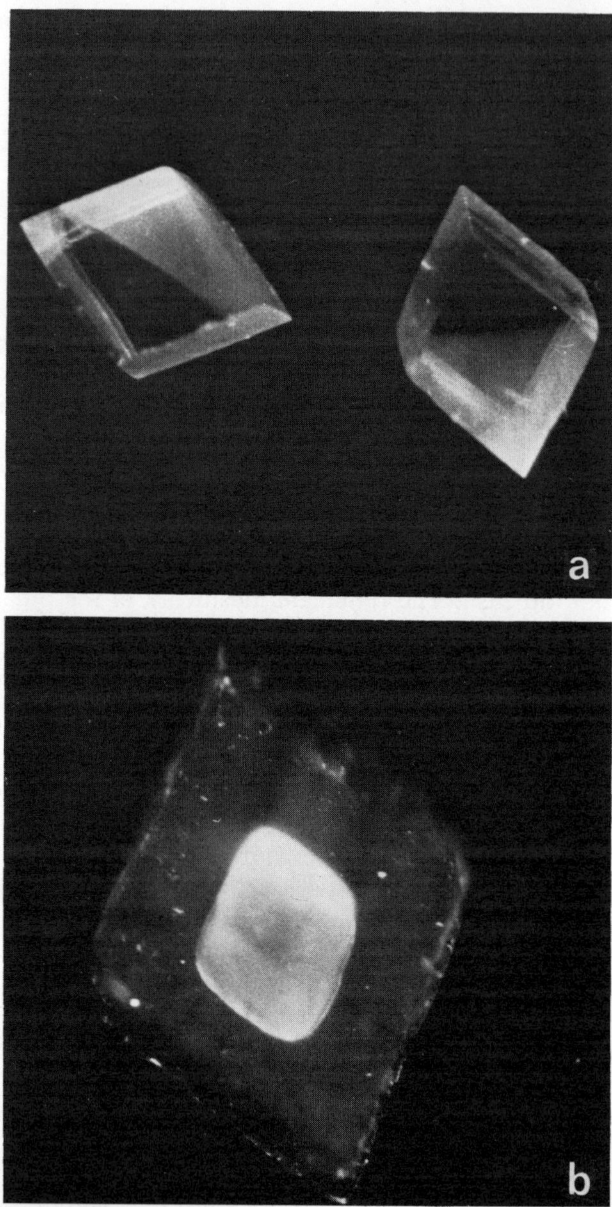


**Figure 4.** A. Calcite rhombs grown at a gel-solution interface. Note the cloudy regions where the single crystals have incorporated some of the gel. B. Partially decalcified gel-grown calcite. The gel trapped in the process of crystal growth retains the form of the original calcite rhomb. The rhombs are about 2.5 mm. on an edge. (after Nickl and Henisch 1969; courtesy of H. K. Henisch).

growth of inorganic calcite crystals in gel media show clearly that this can be readily accomplished. For example, single rhombohedral crystals of calcite grown in silica gels were found to have incorporated significant amounts of the gel during their growth (Figure 4A). Upon decalcification the gel retains the general form of the original enclosing crystal (Figure 4B). Applying this information to biomineralogical systems, any proteins trapped within rather than between single crystals will be more effectively isolated from the effects of exogenous water. It seems likely, then, that diagenetic breakdown of the peptide linkages could be accomplished only with the endogenous water (Hare 1974) simultaneously incorporated with the trapped protein thus limiting the diagenesis from completion. But, interestingly, among the least stable amino acids are the hydroxy amino acids, serine and threonine (Abelson 1954; Vallentyne 1964). Their preferred pathway of decomposition at alkaline pH appears to take place by dehydration (Bada et al., 1978). This process would then release more endogenous water for further peptide hydrolysis. Thus even the intracrystalline protein matrices of carbonate shell material preserved in the most arid or impermeable environments will still be subjected to some diagenetic breakdown in the closed system provided by single crystals.

## CONCLUSIONS

Transmission electron microscopic study of Paleozoic organic materials from both mineralized and unmineralized animal skeletons reveals that physical preservation is often of high quality. The demonstration of ultrastructural detail in both trilobite cuticle and graptolite periderm testifies to this outstanding preservation. However, amino acid analyses of acid hydrolyzates from such material fail to show any significant proteinaceous components, indicating that for these fossils the biochemical preservation is negligible. The use of electron microscopy to demonstrate the physical integrity of ancient fossil proteins will not be a reliable indicator of biochemical integrity. Organic material trapped within single crystals during biomineralization offers the best hope for the study of ancient fossil proteins. Such intracrystalline material is protected from the exogenous water necessary for protein hydrolysis, although even the small amount of endogenous water included with these matrices will permit some diagenesis to take place.

### *ACKNOWLEDGMENTS*

I would like to thank the following people for their assistance: P. E. Hare, D. Teigler, A. Urbanek, and D. Von Endt. H. K. Henisch kindly made available the photographs for Figure 4.

## *REFERENCES*

Abelson, P. H., Paleobiochemistry, *Carnegie Inst. Wash. Year Book 53,* 97–101, 1954.

Bada, J. L., M-Y. Shou, E. H. Man, and R. A. Schroeder, Decomposition of hydroxy amino acids in foraminiferal tests: Kinetics, mechanism and geochronological implications, *Earth Planet. Sci. Lett.* **41,** 67–76, 1978.

Hare, P. E., Amino acid artifacts in organic geochemistry, *Carnegie Inst. Wash. Yearbook 64,* 232–235, 1965.

Hare, P. E., Amino acid dating—a history and an evaluation, *MASCA Newsletter (Applied Science Center for Archaeology),* **10,** (1) 1974.

Henisch, H. K., *Crystal Growth in Gels,* Pennsylvania State Univ. Press, University Park, 1970.

Kozlowski, R., Les Graptolithes et quelques nouveaux groupes d'animaux du Tremadoc de la Pologne, *Palaeontolog. Polon.* **3,** 1–235, 1949.

Nickl, H. J., and H. K. Henisch, Growth of calcite crystals in gels, *J. Electrochem. Soc.,* **116,** 1258–1260, 1969.

Teigler, D. J., and K. M. Towe, Microstructure and composition of the trilobite exoskeleton, *Fossils and Strata,* **4,** 137–149, 1975.

Towe, K. M., Trilobite eyes; calcified lenses *in vivo, Science,* **179,** 1007–1009, 1973.

Towe, K. M., and G. R. Thompson, The structure of some bivalve shell carbonates prepared by ion-beam thinning, *Calcified Tissue Res.,* **10,** 38–48, 1972.

Towe, K. M., and A. Urbanek, Collagen-like structures in Ordovician graptolite periderm, *Nature,* **237,** 443–445, 1972.

Towe, K. M., and A. Urbanek, Fossil organic material: a unique ultrastructure in Silurian graptolites, *8th International Congress on Electron Microscopy,* **2,** 694–695, 1974.

Urbanek, A., and K. M. Towe, Ultrastructural studies on graptolites, 1: The periderm and its derivatives in the Dendroidea and in *Mastigograptus, Smithsonian Contributions to Paleobiology,* No. 20, Washington, D. C., 1974.

Urbanek, A. and K. M. Towe, Ultrastructural studies on graptolites, 2: The periderm and its derivatives in the Graptoloidea, *Smithsonian Contributions to Paleobiology,* No. 22, Washington, D. C., 1975.

Vallentyne, J. R., Biogeochemistry of organic matter, II. Thermal reaction kinetics and transformation products of amino compounds, *Geochim. Cosmochim. Acta,* **28,** 157–188, 1964.

# Osteocalcin: A Specific Protein of Bone with Potential for Fossil Dating

**PETER V. HAUSCHKA**
*Children's Hospital Medical Center and Harvard School of Dental Medicine, Boston, Massachusetts*

---

Bone contains abundant quantities of an unusual small protein called osteocalcin. This protein, with a molecular weight of about 6000 daltons, is distinguished by its content of the $Ca^{2+}$-binding amino acid $\gamma$-carboxyglutamate (Gla; see Fig. 1). It is the purpose of this paper to review our knowledge of osteocalcin and to show how it may be advantageously used in biogeochemical studies of fossilized hard tissues.

Osteocalcin was discovered in our laboratory during the process of tracking down the locus of Gla in whole bone (Hauschka et al., 1975). Osteocalcin purified from chicken bone contains four Gla residues in its total of 57 amino acid residues, and this polypeptide accounts for at least 85% of the Gla content of whole bone (Hauschka and Gallop 1977). Bovine and swordfish osteocalcin show three residues of Gla/49, or 47 total amino acid residues, respectively, and have other compositional differences as shown in Table 1 (see Price et al., 1976a; 1976b; Price 1977). As a tissue, bone contains some 7 to 100 times more Gla than any other whole tissue (Table 2). Osteocalcin comprises about 1% of the total bone protein, or some 10%–20% of the noncollagenous protein. This means that there is about one molecule of osteocalcin for every one to two molecules of bone tropocollagen (Hauschka and Gallop 1977; Hauschka et al., 1978). Surprisingly, osteocalcin is the sixth or seventh most abundant protein in most vertebrate organisms, exceeded on a molar basis only by actin, collagen, hemoglobin, myosin, tropomyosin B, and possibly keratin and serum albumin. The unusual composition and properties of osteocalcin, combined with its exclusive occurrence in osseous tissue, make this protein potentially very useful for biogeochemical studies of all fossilized vertebrate species.

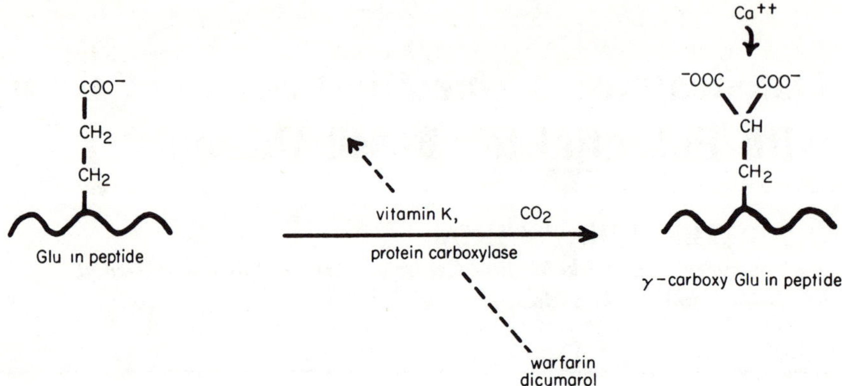


**Fig. 1.** Formation of the $Ca^{2+}$-binding amino acid γ-carboxyglutamate (Gla) occurs by posttranslational protein carboxylation, an enzymatic reaction requiring $CO_2$ and vitamin K. Anticoagulant drugs such as warfarin and dicumarol inhibit the carboxylation reaction *in vivo*.

**Table 1. Species Variation in Osteocalcin Composition**

| Amino acid | Chicken | Bovine[a] | Swordfish[a] |
|---|---|---|---|
| Gla | 4 | 3 | 3 |
| Hyp | — | 1 | — |
| Asp | 6 | 6 | 4 |
| Thr | — | — | 3 |
| Ser | 3 | — | 2 |
| Glu | 7 | 3 | 3 |
| Pro | 7 | 6 | 2 |
| Gly | 6 | 3 | 3 |
| Ala | 6 | 4 | 7 |
| Cys 1/2 | — | 2 | 2 |
| Val | 3 | 2 | 3 |
| Met | — | — | 1 |
| Ile | 1 | 1 | 3 |
| Leu | 4 | 5 | 5 |
| Tyr | 3 | 4 | 3 |
| Phe | 3 | 2 | 1 |
| His | — | 2 | — |
| Lys | — | 1 | — |
| Trp | — | 1 | — |
| Arg | 4 | 3 | 2 |
| | 57 | 49 | 47 |

[a] Price, et al., 1976, 1977.

**Table 2.γ-Carboxyglutamic Acid Content of Various Three Week Chick Tissues**

| Tissue | Relative Gla content[a] |
|---|---|
| Long-bone diaphysis | 100 ± 3 |
| Long-bone metaphysis | 67 |
| Long-bone epiphysis | 23 |
| Long-bone articular cartilage | 18 |
| Long-bone marrow | <1 |
| Mandible | 66 |
| Calvaria | 56 |
| Sternum keel bone | 51 |
| Sternum cartilage | 22 |
| Tracheal cartilage | 17 |
| Tendon | <1 |
| Brain | <1 |
| Kidney | 14 |
| Liver | 10 |

[a] Long-bone diaphysis = 68.9 ± 2.4 residues Gla/$10^5$ amino acid residues.

## DISTRIBUTION AND PROPERTIES OF OSTEOCALCIN AND OTHER GLA-PROTEINS

Osteocalcin is intimately associated with the calcium phosphate mineral phase of bone. This is apparent not only in gross specimens of adult bone, where marrow and cartilage show very low Gla levels (see also Table 2), but also in embryonic bone. Developing long bones (tibiotarsus) of the chick embryo skeleton begin to form osteocalcin coincident with the first observable deposition of bone mineral in the midshaft (Hauschka and Reid 1978a). Even in a rice-grain size 10-day embryonic bone, the Gla level in the mineralized perichondral zone is some seven-fold greater than in the cartilaginous rudimentary core.

The tight association of osteocalcin with bone mineral requires that this mineral be dissolved (using acidic solutions or chelating agents such as EDTA) to release the protein. Such extractions normally solubilize over 90% of the osteocalcin and other noncollagenous proteins while leaving behind the bulk of the insoluble collagen matrix of bone. Purification of osteocalcin has been described for bones of chicken (Hauschka et al., 1975; Hauschka and Gallop 1977), calf (Price et al., 1976b), and swordfish (Price et al., 1977). By inference from direct measurements of Gla, the protein is present in similar quantities in the bones of man, deer, rabbit, rat, mouse, and codfish, and in deer and elk antler (Hauschka 1977; Hauschka, unpublished data). Gla is also found in bovine and rat incisor dentin, while enamel is devoid of Gla (Price et al., 1976; Hauschka 1977; Hauschka and Reddi, 1980).

Within bone, the osteocalcin content varies according to the location, bone tissue type, and degree of mineralization (Hauschka et al., 1978; Hauschka and Reid 1978b). Table 2 shows that Gla is most abundant in the diaphysis (midshaft) of long bones. Less Gla occurs in regions of bone that are more active metabolically or are involved in rapid turnover (e.g., metaphysis and epiphysis). Membranous bones of the skull (e.g., calvaria) contain only about half as much Gla as the long bone diaphyses of the same animal (Table 2). Differences in the Gla content of the two types of bone persist throughout life and are observed in all species so far examined. Whether these differences are related to the presence of endochondral versus intramembranous mineralization is not yet known.

Cartilage contains only 10%–25% as much Gla as long bone diaphysis (Hauschka et al., 1978; Hauschka and Reid 1978). Yet, unlike bone, the Gla in cartilage does not seem to be as closely related to the degree of mineralization (Table 2). The Gla-containing protein(s) in cartilage have not yet been isolated and may be distinctly different from osteocalcin. Calcified cartilage of elasmobranchs was reported by Price, et al (1976b) to have no Gla. However, more recent studies by King, 1978 and in our laboratory (Hauschka, unpublished; Lian et al., 1979) have shown that Gla does occur in both the calcified and uncalcified cartilage and the teeth of several species of sharks.

The first discovery of Gla occurred in 1974 when three laboratories independently described this amino acid as a component of the blood coagulation protein prothrombin (Stenflo et al., 1974; Nelsestuen et al., 1974; Magnusson 1974). Vitamin K plays an essential role in the biosynthesis of Gla (Fig. 1) by acting as a cofactor for the enzymatic posttranslational carboxylation of specific glutamic residues in the polypeptide chain (Esmon et al., 1975; see Hauschka et al., 1978; Stenflo 1977, for reviews). The formation of Gla in the vitamin K–dependent blood coagulation factors (prothrombin; factors VII, IX, and X) endows these proteins with several new properties: (1) ability to specifically bind increased numbers of $Ca^{2+}$ ions; (2) $Ca^{2+}$-mediated adsorption to phospholipid membrane (micelle) surfaces allowing proteolytic activation *in vivo*; and (3) adsorption to barium citrate and barium sulfate (Nelsestuen et al., 1974; Esmon et al, 1975). Gla has recently been found in proteins from sources other than blood and bone. Kidney (Hauschka et al., 1976), urine (Fernlund 1976), kidney stones, atherosclerotic plaque (Lian et al., 1976), and placenta (Friedman et al., 1978) are known to contain Gla-proteins.

Biosynthesis of osteocalcin occurs in bone tissue (Hauschka et al., 1978; Lian et al., 1978; Lian and Friedman 1978) and requires vitamin K (Hauschka and Reid 1978; Lian et al., 1978). Deficiencies in vitamin K or inhibition by anticoagulant drugs such as warfarin and dicumarol cause undercarboxylation of osteocalcin with reduced Gla and increased glutamic acid content (Hauschka and Reid 1978). Osteocalcin biosynthesis has been demonstrated to occur *de novo* in isolated embryonic chick bones cultured *in vitro* as well as in microsomes prepared from chick embryo bone (Lian et al., 1978; Lian and Friedman 1978). Further evidence that osteocalcin is a unique hard-tissue protein, and not merely a fragment of a blood coagulation protein, derives from its primary sequence. Bovine osteocalcin is clearly not homologous to the Gla-containing regions of the bovine blood coagulation proteins (Price et al., 1976; Price 1977).

While the function of osteocalcin in bone is unclear, this protein does exhibit certain features common to other Gla-proteins. At physiological ionic strength each molecule of chicken bone osteocalcin can bind two $Ca^{2+}$ ions noncooperatively with $K_d = 0.8$ m$M$. This affinity for $Ca^{2+}$ is similar to that reported for the Gla-containing blood coagulation proteins (Nelsestuen et al., 1975; Stenflo 1977; Stenflo and Suttie 1977) and for the synthetic peptide L-Gla-L-Gla (Marki et al., 1977). Bovine osteocalcin shows a much lower affinity for $Ca^{2+}$ (Price 1977). Competition studies between $Ca^{2+}$ and other cations for binding to chicken osteocalcin have established the relatively high specificity of the binding site (Hauschka and Gallop 1977). For the alkaline earth cations the affinity is in the order $Ca^{2+} > Mg^{2+} > Sr^{2+} > Ba^{2+}$.

Osteocalcin has a selective affinity for insoluble $Ca^{2+}$ salts. The protein may be purified by adsorption to hydroxyapatite (Hauschka et al., 1975), and it is much more tightly bound to hydroxyapatite than to amorphous calcium phosphate (Price et al., 1976b). Bone contains about one molecule of osteocalcin for every $10^4$ $Ca^{2+}$ ions in hydroxyapatite (Hauschka et al., 1978). Since only one $Ca^{2+}$ in several thousand can be bound to the protein, significant functional processes may alternately involve interaction of osteocalcin with mineral ion clusters, nuclei or crystal surfaces. Osteocalcin is known to inhibit the *in vitro* transition of brushite [$CaHPO_4$ $2H_2O$] → hydroxyapatite [$Ca_{10}(PO_4)_6(OH)_2$] (Hauschka and Gallop, 1977), and it also has an inhibitory effect on hydroxyapatite precipitation (Price et al., 1976b). Whether these properties of the protein *in vitro* relate to its function *in vivo* is speculative.

## OSTEOCALCIN AS A MATERIAL FOR FOSSIL DATING

The extremely slow process of racemization and epimerization of amino acid residues in fossil proteins has provided a means for assessing the age of a fossil over a geological time scale (Hare and Abelson 1968; Bada et al., 1970). Because these reactions are temperature dependent, estimation of paleotemperature has also been attempted (Bada et al., 1973). These methods are complicated by many factors including (1) environmental temperature and fluctuations in temperature; (2) environmental pH; (3) availability of water; (4) position effects on reaction rate: (a) $NH_2$-terminal versus internal versus COOH-terminal versus free amino acid; (b) catalytic or inhibitory effects of neighboring residues; (5) catalytic effects of solutes in the environment: (a) metal ions; (b) racemases produced by putrefying bacteria; (6) differential loss of fossil protein by leaching; (7) contamination by D-amino acids, peptides, and proteins of exogenous origin.

Many of the errors associated with racemization dating could be circumvented by focusing on a fossil peptide of known sequence and composition. Purification of osteocalcin or a smaller Gla-containing peptide from fossil bone would provide an ideal material for dating. First, the presence of Gla would guarantee that the peptide was indigenous to the original living bone tissue and could not have originated by contamination or by secondary bacterial putrefaction. Second, because the amino acid sequence is defined and constant for the species, the extent of racemization would more accurately reflect the age and characteristic paleotemperature for the fossil.

Living matter in which no Gla has been found includes many species of bacteria, plants, and invertebrates (Hauschka 1977; King 1978). These observations are of great importance because they eliminate all the significant sources of a diagenetic contamination that are known to interfere with accurate biogeochemical fossil dating methods.

Although osteocalcin has not yet been isolated from fossil bone, there is no question that fragments of the protein, and possibly even the intact 6000-dalton polypeptide, are preserved for at least 50,000 years. Studies by King (1978) and King and Bada (1978) have shown that Gla levels in ancient bone specimens are dependent on the degree of leaching or exposure to weathering. However, protected specimens such as human mummy (2900 years) or frozen whale bone (52,000 years) are found to have a Gla content of 75%–100% of that in modern bovine bone.

Because of the known instability of Gla in strongly acidic environments (Hauschka et al., 1975; Hauschka 1977), it is important to refine the empirical observation of reduced Gla levels in fossil bone by distingushing true leaching or solubilization of osteocalcin from *in situ* decarboxylation of Gla to glutamic acid. In the latter case, osteocalcin or its peptide fragments with a reduced Gla content and correspondingly increased glutamic acid content should be extractable from the fossil bone. The modern analog of this experiment has been performed with bone from dicumarol-fed chickens (Hauschka and Reid 1978).

Osteocalcin has many features that make it suitable for biogeochemical dating and evolutionary studies: (1) The protein is abundant; each gram of dry bone contains 1–3 mg of osteocalcin. (2) The protein is rich in aspartic acid, alanine, glutamic acid, and leucine (Table 1), which are among the most useful residues for racemization studies. (3) The protein is intimately associated with the calcium mineral phase of bone; thus it should be protected from proteolytic degradation and would be leached only under acidic ground water conditions where mineral leaching also occurs. (4) Osteocalcin is easily isolated and purified, and the Gla content provides a unique index of purity.

## *ACKNOWLEDGMENTS*

The author thanks P. E. Hare, K. King, P. M. Gallop, J. B. Lian and M. J. Glimcher for helpful discussions. Technical assistance was kindly proved by M. L. Reid and J. Komar. Supported by the National Foundation–March of Dimes and by NIH Grants AM 15671 and AM 16754. P.V.H. is a Research Career Development Awardee of the NIDR.

## *REFERENCES*

Bada, J. L., B. P. Luyendyk, and J. B. Maynard, *Science,* **170,** 730–732, 1970.

Bada, J. L., R. Protsch, and R. A. Schroeder, *Nature,* **241,** 394–395, 1973.

Esmon, C. T., J. A. Sadowski, and J. W. Suttie, *J. Biol. Chem.,* **250,** 4744–4748, 1975a.

Esmon, C. T., J. W. Suttie, C. M. Jackson, *J. Biol. Chem,* **250,** 4095–4099, 1975b.

Fernlund, P., *Clin. Chim. Acta,* **72,** 147–155, 1976.

Friedman, P. A., P. V. Hauschka, M. A. Shia, and J. K. Wallace, *Biochim. Biophys. Acta,* **583,** 261–265, 1979.

Gitel, S. N., W. G. Owen, C. T. Esmon, and C. M. Jackson, *Proc. Nat. Acad. Sci. U.S.A..,* **70,** 1344–1348, 1973.

Hare, P. E. and P. H. Abelson, *Carnegie Inst. Wash. Year Book 66,* 526, 1968.

Hauschka, P. V., *Anal. Biochem.,* **80,** 212–223, 1977.

Hauschka, P. V., J. B. Lian, and P. M. Gallop, *Proc. Nat. Acad. Sci. U.S.A.,* **72,** 3925–3929, 1975.

Hauschka, P. V., P. A. Friedman, H. P. Traverso, and P. M. Gallop, *Biochem. Biophys. Res. Commun.*, **71,** 1207–1213, 1976.

Hauschka, P. V. and P. M. Gallop, in *Calcium Binding Proteins and Calcium Function,* Wasserman, R. H., et al., eds., pp. 338–347, Elsevier/North Holland, Amsterdam, 1977.

Hauschka, P. V. and M. L. Reid, *Develop. Biol.,* **65,** 426–434, 1978a.

Hauschka, P. V. and M. L. Reid, *J. Biol. Chem.,* **253,** 9063–9068, 1978b.

Hauschka, P. V., J. B. Lian, and P. M. Gallop, *Trends in Biochem. Sci.,* **3,** 75–78, 1978.

Hauschka, P. V. and A. H. Reddi, *Biochem. Biophys. Res. Commun.,* (in press), 1980.

King, K., *Biochim. Biophys. Acta,* **542,** 542–546, 1978a.

King, K., *Nature,* **273,** 41–43, 1978b.

King, K. and J. L. Bada, *Nature* **281,** 135–137, 1979.

Lian, J. B., E. L. Prien, M. J. Glimcher, and P. M. Gallop, *J. Clin. Invest.,* **59,** 1151–1157, 1977.

Lian, J. B., P. V. Hauschka, and P. M. Gallop, *Fed. Proc.,* **37,** 2615–2620, 1978.

Lian, J. B., M. S. Skinner, M. J. Glimcher, and P. Gallop, *Biochem. Biophys. Res. Commun.,* **73,** 349–355, 1976.

Lian, J. B. and P. A. Friedman, *J. Biol. Chem.,* **253,** 6623–6626, 1978.

Lian, J. B., J. A. Glowacki, and M. J, Glimcher, in *Vitamin K Metabolism and Vitamin K-Dependent Proteins*, J. W. Suttie, ed., pp. 263–268, University Park Press, Baltimore. 1979.

Magnusson, S., L. Sottrup-Jensen, T. E. Peterson, H. R. Morris, and A. Dell, *FEBS Lett.,* **44,** 189–199, 1974.

Marki, W., M. Opplinger, and R. Schwyzer, *Helv. Chim. Acta.,* **60,** 807–815, 1977.

Nelsestuen, G. L., T. H. Zytkovicz, and J. B. Howard, *J. Biol. Chem.,* **249,** 6347–6350, 1974.

Nelsestuen, G. L., Broderius, M., Zytkovicz, T. H. and Howard, J. B. *Biochem. Biophys. Res. Commun.,* **65,** 233–240, 1975.

Price, P. A., J. W. Poser, and N. Raman, *Proc. Natl. Acad. Sci. U.S.A.,* **73,** 3374–3375, 1976a.

Price, P. A., A. S. Otsuka, J. W. Poser, J. Kristaponis, and N. Raman, *Proc. Nat. Acad. Sci. U.S.A.,* **73,** 1447–1451, 1976b.

Price, P. A., A. S. Otsuka, and J. W. Poser in *Calcium Binding Proteins and Calcium Function,* Wasserman, R. H., et al., eds. pp. 333–337, Elsevier/North Holland, Amsterdam, 1977.

Stenflo, J., *Trends in Biochemical Sciences,* **4,** 256–258, 1977.

Stenflo, J., P. Fernlund, W. Egan, and P. Roepstorff, *Proc. Nat. Acad. Sci. U.S.A.,* **71,** 2730–2733, 1974.

Stenflo, J.,and J. W. Suttie, *Annu. Rev. Biochem.,* **46,** 173–200, 1977.

Wehmiller, J., and P. E. Hare, *Science,* **173,** 907–911, 1971.

# Phylogenetic Information Derivable from Fossil Brachiopods

**MARGARET JOPE**
*Queen's University, Belfast, Northern Ireland*

## INTRODUCTION

The Brachiopoda are a small phylum of marine bivalves providing excellent material for the study of shell proteins. They are already present in the L. Cambrian and though they were most prolific in early geological times, some genera have been very long lived, with representatives living today (Figure 1). The genus *Lingula* has an unbroken lineage from Ordovician times to the present day, and living *Crania* differs only in minor morphological details from Ordovician *Petrocrania*. Direct and significant comparisons can thus be made between fossil and living shell proteins, and these do show differences beyond those merely accountable to the effects of fossilization.

Studies on living brachiopod shell proteins have shown a protein taxonomy that amplifies the zoological. The phylum Brachiopoda has been divided into two groups—Inarticulata without a hinge mechanism for opening and closing the shell and Articulata with a hinge. In general, the Inarticulata have shells with phosphatic calcification while the Articulata have shells with carbonate calcification. Among living brachiopods the Craniacea are the exceptions to this generalization in that they are morphologically inarticulate but have carbonate calcification. The main taxonomic distinction given by the shell protein data is the presence of hydroxyproline in phosphatic shells and its absence from carbonate shells including *Crania,* thus placing *Crania* with the articulates with respect to shell protein taxonomy.

It is now clear that proteins can survive as full length polypeptide chains from very early times (Jope 1969; Weiner et al., 1976). Figure 2 shows disc electrophoresis patterns of shell protein from Silurian, Jurassic, and present-day articulate brachiopods. The rates of migration of the three

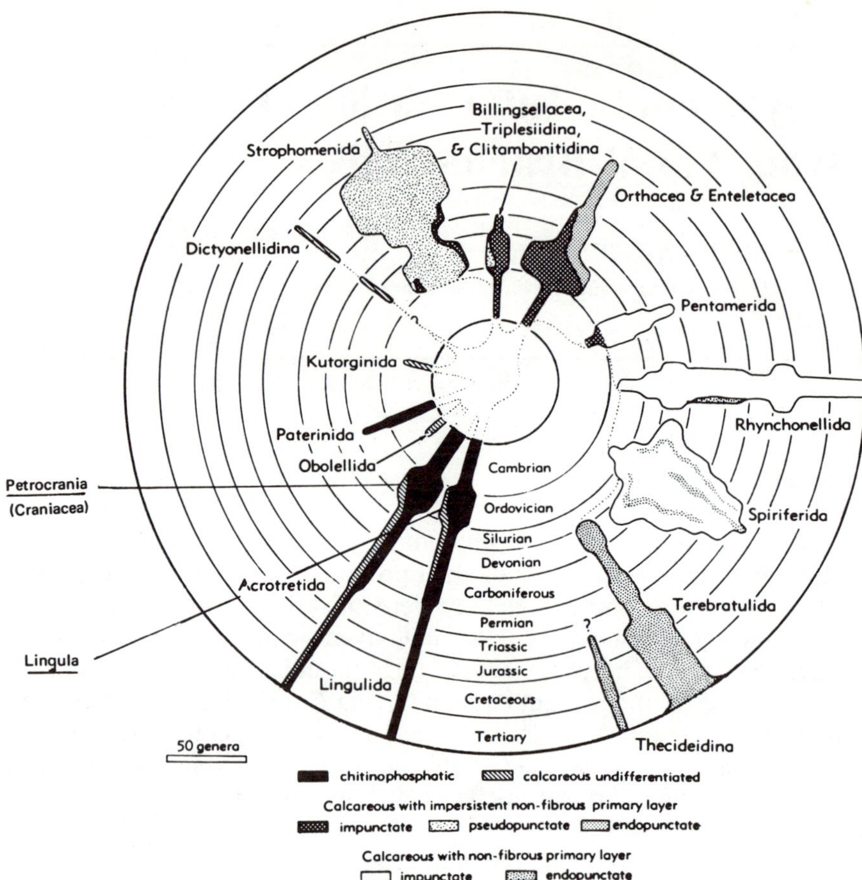


**Figure 1.** Phylogenetic relations within the Brachiopoda, showing rise of carbonate shelled forms within phosphatic shelled groups. (After Williams and Rowell 1965.)

fossil proteins are very close to those of the corresponding present-day proteins, indicating that the protein chains of the fossil genera are comparable in length to those of the living genera. However, the data on the amino acid composition of the ancient protein material differ from those of the present-day protein. Some differences are widespread and must be due to effects of time and environment. The first requirement is to define these fossil effects, to find out what differences remain which could be potentially ascribable to genetic change.

To understand the fossil proteins, the protein systems in living shell must first be understood. The protein in the shell is disposed in a highly

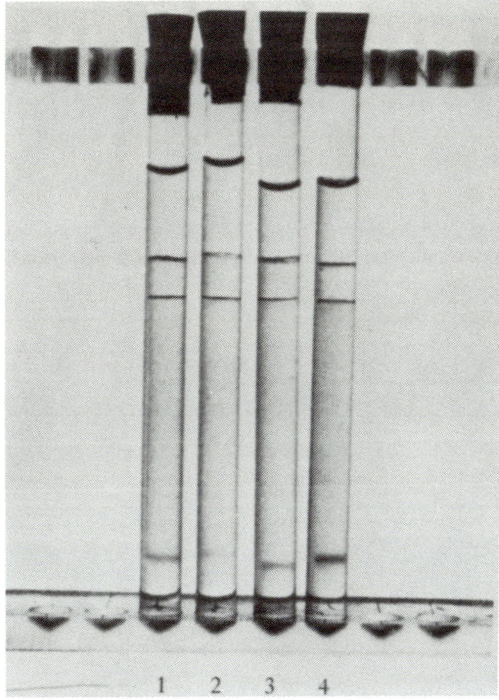


**Figure 2.** Disc electrophoretic patterns (run from top downwards) of calcified shell protein of fossil and present-day articulate brachiopods, in 4% acrylamide gel, 7 *M* urea, Tris-glycine buffer, pH 8.9, for 55 min at 5.5 ma per tube. 1. *Goniorhynchia*, Jurassic. 2. *Camarotoechia*, Mid-Silurian. 3. *Plectothyris*, Jurassic. 4. *Terebratalia*, present day. (From Jope 1969.)

organized way. The main protein is secreted by the outer epithelial cells of the mantle, and in articulate shells it lies around the calcite fibers that make up this part of the shell. Under the electron microscope this protein has a superficially membranous appearance with a triple layered structure 140 Å thick (Williams 1968) though the structure seems not to be a true membrane (Jope 1971).

Each shell protein consists of several distinct protein chains. These are set out for different genera in terms of molecular weights in Figure 3, where the protein fractions have been obtained from SDS-gel electrophoresis (Jope 1979). The molecular weights of the fractions from each shell protein are grouped together in a way suggesting the process of gene splitting, and this is supported by the evidence of homology in amino acid constitution. There is, however, no evidence of supramolecular fibril structure in the shell proteins, such as is seen in the collagen systems

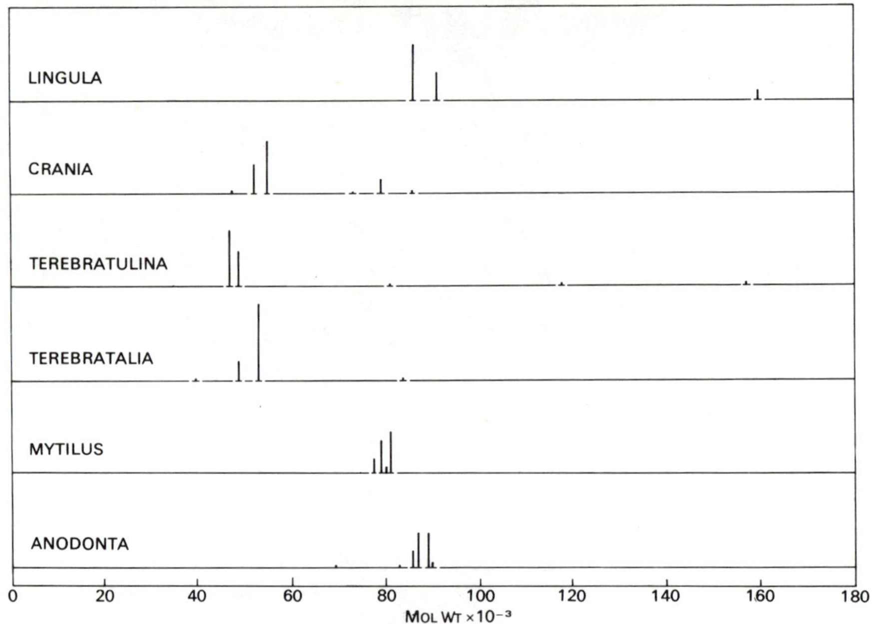


**Figure 3.** Recent brachiopod (and molluscan) shell proteins: molecular weights and relative proportions (0–100%) of fractions from SDS-polyacrylamide gel electrophoresis. 10% gels, acrylamide: crosslinker = 74:1. (From Jope 1979.)

(Miller and Parry 1973; Piez and Miller 1974), and the protein does not seem to contribute a framework of fibrils in shell as it does in bone. Its main function in shell seems to be to provide resilient cushioning between hard masses, sometimes brittle and crystalline. It is still uncertain whether the protein has any directive effect on the initiation or subsequent course of calcification processes, at least some of which seem to be predominantly under cellular control.

The lack of structural constraints needed for complex supramolecular interlocking means that shell protein systems should have a greater tolerance of mutational change than, for example, the collagen systems.

## AMINO ACID COMPOSITIONS OF FOSSIL PROTEINS COMPARED WITH CORRESPONDING LIVING SHELL PROTEINS

The following fossil brachiopods are discussed here: *Lingula* sp., Carboniferous, from Lower Shale Bed, Coal Farm, Pittenween, Fife, Scotland; *Orbiculoidea forbesi* (Davidson) from the Carboniferous limestone, Co. Kildare, Eire; *Petrocrania scabiosa* (Hall) from the Upper Ordovician of Cincinnati, Ohio, U.S.A; *Epithyris oxonica* Arkell, Jurassic, from the Lower Epithyris Bed in White limestone Series, Kirtlington, Oxfordshire, England; *Plectothyris fimbria* (J. Sowerby), Jurassic, from the Lower Marl Band in the Oolitic Marl, Conigyre Wood, Stroud, Gloucestershire, England; *Goniorhynchia boueti* (Davidson), Jurassic, from the Boueti Bed, Forest Marble, Dorset, England; *Camarotoechia* sp., Mid-Silurian, from the Waldron Shale, Hartsville, Indiana, U.S.A.

The present-day brachiopods used for comparison with the fossil are: *Lingula anatina* Larmarck, *Discinisca lamellosa* (Brøderip), *Crania anomala* (Müller), *Terebratulina retusa* (Linné), *Terebratalia transversa* (Sowerby), *Notosaria nigricans* (Sowerby), and *Hemithiris psittacea* (Gmelin).

Though the polypeptide molecular backbone may have remained intact in fossil proteins as indicated by the disc electrophoresis patterns in Figure 2, some changes have evidently taken place in the amino acid side chains because of the effects of geological time and environmental conditions. Some of these alterations are common to most of the fossil proteins and result from the effects of geological time and environmental conditions. They can be allowed for in reconstructing the original amino acid composition of the protein of the shell as living in early times. The differences between the amino acid compositions of shell proteins of fossil brachiopods and those of living descendants are shown diagramatically in Figure 4, which emphasizes those amino acids which are higher or lower in the fossil protein as compared with values for the living protein (i.e., the values for the living protein have been subtracted from those of the fossil protein).

The data in Figure 4 show that all the fossil brachiopod proteins have higher glutamic acid and lysine and lowered pyrrolidines. The leucines are higher in all the fossils except one (*Epithyris*). Histidine is higher in all the fossils except *Petrocrania,* where the fossil is only slightly low. Phenylalanine is usually higher in the fossils. Serine is higher in the articulate fossil proteins but not in the inarticulate. The higher glutamic acid and serine as shown by brachiopod fossil proteins has been widely observed in fossil proteinaceous material, for example, in mollusca (Florkin 1965; Bricteux-Grégoire et al., 1968), and graptolites show a high level of glutamic acid and serine (Foucart et al., 1965). This suggests a tendency for these amino acids to increase in value during geological time; the alterations may be due, for instance, to thermodynamic change in other amino acids in the chain. Higher glutamic acid might be in part connected with the possible presence of $\gamma$-linked glutamyl residues in the peptide chain (this linkage has been shown to occur in collagen by Franzblau et

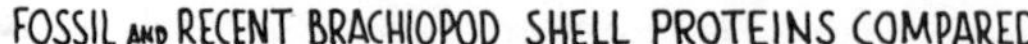


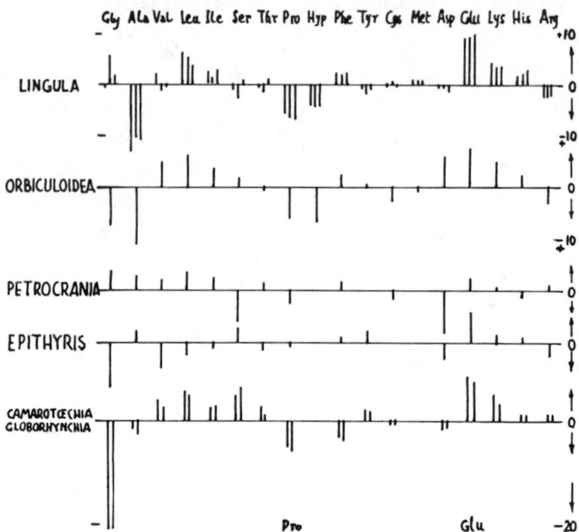


**Figure 4.** Comparison of amino acid compositions of fossil and Recent brachiopod shell proteins showing differences in composition.

al., 1963), where it may have arisen from oxidative transformation of proline (Seifter et al., 1965). If so it could in part explain the rise in glutamic acid in terms of the fall in proline so widely observed in these shell proteins.

Reasons for high serine are not clear. In laboratory experiments serine has been found to be, along with threonine, the most unstable amino acid (Hare and Mitterer 1966), but it is apparently stabilized by peptide bond, and therefore high serine concentration in the insoluble fraction need not be taken to indicate contamination in the older fossils (Akiyama 1971). Factors that may contribute to increased serine are protein aging, hardening, selective degradation, algal penetration (Harington and Toens 1963), bacterial growth, thermodynamic effects or radiation (Alexander and Lett 1967) leading to deamination or decarboxylation. However, the high glutamic and aspartic acid levels generally seen in fossil brachiopod proteins suggest that radiation damage is not a serious disturbance.

## GENETIC DIFFERENCES BETWEEN FOSSIL AND LIVING SHELL PROTEINS

These may be revealed after the general fossilization effects have been allowed for in the amino acid composition differences observed between the fossil proteins and those of their closest living relatives. These are summarized below and in Figure 4.

*Inarticulate Brachiopods. Lingula,* from the Carboniferous, compared with the present day-genus: aspartic acid values are in good agreement, valine, serine and threonine are reasonable; alanine and arginine are low in the fossil, but glycine tends to be high.

*Orbiculoidea.* Carboniferous, compared with present-day *Discinisca:* alanine is again low. Low alanine is not regularly observed throughout fossil brachiopod proteins and could reflect the original composition of the carboniferous protein, altered by the following mutational substitutions: alanine for glycine in *Lingula*; alanine for valine, and glycine for aspartic acid in *Orbiculoidea.* These are simple single-base substitutions; the frequency of mutation leading to an amino acid change is very approximately 1 in $10^3$ million years, depending on how rapidly the gene controlling protein biosynthesis can undergo mutation (Nei 1975).

*Petrocrania.* Ordovician, is unusual in having much lower serine and aspartic acid than living *Crania* (where both these amino acids are exceptionally high; higher serine is more usually observed than lower in fossil proteins. This lower serine might have arisen during geological time by attack from microorganisms resulting in enzymatic cleavage of α and β carbon atoms of serine to form glycine (Schirch and Mason 1963) without breaking the peptide chain, giving the increased glycine seen in Ordovician *Petrocrania.*

Low aspartic acid could be the result of β-decarboxylation of aspartic acid by enzymatic attack from microorganisms during fossilization, giving increased alanine (Mahler and Cordes 1966) though the higher alanine in *Petrocrania* is less than that needed to account for all the loss of aspartic acid. It might be, however, that low serine and aspartic acid of *Petrocrania* are actually those of the Ordovician organism and that they have been increased by mutational change, substituting serine for glycine, and aspartic acid for alanine and valine (Fig. 4) which require only single base changes in the DNA coding.

Glutamic acid is always higher in these fossil brachiopod proteins though less markedly so in *Petrocrania.*

*Articulate brachiopods.* The Jurassic terebratulide, *Epithyris,* compared with living *Terebratulina:* glycine, valine; aspartic acid and (unusually) leucine are low in the fossil; alanine and serine are high. These may represent the true values of the Jurassic protein, altered by mutational changes (involving single base substitutions) at some time since the Jurassic, to give the composition of the present-day protein.

The rhynchonellides, *Camarotoechia* (Silurian) and *Globirhynchia* (Jurassic) compared with the living rhynchonellide *Notosaria:* glycine is much lower, phenylalanine is low (whereas it is high in all other fossil brachiopods), alanine and aspartic acid are low, valine and serine are high. Simple mutational changes would give the present-day proteins (Fig. 4).

## SHELL PROTEIN OF RHYNCHONELLIDE BRACHIOPODS

The shell protein of the rhynchonellides, *Hemithiris* (and also of *Notosaria*) has a tougher, darker layer alternating with the more usual softer protein (of the terebratulides). The amino acid composition of this tougher protein layer of *Hemithiris* shows considerable resemblance to the values for the fossil rhynchonellide proteins (*Camarotoechia* and *Globirhynchia*, especially the latter) by contrast with the softer protein layer of *Hemithiris*, where the values are close to those for the whole protein of present-day *Notosaria*; here the alternating light and dark layers are not so marked as with *Hemithiris* (Figure 5 and Jope 1967a). The data suggest that some of the fossil changes may be due to progressive hardening processes and that in the fossil shell proteins it is the toughened protein which has selectively withstood alteration due to geological time. This toughened protein layer of *Hemithiris* might have arisen during times of stress and could be due to bringing into play a dormant genetic locus. Abnormal protein production of this kind is seen elsewhere in the animal kingdom, for example, with sheep and goat hemoglobins (Wilson et al., 1970).

## TAXONOMIC DISTINCTIONS SHOWN BY FOSSIL PROTEINS

The main taxonomic distinction shown by the shell proteins of living brachiopods is the presence of hydroxyproline in phosphatic shelled inarticulates and its absence from carbonate shelled genera, both inarticulate (*Crania*) and articulate; hydroxyproline thus running with phosphatic calcification as in bone (Jope 1967b). This distinction, which cuts across the zoological classification into Inarticulata and Articulata, is not observable in fossil proteins owing to decomposition of hydroxyproline during geological time. It should be noted that hydroxyproline is formed by post-coding hydroxylation of some proline residues already built into the polypeptide chain as part of the primary coded synthesis.

Another distinction—the high alanine of living phosphatic shelled brachiopods compared with carbonate forms (Jope 1967b)—is less marked

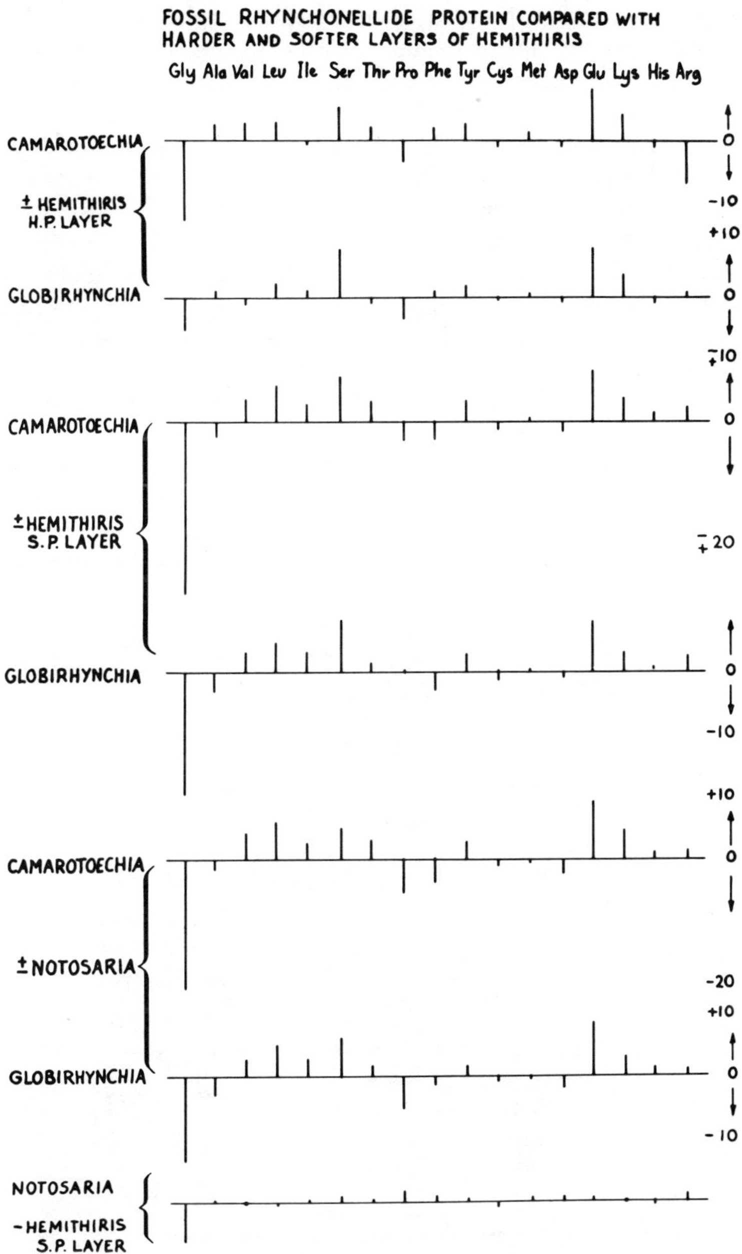


**Figure 5.** Comparison of amino acid compositions of shell proteins of fossil rhynchonellide brachiopods with those of harder and softer layers of the Recent rhynchonellide *Hemithiris*.

in the fossil proteins, where all the inarticulates (phosphatic and carbonate) have somewhat higher alanine than the articulates (Jope 1969). With the fossil proteins aspartic acid tends to be higher in all inarticulates; histidine is lower in carbonate forms.

Figure 6 summarizes the characteristics that distinguish *Petrocrania* from the fossil phosphatic inarticulates and both of those from fossil articulates; the features which distinguish *Crania* from living inarticulates and articulates are also given.

*Petrocrania* is distinguished from the phosphate inarticulates *Orbiculoidea* and Carboniferous *Lingula* in having lower lysine and histidine; it has higher threonine and isoleucine than both the phosphatic inarticulates and the articulates; it is intermediate in position between phosphatic inarticulates and articulates with regard to serine and to the glycine-alanine ratio.

The taxonomic distinctions between fossil groups, though less marked than those shown by present-day proteins (primarily because of the loss of hydroxyproline), set the carbonate shelled inarticulate *Petrocrania* apart from the phosphatic shelled inarticulate fossils, *Orbiculoidea* and

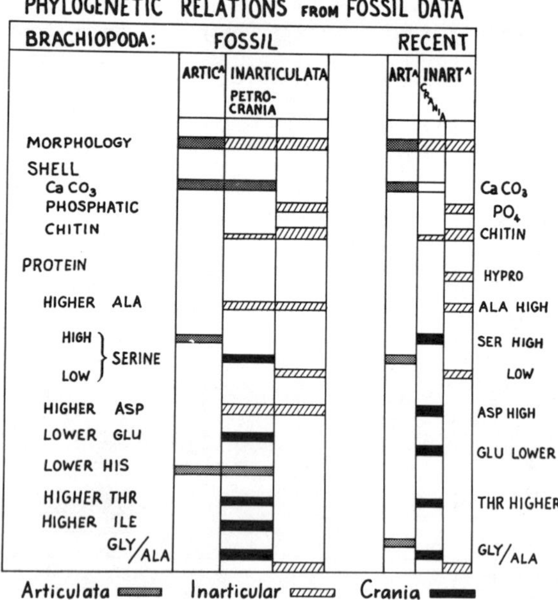


**Figure 6.** Comparison of characteristics of *Petrocrania* with those of fossil inarticulate and fossil articulate brachiopods. Distinctions between Recent *Crania* and Recent inarticulate and articulate brachiopods are also shown.

*Lingula,* just as present-day *Crania* stands apart from present-day *Discinisca* and *Lingula.*

## PHYLOGENETIC IMPLICATIONS OF DATA FROM LIVING SHELL PROTEINS

The study of living material can thus give some basis for a retrospective reconstruction of the shell protein of the original organism, which can clarify the phylogeny and possible evolution of the phylum. With these brachiopod genera of very long stable history, the shell proteins can preserve evidence of phylogenetic relationships even more distant than those apparent in the morphology alone. Gel electrophoresis data showing the molecular size of the protein chains in the shell (Fig. 3) and also end-group data (Jope 1979) reveal evidence of a contribution from a phosphatic shelled protein system in the ancestry of *Crania.* This supports the paleontological view that the Craniacea arose within inarticulate stock in Ordovician times (Fig. 1, Williams and Rowell 1965). The proteins of the living forms *Crania* and *Discinisca* seem, however, too distinct for the Craniacea to have arisen directly out of acrotretid stock.

The terebratulide shell proteins also, though to a lesser extent, reveal a similar element in their ancestry (Fig. 3); this suggests the possibility that the terebratulides (carbonate shelled articulates), too, could have arisen out of phosphatic shelled inarticulate stock, though far back in the history of the phylum (Fig. 1).

## CONCLUSIONS

This preliminary survey shows there is a real possibility of obtaining information of phylogenetic significance from the fossil shell protein data, which may be correlated with the morphological aspects of evolution and with the retrospective implications of protein systems of living brachiopods to clarify the phylogeny of brachiopods as a whole, spanning some 600 million years of evolutionary history.

### *REFERENCES*

Alexander, P. and J. T. Lett, Effects of ionizing radiations on biological macromolecules, In *Comprehensive Biochemistry, Vol. 27: Photobiology, Ionizing Radiations,* M. Florkin and E. H. Stotz, eds. pp. 267–356. Elsevier, Amsterdam, 1967.

Akiyama, M., The amino acid composition of fossil scallop shell proteins and non proteins, *Biomineralization Research Reports*, **3**, 65–70, 1971.

Bricteux-Grégoire, S., M. Florkin, and Ch. Grégoire, Prism conchiolin of modern or fossil molluscan shells. An example of protein paleization, *Comp. Biochem. Physiol*, *24*, 567–572, 1968.

Florkin, M., Paléoprotéines. *Bull. Acad. R. Belg. Cl. Sci.*, **51**, 156–169, 1965.

Foucart, M. F., S. Bricteux-Grégoire, Ch. Jeuniaux, and M. Florkin, Fossil proteins of graptolites, *Life Sci.* **4**, 467–471, 1965.

Franzblau, C., P. M. Gallop, and S. Seifter, The presence in collagen of γ-glutamy peptide linkages, *Biopolymers*, **1**, 79–97, 1963.

Hare, P. E., and R. M. Mitterer, Non protein amino acids in fossil shells, *Carnegie Inst. Wash. Year Book 65*, 362–364, 1966.

Harington, J. A., and P. D. Toens, Natural occurrence of amino acids in dolomitic limestones containing algal growths, *Nature*, **200**, 947–948, 1963.

Jope, M., The protein of brachiopod shell—II. Shell protein from fossil articulates: amino acid composition, *Comp. Biochem. Physiol.* **20**, 601–605, 1967a.

Jope, M., The protein of brachiopod shell—I. Amino acid composition and implied protein taxonomy, *Comp Biochem. Physiol.* **20**, 593–600, 1967b.

Jope, M., The protein of brachiopod shell—IV. Shell protein from fossil inarticulates: Amino acid composition and disc electrophoresis of fossil articulate shell protein, *Comp. Biochem. Physiol.*, **30**, 225–232, 1969.

Jope, M., Constituents of brachiopod shells, in *Comprehensive Biochemistry*, Vol. 26C, M. Florkin and E. H. Stotz, eds., pp. 749–784. Elsevier, Amsterdam, 1971.

Jope, M., The protein of brachiopod shell—VI. C-Terminal end-groups and sodium dodecylsulphate-polyacrylamide gel electrophoresis: Molecular constitution and structure of the protein, *Comp. Biochem. Physiol.*, **63B**, 163–173, 1979.

Mahler, H. R., and E. H. Cordes, *Biological Chemistry*, pp. 1–872, Harper & Row, New York; Weatherhill, Tokyo, 1966.

Miller, A., and D. A. D. Parry, Structure and packing of microfibrils in collagen, *J. Mol. Biol.*, **75**, 441–447, 1973.

Nei, M., *Molecular Population Genetics and Evolution*, pp. 1–283, North-Holland, Amsterdam, 1975.

Piez, K. A., and A. Miller, The Structure of collagen fibrils, *J. Supramol. Structure*, **2**, 121–137, 1974.

Schirch, L. G., and M. Mason, Serine transhydroxymethylase, *J. Biol. Chem.*, **238**, 1032–1037, 1963.

Seifter, S., C. Franzblau, E. Harper, and P. M. Gallop, Special aspects of the primary structure of collagen, in *The Structure and Function of Connective and Skeletal Tissue*, Tristram G. R., ed., pp. 21–26, Butterworths, London, 1965.

Weiner, S., H. A. Lowenstein, and L. Hood. Characterization of 80-million-year-old mollusk shell proteins. *Proc. Natl. Acad. Sci.* (USA), **73**, 2541–2545, 1976.

Williams, A., A history of skeletal secretion among articulate brachiopods, *Lethaia*, **1**, 268–287, 1968.

Williams A., and A. J. Rowell, Classification, in *Treatise on Invertebrate Palaeontology*, Vol. H, Brachiopoda, R. C. Moore, ed., pp. H214–H237, Univ. of Kansas Press, Lawrence, Ks., 1965.

Wilson, J. B., A. Miller, and T. A. J. Huisman, Production of Hemoglobin C in Moufflon (*Ovis musimon* Pallas, 1811) and in Barbary Sheep (*Ammotraqus lervia* Pallas, 1777) during experimental anemia, *Biochem. Genetics*, **4**, 677–688, 1970.

# Well-Preserved Fossil Mollusk Shells: Characterization of Mild Diagenetic Processes

**STEPHEN WEINER**
*Isotope Department, Weizmann Institute of Science, Rehovot, Israel*

**HEINZ A. LOWENSTAM**
*Division of Geological and Planetary Sciences, California Institute of Technology*

*ABSTRACT*

Diagenetic changes in both the bio-inorganic and the organic fractions of well-preserved Upper Cretaceous fossil bivalve shells from Coon Creek, Tennessee and ammonoid shells from Punta San Jose, Baja California, Mexico, have been studied. The mineralogy, trace element contents, and stable isotopic compositions of the shells show only minor diagenetic changes. However, the shell organic matrix components have changed significantly. The alloisoleucine/isoleucine and glycine/alanine ratios of the nondialyzable fractions vary from specimen to specimen and are useful as indicators of the relative states of preservation of the fossil specimens. This combined approach to studying the diagenesis of fossil shells is useful in assessing the state of preservation of relatively well preserved fossil material and is an invaluable tool for paleoecological studies as well as paleobiochemical studies of fossil protein components.

## INTRODUCTION

Well-preserved fossil shells are necessary for studying past evolutionary and environmental changes. An omnipresent problem in studies of this type is the screening of the diagenetic "noise" from the original "signal." The utilization of so-called well-preserved fossil material greatly facilitates this process. However, some diagenetic changes are likely to have occurred even in the best available fossil material. This paper characterizes properties of both the inorganic and the organic fractions of two exceptionally well preserved fossil species from different sites in order to evaluate commonly accepted criteria for determining state of preservation, to monitor diagenetic changes which have occurred, and to determine useful criteria for recognizing mild diagenesis.

Commonly accepted criteria for recognizing the least diagenetically altered fossils are the presence of aragonitic hard parts with original fabric (Boggild 1930) and the similarity in trace element contents of fossil shells with living representatives of the same taxonomic group (e.g., Chave 1954, a, b; Turekian and Armstrong 1961; Pilkey and Goodell 1964; Dodd 1966; Curtis and Krinsley 1965; Hallam and Price 1968; Lorens et al., 1977; Walls et al., 1977). The combination of trace element contents with stable oxygen isotopic composition of the fossil shell carbonate has provided a useful combination of criteria for detecting even minor changes in the chemistry of the bio-inorganic fraction (Lowenstam 1961, 1963; Buchardt 1977).

The organic matrices of fossil mollusk shells have been extensively studied, following the suggestion by Abelson (1954) that this material could provide invaluable information to further our understanding of molecular evolution. The diagenesis of this material has been extensively studied primarily by analyzing the amino acid compositions of fossil shell organic matrices (Abelson 1955; Grégoire 1955, 1959; Bricteux-Grégoire et al., 1968; Matter et al., 1969; Akiyama 1971) by laboratory simulation experiments on extant shells (Jones and Vallentyne 1960; Hare and Mitterer 1968; Vallentyne 1969; Grégoire and Voss-Foucart 1970; Totten et al., 1972; Voss-Foucart and Grégoire 1975), or by determining the extent of racemization of the shell matrix amino acids (Schroeder and Bada 1976; Williams and Smith 1977). Recent studies have identified macromolecular complexes still containing antigenic determinants (de Jong et al., 1974; van der Meide et al., 1978) and preserved characteristic amino acid sequences of fossil shell proteins (Weiner et al., 1976).

In this paper we analyze both the bio-inorganic fraction and the organic fraction of fossil shells in order to evaluate as reliably as possible the state of preservation of the original properties of the fossil material. To do this, we have focused our efforts mainly on the least altered properties of the shells, not only to provide information on the potential usefulness of such fossil shells for evolutionary studies but also as an additional means by which the state of preservation of the bio-inorganic fraction can be evaluated.

## MATERIAL AND METHODS

### Material Selected

Six specimens of the bivalve *Scabrotrigonia thoracica* from the Upper Cretaceous of Coon Creek, Tennessee, were obtained by deep trenching below the oxidized zone. Although the six specimens were obtained within less than 2 square meters of each other, they were not all buried simultaneously but may differ slightly in age. Three specimens of the cephalopod *Baculites inornatus* from the Upper Cretaceous beds at Punta San Jose, Baja California, Mexico, were excavated from among partially exposed material on a sea-cliff which is at present being cut back by the sea. The exposures are all fresh. At both sites the sediments in which the fossils are buried is a clay-sand that is friable at Coon Creek and fairly well consolidated at Punta San Jose. For comparative purposes a recent representative of the Trigonid super family, *Neotrigonia margaritacea* from Victoria, South Australia, was studied, together with *Nautilus pompilius* from the Philippines as a representative of the tetrabranchian cephalopods. Specimens of both of these species were collected alive, and the shells were air-dried after their soft-parts had been removed.

### Physical Appearance

The *S. thoracica* specimens analyzed retained a well-developed pearly luster similar to the recent Trigonid, *N. margaritacea*, and closely resembling that of the pearly oyster, *Pinctada*. The *B. inornatus* specimens show a distinctive ivory luster somewhat similar to the nacreous layer of extant Nautilus. These fossil cephalopod shells, however, tend to develop an "oily irridescence" in areas subject to surficial weathering.

### Preparation of the Shell Material

The *S. thoracica* shells were removed from the poorly consolidated sediment in which they had been buried by immersion in distilled water for about an hour. Periostracal remnants and contaminating foreign material were then mechanically removed. The *B. inornatus* specimens were physically pried loose from their consolidated matrix and the shell fragments carefully examined for adhering sediment and manganese dendrites. Sediment-bearing particles were separated from clean ones.

### Bio-inorganic Fraction

The shell mineralogy was examined using X-ray diffraction (Norelco instrument) with a Debye-Sherrer camera using Ni filtered Cu radiation, together with infrared absorption spectra using a Perkin-Elmer 180 spectrophotometer and KBr pellets. Scanning electron micrographs were obtained with an ETEC Autoscan, and the Sr and Mg contents were determined using a MAC V automated electron microprobe.

### Organic Fraction (Organic Matrix)

The shells were briefly dipped in 5.8% ammonium hydroxide to remove superficially contaminating amino acids and peptides, washed with redistilled water, sonicated for about a second, and then rewashed. The shells were lightly crushed into large (>0.1 mm) fragments* and were dialyzed against 10% (wt/vol) EDTA buffered with 0.05 *M* phosphate to pH 7.0 containing 0.01% (wt/vol) sodium azide to prevent bacterial growth. Cellulose dialysis tubing with approximately 12,000 molecular weight cut-off was used. After decalcification (about 7 days) the contents of the dialysis bag were dialyzed against water and lyophilized; the insoluble fraction was then removed by low speed (about 3000 *g*) centrifugation, and the soluble fraction was desalted on a Sephadex G-25 column (30 × 2.5 cm) in water. The organic matrix of the fossils was almost completely soluble, whereas that of the recent representatives were predominantly insoluble. The fractions included on Sephadex G-25 were exhaustively redialyzed against redistilled water and lyophilized. *S. thoracica* specimen (9) was decalcified in two stages. The EDTA–organic matrix supernatant was removed after about half of the $CaCO_3$ had dissolved. That constituted the first extract. The second extract was obtained after complete dissolution. Aliquots of the different fractions were hydrolyzed with 0.5 ml redistilled 6 *N* HCl under vacuum, after flushing twice with nitrogen, at 108°C for 20 hours. Amino acid compositions were determined using a Durrum 500 amino acid analyzer. The analyses of *S. thoracica* specimen (1) have been previously reported (Weiner et al., 1976).

* A control experiment using extant *N. pompilius* and *Mercenaria mercenaria* showed empirically that if large fragments (>0.1 mm) were used for decalcification, the properties of the extracted organic matrix were more reproducible from one preparation to another than when small fragments are used (Weiner 1976).

Quantitative estimates of the total soluble organic content of the *B. inornatus* fossil shells were made by dissolving material in a molar equivalent amount of redistilled 6 *N* HCl at room temperature and passing the supernatant through an XAD-2 resin column (15 × 0.5 cm) equilibrated with 0.1 *N* HCl, washing the column with water and then eluting the bound organic material with 5% $NH_4OH$ (Mantoura and Riley 1975). The eluted fraction was lyophilized and weighed. The contents of the insoluble and soluble fractions of the extant shells were estimated by weighing the fractions after the normal preparative procedure. Although slight EDTA contamination of the soluble fraction is possible, little of the extant material appears to be lost through the dialysis bag. The protein contents of the fractions were estimated from the yields of amino acids obtained after total hydrolysis of a known amount of sample using norleucine as an internal standard. The total soluble protein and polypeptide of an *S. thoracica* specimen was determined by avoiding dialysis during decalcification and removing the EDTA and salt components by desalting on Sephadex G-25 followed by Sephadex G-10. The total soluble protein and polypeptide of *B. inornatus* specimens was determined using the material obtained from the XAD-2 column.

## RESULTS

### Shell Ultrastructure

Broken surfaces of the fossil shells examined under the scanning electron microscope show fresh, clean, sharp fracture surfaces similar to those seen in the living counterparts. At 3000× magnification the lamellae of the nacreous layer in *S. thoracica* show local thinning resulting in a draping effect over adjacent units. *B. inornatus* nacreous layer shows a similar effect in addition to occasional borings that lack secondary infilling. No evidence of cementation or recrystallization processes was found. These results show that some loss of the carbonate fraction has occurred.

### Mineralogy

In all cases the shells examined were composed entirely of aragonite as determined by both X-ray diffraction and infrared absorption. This is in agreement with the SEM observations in which no cementation or recrystallization was observed.

## Mg and Sr Contents of the Shells

Both the fossil and the extant shells examined have exceedingly low Mg contents, that is, less than 0.1%, which is just above the electron microprobe detection limit (Table 1). This is commonly observed in all recent mollusk aragonitic shells (Lowenstam 1963). The Sr contents of the fossil shells examined are at least twice as high as those of the extant representatives (Table 1) and require some additional comments. The values in Table 1 are averages of the concentrations obtained along traverses in *S. thoracica* normal to the shell's original surfaces and in *B. inornatus* normal to the growth increments from shell chips mechanically dislodged from random sites of the shells. There is considerable variation in concentration levels along the line of traverses in both instances. However, except for the prismatic layer, the highest concentrations found in *S. thoracica* shells are found in the interior of the samples, whereas, in *B. inornatus* they occur, with one exception, in the border zone of the chip surfaces. This points to some differences in the diagenetic history of the *S. thoracica* and *B. inornatus* samples. Mechanical flaking of the more consolidated *Baculites* shells would occur most readily along lines of mechanical weakness. The shells were derived from, or were very close to, the surface of the sedimentary rocks, which are under active erosion by the sea in the cliffs near Punta San Jose. It thus seems likely that a network of permeability channels on the submicroscopic scale could develop through large-scale degradation of the intercrystalline organic matrices. Such a process would account for (1) the lines of preferred mechanical weakness, and (2) for facilitating the introduction of excess Sr from the permeating seawater into the border zones of the adjacent aragonite crystals. This explanation, in principle, is in agreement with the observations of Pilkey and Goodell (1964), who reported that dead, Recent shells have slightly higher Sr contents than conspecific shells from live individuals. The lowest Sr values found in the core of the chips are slightly higher than the values found in extant *Nautilus* shells (Table 1). Although lower than the average Sr values reported by Hallam and Price (1966, 1968) for late Mesozoic ammonite and nautiloid shells, they nevertheless support their conclusion that the Sr contents of the shells in cephalopod groups decreased in the course of their biochemical avolution. Thus there is circumstantial evidence for original as well as diagenetically enriched Sr contents in the *Baculites* shells. On the other hand, the localization of the lowest Sr values along the periphery of the *S. thoracica* shells points to Sr depletion in the contact zone between the shell aragonite and the low Sr connate waters of fresh water origin. The higher values found within the *S. thoracica* shells are considered to be the original concentrations.

**Table 1. Trace Element and Stable Isotopic Composition of the Fossil Shell Carbonate[a]**

| Specimen Fossil | $\delta^{18}O$[c] (‰) | $\delta^{13}C$[c] (‰) | $MgCO_3$[d] (wt %) | $SrCO_3$ (wt %) | Pearly luster |
|---|---|---|---|---|---|
| *S. thoracica* 1 | −0.56 | +1.96 | 0.06 | 0.26 | Very good |
| *S. thoracica* 2 | −0.82 | +1.36 | 0.04 | 0.32 | Good |
| *S. thoracica* 3 | −0.36 | +1.49 | 0.05 | 0.27 | Moderate to good |
| *S. thoracica* 10 | −0.75 | +1.69 | 0.05 (±0.04) | 0.31 (±0.05) | Poor – Sediment filled borings; ferric oxide patches |
| *S. thoracica* 5 | −0.57 | +1.28 | 0.05 (±0.06) | 0.30 (±0.14) | Moderate to poor |
| *S. thoracica* 6 | −0.58 | +2.15 | 0.03 (±0.03) | 0.24 (±0.10) | Moderate |
| *S. thoracica* 7 | −0.51 | +1.05 | 0.05 (±0.02) | 0.52 (±0.30) | Absent |
| Extant | | | | | |
| *N. margaritacea* 1 | — | — | 0.03 (±0.03) | 0.14 (±0.05) | |
| *N. margaritacea* 2 | — | — | 0.02 (±0.02) | 0.14 (±0.07) | |
| Fossil | | | | | |
| *B. inornatus*[b] | −0.25, −0.96 | — | 0.07, 0.07 | 0.67, 0.72 | |
| | −0.90, −0.57 | — | 0.07, 0.07 | 0.69, 0.78 | |
| | | — | 0.07 | 0.67 | |
| Extant | | | | | |
| *N. repertus* | — | — | 0.03 | 0.42 | |

[a] The standard deviations of the trace element contents represent variations within a particular specimen. The standard deviations of the stable isotopic compositions are approximately 0.1%.
[b] Represent analyses of conspecific specimens from the same outcrop. The stable isotopic composition and trace element analyses were not performed on the same set of specimens.
[c] Relative to PDB Chicago Standard.
[d] The $MgCO_3$ contents are close to the limits of detection by the electron microprobe.

## Oxygen and Carbon Isotopic Composition

The range of $\delta^{18}O$ values shown in Table 1 is at most due to only minimal exchange with ground water. Slight variations in temperature and local oxygen isotopic composition of the water cannot be excluded, since all the specimens come from exactly the same stratigraphic horizon. The fact that $\delta^{18}O$ values of the *S. thoracica* specimens analyzed are relatively heavy is illustrated by comparing $\delta^{18}O$ values of *S. thoracica* from the oxidized zone of the same outcrop, which are about 1.2 permil lighter than those obtained by trenching (Lowenstam and Epstein 1954). This observation was previously noted by Weiner et al. (1976). The variations in the $\delta^{13}C$ of the *S. thoracica* do not obviously relate to the corresponding $\delta^{18}O$ values and are not understood.

The above analyses indicate that the bio-inorganic fractions of the fossil shells have undergone only minimal diagenetic alteration. By these criteria the fossils examined are exceptionally well preserved.

## The Organic Matrix

The amino acid compositions of the nondialyzable Sephadex G-25 excluded (G-25-A) and included (G-25-B) fractions of the fossil and extant specimens are shown in Tables 2 and 3, respectively. The fossil organic matrices of both species analyzed are almost entirely soluble after decalcification. A similar observation was made by Van der Meide (1978) for an Upper Cretaceous belemnite, *Gonioteuthis* sp. The amino acid compositions of the G-25-A fractions of the extant representatives differ significantly from their G-25-B fractions (Table 3). The corresponding fossil organic matrix compositions, however, are very similar. Furthermore, the mean amino acid compositions of the *S. thoracica* and *B. inornatus* species closely resemble each other (Table 2) and to a certain extent also resemble, for example, fossil Nautiloids and Ammonoids analyzed by Grégoire and Voss-Foucart (1970). Clearly, diagenetic processes have tended to selectively preserve or alter protein fractions such that their resulting amino acid compositions are similar. The G-25-A and G-25-B fractions of the fossils do, however, differ with respect to their serine contents. In both fossil species the G-25-B fractions contain more serine than the corresponding G-25-A fractions (Table 2), as is found in the extant representatives (Table 3). It is of interest to note that the G-25-B fraction of the extant bivalve *Mercenaria mercenaria* is enriched in polysaccharide relative to protein compared to the G-25-A fraction and that one of the protein-polysaccharide linkages in *M. mercenaria* soluble organic matrix is through serine (Weiner 1976). Thus, despite the gross amino acid com-

**Table 2. Mean and Standard Deviations of the Amino Compositions of the Fossil Organic Matrix Fractions from Different Specimens**

| | *S. thoracica* ($n$ = 8) | | *B. inornatus* ($n$ = 3) | |
|---|---|---|---|---|
| Mole % | G-25-A | G-25-B | G-25-A | G-25-B |
| Asp + Asn | 11.06 ± 0.75 | 9.58 ± 1.10 | 10.28 ± 1.63 | 8.90 ± 1.63 |
| Thr | 5.40 ± 1.03 | 5.26 ± 0.60 | 4.92 ± 0.59 | 5.60 ± 0.93 |
| Ser | 6.54 ± 0.62 | 11.53 ± 5.64 | 6.22 ± 0.55 | 10.57 ± 5.85 |
| Glu + Gln | 9.23 ± 1.93 | 11.41 ± 1.43 | 11.13 ± 0.94 | 11.79 ± 1.25 |
| Pro | 5.21 ± 1.55 | 4.08 ± 1.29 | 4.05 ± 0.16 | 4.74 ± 2.01 |
| Gly | 19.31 ± 8.79 | 16.13 ± 5.29 | 17.39 ± 1.69 | 13.63 ± 4.43 |
| Ala | 10.95 ± 2.23 | 10.70 ± 1.37 | 10.59 ± 1.52 | 11.99 ± 2.32 |
| Cys[a] | 0.30 ± 0.58 | 0.68 ± 0.75 | 0.35 ± 0.60 | 1.65 ± 1.54 |
| Val | 5.36 ± 1.12 | 4.65 ± 0.98 | 7.00 ± 1.76 | 5.53 ± 1.35 |
| Met | 1.51 ± 0.31 | 1.07 ± 0.45 | 1.46 ± 0.78 | 0.61 ± 0.24 |
| Alloile | 0.62 ± 0.90 | 0.23 ± 0.42 | trace | trace |
| Ile | 3.13 ± 0.61 | 2.78 ± 0.56 | 3.71 ± 0.30 | 3.04 ± 0.37 |
| Leu | 6.37 ± 1.40 | 6.17 ± 1.81 | 6.97 ± 0.92 | 6.93 ± 1.60 |
| Tyr | 2.60 ± 1.33 | 2.16 ± 0.93 | 2.26 ± 0.58 | 1.62 ± 0.29 |
| Phe | 3.05 ± 0.64 | 2.76 ± 1.00 | 4.38 ± 0.19 | 3.14 ± 0.38 |
| His | 1.43 ± 0.99 | 2.11 ± 0.89 | 1.61 ± 0.10 | 1.71 ± 0.01 |
| Lys | 3.99 ± 0.68 | 4.79 ± 1.20 | 3.69 ± 1.34 | 5.32 ± 0.59 |
| Arg | 4.10 ± 0.65 | 3.92 ± 1.41 | 3.97 ± 0.96 | 3.77 ± 1.24 |

[a] Might contain aminobutyric acid which co-chromatographs with cysteine.

position similarity between fossil G-25-A and G-25-B fractions, their chromatographic fractionation is still at least partially based on preserved original properties which are responsible for the separation into excluded and included fractions on Sephadex G-25 in extant soluble matrices.

The glycine and alloisoleucine contents of the fossil organic matrix fractions also vary significantly, as is shown by their relatively large standard deviations about the mean (Table 2). The individual analyses are presented for the different specimens examined as the alloisoleucine/isoleucine (Alloile/Ile) and glycine/alanine (Gly/Ala) ratios (Table 4). As alloisoleucine is present in only trace amounts in the nondialyzable fractions of *B. inornatus*, only the *S. thoracica* Alloile/Ile versus Gly/Ala ratios are plotted in Figure 1. In general, where alloisoleucine is present in negligible quantities, the Gly/Ala ratio still varies between 0.76 and 2.09 (Table 4). The Alloile/Ile ratios of G-25-A fractions are inevitably greater than G-25-B fractions, as are the Gly/Ala ratios (with one exception, *S. thoracica* specimen 7).

**Table 3. Amino Acid Compositions of the Organic Matrix Fractions of the Extant Representatives of the Trigoniidae and Tetrabranchian Cephalopods**[a]

| | *N. margaritacea*[b] | | | *N. pompilius* | | |
|---|---|---|---|---|---|---|
| Mole % | G-25-A | G-25-B | Insoluble | G-15-A | G-25-B | Insoluble |
| Asp + Asn | 18.13 | 7.85 | 7.45 | 26.06 | 7.87 | 7.96 |
| Thr | 3.53 | 4.09 | 1.37 | 5.26 | 4.40 | 1.21 |
| Ser | 13.09 | 16.68 | 10.59 | 5.48 | 18.69 | 9.33 |
| Glu + Gln | 8.06 | 15.96 | 3.32 | 5.98 | 17.37 | 4.15 |
| Pro | 3.60 | 3.25 | 1.52 | 3.80 | 0.39 | 1.04 |
| Gly | 23.02 | 18.11 | 32.11 | 22.43 | 21.41 | 33.75 |
| Ala | 9.28 | 11.23 | 23.30 | 3.75 | 7.54 | 24.48 |
| Cys | 1.22 | 0.97 | 0.81 | trace | trace | trace |
| Val | 2.23 | 3.57 | 1.43 | 2.18 | 3.49 | 1.67 |
| Met | 1.94 | 0.65 | 1.98 | 0.89 | 0.88 | 0.69 |
| Ile | 2.09 | 2.34 | 1.45 | 2.18 | 2.15 | 1.60 |
| Leu | 3.53 | 4.02 | 3.72 | 3.08 | 2.97 | 2.53 |
| Tyr | 1.44 | 0.97 | 2.73 | 5.87 | 1.67 | 1.16 |
| Phe | 2.73 | 2.14 | 2.71 | 3.13 | 2.81 | 5.92 |
| His | 0.72 | 2.60 | 0.44 | 3.08 | 3.17 | 0.44 |
| Lys | 2.52 | 3.31 | 1.98 | 3.58 | 3.40 | 0.38 |
| Arg | 2.88 | 2.27 | 3.11 | 3.24 | 1.72 | 3.68 |

[a] The close similarity in amino acid composition of corresponding fractions of these two species is coincidental (e.g., Meenakashi et al., 1971).
[b] Previously published in Weiner et al. (1976).

The nonequilibrium Alloile/Ile ratios reported for the nondialyzable fractions do not imply that the total organic matrix has not been epimerized to a greater extent. An analysis of the total organic matrix of an *S. thoracica* specimen with an exceptionally well preserved pearly luster showed that the Alloile/Ile ratio of this specimen approached equilibrium (P. E. Hare, personal communication). Thus, various organic matrix components appear to exist in different states of epimerization. This had been observed by Hare and Hoering (1977), who found that the Alloile/Ile ratio of the total organic matrix of a Miocene *Mercenaria* shell was close to equilibrium, whereas the humic acid fraction of this shell had an Alloile/Ile ratio of 0.45.

The carbon isotopic compositions of the total shell organic matrix of well-preserved specimens (judging by their pearly luster) of *S. thoracica* and *B. inornatus* are compared to the carbon isotopic compositions of the organic matter in which they were found (Table 5). At both sites the shell

**Table 4. Alloisoleucine/Isoleucine and Glycine/Alanine Ratios of the *S. thoracica* Specimens Analyzed[a]**

| Specimen number | | Alloisoleucine/[b] isoleucine | | Glycine/Alanine | |
|---|---|---|---|---|---|
| | | G-25-A | G-25-B | G-25-A | G-25-B |
| *S. thoracica* | 1 | 0.07 | 0.02 | 1.20 | 1.09 |
| | 2 | 0.10 | 0.02 | 1.21 | 1.02 |
| | 3 | 0.08 | 0.07 | 1.12 | 1.08 |
| | 4 | 1.43 | 0.54 | 6.88 | 3.15 |
| | 5 | 0.02 | trace | 1.22 | 1.10 |
| | 6 | 0.03 | trace | 1.47 | 1.32 |
| | 7 | 0.07 | trace | 1.56 | 1.94 |
| | 8 | 0.39 | 0.10 | 2.07 | 1.89 |
| | 9a Extract 1 | 0.69 | 0.06 | 3.23 | 1.99 |
| | 9b Extract 2 | 0.10 | trace | 1.53 | 1.44 |
| | | — | — | | |
| *B. inornatus* | 1 | trace | trace | 1.66 | 0.91 |
| | 2 | trace | trace | 1.29 | 0.76 |
| | 3 | trace | trace | 2.09 | 2.00 |

[a] The three *B. inornatus* specimens analyzed did not contain detectable quantities of alloisoleucine.
[b] If the alloisoleucine content is less than 0.0005, it is considered as a "trace" quantity.

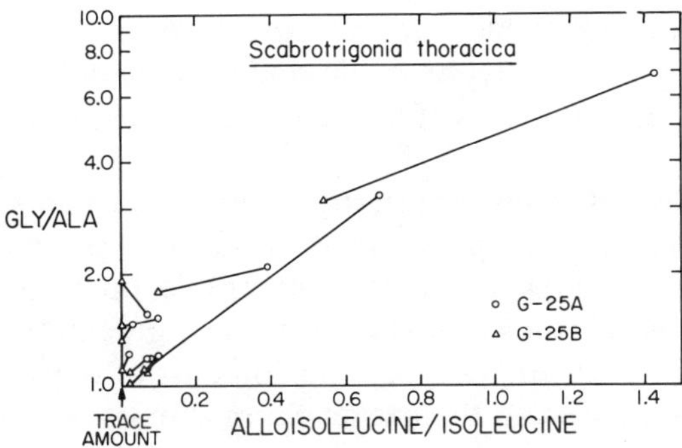


**Figure 1.** The glycine/alanine versus alloisoleucine/isoleucine ratios of the nondialyzable fractions of *S. thoracica* specimens listed in Table 4.

**Table 5. Carbon Isotopic Compositions of the Total Fossil Organic Matrix and the Organic Material of the Sediment in which They Are Buried**

| | $\delta^{13}C$ (‰)[a] | |
|---|---|---|
| | Total Organic Matrix[b] | Sedimentary Organic Material[c] |
| *S. thoracica* | −26.8 | −27.5 |
| *B. inornatus* | −26.3 | −28.2 |

[a] Measurements were performed on a double inlet Atlas M86 mass spectrometer. Estimated error is ±0.2‰. Data are reported in the δ notation relative to the PDB standard.

[b] Prepared with the XAD-2 resin, as described in the Methods section.

[c] Sediment first decalcified in 3 *N* HCl, washed, dried, and then combusted at 900°C in an oxygen atmosphere.

organic material is slightly enriched in $^{13}C$ compared to the sediment organic material. The differences imply that the total shell organic material is at least partially composed of indigenous shell material. As fossil organic material, marine or terrestrial have $\delta^{13}C$ values similar to those reported here (Degens 1969), the carbon isotopic compositions of the fossil shell matrices reported do not necessarily mean that contamination has occurred. The fact that the carbon isotopic compositions of the fossil organic matrices are much lighter than those reported for extant shell organic matrices (Hare and Hoering 1977) could be due to the fossil organic matrices themselves having undergone diagenetic alteration, a process which undoubtedly has occurred. The extent to which the total organic matrix has been degraded is illustrated by the fact that the amino acid composition of the total organic matrix of an *S. thoracica* shell differs from that of the nondialyzable fractions analyzed (Table 2) in that it is greatly enriched in the relatively stable amino acids Gly, Ala, and Glu and contains only small amounts of Ser, Thr and Asp (P. E. Hare, personal communication.)

The total organic matrix content of the *B. inornatus* shells is about 15% of that of the extant *N. pompilius* shell (Table 6). Assuming the fossil ammonoid shell originally contained comparable quantities of organic material as modern *Nautilus*, substantial loss of organic matrix has occurred. The total protein and polypeptide contents of the fossil shells are greatly reduced compared to the extant representatives, and the nondialyzable protein content of the fossils is even smaller (Table 6).

**Table 6. Organic Matrix Contents of the Fossil and Extant Shells (μg/g dry shell)**

| | Bivalvia | | Cephalopoda | |
|---|---|---|---|---|
| | *S. thoracica* (fossil) | *N. margaritacea* (extant) | *B. inornatus* (fossil) | *N. pompili* (extant) |
| Total organic matrix content (excludes an estimated 25% by weight water) | | | | |
| Insoluble | Negligible | 3240 | Negligible | 29,100 |
| Soluble G-25-A | n.d. | 1985 | 5025 ± 860 (n = 3) | 2,025 |
| Total protein and polypeptide[a] content | | | | |
| Insoluble | Negligible | 3280[b] | Negligible | 29,100[b] |
| Soluble | G-25-A 31 | 96 | 50 ± 24[d] | 850 |
| | G-25-B 29 | 40 | (n = 3) | n.d. |
| Non-dialyzable protein content[a] | | | | |
| Insoluble | Negligible | 3280[c] | Negligible | 29,100 |
| Soluble G-25-A | 4.8 ± 1.3 (n = 6) | 96 | 6.8 ± 3.6 (n = 3) | 850 |
| G-25-B | 6.9 ± 1.2 (n = 6) | 40 | 6.0 ± 2.3 | n.d. |

n.d. = not determined.

[a] Calculated from weights of amino acid per unit weight of organic matrix fraction obtained on complete hydrolysis and then normalized to original shell weight.

[b] The values are assumed to be the same as those for the non-dialysable protein content as not much material is thought to be lost during dialysis.

[c] Previously reported in Weiner et al. (1976).

[d] Represents the quantities of amino acids in the samples purified by XAD-2 chromatography.

## DISCUSSION

The analyses of the bio-inorganic fraction show that the diagenetic changes are limited to losses of mineral phase, especially around the margins of the microarchitectural structural units. No additions to, or alterations of, the biogenic aragonite appear to have occurred. The observed significant variations in stable isotopic composition and trace element contents between specimens of the same burial assemblage are considered to represent to a large extent original paleoenvironmental and/or individual physiological differences. The preservation of the differences within the burial assemblage is in itself an indication of minimal diagenetic alteration (Curtis and Krinsley 1965).

The organic matrix, however, has undoubtedly been subject to severe diagenetic change. Protein, which is the major constituent of extant organic matrices, has been reduced to a very minor component, as shown by the amino acid content of nondialyzable fossil matrix components (Table 6). Similar reductions of the protein content of fossil shells have been observed many times (e.g., Abelson 1955; Ho 1966; Matter et al., 1969). The bulk of the nondialyzable preserved organic matrix is a dark-brown material that partially precipitates on acidification. It has been described in other fossil shells as "humic acid" and is considered to be a diagenetic alteration product of an indigenous constituent of the shell (Hare and Hoering 1977; van der Meide et al., 1978). The amino acid composition of the nondialyzable fraction has changed to such an extent that it has lost its species-specific character, well known in extant mollusk organic matrices (e.g., Degens et al., 1967), and even the fossil bivalve and cephalopod protein remnants resemble each other. An additional indication of major breakdown of the protein is that most of the protein/polypeptide content of an *S. thoracica* shell is in the dialyzable fraction (Table 6), indicating that it has been degraded into low molecular weight components.

Quantitative loss of organic matrix material from the shell appears to have occurred in *B. inornatus* (Table 6), assuming that the original *B. inornatus* shell contained the same amount of organic matrix as a modern *Nautilus* shell. Thus, one diagenetic process that has affected these fossil shells is a minor loss of bio-inorganic material and a major loss of organic material. Contamination of the nondialyzable matrix fractions from the surrounding sediment seems unlikely for the following reasons: (1) the contaminating material would have to selectively permeate some specimens but not others, even though they are buried within a few square meters of each other, in order to account for the range of Alloile/Ile ratios observed in the suite of *S. thoracica* specimens (Table 4); (2) the shells with high alloisoleucine and glycine contents in their nondialyzable fractions are also those with poorly preserved pearly luster; (3) the contaminating material would have to be of high molecular weight, as the total organic matrix in one of the fossil shells (and presumably in all) is close to equilibrium with respect to the epimerization state of isoleucine; (4) the two-stage decalcification experiment showed that the outer surfaces of the crushed grains of one *S. thoracica* specimen contain more alloisoleucine and glycine than their inner cores (Table 4); (5) the carbon isotopic compositions of the total shell organic matrix at both sites differ from the organic matter of the sediment in which they are buried (Table 5) implying that at least part, if not all, of the shell organic matrix is indigenous.

To assess the relative states of preservation of these fossil shells, we have studied primarily the least altered fraction of the organic matrix, namely the nondialyzable, presumably high molecular weight fraction. Even though the amino acid compositions of all the fractions studied at both sites are similar, they vary considerably in their Alloile/Ile and Gly/Ala ratios (Table 4, Fig. 1). As alloisoleucine is not present in extant proteins, the increase in the Alloile/Ile ratio implies that matrix fractions containing more alloisoleucine have been more diagenetically altered than those containing less. The concomitant increase in the Gly/Ala ratio with increasing Alloile/Ile ratio in some of the *S. thoracica* shells is therefore also a diagenetic phenomenon. In addition, variations in the Gly/Ala ratio can result from differential preservation of the various fractions of the original organic matrix, as these differ greatly with respect to their Gly/Ala ratios (Table 3). Both the Alloile/Ile and Gly/Ala ratios are thus considered useful indicators of diagenetic change in the organic matrix.

The factors that promote increase in the ratios of these two parameters are not necessarily the same, but they are both probably strongly influenced by temperature, pH fixed in a carbonate matrix, age of sample, and availability of connate water (Hare 1966, 1974; Schroeder and Bada 1976; Williams and Smith 1977). It is of interest to note that the increase in glycine content is not accompanied by a corresponding increase in the other, more stable amino acids such as alanine or valine. Thus the mechanism responsible for increasing the glycine content of the nondialyzable fraction does not involve the selective removal or in situ destruction of the less stable shell protein components, as has been observed in the case of other fossil proteins (King and Hare 1972). Possible mechanisms by which the glycine content can increase include selective removal of the reactive side groups of other amino acids without coincident breakage of peptide bonds or the destruction of free amino acids to glycine (Vallentyne 1964) and their subsequent adsorption or inclusion into high molecular weight humic acid polymers (Hare and Hoering 1977).

The fact that both the fossil *S. thoracica* and *B. inornatus* organic matrices are almost entirely soluble after decalcification may be due to some diagenetic solubilization process of what was originally the insoluble fraction. However, preferential preservation of the soluble fraction is considered more likely, since ion exchange chromatography of the *Baculites* soluble matrix shows that most of the preserved protein fractions are analogous to extant soluble matrix fractions. The soluble matrix of extant mollusk shells contains aspartic acid rich proteins. These proteins are enriched on the surfaces of the intercrystalline organic matrices of *Nautilus repertus* nacreous layer relative to other soluble matrix proteins, by an in situ immunochemical mapping technique (in preparation). Soluble matrix components are thus in intimate association with the mineral phase and are therefore presumably preferentially stabilized relative to the insoluble components, and for this reason they might have been differentially preserved.

The two deposits in which the fossil Trigonids and the Ammonoids are found are both composed of clay-sand with extremely low permeability. At both localities, rainwater collecting in depressions made in freshly exposed sediment penetrated the clay-sand but only surficially. The overburden at the Coon Creek site never exceeded 1000 feet and probably was no more than 500 feet thick (W. A. Pryor, personal communication). Elevated temperatures due to the geothermal gradient did not exceed about 10°C and could have been as low as 6°C. Although reliable estimates of the overburden at Punta San Jose do not exist, it too was probably never excessive. These factors have undoubtedly contributed to the excellent state of preservation of the fossils.

The differences observed in the organic matrix properties within the burial assemblage of shells at each site (in particular the Alloile/Ile and Gly/Ala ratios) are probably due to local sedimentary heterogeneity reflected in permeability differences, together with variations in the degree of protection afforded the organic matrix components by the mineral phase. Furthermore, these differences exist within subfractions of the organic matrix components; for example, the Gly/Ala ratios of the G-25-B fractions are lower than the G-25-A fractions in every specimen examined (Table 4, Fig. 1), suggesting that the G-25-B fractions are better preserved than the G-25-A fractions. Similarly, the second extract of *S. thoracica* specimen (9) appears to be better preserved than the first extract (Table 4, Fig. 1). These observed heterogeneities within a specimen and between specimens of the same burial assemblage need to be taken into account when using the properties of fossil organic matrices for stratigraphic correlations or age determinations.

The Alloile/Ile and Gly/Ala ratios of the fossil organic matrices are useful indicators of the state of preservation of the organic matrix and in turn the fossil shell itself. The fossil shells analyzed in this study appear to be exceptionally well preserved, so most of the variations in the properties of the bioinorganic phase are probably not diagenetic. The Alloile/Ile and Gly/Ala ratios however, provide an independent assessment of the state of preservation of the bioinorganic phase and could be especially useful in studies where the fossil material is not as well preserved as the specimens from those two locations.

## CONCLUSION

The study of properties of both the bio-inorganic and organic phases of assemblages of fossil shells provides a reliable means of assessing the state of preservation of individual fossil specimens. This approach is especially important when apparently well-preserved material is being examined and where the diagenetic changes are small but still of sufficient magnitude to seriously affect the interpretation of the ecologic, evolutionary, and diagenetic history of the samples. Furthermore, the identification of two properties of the organic matrix that are sensitive indicators of the preservational state of the organic matrix; namely, Alloile/Ile and Gly/Ala ratios, can greatly facilitate the selection of suitably preserved material for fossil protein extraction and characterization. In fact, Weiner et al. (1976) reported that one of these specimens (*S. thoracica* specimen 1) has preserved a protein fraction with a characteristic original amino acid sequence.

### *ACKNOWLEDGMENTS*

We thank S. Stevenson for the shells of *Neotrigonia margaritacea*; A. Nissenbaum, S. Savin, and C. Emiliani for the oxygen and carbon isotopic determinations; P. E. Hare for the total organic matrix amino acid composition analyses; W. A. Pryor for the overburden estimate; L. Mortin and B. Taborek for their help in analyzing the organic matrices; and L. Hood for use of his laboratory facilities and for his helpful suggestions. A. Chodos performed the electronprobe analyses of the trace elements. This study was supported in part by NSF Grant DE 575-08659 to H.A.L., NSF Grant PCM71-00770 to Professor L. Hood and by a United States–Israel Bionational Science Foundation (BSF) grant to S.W.

## *REFERENCES*


Abelson, P. H., Organic constituents of fossils, *Carnegie Inst. Wash. Year. Book 53,* 97–101, 1954.

Abelson, P. H., Organic constituents of fossils, *Carnegie Inst. Wash. Year. Book 54,* 107–109, 1955.

Akiyama, M., The amino acid composition of fossil scallop shell proteins and non-proteins, *Biomineralisation Res. Rept.,* **3,** 65–70, 1971.

Buchardt, B., Oxygen isotope ratios from shell material from the Danish Middle Paleocene (Selandian) deposits and their interpretation as paleotemperature indicators, *Palaeogeogr. Palaeoclimatol. Palaeoecol.,* **22,** 209–230, 1977.

Boggild, O. B., The shell structure of the mollusks, *Kgl. Dan. Vidensk. Selsk. Skr., Naturv. Mat.,* **9**(33) 1–15, 1930.

Bricteux-Grégoire, S., M. Florkin, and Ch. Grégoire, Prism conchiolin of modern or fossil molluscan shells: An example of protein paleization, *Comp. Biochem. Physiol.,* **24,** 567–572, 1968.

Chave, K. E., Aspects of biochemistry of magnesium, (1) calcareous marine organisms, *J. Geol.,* **62,** 266–283, 1954a.

Chave, K. E., Aspects of biochemistry of magnesium, (2) calcareous sediments and rocks, *J. Geol.,* **62,** 587–599, 1954b.

Curtis, J. D., and D. Krinsley, The detection of minor diagenetic alteration in shell material, *Geochim. Cosmochim. Acta,* **29,** 71–84, 1965.

De Jong, E. W., P. Westbroek, J. F. Westbroek, J. W. Bruning, Preservation of antigenic properties of macromolecules over 70 Myr. *Nature,* **252,** 63–64, 1974.

Degens, E. T., Biogeochemistry of stable carbon isotopes, In *Organic Geochemistry,* G. Eglinton and M. T. J. Murphey, eds., pp. 304–328, Longman Springer Verlag, New York, 1969.

Degens, E. T., D. W. Spencer, and R. H. Parker, Paleobiochemistry of molluscan shell proteins, *Comp. Biochem. Physiol.,* **20,** 553–579, 1967.

Dodd, J. R., Diagenetic stability of temperature sensitive skeletal properties in Mytilus from the Pleistocene of California, *Bull. Geol. Soc. Am.,* **77,** 1213–1224, 1966.

Gregoire, Ch., Gh. Duchateau, and M. Florkin, La trame protidique des nacres et des perles, *Ann. Inst. Oceanogr.,* **31,** 1–36, 1955.

Grégoire, Ch., Conchiolin remnants in mother-of-pearl from fossil Cephalopoda, *Nature,* **184,** 1157–1158, 1959.

Grégoire, Ch., and M. F. Voss-Foucart, Proteins in shells of fossil Cephalopods (Nautiloids and Amnonoids) and experimental simulation of their alterations, *Arc. Int. Physiol. Bioch.,* **78,** 191–203, 1970.

Hallam, A., and N. B. Price, Strontium contents of recent and fossil aragonitic cephalopod shells, *Nature,* **212,** 25–27, 1966.

Hallam, A., and N. B. Price, Further notes on the strontium contents of unaltered Cephalopod shells, *Geol. Mag.,* **105,** 52–55, 1968.

Hare, P. E., Amino acid dating of bone—the influence of water, *Carnegie Inst. Wash. Year Book 73,* 576–581, 1974.

Hare, P. E., and R. M. Mitterer, Nonprotein amino acids in fossil shells, *Carnegie Inst. Wash. Year. Book 65,* 362–365, 1966.

Hare, P. E., and R. M. Mitterer, Laboratory simulation of amino acid diagenesis in fossils, *Carnegie Inst. Wash. Year. Book 67,* 205–208, 1968.

Hare, P. E. and T. C. Hoering, The organic constituents of fossil mollusc shells, *Carnegie Inst. Wash. Year. Book 76,* 625–631, 1977.

Ho, T., Stratigraphic and paleoecologic applications of water-insoluble fraction of residual shell-proteins in fossil shells, *Bull. Geol. Soc. Amer.,* **77,** 375–392, 1966.

Jones, J. D., and J. R. Vallentyne, Biogeochemistry of organic matter—I, *Geochim. Cosmochim. Acta,* **21,** 1–34, 1960.

King, K., and P. E. Hare, Amino acid composition of planktonic foraminifera: A paleobiochemical approach to evolution, *Science,* **175,** 1461–1463, 1972.

Lorens, R. B., D. F. Williams, and M. L. Bender, The early nonstructural chemical diagenesis of foraminiferal calcite, *J. Sed. Pet.,* **47,** 1602–1609, 1977.

Lowenstam, H. A., Mineralogy, $^{18}O/^{16}O$ ratios, and strontium and magnesium contents of recent and fossil brachiopods and their bearing on the history of the oceans, *J. Geol.,* **69,** 241–260, 1961.

Lowenstam, H. A., Biologic problems relating to the composition and diagenesis of sediments, In *The Earth Sciences,* T. W. Donnelly, ed., pp. 137–195, Univ. Chicago Press, 1963.

Lowenstam, H. A., and S. Epstein, Paleotemperatures of the post-Aptian Cretaceous as determined by the oxygen isotope method, *J. Geol.,* **62,** 207–248, 1954.

Mantoura, R. F. C., and J. P. Riley, The analytical concentration of humic substances from natural waters, *Anal. Chim. Acta,* **76,** 97–106, 1975.

Matter, P., F. D. Davidson, and R. W. G. Wyckoff, The composition of fossil oyster shell proteins, *Proc. Nat. Acad. Sci. U.S.A.,* **64,** 970–972, 1969.

Meenakaski, V. R., P. E. Hare, and K. M. Wilbur, Amino acids of the organic matrix of Neogastropod shells, *Comp. Biochem. Physiol.,* **40B,** 1037–1043, 1971.

Pilkey, O. H., and H. G. Goodell, Comparison of the composition of fossil and recent mollusk shells, *Bull. Geol. Soc. Amer.,* **75,** 217–228, 1964.

Schroeder, R. A., and J. L. Bada, A review of the geochemical applications of the amino acid racemization reaction, *Earth Sci. Rev.,* **12,** 347–391, 1976.

Totten, D. K., F. D. Davidson, and R. W. G. Wyckoff, Amino acid composition of heated oyster shells, *Proc. Nat. Acad. Sci. U.S.A.,* **69,** 784–785, 1972.

Turekian, K. K., and R. L. Armstrong, Chemical and mineralogical composition of fossil molluscan shells from the Fox Hills Formation, South Dakota, *Bull. Geol. Soc. Amer.,* **72,** 1817–1828, 1961.

Vallentyne, J. R., Biogeochemistry of organic matter, *Geochim. Cosmochim. Acta,* **28,** 157–188, 1964.

Vallentyne, J. R., Pyrolysis of amino acids in Pleistocene *Mercenaria* shells, *Geochim. Cosmochim. Acta,* **33,** 1453–1458, 1969.

Van der Meide, P. H., P. Westbroek, and E. W. de Jong, Characterization ofmacromolecules from fossil shells by immunology and Curie-point pyrolysis mass spectrometry, in *The 3rd International Symposium on the Mechanism of Biomineralization in the Invertebrates and Plants,* in press, 1979.

Voss-Foucart, M. F., and Ch. Grégoire, On biochemical and structural alterations of the nacre conchiolin in the Nautilus shell under conditions of protracted, moderate heating and pressure, *Arch. Internat. Physiol. Biochim.,* **83,** 43–52, 1975.

Walls, R. A., P. C. Raglund, E. L. Crisp, Experimental and natural early diagenetic mobility of Sr and Mg in biogenic carbonates, *Geochim. Cosmochim. Acta,* **41,** 1731–1737, 1977.

Weiner, S., Aspects of the biochemistry of the organic matrix of extant and fossil mollusks, 114 pp., Ph.D. dissert., California Institute of Technology, 1976.

Weiner, S., and L. Hood, Soluble protein of the organic matrix of mollusk shells: A potential template for shell formation, *Science,* **190,** 987–989, 1975.

Weiner, S., H. A. Lowenstam, L. Hood, Characterization of 80-million-year-old mollusk shell proteins, *Proc. Nat. Acad. Sci. U.S.A.,* **73,** 2541–2545, 1976.

Williams, K. M., and G. G. Smith, A critical evaluation of the application of amino acid racemization to geochronology and geothermometry, *Origins of Life,* **8,** 91–144, 1977.

# Diagenetic Decomposition of Peptide-Linked Serine Residues in the Fossil Scallop Shells

**MASAHIKO AKIYAMA**
*Department of Geology and Mineralogy, Faculty of Science, Hokkaido University, Sapporo, Japan*

*ABSTRACT*

Fossil pelecypod shells contain less serine in peptides and proteins than do the modern shells. It is suggested that serine undergoes decomposition while still bound in a polypeptide chain. This alteration was confirmed with Pleistocene scallop shell and heated modern scallop, using a method of partial hydrolysis with acetic acid.

## INTRODUCTION

Proteins of ancient organisms undergo hydrolysis yielding peptides of various chain lengths and amino acids. Degradation of amino acids by decarboxylation and/or deamination results in amines or carboxylic acids and finally hydrocarbons. Recently, compounds in the hydrolysate of proteins in fossils were noted to polymerize into macromolecules similar to humic acids and kerogen, which are widely found in sediments (Hare and Hoering 1977).

Organic remnants in fossil hard tissues such as bones, teeth, and shells may provide a clue to the reconstruction of the proteins that originally constituted ancient organisms. However, a close examination of the amino acid compositions of the modern and fossil conchiolin proteins in shells has revealed that the fossil proteins differ from the modern ones in having less aspartic acid, serine, and glycine (Akiyama 1971; Matter et al., 1969; Totten et al., 1972). The difference may be due to the following reasons: (1) The conchiolin protein consisted of various proteins with different stability to diagenetic alteration. Only the stable ones were preserved in the fossil scallop shell. (2) Degradation of the side chains of certain amino acids might have yielded the different amino acid compositions.

By experiments with heated scallop shells, the writer has demonstrated (Akiyama 1978) that the conchiolin consists of proteins with varying stability to diagenetic alteration. It is also known that serine decomposes thermally into glycine and alanine. This conversion has been shown by the pyrolysis of serine in *Mercenaria* shells as well as by the pyrolysis of serine as the free amino acid (Vallentyne 1964, 1969). This indicates that serine is thermally unstable and should be considered as a cause of the change of amino acid composition of fossil conchiolin proteins.

Hydrolysis with acetic acid indicates that the soluble protein in some modern and fossil shells consists mainly of a repeating sequence of $(\text{Asp} - \text{Y})_n$, where Asp is aspartic acid and Y is predominantly glycine and serine (Weiner and Hood 1975; Weiner et al., 1976). Acetic acid hydrolysis seems to preferentially cleave the peptide bond at both sides of aspartic acid. A comparison of the amino acid composition obtained by acetic acid hydrolysis of modern and fossil shells may show a difference in the serine content at the Y position, which would indicate the decomposition of serine during diagenesis.

For this study the fossil protein with high aspartic acid content is more suitable, though most of the pelecypod fossil shells contain an aspartic acid–poor protein.

## SAMPLES AND ANALYTICAL PROCEDURES

The fossil scallop shells (*Patinopecten yessoensis*) favorable for this study were collected from the Pleistocene Shishinai Formation in an eastern suburb of Sapporo, Hokkaido.

The right valves of the fossils were pulverized after removing the surface contamination. After one half of the pulverized shells were decalcified with 2 *N* HCl, the resulting solution was dialyzed with distilled water in Visking seamless cellulose tubing to recover proteins and polypeptides, thereby removing various inorganic ions, peptides of small chain lengths, and free amino acids dissolved in the solution. The soluble (S) and insoluble (I) fractions recovered after centrifugation were individually subjected to 6 *N* HCl hydrolysis at 108°C for 22 hours under $N_2$ atmosphere. The soluble and insoluble fractions obtained from the other half of the shells were hydrolyzed with 0.25 *N* $CH_3COOH$ at 108°C for

48 hours to cleave preferentially on both sides of aspartic acid (Weiner and Hood 1975).

For comparison with the fossil specimen the modern shell of the scallop was obtained from the shore of the Uchiura Bay, southwest of Hokkaido. Several right valves were pulverized after thorough washing and removal of the periostracum. One-gram portions of the pulverized shells were heated in open glass ampoules at 138°C for 50 hours or 160°C for 47 hours. This heating experiment was carried out in an open ampoule because less hydrolysis of proteins takes place during dry heating. The decalcification and hydrolyses of the soluble fraction of the heated shell and the insoluble one of the unheated shell were performed in the same manner as in the fossil sample.

The resulting hydrolysates were dried by evaporation at 50°C under reduced pressure and were redissolved in an appropriate amount of 1/100 *N* HCl solution for amino acid analysis. The analysis was performed by an automatic amino acid analyzer (JLC – 6AH).

## RESULTS AND DISCUSSIONS

The amino acid compositions obtained by 6 *N* HCl hydrolysis are shown in Table 1. The heating treatment of the shell at 138°C and 160°C solubilized most of the conchiolin protein. The total amount (S + I) of the amino acids in the heated shells indicates no remarkable difference from the conchiolin protein of the unheated shells.

Approximately 93% of the conchiolin protein was dissolved by the acetic acid hydrolysis. The amino acid composition of the dissolved protein consists mainly of aspartic acid, glycine, and serine, similar to the total amino acid composition in the original conchiolin protein.

Examination of the amino acid composition at the Y position in the repeating sequence shows smaller serine and greater glycine and alanine contents in the fossil than in the modern scallop shell (Table 2). This result shows that serine undergoes decomposition while still bound in polypeptide chains. If the bond between aspartic acid and serine decomposed preferentially, the result would be a decrease of serine content in the $(Asp - Y)_n$ sequence.

To examine this possibility the heating experiment of the modern shell was carried out, simulating a diagenetic change of amino acids in the fossil shell. After heating, the acetic acid hydrolysates of the soluble fraction showed the decrease of serine and increase of glycine and alanine. This result clearly supports the decomposition of serine residues bound at the Y position in the repeating sequence with the additional finding of no

**Table 1. Amino Acid Compositions, Released by 6 *N* HCl Hydrolysis, of the Insoluble (I) and Soluble (S) Fractions in the Modern and Pleistocene Fossil Shells (*Patinopecten yessoensis*) (residues/1000 residues)**

| | Moden Shell | | | | | | |
|---|---|---|---|---|---|---|---|
| | Unheated | Heated at 138°C for 50 hrs | | Heated at 160°C for 47 hrs | | Fossil Shell Unheated | |
| Fraction | I | I | S | I | S | I | S |
| Lysine | 26 | 75 | 28 | 69 | 30 | 35 | 29 |
| Histidine | 3.2 | 13 | 3.3 | 9.4 | 3.9 | 13 | 8.5 |
| Arginine | 22 | 35 | 9.6 | 43 | 8.1 | 12 | 2.6 |
| Aspartic acid | 262 | 79 | 303 | 106 | 348 | 422 | 447 |
| Threonine | 23 | 68 | 7.1 | 47 | + | 20 | 18 |
| Serine | 216 | 119 | 127 | 78 | 85 | 97 | 35 |
| Glutamic acid | 50 | 96 | 52 | 87 | 67 | 100 | 115 |
| Proline | 26 | 58 | 22 | 53 | 28 | 19 | 18 |
| Glycine | 265 | 170 | 338 | 202 | 318 | 144 | 137 |
| Alanine | 52 | 80 | 82 | 78 | 76 | 53 | 81 |
| Valine | 13 | 56 | 5.2 | 71 | 8.7 | 17 | 22 |
| Methionine | 2.3 | 35 | – | – | – | + | – |
| allo-Isoleucine | – | – | – | – | – | 0.5 | 1.8 |
| Isoleucine | 6.8 | 32 | 1.0 | 42 | 2.6 | 12 | 10 |
| Leucine | 17 | 56 | 16 | 57 | 20 | 38 | 63 |
| Tyrosine | 8.7 | + | 2.0 | 21 | 1.3 | 4.0 | – |
| Phenylalanine | 7.6 | 27 | 2.9 | 37 | 3.1 | 12 | 13 |
| Total (μM/g of shell) | 14.54 | 0.08 | 11.21 | 0.32 | 11.45 | 1.21 | 0.27 |

remarkable change in total amino acid quantities and in those of $(Asp-Y)_n$ sequence after the heating experiment.

Recent amino acid contamination is sometimes significant in the fossil shells (Hare 1969; Mitterer 1972). In the fossil scallop studied, D-alloisoleucine/L-isoleucine ratios are 0.04 in the insoluble fraction. Fossil shells of the Pleistocene Narita Formation in the Kanto district equivalent to the Shishinai Formation in geological age showed a ratio between 0.05 and 0.07 (Akiyama and Ujiié 1976). These values were considered to be contamination free by a systematic analysis of the shells collected from the Narita Formation and the several underlying other formations. Since the Shishinai Formation is in the northern part of Japan where the annual temperature is lower than in the Kanto district, these two ratios from the

**Table 2. Amino Acids Released from the Modern and Fossil Scallop Shells by 0.25 *M* Acetic Acid Hydrolysis at 108°C for 48 hrs**

| | Minimum estimate of $(Asp–Y)_n$ sequence (%)[a] | Molar ratios of the amino acids at the Y position to aspartic acid | | | | |
|---|---|---|---|---|---|---|
| | | Glycine | Serine | Alanine | Leucine | Glutamic acid |
| Unheated modern shell (I)[b] | 31.5 | 0.25 | 0.26 | 0.08 | 0.02 | 0.01 |
| Heated modern shell at 138°C (S)[c] | 34.9 | 0.32 | 0.16 | 0.12 | 0.02 | 0.01 |
| Heated modern shell at 160°C (S) | 32.5 | 0.31 | 0.11 | 0.10 | 0.03 | 0.01 |
| Pleistocene fossil shell | | | | | | |
| (I) | 11.5 | 0.28 | 0.10 | 0.12 | 0.08 | 0.01 |
| (S) | 9.1 | 0.27 | 0.05 | 0.14 | 0.06 | 0.01 |

[a] Values are calculated from the amino acids released by the acetic acid hydrolysis to the total amino acids recovered from complete hydrolysis by 6 *N* HCl.
[b] Since a modern scallop shell contains no soluble proteins nor soluble peptides (S), the insoluble protein fraction (I) means total protein components included in the shell.
[c] Most of the insoluble shell protein changed to the soluble fraction by heating. This soluble fraction is suitable for the examination of the repeating sequence of $(Asp–Y)_n$ because of the high content of aspartic acid.

two formations are very similar and the samples analyzed have not undergone significant amino acid contamination.

In conclusion, some of the serine residues in peptide linkage may be preferentially altered to glycine and alanine in a polypeptide chain during diagenesis, as observed in degradation of free molecules of serine by heating experiments (Vallentyne 1964). The determination of the amino acid composition of the original paleoprotein from the material in the fossil shell must be carefully studied. The degradation of the unstable amino acids that occurs while in polypeptides should be considered in the study of molecular evolution based on fossil organic molecules.

## *ACKNOWLEDGMENTS*

The writer wishes to thank Dr. A. Shimoyama of the University of Maryland for reading the manuscript and making several valuable suggestions.

## *REFERENCES*

Akiyama, M., The amino acid composition of fossil scallop shell proteins and non-proteins, *Biomineralisation,* **3,** 65–70, 1971.

Akiyama, M., Amino acid composition of heated scallop shells, *J. Fac. Sci., Hokkaido Univ., Ser. 4,* **18,** 117–121, 1978.

Akiyama, M. and Y. Ujiié, Racemization of isoleucine and a possible application to geology (Pt. 2). In case of Pleistocene molluscan fossils from the Kanto district, Japan, *Earth Sci. (Chikyu Kagaku),* **30,** 186–190, 1976.

Hare, P. E., Geochemistry of proteins, peptides and amino acids, in *Organic Geochemistry,* G. Eglinton and M. T. J. Murphy, eds., Springer-Verlag, New York, pp. 438–463, 1969.

Hare, P. E., and T. C. Hoering, The organic constituents of fossil mollusc shells, *Carnegie Inst. Wash. Year Book 76,* 625–631, 1977.

Matter, P. III, F. D. Davidson, and R. W. G. Wyckoff, The composition of fossil oyster shell proteins, *Proc. Nat. Acad. Sci. U.S.A.,* **64,** 790–792, 1969.

Mitterer, R. M., Calcified proteins in the sedimentary environment, in *Advances in Organic Geochemistry,* H. R. Gaertner and H. Wehner, eds., Pergamon, Elmsford, NY, pp. 441–451, 1972.

Totten, D. K., F. D. Davidson, and R. W. G. Wyckoff, Amino acid composition of heated oyster shells, *Proc. Nat. Acad. Sci. U.S.A.,* **69,** 784–785, 1972.

Vallentyne, J. R., Biogeochemistry of organic matter II—Thermal reaction kinetics and transformation products of amino acid compounds, *Geochim. Cosmochim. Acta,* **28,** 157–188, 1964.

Vallentyne, J. R., Pyrolysis of amino acids in Pleistocene *Mercenaria* shells, *Geochim. Cosmochim. Acta,* **33,** 1453–1458, 1969.

Weiner, S. and L. Hood, Soluble protein of the organic matrix of mollusc shells. A potential template for shell formation, *Science,* **190,** 987–989, 1975.

Weiner, S., H. A. Lowenstam, and L. Hood, Characterization of 80-million-year-old mollusk shell proteins, *Proc. Nat. Acad. Sci. U.S.A.,* **73,** 2541–2545, 1976.

# Amino Acid Stereochemistry in Antarctic Marine Sediments: Islas Orcadas Piston Cores 12-77-41 and 12-77-24

**DAVID J. BLUNT**

**DETLEF A. WARNKE**

*U.S. Geological Survey, Menlo Park, California, and Department of Geological Sciences, California State University, Hayward*

*ABSTRACT*

The extents of isoleucine epimerization in Antarctic marine sediments from two high-latitude piston cores are compared. The percentages of D-alloisoleucine in a bound fraction isolated from sediments in Islas Orcadas core 12-77-24 older than 0.7 m.y. are anomalously high relative to the percentages of D-alloisoleucine isolated from younger than 0.7 m.y. age sediments in Islas Orcadas core 12-77-41. The source of the additional D-alloisoleucine is suggested to originate from the foraminiferal (carbonate) fraction. Percentages of D-amino acids in the foraminiferal fraction range from a low of 31% in valine to a high of 41% in alanine and aspartic acid. Rapid sedimentation and slow racemization rates contribute to the lack of increase in foraminiferal D-amino acid percentages with core depth.

The stereochemistry of amino acids in marine sediments is complexly related to the biologic and lithologic composition of the marine sediment and to the diagenesis of organic compounds in the particular fraction of sediment isolated for study. In calcareous marine sediment, the interconversion of the amino acid L-configuration to the D-configuration by racemization (epimerization) proceeds at a faster rate in the foraminiferal fraction than in the bound (residue) fraction (Bada and Schroeder 1972; and Bada and Man 1973). Different species of fossil planktonic foraminifera have significant differences in the extent of epimerization (King and Hare 1972; King and Neville 1977). A reproducible fraction of marine sediment isolated for amino acid analysis is thus an important requirement for correlation and geochronology by amino acid stereochemistry.

Antarctic glacial marine sediments provide a unique and complex environment in which to study amino acid stereochemistry. In Antarctic waters, solution of foraminiferal (carbonate) tests is both a temporal and a spatial phenomenon related to the local glacial regime (Anderson 1975). Correlation and geochronology of marine sediment using amino acids from foraminifera, although the most promising technique, is not applicable when the fossil test is not preserved. Thus, the present study was undertaken to isolate and compare the extent of isoleucine epimerization in the bound fractions of calcareous and noncalcareous Antarctic marine sediments that contain few siliceous fossil tests. Isoleucine epimerization was measured as the percentage of D-alloisoleucine.

The marine sediments were recovered during cruise 12-77 of A.R.A. *Islas Orcadas* (see Gordon and La Brecque 1977). Islas Orcadas (I.O.) piston core 12-77-41, taken at Lat. 69°59.9′S, Long. 05°04.6′W, is approximately 11.73 m long and contains olive-gray (5Y 3/2) mud. This core was recovered from a water depth of 1873 meters. Smear slide analyses of this essentially noncalcareous core show that the abundance of calcareous and siliceous tests is less than 1%. Core I.O. 12-77-24 was recovered from a water depth of 1862 meters at Lat. 68°10.0′S, Long. 11°58.8′E. This calcareous 11.80-m core is composed of light-olive-gray (5Y 5/2) foraminiferal mud with about 1% siliceous material.

Paleontologic examination of the few diatoms present in these sediments (David R. DeFelice, personal communication 1978) indicates a basal age of less than 0.7 m.y. for core I.O. 12-77-41 and between 0.7 m.y. and 1.6 m.y. for core 12-77-24 based on the occurrence of diatoms characteristic of the *Coscinodiscus lentiginosus* Zone and the *Coscinodiscus elliptipora/Actinocyclus ingens* Zone of McCollum (1975).

## SAMPLING AND ANALYSIS

Piston cores were kept refrigerated after recovery and were opened just before sampling. Approximately 60 g of sediment was carefully removed from the center of the core over intervals of about 15 cm. Each sediment sample was homogenized by stirring before chemical treatment.

Reagents used in the analytical procedure were made up by bubbling the appropriate gas (HCl, $NH_4OH$) into water distilled in a quartz still and diluted to lower concentrations. Sodium hydroxide pellets were dissolved in distilled water to a concentration of 10 *N* from which dilutions were used for cation precipitation and cation exchange resin regeneration. The 5 *N* HF–0.1 *N* HCl solution was made up according to the procedure in Stevenson and Cheng (1970).

To isolate the bound (residue) amino acid fraction, 10 g of homogenized sediment was placed in a 100-ml polypropylene tube. To the tube was added 40 ml of 2 *N* HCl to dissolve carbonate and remove extractable amino acids. The tube was centrifuged and the supernatant removed. This operation was performed three times to assure complete dissolution and removal of all calcareous material. The residue in the tube was than treated with 40 ml of 5 *N* HF–0.1 *N* HCl solution for 24 hours on a reciprocating shaker to break down the silicate matrix (Stevenson and Cheng 1970). The mixture was then evaporated to dryness on a rotary evaporator, taken up in 40 ml 6 *N* HCl and hydrolyzed in an oil bath at 110°C for 20 hours. The tube was then centrifuged and the hydrolysate was removed with a 6 *N* HCl rinse. Two additional rinses in 6 *N* HCl followed by centrifugations were added to the initial rinse, and the combined mixture was evaporated to dryness.

The foraminiferal fraction was isolated by sieving in a 0.062-mm stainless steel sieve. The collected foraminiferal sand was then washed with 0.05 *N* HCl and briefly sonicated. The foraminiferal sand was placed in a screw-cap hydrolysis tube into which 5 ml 6*N* HCl was added. The tube was purged with $N_2$ gas, sealed, and hydrolyzed at 110°C in a block heater for 20 hours. The hydrolysate was separated from the undissolved clastic material with two additional 6 *N* HCl rinses and evaporated to dryness.

Cations were removed from the bound-fraction hydrolysate by the precipitation method of Pollock et al. (1970), with the exception that pH 9.5 was used instead of pH 10. The combined filtrates from precipitation were acidified and evaporated to dryness. The remaining cations in the dried filtrate and in the dried foraminiferal hydrolysate were adjusted to pH 1 and carefully applied to the top of a freshly regenerated 14-ml column of AG 50 W-X8 (H+) 50-100 mesh cation exchange resin and eluted with 2 *N* $NH_4OH$ at a flow rate of 0.5–1.0 ml/min. The eluates were dried, taken up in pH 2.2 sodium citrate dilution buffer and applied in duplicate to an automatic ion-exchange chromatograph to determine the extent of isoleucine epimerization. The extent of racemization of other amino acids was determined by gas chromatography of their N-pentafluoropropionyl-

(+)-2-butyl ester derivatives and were resolved on the stationary phases of UCON 75H-90,000 and Carbowax 20 M. The ion-exchange and gas chromatographic data are not corrected for effects of hydrolysis. Also, no correction is made for the approximately 1% impurity in the (+)-2-butanol used for derivatization.

## RESULTS AND DISCUSSION

The results of the analysis on isoleucine epimerization in the bound fraction are recorded in Table 1. The percentages of D-alloisoleucine in core I.O. 12-77-24 are clearly greater than the percentages determined for core I.O. 12-77-41. The percentages of D-alloisoleucine do not increase uniformly with depth in either core. Because core I.O. 12-77-24 contains older sediments, epimerization of isoleucine over a longer time may have resulted in the apparent extent of isoleucine epimerization observed.

The extent of epimerization of isoleucine in the bound or residue fraction has been reported in marine sediments whose ages span the Quaternary. Kvenvolden et al. (1970) reported 9000-year-old sediments from Saanich Inlet, Canada, that contained 3% D-alloisoleucine. Aizenshtat et al, (1973) reported 14% D-alloisoleucine for Pleistocene sediments. Pliocene marine sediments having 10 to 12% D-alloisoleucine were reported by Bada and Schroeder (1972) and Bada and Man (1973). Clearly the amounts of D-alloisoleucine in I.O. 12-77-24 (19–38%) cannot be accounted for by epimerization of isoleucine in a bound amino acid fraction.

There are at least three possible sources for the high percentages of D-alloisoleucine in core I.O. 12-77-24: (1) recycled organic matter; (2) bacteria; and (3) fossil siliceous and calcareous skeletal material. Each of these possibilities will be considered in turn. Sackett et al. (1974) gave

**Table 1. Percent D-alloisoleucine in the Bound Fraction of Marine Sediment from Two Antarctic Piston Cores**

| I.O. 12-77-41 | | I.O. 12-77-24 | |
|---|---|---|---|
| Depth (m) | % D-alloisoleucine | Depth (m) | % D-alloisoleucine |
| 0.42–0.57 | 2 | 0.82–0.98 | 30 |
| 1.62–1.77 | 11 | 2.02–2.18 | 19 |
| 3.02–3.17 | 2 | 3.02–3.17 | 34 |
| | | 4.42–4.57 | 38 |

organic carbon isotopic evidence for recycling of kerogen in Antarctic marine sediments in the Ross Sea. These data suggest that recycling of organic material is a real factor that must be evaluated in organic geochemical studies. Amino acids, however, would undergo biological attack upon their release into the water column (Lee and Bada 1977).

In a study of possible microbial influences on Recent marine sediments, Pollock and Kvenvolden (1978) attributed the small amount (1–2%) of D-alloisoleucine to racemization, and they also considered bacterial sources from the overlying water column. Bada and Lee (1977) reported a dissolved alloisoleucine/isoleucine ratio as high as 0.22 (18% D-alloisoleucine) in seawater from the Sargasso Sea and attributed the D-amino acids to bacterial sources. Holm-Hansen et al. (1977) reported that at least 10% of the total biomass in the surface water of McMurdo Sound, Antarctica, is composed of bacteria-size cells. Bacteria, although not directly isolated in this study, cannot be excluded as a source of D-amino acids in sediments from cores I.O. 12-77-24 and 12-77-41.

Fossil siliceous and calcareous skeletal material not extracted in the laboratory procedure could be a source of D-amino acids. Siliceous matter containing amino acids (King 1975, 1977) that have undergone racemization would not be extracted in a 2 *N* HCl pretreatment, but the 1% or less abundance makes this a minor component of these sediments. Identical amino acid patterns between a foraminiferal fraction and a total sediment fraction in Caribbean core P6304-9 were attributed by Bada and Schroeder (1972) to the influence of the calcareous component. To test for this possibility, a foraminiferal fraction (>0.062 mm) was isolated from sediments in core I.O. 12-77-24. Measurements of this fraction by ion-exchange chromatography could not be made because of low concentrations, but the concentrations of alanine, aspartic acid, leucine, glutamic acid, and valine were detectable by gas chromatography. These five amino acids are reported in Table 2 as the percentage of D-amino acid.

$$\%\ \text{D-amino acid} = \frac{\text{D-amino acid} \times 100}{\text{D-amino acid} + \text{L-amino acid}}$$

Percentages of D-amino acid in the foraminiferal fraction range from 31% for valine to 41% for alanine and aspartic acid. The rates of racemization are in the order: aspartic acid ≈ alanine > leucine > glutamic acid > valine, and are consistent with the values found by Kvenvolden et al. (1973).

The extent of isoleucine epimerization in the bound fraction and the extent of amino acid racemization in the foraminiferal fraction show similarities in magnitude. Kvenvolden et al. (1973) showed that the extent of isoleucine epimerization was roughly between the extents of valine and

**Table 2. Percent D-amino Acids in the Foraminiferal Fraction (>.062 mm) Isolated from Sediments in Core I.O. 12-77-24**

| | Depth (m) | | | |
|---|---|---|---|---|
| Amino acids | 0.82–0.98 | 2.02–2.18 | 3.02–3.17 | 4.42–4.57 |
| Alanine | 41 | 37 | 41 | 40 |
| Aspartic acid | 41 | 39 | 40 | 41 |
| Leucine | 38 | 37 | 37 | 38 |
| Glutamic acid | 36 | 34 | 36 | 35 |
| Valine | 33 | 34 | 32 | 31 |

aspartic acid racemization. The bound D-alloisoleucine percentages in core I.O. 12-77-24, which are between the D-amino acid percentages of valine and aspartic acid in the foraminiferal fraction, suggest incomplete isolation of the foraminiferal fraction and a calcareous source for the anomalously high bound D-alloisoleucine percentages. The bound fraction in this study, therefore, more likely represents a mixture of amino acids having a complex origin.

The high percentages of D-amino acids in the foraminiferal fraction near the top of core I.O. 12-77-24 is a result of sediment erosion or nondeposition above the top sampled interval. The absence of any significant increase in extent of racemization down-core may result from rapid sedimentation and slow racemization rates in this fraction, which consists mainly of *Neogloboquadrina pachyderma*. King and Neville (1977) show that two species of planktonic foraminifera have little increase in extent of isoleucine epimerization with depth beyond an age of 0.4 m.y. at a sedimentation rate of 1.71 cm/$10^3$ yr. Rates of Antarctic shelf and slope marine sedimentation greater than this are common in Quaternary time (Piper et al., 1975; Tucholke et al., 1976). Because the estimated age for core I.O. 12-77-24 is greater than 0.4 m.y., the slow racemization rates and rapid sedimentation result in little increase in apparent extent of racemization with depth, even in a long piston core (see Blunt et al., 1978).

The apparent stereochemistry of amino acids in marine sediments is complexly related to the sediment composition, diagenesis of the organic compounds, and the methodology used to extract the amino acids. In this study, the calcareous foraminiferal fraction was considered to be the most probable source for the apparent extent of epimerization found in the bound fraction. Future studies of the amino acid stereochemistry of a true bound fraction in Antarctic marine sediments may provide a means of improving geochronology and increasing understanding of local marine environmental conditions during sedimentation.

## *ACKNOWLEDGMENTS*

We thank D. S. Cassidy, curator of the Antarctic Marine Geology Research Facility at Florida State University, and A. Kaharoeddin for generous help in sampling. Paleontologic information was provided by D. DeFelice at Florida State University. Comments and helpful discussion were provided by K. Kvenvolden, U.S.G.S., Menlo Park, California, and by G. Pollock, NASA–Ames Research Center, Moffett Field, California.

## *REFERENCES*

Aizenshtat, Z., M. J. Baedecker, and I. R. Kaplan, Distribution and diagenesis of organic compounds in JOIDES sediment from Gulf of Mexico and western Atlantic, *Geochim. Cosmochim. Acta,* **37,** 1881–1898, 1973.

Anderson, J. B., Factors controlling $CaCO_3$ dissolution in the Weddell Sea from foraminiferal distribution patterns, *Marine Geology,* **19,** 315–332, 1975.

Bada, J. L., Kinetics of the non-biological decomposition and racemization of amino acids in natural waters, *Amer. Chem. Soc., Adv. Chem. Ser.,* **106,** 309–331, 1971.

Bada, J. L. and C. Lee, Decomposition and alteration of organic compounds dissolved in seawater, *Marine Chemistry,* **5,** 523–534, 1977.

Bada, J. L. and E. H. Man, Racemization of isoleucine in cores from Leg 15, Site 148, in *Initial Reports of the Deep Sea Drilling Project 20,* B.C. Heezen et al., eds., U.S. Government Printing Office, Washington, D.C., 947–951, 1973.

Bada, J. L. and R. A. Schroeder, Racemization of isoleucine in calcareous marine sediments: Kinetics and mechanism, *Earth Planet. Sci. Lett.,* **15,** 1–11, 1972.

Bada, J. L. and R. A. Schroeder, Amino acid racemization reactions and their geochemical implications, *Naturwissenschaften,* **62,** 71–79, 1975.

Blunt, D. J., D. A. Warnke, and D. R. DeFelice, Amino acid racemization in Islas Orcadas piston core 12-77-24, *Antarctic Journal of the U.S.,* **13,** 99, 1978.

Gieskes, J. M. and J. R. Lawrence, Interstitial water studies, Leg 35, in *Initial Reports of the Deep Sea Drilling Project 35,* C. D. Hollister, C. Craddock et al., eds., U.S. Govt. Printing Office, Washington, D.C., 407–424, 1976.

Gordon, A. L. and J. La Brecque, Islas Orcadas cruise 12: Cape Town to Buenos Aires, *Antarctic Journal of the U.S.,* **12,** 60–63, 1977.

Holm-Hansen, O., F. Azam, A. F. Carlucci, R. E. Hodson, and D. M. Karl, Microbial distribution and activity in and around McMurdo Sound, *Antarctic Journal of the U.S.,* **12,** 29–32, 1977.

King, K., Jr., Amino acid composition of the silicified organic matrix in fossil polycystine radiolaria, *Micropaleontology,* **21,** 215–226, 1975.

King, K., Jr., Amino acid survey of recent calcareous and siliceous deep-sea microfossils, *Micropaleontology,* **23,** 180–193, 1977.

King, K., Jr., and P. E. Hare, Species effects in the epimerization of L-isoleucine in fossil planktonic foraminifera, *Carnegie Inst. Wash. Year Book 71,* 596–598, 1972.

King, K., Jr. and C. Neville, Isoleucine epimerization for dating marine sediments: Importance of analyzing monospecific foraminiferal samples, *Science,* **195,** 1333–1335, 1977.

Kvenvolden, K. A., E. Peterson, and F. S. Brown, Racemization of amino acids in sediments from Saanich Inlet, British Columbia, *Science,* **169,** 1079–1082, 1970.

Kvenvolden, K. A., E. Peterson, J. Wehmiller, and P. E. Hare, Racemization of amino acids in marine sediments determined by gas chromatography, *Geochim. Cosmochim Acta,* **37,** 2215–2225, 1973.

Lee, C. and J. L. Bada, Dissolved amino acids in the Equatorial Pacific, Sargasso Sea and Biscayne Bay, *Limnol. Oceanogr.,* **22,** 502–510, 1977.

McCollum, D. W., Antarctic Cenozoic diatoms: Leg 28, Deep Sea Drilling Project, in *Initial Reports of the Deep Sea Drilling Project 28,* L. A. Frakes et al., eds., U.S. Govt. Printing Office, Washington, D.C., 515–572, 1975.

Piper, D. J. W. and C. D. Brisco, Deep-water continental margin sedimentation, DSDP Leg 28, Antarctica, in *Initial Reports of the Deep Sea Drilling Project 28,* L. A. Frakes et al., eds., U.S. Govt. Printing Office, Washington, D.C., 727–756, 1975.

Pollock, G. E. and K. A. Kvenvolden, Stereochemistry of amino acids in surface samples of a marine sediment, *Geochim. Cosmochim. Acta,* **42,** 1903–1905, 1978.

Pollock, G. E., A. K. Miyamoto, and V. I. Oyama, The detection of optical asymmetry in biogenic molecules by gas chromatography for extraterrestrial space exploration: Sample processing studies, *Life Sciences and Space Research,* **8,** 99–107, 1970.

Sackett, W. M., C. W. Poag, and B. J. Eadie, Kerogen recycling in the Ross Sea, Antarctica, *Science,* **185,** 1045–1047, 1974.

Stevenson, F. J. and C. N. Cheng, Amino acids in sediments: Recovery by acid hydrolysis and quantitative estimation by a colorimetric procedure, *Geochim. Cosmochim. Acta,* **34,** 77–88, 1970.

Tucholke, B. E., C. D. Hollister, and W. R. Vennum, Continental rise and abyssal plain sedimentation in the Southeast Pacific Basin—Leg 35 Deep Sea Drilling Project, in *Initial Reports of the Deep Sea Drilling Project 35,* Hollister, C. D., Craddock, C. et al., eds., U.S. Govt. Printing Office, Washington, D.C., 359–400, 1976.

# Metal Binding Study of Fulvic Acids from Carbonate Sediments Using Manganese as a Magnetic Resonance Probe

**ROBERT CUNNINGHAM**
**RICHARD M. MITTERER**
*Program in Geosciences, The University of Texas at Dallas*

*ABSTRACT*

The metal-binding ability of a fulvic acid fraction (>10,000 MW) extracted from carbonate sediments was studied by water proton relaxation enhancement and electron paramagnetic resonance spectroscopy. The transition metal $Mn^{2+}$ was used as the paramagnetic probe for the binding study. Fulvic acids extracted from two sediment samples from the Florida Keys were each found to have two classes of $Mn^{2+}$ binding sites, a high affinity class with a stability constant approximately equal to 840 $M^{-1}$ (pH = 6.5, $\mu$ = 0.4, T = 25°C) and a low affinity class with a stability constant approximately equal to 110 $M^{-1}$. Fulvic acid extracted from Bahamian ooids has an affinity for $Mn^{2+}$ similar to that of the low affinity sites of the other fulvates, yet the ooid organic matter binds about twice as much total $Mn^{2+}$ as the organic matter from the other sediments. The fulvic acids, high in the acidic amino acids, generally were found to bind $Mn^{2+}$ in proportion to their aspartic acid content. The metal-binding ability of asparatic acid–rich organic matter absorbed onto carbonate sediment grains may be an important factor in further precipitation of $CaCO_3$ on the sediment grains.

## INTRODUCTION

Sedimentary organic matter may participate in two important processes in shallow marine carbonate environments. One of these is adsorption onto carbonate grain surfaces, a process by which the organic matter may coat the mineral with an unreactive sheath precluding equilibration with seawater (Chave 1965; Chave and Suess 1970). Recently, more active roles have been ascribed to the organic-carbonate interaction. For example, carbonate surfaces exert considerable influence on the composition of the adsorbed organic molecules. Experiments indicate that aspartic acid-rich organic matter is preferentially adsorbed by calcite grains but not by quartz grains (Mitterer 1972; Carter 1978). The enrichment of aspartic acid in the sorbed fraction has also been noted for natural carbonate sediments (Mitterer and Carter 1977; Carter and Mitterer 1978; Müller and Suess 1977). Calcareous ooids from various localities are found to have similar amino acid compositions dominated by aspartic acid (Mitterer 1968, 1972; Trichet 1968). Artificial ooids have been experimentally precipitated after adsorption of humic acids onto grain surfaces (Suess and Fütterer 1972; Davies et al., 1978; Ferguson et al., 1978).

A second important process in which sedimentary organic matter may participate is metal ion complexation. Metal ions may be complexed by dissolved, particulate, and adsorbed organic matter. There is an extensive literature on metal complexes with terrestrial humic substances. Much less is known about metal complexes with marine humic substances (Rashid 1970; Nissenbaum and Swaine 1976). A study of the metal-binding ability of organic matter associated with marine carbonate sediments is significant for two reasons. First, it adds to the scant body of data on metal-marine humic interactions. Second, it is a necessary step in understanding the influence of organic matter on the inorganic precipitation of $CaCO_3$ in the marine environment.

Aspartic acid-rich organic matter is a feature of the calcified protein of skeletal carbonates as well as the adsorbed phase on carbonate grains. One of the common features of calcified tissues is the presence of organic components that bind calcium (Simkiss and Tyler 1958; Simkiss 1960; Crenshaw 1972; Vittur et al., 1972). The binding of calcium to a macromolecule may be the first and most important event in the process of biological calcification (Waddell 1972). In a similar fashion, if sedimentary organic matter is important in geological calcification processes, metal-binding ability by the adsorbed organic phase may be a prerequisite.

The present investigation utilizes an analysis of the water proton relaxation rate enhancements to investigate the metal-binding ability of a fulvic acid fraction extracted from carbonate sediments. Divalent manganese ($Mn^{2+}$) has been used as a paramagnetic probe for nuclear relaxation studies of macromolecular systems (Cohn and Leigh 1962). Manganese-binding to a macromolecule can cause large enhancements in the solvent water proton relaxation rates. These enhancements depend on the amount of metal ion bound to the macromolecule. By varying the $[Mn^{2+}]/[\text{macromolecule}]$ ratio the measured enhancements can be used to determine stability constants for the Mn-macromolecule complex (Dwek 1972). This method has been used extensively for the study of $Mn^{2+}$ binding to biological macromolecules (Reuben 1971; Valentine and Cottom 1973; Sherry and Cottom 1973; Epstein et al., 1974) but apparently has not been used previously to study $Mn^{2+}$ binding with sedimentary organic matter. With this method it is possible to make rapid determinations of stability constants and to obtain structural information about the metal-fulvate complex.

Electron paramagnetic resonance spectroscopy is used as an alternative means of determining the degree of $Mn^{2+}$ binding. This method is based on the observation that free $Mn^{2+}$ gives a characteristic first derivative absorption spectrum from which the free $Mn^{2+}$ concentration can be determined. Bound $Mn^{2+}$ gives a different and much weaker signal (Cohn and Townsend 1954). Bound $Mn^{2+}$ is taken as the difference between the known total $Mn^{2+}$ and free $Mn^{2+}$ determined at several point in an $Mn^{2+}$ organic ligand titration. Using suitable graphic techniques the intrinsic association constant and number of binding sites can be obtained (Scatchard 1949). This spectroscopic technique has recently been used to study $Mn^{2+}$ binding to soil fulvic acid (McBride 1977; Gamble et al., 1977). Although the present study pertains specifically to manganese binding rather than calcium binding, and to organic matter extracted from carbonate sediments, the results are intended to demonstrate and quantitate the general metal-binding ability of this organic mattter.

## MATERIALS AND METHODS

### Fulvic Acids

Fulvic acids were obtained from: (1) carbonate mud collected from Cross Bank in Florida Bay, (2) carbonate sand from White Bank in the back-reef platform of the Florida Keys, and (3) ooids from Joulters Cay, Great Bahama Bank, Bahama Islands. Raw sediments were sieved through U.S. Standard No. 45 (355 μm) sieves to remove as much coarse organic and inorganic debris as possible. Humic substances were extracted from the <355 μm fraction with 0.1 *N* NaOH under nitrogen for 24 hours. Bahamian ooids were dissolved in enough cold 0.4 *N* HCl to dissolve the outer half of the ooid cortex. Humic acids were precipitated by acidification to pH 2 and removed by centrifugation. The fulvic acid remaining in solution was exhaustively dialyzed against deionized water and concentrated with an Amicon model 202 ultrafiltration system using PM-10 ultrafilter. The effective retentive capacity of this membrane is approximately 10,000 molecular weight. The >10,000 M.W. fraction was used throughout this study. The extract was shaken over a slurry of Chelex-100 resin in the Na-form to remove previously bound metals. The slurry was filtered through acid-washed glass fiber filters to remove the resin beads, and the Na-salt of the fulvic acid was again exhaustively dialyzed against deionized water and lyophilized.

## Apparatus

The longitudinal water proton relaxation rates were measured by the pulsed nuclear magnetic resonance method of Carr and Purcell (1954) with a Nuclear Magnetic Resonance Specialties PS 60-AW pulsed spectrometer operating at 24.3 MHz. The concentration of free $Mn^{2+}$ for the ooid fulvic acid experiment was determined by electron paramagnetic resonance (Mildvan and Cohn 1963) using a Varian E4 spectrometer at 9.12 GHz. Amino acid analyses were obtained on a custom-made analyzer similar to that described by Hare (1969).

## Procedure

The water proton relaxation rate experiments were carried out by varying the ratio of [$Mn^{+2}$] to [macromolecule]. The enhancement of the relaxation rate ($\varepsilon^*$) is the ratio of the paramagnetic contribution to the relaxation rate in the presence (*) and absence of ligand (Mildvan and Cohn 1963; Dwek 1972).

$$\varepsilon^* = (T_{1p})^* - 1/(T_{1p})^{-1} \qquad (1)$$

The paramagnetic contribution to the relaxation rate, $(T_{1p})^{-1} = (T_1)^{-1} - (T_{1(0)})^{-1}$, is defined as the difference of the longitudinal relaxation rate in the presence and absence (0) of paramagnetic ion. Observation of an enhancement ($\varepsilon^* > 1$) is evidence for the formation of a metal-macromolecule complex. The characteristic enhancement of the fully formed metal-macromolecule complex ($\varepsilon_b$) was determined by titrating a fixed concentration of $Mn^{2+}$ with fulvic acid. From $\varepsilon_b$ and the observed enhancements obtained by titrating $Mn^{2+}$ with a fixed concentration of fulvic acid, the free and the bound metal concentration in each sample are calculated using the following equation (Dwek 1972)

$$[M_b] = \frac{\varepsilon^* - 1}{\varepsilon_b - 1} \cdot [M_t] \qquad (2)$$

where $[M_b]$ and $[M_t]$ are the bound and the total metal concentrations, respectively. Free manganese, $[M_f]$, was determined by difference. The results were analyzed according to Scatchard (1949) methodology,

$$\frac{\bar{\mu}}{[M_f]} = K_a \cdot (n - \bar{\mu}) \qquad (3)$$

where $\bar{\mu}$ is the moles of bound metal ion per gram of fulvic acid, and n is the number of metal binding sites per gram fulvic acid. From a plot of $\bar{\mu}$ versus $\bar{\mu}/[M_f]$ the intrinsic association constant or stability constant, $K_a$, and n can be obtained as the negative value of the slope and x-intercept, respectively.

Sample solutions, never exceeding 200 μl, contained 0.3 *M* NaCl and were buffered at pH 6.5 with 0.05 *M* sodium cacodylate. The pulsed NMR probe was thermostated with a stream of dry nitrogen at 25°C.

For amino acid analyses approximately 1 mg of lyophilized sample was hydrolyzed for 22 hours with 6 *N* HCl. All hydrolyses were carried out at 110°C under $N_2$ in sealed tubes.

## RESULTS AND DISCUSSION

### Binding Experiments

The measurements of water proton relaxation rate enhancement from the titration of $Mn^{2+}$ with each of the fulvic acid samples indicate quantitative binding of the metal by the polymeric material. Although not large, the observed enhancements ($\varepsilon^*$) display a relationship to the [metal]/[fulvic acid] suggesting the formation of a $Mn^{2+}$-fulvic acid complex (Fig. 1). The plots of $[Mn]_t$ versus $\varepsilon^*$ for the Cross Bank fulvic acid (CBFA) and White Bank fulvic acid (WBFA) show an inflection point in the manganese concentration range of $2 \times 10^{-3}$ *M* to $4 \times 10^{-3}$ *M*. This inflection represents a transition during the titration from $Mn^{2+}$-binding by one class of metal binding sites to $Mn^{2+}$-binding by another. The Bahamian ooid fulvic acid (BOFA) plot has no inflection point, indicating only one class of metal binding sites active in this $(Mn)_t$ range.

The enhancement of the bound form ($\varepsilon_b$) for the $Mn^{2+}$-fulvates was determined by extrapolation of a double reciprocal plot obtained by ti-

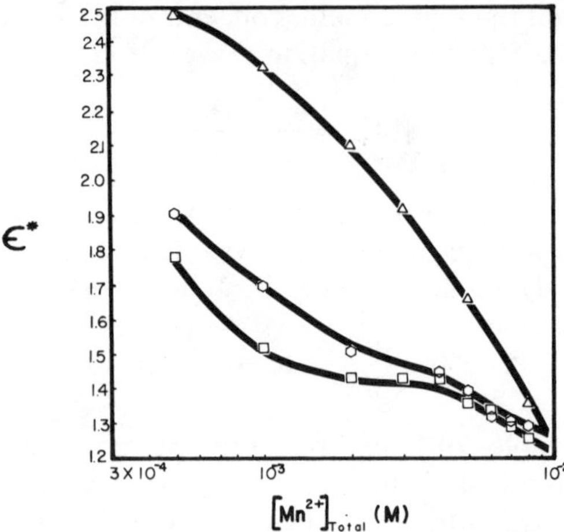


**Figure 1.** Variation of the enhancement of water proton relaxation ($\varepsilon^*$) with change in the total $Mn^{2+}$ concentration during titrations of $Mn^{2+}$ with constant fulvic acid concentration. The reaction mixtures contained 0.05 *M* sodium cacodylate buffer, pH 6.5, 0.30 *M* sodium chloride and 4.00 g/l Bahamian ooid fulvic acid (△), 3.01 g/l Cross Bank fulvic acid (○), 4.62 g/l White Bank fulvic acid (□), and the noted concentrations of $Mn^{2+}$. Temperature = 25°C.

trating different concentrations of fulvic acid with a constant concentration of $Mn^{2+}$ (Fig. 2). The $\varepsilon_b$ values for CBFA and WBFA are identical (4.8). For protein systems $\varepsilon_b$ values range from 1.4 to 12 (Reuben 1975). Using the $\varepsilon_b$ value, the enhancements ($\varepsilon^*$) from the $Mn^{2+}$ titration, and Eq. 2, the concentrations of free and bound $Mn^{2+}$ at each point of the titration are calculated. The resulting Scatchard plots for the two samples (CBFA and WBFA) further indicate two classes of metal binding sites (Figs. 3 and 4). The CBFA contains 0.13 μmoles/mg fulvic acid of high affinity sites with $K_a$ = 1000 $M^{-1}$ and 0.42 μmoles/mg fulvic acid of low affinity sites with $K_a$ = 120 $M^{-1}$. The two classes of binding sites for WBFA are 0.10 μmoles/mg fulvic acid of high affinity sites with $K_a$ = 670 $M^{-1}$ and 0.24 μmoles/mg fulvic acid of low affinity sites with $K_a$ = 98 $M^{-1}$. Two classes of binding sites have also been demonstrated to occur in peat and lake water humic acids (Mantoura and Riley 1975), river water humic material (Buffle et al., 1977), and soil fulvic acid (Gamble et al., 1970).

The two carbonate sediment fulvates show close agreement in their ability to bind manganese. The stability constants for the low affinity sites of WBFA and CBFA are similar (98 $M^{-1}$ and 120 $M^{-1}$, respectively), and the overall stability constants for the two samples are almost identical

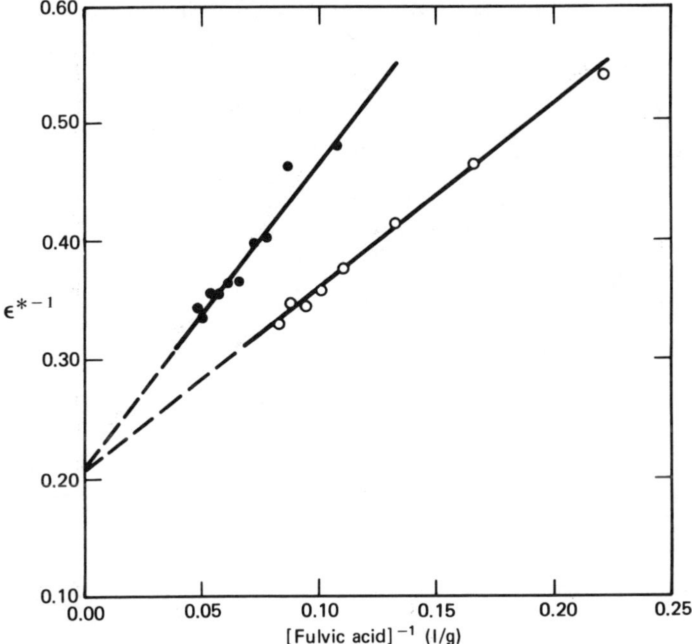


**Figure 2.** Double reciprocal plot of the observed enhancement versus total concentration of fulvic acid obtained during titrations of fulvic acid with constant $Mn^{2+}$ concentration. Solutions contained $1 \times 10^{-3}$ *M* $Mn^{2+}$, 0.05 *M* sodium cacodylate buffer, pH 6.5, 0.30 *M* sodium chloride, and the noted concentration of White Bank fulvic acid (●) and Cross Bank fulvic acid (○).

(log $K_a \cong 2.5$). The latter value is consistent with those quoted by Schnitzer and Khan (1972) for $Mn^{2+}$ complexes with soil fulvic acid. Recent studies of soil $Mn^{2+}$-fulvate complexes using electron paramagnetic resonance and nuclear magnetic resonance techniques indicate that $Mn^{2+}$ is bound electrostatically as $Mn(OH_2)_6^{2+}$, with the fulvic acid donor groups in outer sphere complexing sites (Gamble et al., 1976, 1977; McBride 1978). It is possible that the low affinity sites of the carbonate sediment fulvic acids (average $K_a \cong 110\ M^{-1}$) act in similar fashion and electrostatically bind $Mn^{2+}$ in its fully hydrated form.

The stability constants for the high affinity sites of the CBFA and WBFA samples do not show such a close resemblance as do the low affinity sites. They are, however, approximately eight times stronger in metal-binding ability than the low affinity sites. This indicates that the high affinity sites may bind the metal in more rigid inner sphere or chelate complexes, with differences in site composition and/or geometry between

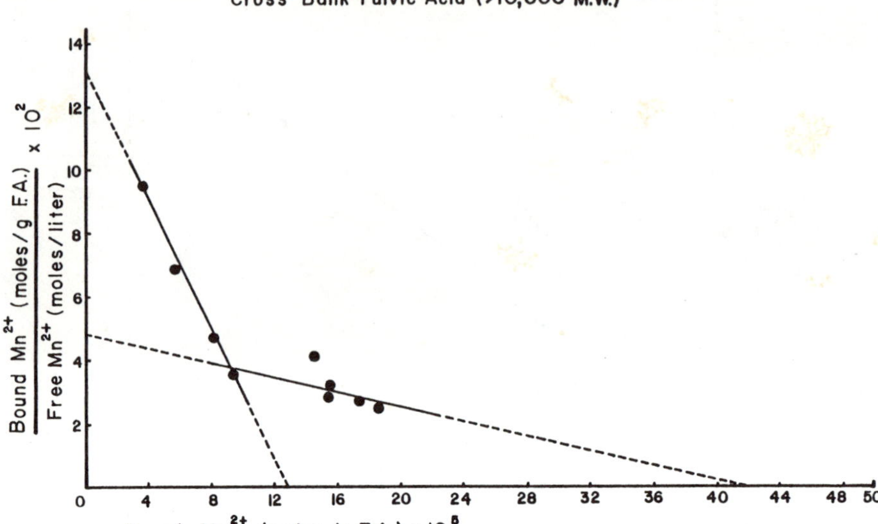


**Figure 3.** Scatchard plot resulting from a titration of $Mn^{2+}$ with a constant concentration of Cross Bank fulvic acid. The sample solutions contained 3.01 g/l fulvic acid, 0.5–8 m*M* $Mn^{2+}$, 0.05 *M* sodium cacodylate buffer, pH 6.5, and 0.30 *M* sodium chloride. Temperature = 25°C.

the samples resulting in the difference in $Mn^{2+}$ affinity. The results are analogous to those of Senesi et al. (1977) in which $Fe^{3+}$ was found to bind to humic material in high affinity, well-structured sites and also to bind weakly to the external surfaces of the molecules.

The Scatchard plot for the interaction of $Mn^{2+}$ with Bahamian ooid fulvic acid as detected by electron paramagnetic resonance spectroscopy is seen in Figure 5. This fulvic acid exhibits only one class of binding sites having a concentration of 1μ mole of sites/mg fulvic acid and $K_a$ = 200 $M^{-1}$. The stability constant is intermediate between those of the high and the low affinity sites of the sediment extracts; but the number of binding sites is two to three times as great. When compared to the stability constants of the other carbonate sediment fulvic acids, the BOFA stability constant is most similar to those of the low affinity sites. The BOFA binding sites may also electrostatically bind $Mn^{2+}$ in its fully hydrated form. It is possible that a greater negative charge density exists on the BOFA molecules than on the CBFA or WBFA molecules, resulting in the greater amount of $Mn^{2+}$ binding.

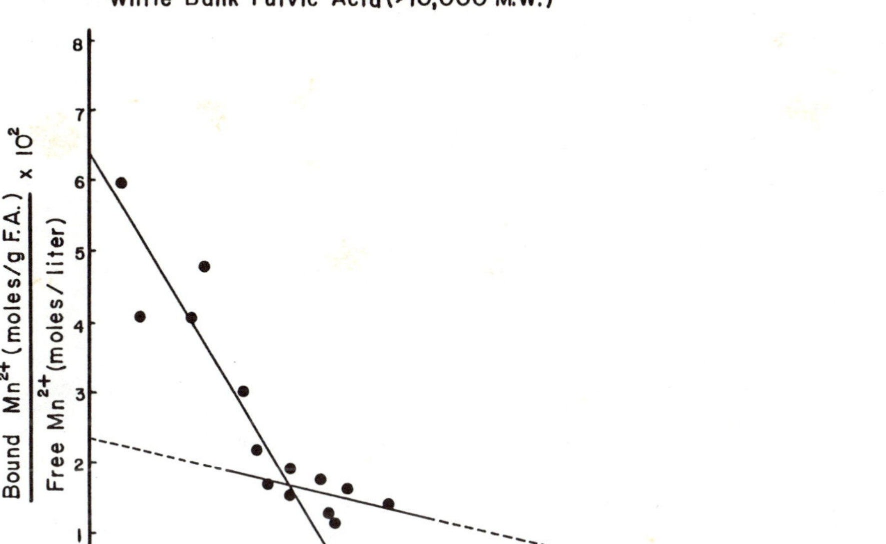


**Figure 4.** Scatchard plot obtained from a titration of $Mn^{2+}$ with a constant concentration of White Bank fulvic acid. The reaction mixtures contained 4.62 or 4.06 g/l fulvic acid, 0.5–8 m*M* $Mn^{2+}$, 0.05 *M* sodium cacodylate buffer, pH 6.5, and 0.30 *M* sodium chloride. Temperature = 25°C.

## Amino Acid Composition

A knowledge of the chemical composition and structure of the carbonate sediment fulvic acids is necessary to understand the metal-binding process. The chemical structure of soil humic substances has been suggested to be a complex moiety of benzenecarboxylic and phenolic acids polymerized by hydrogen-bonding, with inorganic and other organic substances bound within and on the periphery of the structure (Haworth 1971; Khan and Schnitzer 1971). Carboxyl functions, especially those ortho to phenolic OH groups on aromatic rings, are reported to be the major metal-binding ligand of soil fulvic acids (Gamble et al., 1970). Marine humic acids contain similar components but have more aliphatic and heterocyclic structures than soil humates (Nissenbaum and Swaine 1976). Little is known about the functional groups that are important to metal binding in marine humic substances, but there are fewer oxygenated and more

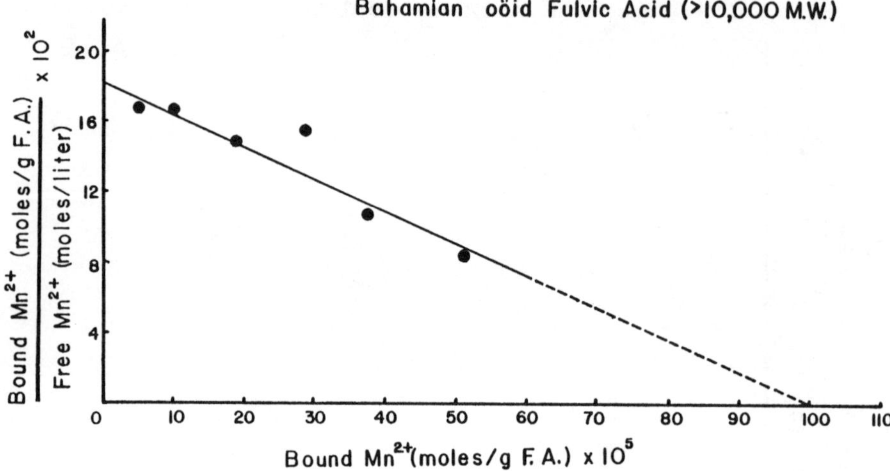


**Figure 5.** Scatchard plot for the binding of $Mn^{2+}$ by Bahamian ooid fulvic acid. The concentration of free $Mn^{2+}$ was determined by electron paramagnetic resonance using a Varian E4 spectrometer. The sample solutions contained 4.00 g/l fulvic acid, 0.05 *M* sodium cacodylate buffer, pH 6.5, 0.30 *M* sodium chloride, and 0.5–8 m*M* $Mn^{2+}$. Temperature = 25°C.

nitrogen and sulfur containing functional groups than in soil humates (Nissenbaum and Swaine 1976).

Organic matter associated with shallow marine carbonate sediments is distinctly different from that associated with noncarbonate marine sediments with respect to amino acid composition (Mitterer and Carter 1977; Carter and Mitterer 1978). The amino acid composition of carbonate sediments is dominated by the acidic amino acids, (aspartic acid and glutamic acid), whereas that of noncarbonate sediments is comprised primarily of glycine and alanine. Generally, humic substances extracted from both carbonate and noncarbonate sources contain from 15% to 36% amino acids by dry weight. Side chain carboxyl groups of the acidic amino acids may account for a minimum of 15–30% of the total carboxyl groups present in humic substances from carbonate sediments (Carter and Mitterer 1978).

Monomeric or peptide-bound amino acids incorporated into the humic structure may possess ionizable functional groups, most probably side-chain and terminal carboxyl groups, which could participate in metal complexing. Representative acid pK values for terminal and side-chain carboxyl groups on proteins are 3.1–3.5 and 4.4–4.7, respectively (Breslow 1973). These are in the same range as the acid pK values for carboxyl groups reported by Gamble et al. (1970) to be involved in soil fulvic acid–metal ion complexing. Calcium-binding to a single carboxylate group on a protein does not yield a complex with significant stability ($K_a < 10^2$ $M^{-1}$) (Williams 1977). However, one or more nearby carboxyl groups on the protein can coordinate the metal ion, resulting in a chelate with considerable stability. Where $M$ is the metallic ion and $L$ is the ligand, complexes could have the form $M_nL$ ($n > 1$), with several metal ions bound per molecule (Buffle et al., 1977). It is also possible that the acidic amino acids may be crosslinked via a bound metal ion to a ligand of another humic acid component, for example, a benzene-carboxylate, thus forming an $ML_n$ ($n > 1$) complex.

Recognizing the compositional importance of amino acids in the organic matter extracted from carbonate sediments, it is interesting to note the correlation between the amino acid composition and the $Mn^{2+}$-binding ability of the fulvic acid samples used in this study (Table 1). The CBFA and WBFA contain 16% and 27% amino acids by weight, respectively. The acidic amino acids constitute 30% of the total amino acids in both samples. The Bahamian ooid fulvic acid contains 16% total amino acids but has the greatest percentage of acidic amino acids (60%) and aspartic acid (48%). The ooid fulvic acid, which contains about twice as much aspartic acid as the CBFA and WBFA samples, binds approximately twice as much $Mn^{2+}$ as the latter fulvic acid samples. The greater abundance of acidic amino acids, in particular aspartic acid, in the BOFA could have enhanced its negative charge density, resulting in an increased number of metal binding sites.

These results have a bearing on the theories for the formation of ooids. Mitterer (1972) has suggested that the aspartic acid-rich organic matter within ooids may enhance the concentration of calcium ions sufficiently to permit precipitation of $CaCO_3$. More recently, Davies et al. (1978) and Ferguson et al. (1978) have suggested that humic material adsorbed on the surface of the ooid grain may concentrate calcium and magnesium ions through complexation with carboxyl functions, resulting in precipitation of a layer of $CaCO_3$. From the present study it appears that organic material associated with carbonate sediments binds metal ions in proportion to the amount of aspartic acid or total acidic amino acids. Calcium ions bound to an organic layer adsorbed on the surface of an ooid may be able to further coordinate carbonate ions, thus inducing $CaCO_3$ precipitation. A high concentration of negatively charged functional groups on the surface of the humic molecules, including the carboxyls of the acidic amino acids, may be required if these molecules are to participate in sedimentary calcification processes.

**Table 1. Amino Acid Composition of Carbonate Sediment Fulvic Acids (>10,000 MW fraction) in Residues/1000 Amino Acids**

| | Florida Keys Carbonate Sediments | | |
|---|---|---|---|
| | Cross Bank | White Bank | Bahamian Ooids |
| Asp | 176 | 172 | 478 |
| Thr | 77 | 77 | 37 |
| Ser | 54 | 59 | 30 |
| Glu | 131 | 139 | 126 |
| Gly | 154 | 132 | 72 |
| Ala | 105 | 105 | 57 |
| Val | 74 | 80 | 41 |
| Met | 18 | 16 | 5 |
| Ile | 35 | 48 | 28 |
| Leu | 43 | 64 | 41 |
| Tyr | 22 | 17 | 5 |
| Phe | 23 | 35 | 23 |
| His + Lys | 61 | 37 | 37 |
| Arg | 29 | 18 | 19 |
| Total (μM/mgF.A.) | 1.289 | 2.233 | 1.296 |
| Asp (μM/mgF.A.) | 0.226 | 0.385 | 0.620 |
| Bound $Mn^{2+}$ (μM/mgF.A.) | 0.55 | 0.34 | 1.00 |

## SUMMARY AND CONCLUSIONS

Two classes of $Mn^{2+}$ binding sites (high and low affinity) exist on the fulvic acid fraction extracted from carbonate sediments from the Florida Keys. Only one class of binding sites exists on fulvic acid extracted from Bahamian ooids. This class of binding sites has an affinity for $Mn^{2+}$ similar to that of the low affinity sites of the other sediment extracts but binds two to three times as much $Mn^{2+}$ as the total $Mn^{2+}$ bound by the other sediment extracts. The high affinity sites may represent inner sphere metal ion chelation by the fulvic acid, whereas the low affinity sites may be outer sphere electrostatic interaction of the aquo-$Mn^{2+}$ with the fulvic acid.

The fulvic acid extracts are enriched in the acidic amino acids, as is the organic matter associated with other skeletal and nonskeletal carbonates. With their free carboxyl function, the acidic amino acids may participate in binding metal ions. Consequently, the metal-binding ability of organic matter adsorbed on the surfaces of carbonate grains may be especially important to the precipitation of $CaCO_3$.

## ACKNOWLEDGMENT

We are grateful to Drs. A. D. Sherry and G. L. Cottam for use of the pulsed NMR and EPR spectrometers and for their advice during this study. This study was funded by NSF Grant OCE77-19452. Contribution No. 367, the Geosciences Program, The University of Texas at Dallas.

## REFERENCES

Breslow, E., Metal-Protein Complexes, in *Inorganic Biochemistry,* Vol. 1, G. L. Eichhorn, ed., Elsevier, New York, pp. 227–247, 1973.

Buffle, J., F. Greter, and W. Haerdi, Measurement of complexation properties of humic and fulvic acids in natural waters with lead and copper ion-selective electrodes, *Anal. Chem.,* **49,** 216–222, 1977.

Carr, H. Y. and E. M. Purcell, Effects of diffusion on free precession in nuclear magnetic resonance experiments, *Phys. Rev.,* **94,** 630–638, 1954.

Carter, P. W., Adsorption of amino acid–containing organic matter by calcite and quartz, *Geochim. Cosmochim. Acta,* **42,** 1239–1242, 1978.

Carter, P. W. and R. M. Mitterer, Amino acid composition of organic matter associated with carbonate and non-carbonate sediments, *Geochim. Cosmochim. Acta,* **42,** 1231–1238, 1978.

Chave, K. E., Carbonates: Association with organic matter in surface seawater, *Science,* **148,** 1723–1724, 1965.

Chave, K. E. and E. Suess, Calcium carbonate saturation in seawater: Effects of dissolved organic matter, *Limnol. Oceanogr.,* **15,** 633–637, 1970.

Cohn, M. and J. Townsend, A study of manganous complexes by paramagnetic resonance absorption, *Nature,* **173,** 1090–1091, 1954.

Cohn, M. and J. S. Leigh, Magnetic resonance investigations of ternary complexes of enzyme-metal-substrate, *Nature,* **193,** 1037–1040, 1962.

Crenshaw, M. A., The soluble matrix from *Mercenaria mercenaria* shell, *Biomineralization,* **6,** 6–11, 1972.

Davies, P. J., B. Bubela, and J. Ferguson, The formation of ooids, *Sedimentology,* **25,** 703–730, 1978.

Dwek, R. A., Proton relaxation enhancement probes—applications and limitations to systems containing macromolecules, *Advan. Mol. Relaxation Processes,* **4,** 1–53, 1972.

Epstein, M., A. Levitzki, and J. Reuben, Binding of lanthanides and of divalent metal ions to porcine trypsin, *Biochemistry,* **13,** 1777–1782, 1974.

Ferguson, J., B. Bubela, and P. J. Davies, Synthesis and possible mechanism of formation of radial carbonate ooids, *Chem. Geol.,* **22,** 285–308, 1978.

Gamble, D. S., M. Schnitzer, and I. Hoffman, $Cu^{2+}$-fulvic acid chelation equilibrium in 0.1 m KCl at 25.0°C, *Can. J. Chem.*, **48,** 3197–3204, 1970.

Gamble, D. S., C. M. Langford, and J. P. K. Tong, The structure and equilibria of a manganese (II) complex of fulvic acid studies by ion exchange and nuclear magnetic resonance, *Can. J. Chem.*, **54,** 1239–1245, 1976.

Gamble, D. S. M., Schnitzer, and D. S. Skinner, $Mn^{II}$-fulvic acid complexing equilibrium measurements by electron spin resonance spectrometry, *Can. J. Soil Sci.*, **57,** 47–53, 1977.

Greenfield, L. J., R. O. Hamilton, and C. Weiner, Nondestructive determination of protein, total amino acids, and ammonia in marine sediments, *Bull. Mar. Sci.*, **20,** 289–304, 1970.

Hare, P. E., Geochemistry of proteins, peptides, and amino acids, in *Organic Geochemistry: Methods and Results*, G. Eglington and M. T. J. Murphy, eds., Springer-Verlag, New York, pp. 438–463, 1969.

Haworth, R. D., Chemical nature of humic acid, *Soil Sci.*, **111,** 71–79, 1971.

Khan, S. U. and M. Schnitzer, Chemistry of fulvic acid, a soil humic fraction, *Can. J. Chem.*, **49,** 2302–2309, 1971.

Mantoura, R. F. C. and J. P. Riley, The use of gel filtration in the study of metal binding by humic acids and related compounds, *Anal. Chim. Acta*, **78,** 193–200, 1975.

McBride, M. B., Transition metal bonding in humic acid: An ESR study, *Soil Sci.*, **126,** 200–209, 1978.

Mildvan, A. S. and M. Cohn, Magnetic resonance studies of the interaction of the manganous ion with bovine serum albumin, *Biochemistry*, **2,** 910–919, 1963.

Mitterer, R. M., Amino acid composition of organic matrix in calcareous oolites, *Science*, **162,** 1498–1499, 1968.

Mitterer, R. M., Biogeochemistry of aragonite mud and oolites, *Geochim. Cosmochim. Acta*, **36,** 1407–1422, 1972.

Mitterer, R. M. and P. W. Carter, Some analytical and experimental data on organic-carbonate interaction, *Proceedings, Third International Coral Reef Symposium*, University of Miami, FL, 541–547, 1977.

Müller, P. J. and E. Suess, Interaction of organic compounds with calcium carbonate—III. Amino acid composition of sorbed layers, *Geochim. Cosmochim. Acta*, **41,** 941–949, 1977.

Nissenbaum, A. and D. J. Swaine, Organic matter–metal interactions in Recent sediments: The role of humic substances, *Geochim. Cosmochim. Acta*, **40,** 809–816, 1976.

Rashid, M. A., Role of humic acids of marine origin and their different molecular weight fractions in complexing di- and tri-valent metals, *Soil Sci.*, **III,** 298–306, 1970.

Reuben, J., Gadolinium (III) as a paramagnetic probe for a proton relaxation studies of biological macromolecules. Binding to bovine serum albumin. *Biochemistry*, **10,** 2834–2838, 1971.

Reuben, J., The lanthanides as spectroscopic and magnetic resonance probes in biological systems, *Naturwissenschaften*, **62,** 172–178, 1975.

Scatchard, G., The attraction of proteins for small molecules and ions, *Ann. N.Y. Acad. Sci.*, **51,** 660–672, 1949.

Schnitzer, M. and S. U. Khan, *Humic Substances in the Environment*, M. Dekker, New York, 1972.

Senesi, N., S. M. Griffith, and M. Schnitzer, Binding of $Fe^{3+}$ by humic materials, *Geochim. Cosmochim. Acta*, **41,** 969–976, 1977.

Sherry, A. D. and G. L. Cottam, Proton relaxation rate and fluorometric studies of manganese and rare earth binding to Concanavalin A, *Arch. Biochem. Biophys.*, **156,** 665–672, 1973.

Simkiss, K. and C. Tyler, Reactions between eggshell matrix and metallic ions, *Quart. J. Microsc. Sci.*, **99,** 5–19, 1958.

Simkiss, K., Some properties of the organic matrix of the shell of the cockle, *Proc. Malac. Soc. Lond.*, **34,** 89–95, 1960.

Suess, E. and D. Fütterer, Aragonitic ooids: Experimental precipitation from sea water in the presence of humic acid, *Sedimentology,* **19,** 129–139, 1972.

Trichet, J., Étude de la composition de la fraction organique des oolites. Comparison avec celle des membranes des bacteries et des cyanophycées, *Compt. Rend.*, **267,** 1492–1494, 1968.

Valentine, K. M. and G. L. Cottam, Gadolinium as a probe of the alkaline earth and ATP-metal binding sites in pyruvate kinase, *Arch. Biochem. Biophys.*, **158,** 346–354, 1973.

Vittur, F., M. C. Pugliarello, and B. deBernard, The calcium binding properties of a glycoprotein isolated from pre-osseous cartilage, *Biochem. Biophys. Res. Comm.*, **48,** 143–152, 1972.

Waddell, W. J., A molecular mechanism for biological calcification, *Biochem. Biophys. Res. Comm.*, **49,** 127–132, 1972.

Williams, R. J. P., Calcium chemistry and its relation to protein binding, in *Calcium-Binding Proteins and Calcium Function,* R. H. Wasserman et al., eds., pp. 3–12, 1977.

# Adsorption of Some Amino Acids on Na-Montmorillonite: Implication of the Adsorption for Chemical Evolution

AKIRA SHIMOYAMA
CYRIL PONNAMPERUMA
*Laboratory of Chemical Evolution, University of Maryland*

## INTRODUCTION

The hypothesis of chemical evolution has been substantiated by extensive laboratory experiments on the abiotic synthesis of molecules of biological importance (see reviews by Lemmon 1970; Ponnamperuma 1972; Miller and Orgel 1974; Ponnamperuma and Gabel 1974; Ponnamperuma 1978). Although there are many examples of experiments simulating conditions on the primitive earth, there have been very few investigations of the effect of surfaces. Such experiments could recreate a truer prebiotic picture if material relevant to the constituents of the primitive earth's surface are also included.

Clay minerals on the primitive earth may have contributed to chemical evolution because of their wide distribution and their high affinity for organic compounds. Bernal (1951) proposed that clays at the interface of the hydrosphere and lithosphere might have adsorbed organic micromolecules, providing enhanced local concentrations of the reactants needed to form certain biologically important molecules.

We have extended Bernal's postulate to several other roles which clay minerals may have performed in the different stages of chemical evolution. They may be outlined as follows: (1) the catalytic effects in the synthesis of monomers from gaseous constituents of the primordial atmosphere; (2) the concentration of monomers by adsorption on a clay surface; (3) the specific orientation of monomers on a clay surface giving rise to a particular sequence of monomers; (4) the adsorption of certain organic compounds by the clay surface, in preference to others; (5) the condensation of monomers to polymers of biological importance by clay catalysis; (6) the protection of organic compounds (monomers and polymers) by adsorption from possible degradation by the high energy required for the formation of these compounds.

We have investigated the first role, catalytic effects, on the abiotic formation of amino acids from the constituents of a simulated primordial atmosphere (Shimoyama et al., 1978). All the other roles (2 to 6) involve mechanisms of organic adsorption.

The adsorption of organic compounds on the clay surface takes place through various mechanisms (Mortland 1970). The mechanisms on the adsorption of ionizable molecules, such as amino acids, depend largely upon the acidity-basicity, dipole moment, physico-chemical nature of the molecules, and the solution pH. The adsorption is affected also by the properties of clay minerals. The major factors controlling the adsorption are the charge density in octahedral and tetrahedral sheets and the interlayer cations.

Studies of amino acid adsorption on the clay surface (McLaren et al., 1958; Sieskind and Wey 1959; Sieskind 1960; Cloos et al., 1966) have been reported in a context unrelated to chemical evolution. In these studies, the adsorption experiments were performed with single amino acids (not mixtures), since the nitrogen determination by Kjeldahl was used to determine the amount of amino acids. Using ion exchange chromatography, the quantity of each amino acid can be well estimated directly from a mixture of amino acids. This enables us to examine a mixture of amino acids adsorbed and then released from the clay surface.

Our current study was designed to investigate whether a preferential adsorption of amino acids by clay minerals can be demonstrated. Our particular interest is in the adsorption of protein versus nonprotein amino acids by montmorillonite, which is a common clay mineral and has a high adsorption capacity of organic compounds. Is there any prebiotic process which sorted some or all of 20 protein amino acids out of the large number of amino acids that may have been produced under prebiotic conditions? Can clay minerals selectively adsorb protein amino acids over nonprotein amino acids or vice versa?

## EXPERIMENT

The clay mineral used in this experiment was Na-montmorillonite from the Source Clay Repository of the Clay Minerals Society located at the University of Missouri at Columbia. The clay mineral was size-fractionated (less than 2 μm) by centrifugation, cleaned for organic contaminants with distilled methanol, and washed with distilled water. The montmorillonite was further treated with NaCl solution to obtain the Na-saturated form. The cation exchange capacity (CEC) of the clay mineral was 101 millieq $Na^+$ per 100 grams dry weight of the clay as determined by atomic adsorption spectroscopy.

Adsorption experiment was performed with 10 ml of α-amino-n-butyric acid (0.4 m moles) mixed with approximately 10 ml of the clay suspension containing 400 mg Na-montmorillonite. The quantity of the amino acid used corresponds to 100% CEC of the amount of the clay. For a mixture of valine and norvaline, 10 ml of the amino acid mixture (0.1 moles each) was mixed with about 10 ml of the clay suspension containing 200 mg Na-montmorillonite. Experiments at 25% and 100% CEC levels were performed with 5 ml of a mixture of α-alanine and β-alanine (0.05 m moles each) and 10 ml of the mixture (0.20 m moles each), respectively, mixed with 10 ml of the clay suspension containing 400 mg Na-montmorillonite. After mixing the two solutions, the pH of the clay-amino acid solution was adjusted to 3, 7, or 10 by the addition of pH-preadjusted water and a few milliliters of HCl or NaOH solution. The final volume of the mixed solution was approximately 40 ml. The mixture was then shaken continuously for 48 hours to ensure equilibrium adsorption. After the shaking, the clay fraction was centrifuged (12000 rpm, 90 min.) from the supernatant. The clay deposit was then gently washed with a solution of the corresponding pH and certrifuged again. Finally, the clay deposit was treated with 1 *N* HCl solution (continuous shaking for 24 hours) to release amino acids from the clay surface. After this treatment, the suspension was centrifuged and the supernatant was recovered and lyophilized to remove HCl. The residue containing released amino acids was redissolved with water. Thus, we prepared three fractions, the supernatant (nonadsorbed), wash (weakly adsorbed), and clay (adsorbed on the surface), for the quantitative estimation of each amino acid by a Durrum 500 amino acid analyzer.

## RESULTS AND DISCUSSIONS

There have been some problems in reported quantities of amino acids adsorbed on the clay surface. These problems stem mainly from an indirect method of estimation (i.e., measuring the difference between the initial and the nonadsorbed concentration of the solution). Beside experimental error, there is always a possibility of microbial degradation of amino acids, especially the protein amino acids, during adsorption experiments. Generally, indirect methods give rise to a higher estimate of amino acid adsorbed. To overcome this problem and to establish a method for the estimation of the amount adsorbed, we have investigated the quantities of α-amino-n-butyric acid, a nonprotein amino acid in the three fractions. These estimates are shown in Table 1. The total recovery was between 99% and 105%. These differences are within experimental error. However, these estimates appear quite satisfactory when we consider the analytical process involving several steps. The 1 *N* HCl extraction of amino acids from the clay seems to cause no problem. Presumably, a large quantity of hydrogen ions introduced by the HCl treatment easily replaces almost all of the cationic amino acid in the interlayer positions.

The trend with pH in the amounts adsorbed by the clay clearly indicates that the adsorption took place through cationic exchange mechanisms. However, even at pH 3 (the amino acid molecules are cationic under this solution pH), the percentage adsorbed is rather low. This is probably because of the competition for adsorption between the amino acid cations and the hydrogen ions. There is no clear boundary for the separation of the nonadsorbed molecules and the weakly adsorbed molecules for Na-montmorillonite which can adsorb water molecules unlimitedly and gives no measurable interlayer spacing in water. However, the two trends in quantities of the weakly adsorbed and the nonadsorbed with pH are opposite and support the technique and estimate for the weakly adsorbed quantity. It seems from the trend that two adsorption mechanisms are involved in the weak adsorption. One is the ion-dipole interaction (ion of the Na ion in the interlayer position of the Na-montmorillonite and dipole of the amino acid molecules). The other is the hydrogen bonds of the weakly adsorbed molecules to the cationically adsorbed molecules and to the surface oxygen of the interlayer surface.

Results of the adsorption study of the equimolar mixture of valine and norvaline are shown in Table 2. Both valine and norvaline are α-amino acids and structural isomers. However, valine is a protein amino acid, whereas norvaline is a nonprotein amino acid. Apparently there is no significant difference between the percentages of the two amino acids

**Table 1. Percent Adsorption of α-Amino-n-Butyric Acid on Na-Montmorillonite[a]**

| pH | Supernatant | Wash | Clay | Total |
|---|---|---|---|---|
| 3.0 | 72.1 | 18.1 | 9.5 | 99.7 |
| 7.0 | 79.4 | 16.4 | 5.0 | 10.8 |
| 10.0 | 86.3 | 15.6 | 3.2 | 105.0 |

[a] Initial amount in solution was equivalent to 100% CEC.

**Table 2. Percent of Valine and Norvaline Adsorbed on Na-Montmorillonite[a]**

| pH | | Supernatant | Wash | Clay | Total |
|---|---|---|---|---|---|
| 3.0 | Valine | 68.6 | 19.7 | 8.3 | 96.6 |
| | Norvaline | 65.9 | 20.6 | 8.9 | 95.4 |
| 7.0 | Valine | 89.4 | 9.1 | 1.6 | 100.1 |
| | Norvaline | 87.2 | 7.2 | <0.1 | 94.4 |
| 10.0 | Valine | 95.7 | 6.6 | 1.2 | 103.5 |
| | Norvaline | 95.8 | 5.8 | <0.1 | 101.6 |

[a] Initial amount in solution was equivalent to 100% CEC, 50% of each amino acid.

found in the supernatant, wash, and clay. Both amino acids behaved almost identically at each solution pH. The molecules found in the adsorbed and weakly adsorbed fractions show similar adsorption mechanisms as those for α-amino-n-butyric acid. The similarity of these behaviors is due to their closeness in isoelectric points (6.06 for α-amino-n-butyric acid and 5.97 for valine) and their similar molecular sizes and structures. Probably the difference between valine and norvaline structures (iso and normal, respectively) is too small to be recognized by the interlayer environment of the Na-montmorillonite.

Table 3 shows that the results of an adsorption experiment with another equimolar mixture consisting of α-alanine (protein amino acid) and β-alanine (nonprotein amino acid). The percentages adsorbed are nearly the same at both 25% CEC and 100% CEC levels of the initial solution for the given pH. The average percents of the two amino acids adsorbed at pH 3 are 8.7% at the 25% CEC, 8.2% at the 100% CEC, level and are similar to those percentages observed in the experiments with α-amino-n-butyric acid and the mixture of valine and norvaline. However, the comparison between the percentage adsorbed for α-alanine and β-alanine in-

**Table 3. Percent of α-alanine and β-alanine Adsorbed on Na-Montmorillonite**

| pH | | Total initial amounts in solution equivalent to: 25% CEC | 100% CEC |
|---|---|---|---|
| 3.0 | α-alanine | 4.2% | 4.8% |
| | β-alanine | 13.1% | 11.6% |
| 7.0 | α-alanine | 1.2% | 0.9% |
| | β-alanine | 1.2% | 1.1% |
| 10.0 | α-alanine | 0.5% | 0.5% |
| | β-alanine | 0.6% | 0.7% |

dicates a preferential adsorption at pH 3. The difference cannot be accounted for by the difference in their isoelectric points (6.02 for α-alanine and 6.90 for β-alanine) because the solution pH was 3. The only plausible explanation is the difference in their dipole moments. Molecules of larger dipole moment are more readily adsorbed.

The isoelectric points, physical forces, and hydrogen bonding of amino acids, major controlling factors for adsorption, do not give any clue for the differentiation of the protein amino acids from the nonprotein ones. On the other hand, the dipole moment of amino acids affects the separation of the α-amino acids from non–α-amino acids. Note that all protein amino acids are α-amino acids. The ion-dipole interaction on the clay surface is the least for α-amino acids.

Although the presence of clay minerals on the prebiotic earth is yet to be confirmed by observation, it is most likely that they did exist on the juvenile earth. We have found evidence of clay minerals (chlorite and illite) in the Fig Tree Shale (details will be published elsewhere). The Fig Tree Shale is known to be one of the oldest unmetamorphosed sediments available on the earth. It is dated at 3.1 billion years and is reported to contain microfossils of early life (Barghoorn and Schopf 1966; Knoll and Barghoorn 1977). Clay particles may have occurred as suspensions in the primordial ocean and in the early atmosphere. The abiotic synthesis (Shimoyama et al., 1978) and the adsorption experiments of this study have both demonstrated that clay may have played an important role in chemical evolution.

## *REFERENCES*


Barghoorn, E. S. and J. W. Schopf, Microorganisms three billion years old from the Precambrian of South Africa, *Science,* **152,** 758–763, 1966.

Bernal, J. D., *The Physical Basis of Life,* Routledge and Kegan Paul, London, 1951.

Cloos, P., B. Calicis, J. J. Fripiat, and K. Makay, Adsorption of amino-acids and peptides by montmorillonite, I. Chemical and x-ray diffraction studies, *Proc. Intern. Clay Conf.,* Jerusalem, Vol. I, pp. 223–232, 1966.

Knoll, A. H. and E. S. Barghoorn, Archean microfossils showing cell divisions from the Swaziland system of South Africa, *Science,* **198,** 396–398, 1977.

Lemmon, R. M., Chemical Evolution, *Chem. Rev.,* **70,** 95–109, 1970.

MacLaren, A. D., G. H. Peterson, and I. Barshad, The adsorption and reaction of enzymes and proteins on clay minerals, IV. Kaolinite and montmorillonite, *Soil Sci. Soc. Amer. Proc.,* **22,** 239–224, 1958.

Miller, S. L. and L. E. Orgel, *The Origins of Life on the Earth,* Prentice-Hall, Englewood Cliffs, NJ, 1974.

Mortland, M. M., Clay organic complexes and interactions, *Adv. Agron.,* **22,** 75–117, 1970.

Ponnamperuma, C., *The Origins of Life,* Thames and Hudson, London, 1972.

Ponnamperuma, C., Prebiotic Molecular Evolution, *Origin of Life,* H. Noda, ed., Center for Academic Publication, Tokyo, pp. 67–82, 1978.

Ponnamperuma, C. and N. N. Gabel, The precellular evolution and organization of molecules, *Symposia of the Society for General Microbiology, Number XXIV, Evolution in the Microbial World,* 393–413, 1974.

Shimoyama, A., N. Blair, and C., Ponnamperuma, Synthesis of amino acids under primitive earth conditions in the presence of clay, *Origin of Life,* H. Noda, ed., Center for Academic Publication, Tokyo, pp. 95–100, 1978.

Sieskind, O., Etude des complexes d'adsorption formés entre la montmorillonite-H. et certains acides aminés: Leur structure, *C. R. Acad. Sci.,* **250,** 2228–2230, 1960.

Sieskind, O. and R. Wey, Sur l'adsorption d'acides aminés par la montmorillonite-H. Influence de la position relative des deux fonctions $NH_2$ et COOH, *C. R., Acad. Sci.,* **248,** 1652–1655, 1959.

# Amino Acids of the Murchison Meteorite

**JOHN R. CRONIN**
**WILLIAM E. GANDY**
**SANDRA PIZZARELLO**
*Department of Chemistry and Center for Meteorite Studies, Arizona State University*

## INTRODUCTION

Carbonaceous chondrites have been known for almost 175 years and for the last 100 years have been recognized as a distinct meteorite class based on their unusual and characteristic chemical composition. They are unique among meteorites in having a high content of water and carbon, and in the occurrence of much of this carbon in the form of organic (in the chemical sense) compounds. The carbonaceous chondrites are alone among the chondritic meteorites in having representation in petrologic types I and II, the categories reserved for chondrites lacking, or nearly lacking, signs of metamorphic effects (Van Schmus and Wood 1967). Much evidence, including compaction ages (≥4.3 Gy), now suggests that the types I and II carbonaceous chondrites (C1 and C2 chondrites) are a relatively undifferentiated and unaltered condensate of the primordial solar nebula.

The presence of organic compounds in carbonaceous chondrites was noted by Berzelius in his analysis of the Alais (C1, 1806) meteorite, the first recognized carbonaceous chondrite (Berzelius 1834). The existence of complex organic compounds in such primitive materials remains one of their more intriguing characteristics. Analyses of these compounds in several of the nearly 40 carbonaceous chondrites that have been recovered during the subsequent 173 years now allow at least a partial accounting in terms of classes of compounds, and in some cases, specific compounds.

About 70% of the organic carbon in C1 and C2 chondrites is present in the form of an insoluble, extensively substituted aromatic polymer. The remaining 30% can be extracted by water or organic solvents. The soluble fraction includes aliphatic and aromatic hydrocarbons, phenols, low molecular weight alcohols, aldehydes and ketones, mono- and dicarboxylic acids, nitrogen heterocycles, and amino acids. Analyses of meteorite organic constituents have been reviewed by Hayes (1967) and by Nagy (1975).

Berzelius raised the question whether the organic compounds in Alais "hint of organic formations on other planets." The question of the relationship between meteorite organic compounds and extraterrestrial life has since been raised periodically and, although it cannot be answered with absolute certainty, analyses of organic compounds in the Murchison meteorite (C2, 1969) carried out during the last 10 years yield little if any evidence to suggest the involvement of life in the synthesis of these compounds. To the contrary, these data can be readily used to support the view that the meteorite organic compounds are the products of an exclusively chemical synthesis from simple precursors, that is, of a process of organic chemical evolution. Organic chemical evolution is an essential postulate of scientific theories of the origin of life. Although the mechanisms are unknown by which the organic compounds of the primitive Earth and the meteorites were synthesized, the presence of ubiquitous biomolecules among the meteorite organics suggests at least some analogy between these processes. It is possible that a direct relationship exists between meteoritic organic compounds and the origin of terrestrial life. The heavy meteoritic bombardment that the Earth received prior to about 3.3 Gy ago very likely contributed large quantities of organic compounds to the primordial surface. Whatever the relationship to terrestrial processes may be, the carbonaceous chondrites appear to carry the clearest evidence yet found in nature for the occurrence of organic chemical evolution.

## METEORITE AMINO ACIDS

Several reports of the occurrence of amino acids in carbonaceous chondrites appeared in the literature in the period 1960–1965. Serious doubt was cast on the significance of these early findings when it was recognized that there were striking similarities between some of these data and the amino acid content of potential contaminants such as fingerprints and laboratory dust (Hayes 1967). In late 1969, specimens from two recent carbonaceous chondrite falls (Murchison; Allende, C3) became available for analysis. Several laboratories were at that time equipped to carry out amino acid analyses with both high sensitivity and effective contamination control in preparation for the analysis of returned lunar samples. Their analytical capability included quantitation of the individual amino acid enantiomers by combined gas chromatography–mass spectrometry (GC-MS). This capability provided a new and powerful tool for evaluating the extent of terrestrial contamination, since the biological occurrence of amino acids is predominantly as the L-enantiomer. This technique along with conventional ion exchange chromatographic methods was applied to the analysis of aqueous extracts of the Murchison meteorite, and it was soon apparent that the amino acids found could not readily be explained in terms of terrestrial contamination. The presence of amino acids that do not occur or occur but rarely in biological materials and the finding of amino acids with a chiral center in nearly racemic proportions were strong arguments favoring a meteoritic origin (Kvenvolden et al., 1970, 1971). Further studies of Murchison amino acids done in various laboratories have confirmed and extended these results (Cronin and Moore 1971; Oro' et al., 1971; Lawless 1973; Pollock et al., 1975; Buhl 1975). Similar analyses have now been carried out with a C1 chondrite, Orgueil (Lawless et al., 1972; Cronin, unpublished), other C2 chondrites, Murray (Cronin and Moore 1971; Lawless et al., 1971), Mighei (Buhl 1975; Cronin, unpublished), Nogoya and Mokoia (Cronin and Moore 1976), Renazzo, Cold Bokkeveld, and Santa Cruz (Cronin, unpublished), and a C3 chondrite, Allende (Cronin and Moore 1971; Harada and Hare 1972). The C1 chondrites are of particular interest because they represent the most primitive and least altered meteoritic material. The carbon (average 3.5% by weight) and water (about 20%) contents of the C1 chondrites exceed those of all other carbonaceous meteorites, and it is reasonable to expect that the same relationship would hold for organic constituents such as amino acids. However, this does not seem to be born out by recent results on Orgueil meteorite which indicate only a low level of indigenous amino acids overlaid by substantial amounts of terrestrial contaminants (Lawless et al., 1972; Cronin, unpublished). Contamination of this meteorite is not surprising in view of its long terrestrial history (115 yr) and the decrepit state of the samples available for analysis. However, the low level of indigenous amino acids is unexpected and suggests the possibility of considerable microbial alteration of the original organic constituents.

The single C3 chondrite analyzed to date was initially found to be devoid of amino acids (Cronin and Moore 1971). However, other workers analyzed this meteorite using an analytical system of much higher sensitivity and found a low but reproducible level of amino acids in hydrolyzed aqueous extracts of interior samples (Harada and Hare 1972). Thus it appears that amino acids (or acid-labile precursors) are indigenous to this meteorite, but at levels two to three orders of magnitude lower than found in C2 chondrites.

The results of analyses of C2 chondrites will not be individually described except in the case of Murchison. The suite of amino acids found in this meteorite has been analyzed most extensively and can be considered qualitatively typical of C2 chondrites with the exception of Mokoia and possibly Renazzo. The latter are sufficiently distinct chemically to be grouped separately in the Vigarano subtype (C2V), whereas the other C2 chondrites analyzed are placed in the Mighei subtype (C2M) (Wasson 1974). Although the C2M chondrites are qualitatively similar, there is considerable quantitative variation among those analyzed thus far, with decreases in total amino acid content relative to Murchison by factors ranging from 0.5 to 0.1 (Cronin, unpublished).

## Methodology

Whenever possible it is desirable to select for analysis interior fragments from a larger stone broken in the laboratory. When this was not possible, acceptable samples were obtained either by removing a surface layer from the specimen prior to crushing, or by drilling into the interior of a stone and collecting the grindings obtained after the first few millimeters of surface were removed. In the former cases, the clean fragment(s) thus obtained are then pulverized with an agate or steel mortar and pestle. Care is taken to avoid sample contamination by handling with heat-cleaned implements and using heat-cleaned aluminum foil for collection of fragments, grindings, crushed samples, etc.

The pulverized sample is extracted with hot water for 20–24 hours either under reflux conditions or in a sealed Pyrex ampule (110°). The meteorite residue is removed from the extract by vacuum filtration through a fine frit (4–5.5 μ) glass funnel and the filtrate dried by rotary evaporation or in a vacuum desiccator over NaOH pellets.

This extract may be used directly for analysis of *free* amino acids by ion exchange chromatographic methods or by gas chromatographic methods after desalting (Pollock et al., 1977) and derivatization (Kvenvolden et al., 1971). Alternatively the extract may be hydrolyzed in 6 *N* (constant boiling) HCl for 24 hours at 110°C, dried as before, and analyzed for *total* amino acids by either ion exchange chromatography (directly) or gas chromatography (after desalting and derivatization).

The water used for extraction and reagent preparation is first distilled from glass and then redistilled from a quartz subboiling distillation apparatus. All glassware is carefully washed, rinsed, and taken to the annealing temperature to burn off any organic residue adhering to the surface.

## Murchison Amino Acids

A hydrolyzed hot-water extract of Murchison obtained as described above contains a large number of components that behave chromatographically as amino acids. On the chromatogram obtained from a ninhydrin-based ion exchange system it is possible to identify about 70 distinct features (resolved and partially resolved peaks) in the region in which amino acids elute. Twenty-two amino acids have been positively identified by GC-MS, and 20 or 21 others have been tentatively identified from coincidence of retention times in one or more chromatographic systems or have been characterized in a general way by GC-MS without making a specific identification. The latter group includes six dicarboxylic amino acids isomeric with glutamic acid or higher homologs, one proline and two pipecolic acid isomers, and five or six noncyclic 5-carbon saturated aliphatic monoaminocarboxylic acids (Lawless 1973). The specific amino acids that have been either positively or tentatively identified in Murchison extracts are listed in Table 1.

The Murchison amino acids identified thus far fall into one of two general categories. There is a large group of cyclic and noncyclic saturated aliphatic monoamino monocarboxylic acids that includes both N-alkyl forms and various amino position isomers, that is, α-, β-, γ, and δ-amino acids. There is a second smaller group of monoamino dicarboxylic acids. These two groups include many amino acids that are found in biological materials. There are eight protein amino acids (glycine, alanine, valine, leucine, isoleucine, proline, aspartic acid, and glutamic acid) that are common to all terrestrial life and several others that range from wide (e.g., β-alanine) to quite restricted (e.g., α-amino-isobutyric acid) biological distribution. It is important to note that 12 of the ubiquitous protein amino acids have *not* been found in Murchison extracts. These amino acids have in common the presence of oxygen-, sulfur-, or nitrogen-containing functional groups or aromatic groups in their side chains. The hydroxy amino acids, serine, and threonine, have been detected at low levels by ion exchange chromatographic analysis of Murchison extracts (Cronin 1976). Whether these amino acids are indigenous to the meteorite remains to be established. In evaluating the extent to which meteorite samples have been contaminated during their terrestrial history it is useful to note the level at which extracts contain the nonmeteoritic protein amino acids.

It is interesting that several of the Murchison amino acids have not been reported as occurring biologically, in which case, carbonaceous

**Table 1.**

| Amino Acid | Amount (nmole/g) | Number of chiral centers | Biological occurrence[b] | Identification[c] | Identification reference |
|---|---|---|---|---|---|
| 2 Carbon | | | | | |
| Glycine | 98 | 0 | BP | C-MS | Kvenvolden, et al., 1970 |
| 3 Carbon | | | | | |
| Alanine | 44 | 1[a] | BP | C-MS | Kvenvolden, et al., 1970 |
| β-Alanine | 14 | 0 | B | C-MS | Kvenvolden, et al., 1971 |
| Sarcosine | 19 | 0 | B | C-MS | Kvenvolden, et al., 1970 |
| 4 Carbon | | | | | |
| α-Amino-n-butyric acid | 18 | 1[a] | B | C-MS | Kvenvolden, et al., 1971 |
| β-Amino-n-butyric acid | 4 | 1 | B | C-MS | Kvenvolden, et al., 1971 |
| γ-Amino-n-butyric acid | 24 | 0 | B | C-MS | Kvenvolden, et al., 1971 |
| α-Aminoisobutyric acid | 117 | 0 | B | C-MS | Kvenvolden, et al., 1970 |
| β-Aminoisobutyric acid | 4 | 1[a] | B | C-MS | Lawless, 1973 |
| N-Methyl-alanine | — | 1 | ? | C-MS | Lawless, 1973 |
| N-Ethyl-glycine | — | 0 | ? | C-MS | Lawless, 1973 |
| Aspartic acid | 5 | 1[a] | BP | C-MS | Lawless, 1973 |
| 5 Carbon | | | | | |
| Norvaline | 3 | 1[a] | ? | C-MS | Lawless, 1973 |
| Isovaline | 28 | 1[a] | ? | C-MS | Kvenvolden, et al., 1971 |
| Valine | 10 | 1[a] | BP | C-MS | Kvenvolden, et al., 1970 |

| | | | | | |
|---|---|---|---|---|---|
| δ-Aminovaleric acid | — | 0 | B | C-MS | Buhl, 1975 |
| Proline | 16 | 1[a] | BP | C-MS | Kvenvolden, et al., 1970 |
| Glutamic acid | 18 | 1[a] | BP | C-MS | Kvenvolden, et al., 1970 |
| 6 Carbon | | | | | |
| Leucine | 4 | 1 | BP | C-MS | Pereira, et al., 1975 |
| Isoleucine | 4 | 2 | BP | C-MS | Pereira, et al., 1975 |
| *Allo*-isoleucine | 4 | 2 | B | C | Buhl, 1975 |
| Norleucine | 2 | 1 | ? | C | Buhl, 1975 |
| 2-Amino-3,3-dimethylbutyric acid (pseudoleucine) | 4 | 1 | ? | C | This paper |
| 6 Carbon | | | | | |
| 2-Amino-2-ethylbutyric acid | 11 | 0 | ? | C | This paper |
| 2-Amino-2,3-dimethylbutyric acid | 17 | 1 | ? | C | This paper |
| 2-Amino-2-methylvaleric acid (2-methyl-norvaline) | 4 | 1 | ? | C | This paper |
| Pipecolic acid | — | 1[a] | B | C-MS | Kvenvolden, et al., 1971 |
| α-Aminoadipic acid | — | 1 | B | C-MS | Buhl, 1975 |

[a] Determined to be racemic.

[b] B: known biological occurrence.
BP: protein amino acid.
?: no known biological occurrence.

[c] C: identification by chromatographic retention time (gas and/or ion exchange).
C-MS: identification by chromatographic criteria and mass spectrum (GC-MS).

chondrites may be their only natural source and an amino acid suite indigenous to carbonaceous chondrites uniquely recognized by their presence. Isomers of valine (norvaline, isovaline) and leucine (e.g., norleucine, pseudoleucine) are characteristic in this regard. The presence of several of these amino acids has been helpful in establishing that extracts of a C2 carbonaceous chondrite recently recovered from Antarctica contain amino acids of meteoritic origin (Cronin et al., 1979).

Twenty of the 28 Murchison amino acids listed in Table 1 have at least one asymmetric carbon atom and thus can occur as enantiomeric pairs. Of these 20, ten have been sufficiently resolved by gas chromatography of their diastereomeric N-trifluoroacetyl (TFA) D-2-butyl or D-2-pentylesters to determine that they are racemic within experimental error (Kvenvolden et al., 1970; Kvenvolden, et al., 1971; Lawless 1973; Pollock et al., 1975). This has been taken as compelling evidence against a significant contribution from terrestrial contaminants.

Aqueous extracts of the Murchison meteorite have been found to contain both free amino acids and combined forms that can be converted to free amino acids by acid hydrolysis (Cronin 1976). Overall, the combined forms are found in an amount about equal to that of the free amino acids; however, there is considerable variation among the individual amino acids in the ratio of combined to free forms. The aspartic acid content increases eight-fold on hydrolysis of the aqueous extract, whereas the α-aminoisobutyric acid content increases by only about 40%. These acid-labile derivatives have been shown in experiments with ion exchange resins to have lost the basicity but to have retained the acidic properties of the amino acids (Cronin 1975). The nature of these compounds is not known, but their properties are consistent with those of N-acyl amino acids.

The 2,3, and 4 carbon amino acids of Murchison show nearly the full range of structural diversity available within the noncyclic saturated aliphatic monoamino monocarboxylic acid class. The number of structural isomers possible for 2, 3, and 4 carbon amino acids of this class are 1, 3, and 9, respectively. In Table 1 it can be seen that all of the 1 and 2 carbon isomers have been identified as have seven of the nine possible 4 carbon amino acids. Of this latter group only N-dimethyl glycine and N-methyl-β-alanine have not been found (Lawless 1973). The total isomeric possibilities increase to 24 for 5 carbon, and to 63 for 6 carbon amino acids of this class.

From the quantitative data given in Table 1 it is apparent that primary amino acids are favored over secondary, that is, the N-alkyl forms. It also appears that α-amino acids are favored over the corresponding β-, γ-, etc., forms. Excluding the N-alkyl forms from consideration, the isomeric possibilities through 6 carbon amino acids are shown in Table 2.

**Table 2. Noncyclic Saturated Aliphatic Primary Monoamino Monocarboxylic Acid Isomers**

| Carbon atoms | α | β | γ | δ | ε |
|---|---|---|---|---|---|
| 2 | 1(1) | — | — | — | — |
| 3 | 1(1) | 1(1) | — | — | — |
| 4 | 2(2) | 2(2) | 1(1) | — | — |
| 5 | 3(3) | 5(0) | 3(1) | 1(1) | — |
| 6 | 7(7) | 10(0) | 9(0) | 4(0) | 1(1) |

The number of isomers that have been positively or tentatively identified are given in parentheses. It can be seen that the tentative identification of the 6 carbon, noncyclic, saturated aliphatic primary α-monoamino monocarboxylic acids (leucine isomers) to be described completes the identification of all members of this series. Complete identification of *all* isomers through 6 carbon atoms will require a search for 30 additional 5 carbon and 6 carbon β-, γ-, and δ-isomers. Many of these are undoubtedly present. As mentioned above GC-MS has allowed the recognition of five or six of the seven remaining 5-carbon isomers and ion exchange chromatographic methods show a large, but incompletely resolved group of amino acids in the region where these isomers would elute (see Fig. 1a, b; 95–120 min elution period).

## Identification of Leucine Isomers

Gas chromatography (Oró et al., 1971; Buhl 1975), ion exchange chromatography (Cronin 1976), and GC-MS (Lawless 1973; Pereira et al., 1975) provide evidence for the presence of three (leucine, norleucine, and isoleucine) of the seven possible leucine isomers. To further test the proposition that Murchison contains a total isomeric distribution of amino acids within the saturated aliphatic monoamino monocarboxylic acid group, analysis of the remaining four isomers was undertaken. The isomers are: 2-amino-3,3-dimethylbutyric acid (pseudoleucine), 2-amino-2-ethylbutyric acid, 2-amino-2,3-dimethyl butyric acid, and 2-amino-2-methylvaleric acid (2-methyl norvaline). The identification of these isomers adds several members to the nonbiological group of meteorite amino acids. They are particularly useful in that they are easily analyzed by high sensitivity fluorimetric ion exchange chromatographic methods (Cronin et al., 1979). Furthermore, three of the isomers lack an α-hydrogen atom, a condition that makes them of special interest in terms of elucidating the original enantiomeric distribution of the Murchison amino acids (Pollock et al., 1975).

Since these leucine isomers are not commercially available, they were obtained by synthesis of the corresponding hydantoins (Henze and Speer 1942) followed by alkaline hydrolysis (Upjohn and Dermer 1957). The products were used as comparison standards in both gas and ion exchange chromatographic analyses.

Gas chromatography was carried out with both N-TFA-D(+)-2-butyl and N-TFA-D(+)-2-pentyl esters of the amino acids employing steel capillary columns (0.02 in × 150 ft) coated with Carbowax 20M. The results can be summarized as follows.

**2-Amino-2-ethylbutyric acid.** This amino acid does not have a chiral center and appears as a single peak resolved from all other known meteorite amino acids in the case of both derivatives. A corresponding peak, identical by cochromatography, appears in the Murchison extract.

**2-Methyl norvaline.** The diastereomeric N-TFA esters of DL-2-methylnorvaline are unresolved in the butanol system but are separated with pentanol. In both cases they are resolved from other known Murchison amino acids. When the Murchison extract was run a corresponding single peak was seen with butanol and two approximately equal-sized peaks were found with D(+)-2-pentanol. Since 2-methyl norvaline lacks an α-hydrogen and cannot undergo racemization readily, this finding further supports the contention that the Murchison amino acids were synthesized as racemic mixtures (Pollock et al., 1975).

**Pseudoleucine.** The diastereomeric esters of DL-pseudoleucine are resolved in both systems, but in both cases the peak corresponding to the D-enantiomer coelutes with isovaline. A peak corresponding to L-pseudoleucine is seen in the Murchison extract with both esters. Thus the amino acid is tentatively identified, although the enantiomeric distribution cannot be determined.

**2-Amino-2,3-dimethylbutyric acid.** The diastereomeric derivatives of this amino acid are resolved in both cases although somewhat better as the D(+)-2-butyl esters than as the D(+)-2-pentyl esters. The resolution from other Murchison amino acids is rather poor, with the L-enantiomer overlapped by D-valine. The D-enantiomer appears as a partially resolved peak or shoulder (depending on column condition) on the leading edge of the D-valine peak. The derivatized meteorite extract shows this feature with either ester.

To gain additional evidence for the presence of these amino acids in Murchison their analysis was attempted by ion exchange chromatography. Detection of these amino acids with a ninhydrin-based analytical system is difficult because of low color yield. The analogous problem, low fluorescence yields (0.01–0.12 relative to glycine), hampers their detection with an o-phthalaldehyde (OPA)-based system. However, it was found that detection sensitivity with OPA could be substantially improved by making two minor modifications: (1) substitution of methanethiol or ethanethiol for mercaptoethanol in the OPA reagent, and (2) placing a short, 100° reaction coil between the reagent mixer and the fluorimeter (Cronin et al., 1979). In Figure 1 are shown chromatograms obtained with a hydrolyzed Murchison extract with fluorogen development at 100° as described above (Fig. 1a) and at 25° (Fig. 1b). It can be seen that peaks corresponding to 2-amino-2-ethylbutyric acid (2-A-2-EBA), pseudoleucine (PLEU) plus 2-amino-2,3-dimethylbutyric acid (2-A-2,3-DMBA), and 2-methylnorvaline (2-MNVAL) plus *allo*-isoleucine (ALILE) are readily apparent at 100° and are greatly diminished at 25° reaction temperature.

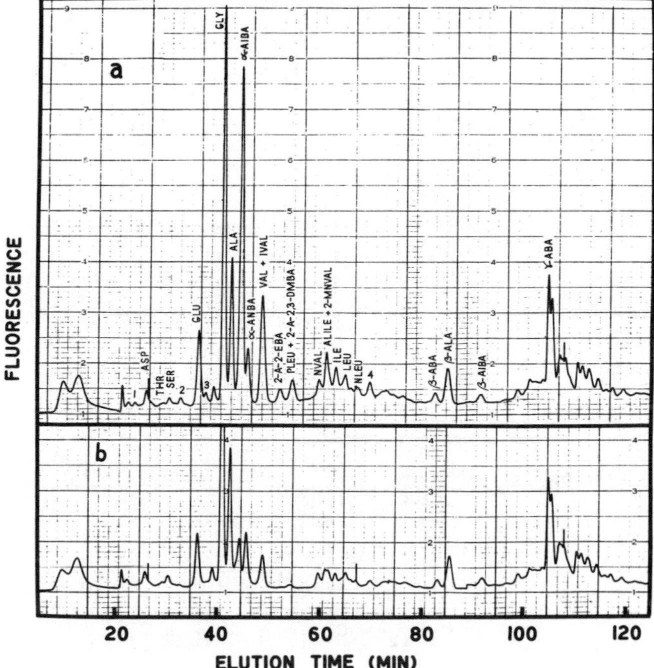


**Figure 1.** Chromatograms of hydrolyzed aqueous extracts from the Murchison meteorite, 4.4 mg. Fluorogen development: (a) 100°, (b) 25°.

Not only retention times but also the characteristic temperature dependence of their fluorescence reactions give further support to the identification of these amino acids in the Murchison extract.

Amino acids showing a marked increase in fluorescence developed at 100° compared with that at 25° are typically those lacking an α-hydrogen atom, but pseudoleucine may show a similar effect because of the presence of the bulky t-butyl side-chain. In common with the leucine isomers, isovaline and α-aminoisobutyric acid behave in this way. It can be seen by comparing Figures 1a and 1b that in addition to these amino acids there are at least four (peaks numbered 1–4) other components of the Murchison extract that show a reaction temperature–dependent fluorescence and thus may lack an α-hydrogen atom.

The difference in fluorescence yield for these amino acids at 25° and 100° allows their quantitation when they are unresolved from other components, for example, PLEU plus 2-A-2,3-DMBA and 2-MNVAL plus ALILE. The primary requirement is that the unresolved amino acids be substantially different with respect to difference in fluorescence yield at

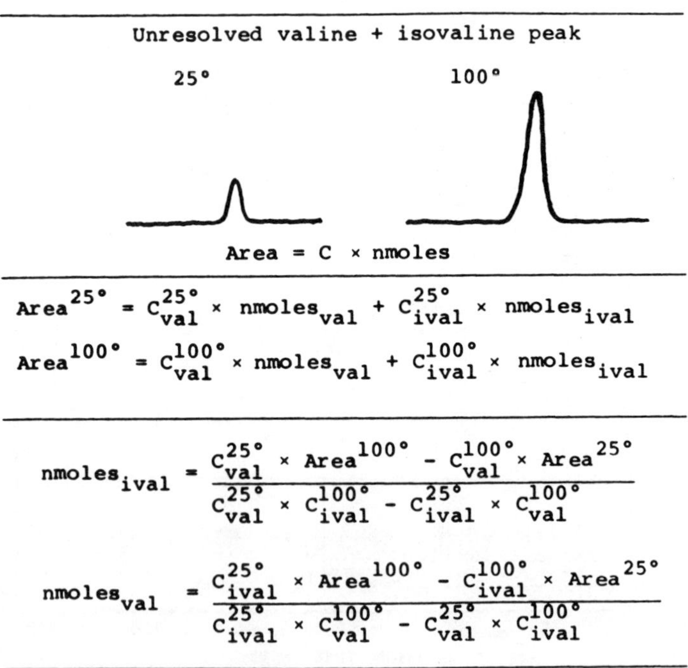


**Figure 2.** Calculation of amounts of chromatographically unresolved amino acids from OPA fluorescence at two temperatures (25° and 100°).

25° and 100°. Analyses of standards and samples at 25° and 100° allows the construction of simultaneous equations and their solution for individual amounts of the two unresolved components. This is illustrated in Figure 2 for valine and isovaline. This technique allowed determination of the amounts of valine, isovaline, pseudoleucine, 2-amino-2,3-dimethylbutyric acid, 2-methyl norvaline, and *allo*-isoleucine given in Table 1.

## DISCUSSION AND SUMMARY

The results described above allow identification of 2-amino-2-ethylbutyric acid, 2-amino-2,3-dimethylbutyric acid, pseudoleucine, and 2-methylnorvaline as Murchison constituents and completes at least a tentative identification of all seven leucine isomers (eight, including *allo*-isoleucine). Positive identification of each by GC-MS is under way.

Quantitative results have now been obtained for all primary α-amino acids through 6 carbon atoms. Plots of these data are given in Figure 3. If only the straight chain (*normal*) isomers are considered, there is a nearly exponential decline in content with increasing carbon number. However, if total amounts of primary α-amino acids are considered, there is a pronounced departure from the smooth decline seen within the *normal* homologous series. The preference for branched chain forms and the pos-

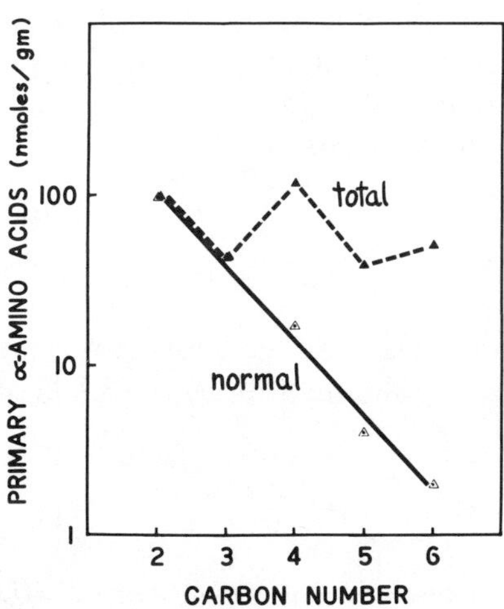


**Figure 3.** Amounts of 2 through 6 carbon amino acids in the Murchison meteorite: Dashed line, total noncyclic saturated aliphatic primary α-amino acids; solid line, straight chain (normal) noncyclic saturated aliphatic primary α-amino acids.

sibilities for chain branching at carbon numbers greater than 3 appear to partially overcome the tendency for declining content with increasing carbon number. Although little is known about the mechanism of meteorite amino acid synthesis, these data are consistent with the concept of a synthesis from a single carbon atom precursor in which radical forms are important intermediates.

The question might be raised as to how much further analytical studies can profitably be pursued. Identification of the 5 and 6 carbon β-, γ-, and δ-isomers seems feasible, as does the identification of additional dicarboxylic amino acids through 6 carbons. In view of the lower content of N-alkyl amino acids and the lower sensitivity of ion exchange chromatographic detection systems for these isomers, a systematic identification and quantification of 5 and 6 carbon isomers of this type appears to be a difficult undertaking at present. Nevertheless quantitative determination of the other isomers mentioned above, along with the monocarboxylic acids, (Lawless and Yuen 1979), dicarboxylic acids (Lawless et al., 1974), and amines (Jungclaus et al., 1976) of Murchison will establish a useful data base to guide theoretical and chemical modelling studies of the amino acid synthetic mechanism.

In summary, the Murchison amino acids can be presently described as a mixture of 2 carbon through 6 carbon cyclic and noncyclic saturated aliphatic monoamino-, mono- and dicarboxylic acids of total or nearly total structural diversity, in which (1) branched carbon chain, primary amino-, and α-amino forms predominate, (2) there is a declining content with increasing carbon number within homologous series, and (3) those forms with a single chiral center are racemic. They occur as both free and acid-hydrolyzable combined forms and are apparently products of an extraterrestrial chemical synthesis.

## *ACKNOWLEDGMENTS*

The authors gratefully acknowledge support received from the National Aeronautics and Space Administration, NASA Grant NSG-7255.

## *REFERENCES*

Berzelius, J. J., Ueber Meteorsteine, *Ann. Phys. Chem.*, **33**, 113–148, 1834.

Buhl, P., An investigation of organic compounds in the Mighei meteorite, Ph.D. Thesis, University of Maryland, College Park, MD, 1975.

Cronin, J. R., Amino acids and their derivatives in carbonaceous chondrites, *Carnegie Inst. Wash. Year Book 74*, 617–619, 1975.

Cronin, J. R., Acid-labile amino acid precursors in the Murchison meteorite. I. chromatographic fractionation, *Origins of Life*, **7**, 337–342, 1976.

Cronin, J. R., Acid-labile amino acid precursors in the Murchison meteorite, II. a search for peptides and amino acyl amides, *Origins of Life*, **7**, 343–348, 1976.

Cronin, J. R. and C. B. Moore, Amino acid analyses of the Murchison, Murray, and Allende carbonaceous chondrites, *Science*, **172**, 1327–1329, 1971.

Cronin, J. R. and C. B. Moore, Amino acids of the Nogoya and Mokoia carbonaceous chondrites, *Geochim. Cosmochim. Acta*, **40**, 853–857, 1976.

Cronin, J. R., S. Pizzarello, and W. E. Gandy, Amino acid analysis with o-phthalaldehyde detection: effects of reaction temperature and thiol on fluorescence yields, *Anal. Biochem.*, **93**, 174–179, 1979.

Cronin, J. R., S. Pizzarello, and C. B. Moore, Amino acids in the Allan Hills, Antarctica C2 carbonaceous chondrite ALHA 77306 (abstract), *Lunar and Planetary Science*, **X**, pp. 251–253, Lunar and Planetary Institute, Houston, 1979.

Harada, K. and P. E. Hare, Analyses of amino acids from the Allende chondrite, presented at a meeting of the Committee for the Analysis of Carbon Compounds in Meteorites and Lunar Samples, Houston, 1972.

Hayes, J. M., Organic constituents of meteorites, *Geochim. Cosmochim. Acta*, **31**, 1395–1440, 1967.

Henze, H. R. and R. J. Speer, Identification of carbonyl compounds through conversion into hydantoins, *J. Amer. Chem. Soc.*, **64**, 522–523, 1942.

Jungclaus, G., J. R. Cronin, C. B. Moore, and G. Yuen, Aliphatic amines in the Murchison meteorite, *Nature*, **261**, 126–128, 1976.

Kvenvolden, K., J. G. Lawless, and C. Ponnamperuma, Nonprotein amino acids in the Murchison meteorite, *Proc. Nat. Acad. Sci. U.S.A.*, **68**, 486–490, 1971.

Kvenvolden, K. A., E. Peterson, and G. E. Pollock, Geochemistry of amino acid enantiomers: Gas chromatography of their diastereomeric derivatives, *Adv. Org. Geochem.*, **1971**, 387–401, 1972.

Kvenvolden, K., J. Lawless, K. Pering, E. Peterson, J. Flores, C. Ponnamperuma, I. R. Kaplan, and C. Moore, Evidence for extraterrestrial amino acids and hydrocarbons in the Murchison meteorite, *Nature*, **228**, 923–926, 1970.

Lawless, J. G., Amino acids in the Murchison meteorite, *Geochim. Cosmochim. Acta*, **37**, 2207–2212, 1973.

Lawless, J. G. and G. U. Yuen, Quantification of monocarboxylic acids in the Murchison meteorite, *Nature*, **282**, 396–398, 1979.

Lawless, J. G., K. Kvenvolden, E. Peterson, C. Ponnamperuma, and E. Jarosewich, Evidence for amino acids of extraterrestrial origin in the Orgueil meteorite, *Nature*, **236**, 66–67, 1972.

Lawless, J. G., K. A. Kvenvolden, E. Peterson, C. Ponnamperuma, and C. Moore, Amino acids indigenous to the Murray meteorite, *Science*, **173**, 626–627, 1971.

Lawless, J. G., B. Zeitman, W. E. Pereira, R. L. Summons, A. M. Duffield, Dicarboxylic acids in the Murchison meteorite, *Nature*, **251**, 40–41, 1974.

Nagy, B., *Carbonaceous Meteorites*, Elsevier, Amsterdam, 1975.

Oro', J., J. Gibert, H. Lichtenstein, S. Wikstrom, and D. A. Flory, Amino-acids, aliphatic and aromatic hydrocarbons in the Murchison meteorite, *Nature*, **230**, 105–106, 1971.

Pereira, W. E., R. E. Summons, T. C. Rindfleisch, A. M. Duffield, B. Zeitman, and J. G. Lawless, Stable isotope mass fragmentography: Quantitation and hydrogen-deuterium exchange studies of eight Murchison meteorite amino acids, *Geochim. Cosmochim. Acta,* **39,** 163–172, 1975.

Pollock, G. W., C-N. Cheng, S. E. Cronin, Determination of the D and L isomers of some protein amino acids present in soils, *Anal. Chem.,* **49,** 2–7, 1977.

Pollock, G., C-N. Cheng, S. E. Cronin, and K. A. Kvenvolden, Stereoisomers of isovaline in the Murchison meteorite, *Geochim. Cosmochim. Acta,* **39,** 1571–1573, 1975.

Upjohn, S. D. and O. C. Dermer, A new series of anticonvulsant drugs: Branched-chain α-amino acetamides, *J. Org. Chem.,* **22,** 799–802, 1957.

Van Schmus, W. E. and J. A. Wood, A chemical-petrologic classification for the chondritic meteorites, *Geochim. Cosmochim. Acta,* **31,** 747–765, 1967.

Wasson, J. T., *Meteorites, Classification and Properties,* Springer-Verlag, New York, 1974.

# Analyses of Amino Acids from the Allende Meteorite

**KAORU HARADA***
*Institute for Molecular and Cellular Evolution and Department of Chemistry, University of Miami*

**P. E. HARE**
*Geophysical Laboratory, Carnegie Institution of Washington, Washington, D.C.*

* Present Address: Department of Chemistry, the University of Tsukuba, Niihari-Gun, Ibaraki-Ken, 305, Japan

*ABSTRACT*

Amino acid analyses of hydrolyzed hot water extracts of successively drilled sample aliquots from three specimens of the Allende meteorite were carried out. The analytical results reveal a profound difference in the distribution of amino acids (in quantity and in species) found at the specimen surface compared to the pattern found within the specimen. The surface sample had between one and two orders of magnitude higher concentration of amino acids as well as a more complex mixture containing virtually all of the common amino acids. In contrast, the samples from the interior of the meteorite specimens showed a simple mixture of only six or seven amino acids. Analytical results indicate that the amino acids from the surface of the meteorite are largely terrestrial contamination and that the amino acids from the interior are probably indigenous to the meteorite. Comparison to other meteorites and to lunar samples suggests a common mode of origin for these suites of amino acids.

## INTRODUCTION

The stony meteorites that contain organic compounds are called carbonaceous chondrites. A few more than 30 of this type of meteorite have been known since the fall at Alais, France, in 1806. The famous Swedish chemist Berzelius examined the organic compounds of the Alais meteorite. By 1900, several scientists had analyzed the organic compounds of carbonaceous chondrites. After a half-century recess in analytical study, Mueller (1953) isolated organic compounds of the Cold Bokkeveld meteorite. Since then, numerous chemical analyses of organic compounds in meteorites have been carried out. The results are summarized by Hayes (1967).

It would be very important for concepts of the chemical evolution of carbon compounds if amino acids were found to be indigenous to carbonaceous chondrites. Degens and Bajor (1962) found amino acids in the extracts of Murray and Bruderheim meteorites. Kaplan et al. (1963) analyzed amino acids from Orqueil and other meteorites. Anders (1964) and Vallentyne (1965) analyzed amino acids in several meteorites for possible contaminations. The meteorites used for these analytical studies were from museum specimens, and terrestrial contamination would be inevitable. Hamilton (1965) pointed out the similarity of amino acid patterns in meteorites to those of human fingerprints. Oró and Skewes (1965) also published the amino acid composition of human fingerprints. These findings indicate that most of the amino acids found in meteorites in the previous studies could be due to contamination during human handling. In addition, contamination during the fall, in the soil after the impact, and during storage or exhibition should be considered.

In 1969, two major falls of meteorites were recorded. One is the Allende carbonaceous chondrite (Type III) in Mexico on 8 February, and the other is the Murchison carbonaceous chondrite (Type II) in Australia on 28 September.

The Murchison meteorite was collected soon after the impact. The outside layer of the specimen was removed and the interior analyzed by Kvenvolden et al. (1970). The sample was extracted with hot water, and the extract was hydrolyzed. This method is similar to that (Fox et al., 1970, 1972b) developed and successfully employed in the amino acid analyses from lunar fines. The five abundant amino acids found by Kvenvolden et al. with the use of gas-liquid chromatography were glycine (6 mg/g), alanine (3 mg/g), glutamic acid (3 mg/g), valine (2 mg/g), and proline (1 mg/g). The gas chromatogram of optically active diastereomers showed that these amino acids were close to the racemic form. In addition, the presence of nonproteinous amino acids (Kvenvolden et al., 1971) suggests that these amino acids are derived from indigenous precursors. Later, Cronin and Moore (1971) analyzed the Murchison and Murray meteorites by ion exchange chromatography and found glycine, alanine, aspartic acid, glutamic acid, and other amino acids.

The Allende meteorite, as a Type III chondrite, contains less carbonaceous materials. The carbon content of the Allende meteorite was reported as 0.29% (Clarke, Jr., et al., 1970). Oró and Gelpi (1969) analyzed organic compounds in the Allende meteorite. Ponnamperuma et al. (1970) examined the N-trifluoroacetyl-n-butyl ester derivatives of the direct hydrolyzate of the Allende meteorite with 1 *N* hydrochloric acid. Cronin and Moore (1971) also analyzed amino acids in the hydrolyzed sample of a hot water extract of the Allende meteorite by the use of a sensitive amino acid analyzer. In all cases, no amino acids were identified by the method or instrument they employed. However, organic compounds such as formaldehyde were found in the Allende meteorite by Breger et al. (1972).

The failure to identify amino acids in the meteorite could have been due to the low concentration of amino acid precursors. From the results obtained by Cronin and Moore, the amino acid content of the Allende meteorite, if any, is probably less than 1/100 of that of the Murray and Murchison meteorites. Therefore, the sample preparation for amino acid analysis of the Allende meteorite must be carried out properly (direct hydrolysis should, for example, be avoided), and a much more sensitive amino acid analyzer is needed.

## MATERIALS AND PROCEDURES

We have analyzed three specimens of the Allende meteorite.

| Institutional Source | Specimen number | Weight, g |
|---|---|---|
| Smithsonian Institution | 3577 | 429 |
| Arizona State University | 818 | 459.3 |
| University of Miami | — | 104 |

These meteoritic specimens were already exposed to human hands; the surfaces were inevitably contaminated. However, it was possible to assume that, since the Allende meteorite is a Type III carbonaceous chondrite and the structure is tighter than that of other carbonaceous chondrites, the penetration of terrestrial contamination would be slower. Accepting this, if we drill the meteorite from the outside to the interior, the analytical data could indicate the indigenous amino acids or their precursors in the meteorite, depending on the distribution of quantities and species of analyzed amino acids. Such step-by-step drilling analysis was expected to be more informative than a single analysis of an interior part of the meteorite.

The description of analytical procedures (Fox et al., 1972) is that reported for the lunar amino acid analyses (Fox et al., 1970; Harada et al., 1971; Fox et al., 1972a; Fox et al., 1972b; Fox et al., 1973; Fox et al., 1976). All extracts were done in a laminar flow box in a cleanroom. Water, hydrochloric acid, and glassware were specially prepared. The meteorites were brushed thoroughly to remove dust on the surface. Then the meteorites were drilled in an area of 1.3–1.6 cm × 1.8–2.3 cm. In two cases, three layers were collected; in one case five layers of sample were subjected to analysis. The actual amount of sample in each analysis is shown in the Results section. Pictures of the drilled meteorites are shown in Figure 1. The drilled samples were extracted with three to ten times their weight of hot water under gentle reflux for 20 hours. The aqueous extracts were separated by centrifugation. Then the aqueous layers were acidified with five drops of 6 *N* hydrochloric acid, and each layer was dried under reduced pressure over sodium hydroxide (Kjeldahl grade). The dried residue was washed with 6 *N* hydrochloric acid (3 × 1 ml) into a sealed tube. The tube was evacuated under a water pump and sealed. The hydrochloric acid solution was hydrolyzed at 110°C for 24 hours in an oil bath. The sealed tube was opened and dried under reduced pressure over sodium hydroxide. The residual material was diluted with 0.01 *N* hydrochloric acid and applied to the ultrasensitive automatic amino acid analyzer (Hare 1966). Usually, hot water extractions were repeated. Overall blanks without using meteoritic samples were carried out.

(A)

**Figure 1.** Drilled Allende meteorites. A, University of Miami sample; B, Arizona State University sample; C, Smithsonian Institution sample.

(B)

(C)

## RESULTS AND DISCUSSION

An Allende meteorite (University of Miami sample) was drilled from the outside to the inside in five layers. The collected powder from each stage of the drilling was extracted with hot water three times. Amino acid compositions were determined by the use of an ultrasensitive amino acid analyzer. The amino acid content of each sample relative to the depth of the layer is shown in Figure 2. The total amino acid content from the first layer was rather high (40 nanomoles/g) but decreased rapidly as we drilled further. At a depth of 1 cm, the amino acid content was about 0.5 nanomole/g, and the level of the content was almost constant in value as

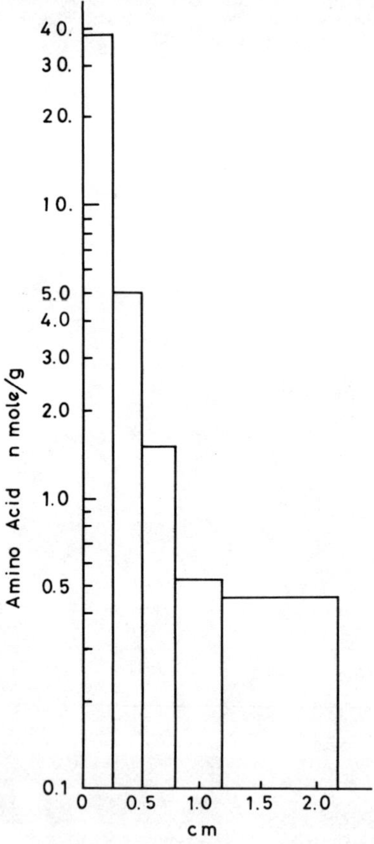


**Figure 2.** Total amino acid content of each layer relative to the depth of the sample (University of Miami).

we drilled further. These results indicate that the sample is heavily contaminated outside; however, the penetration of the contamination appears to be only 0.5–0.8 cm from the surface. The amounts of total amino acids from the second and third extractions were all less than half that of amino acids from the first and the second extractions of the drilled layers.

Table 1 shows the amino acid analyses of hydrolyzates of five layers of the first extracts of the Allende meteorite (University of Miami sample). The hydrolyzate of the extract of the first layer contains almost all the common amino acids. Glycine, glutamic acid, and serine are the constituents present in the highest proportion in the hydrolyzate on a molar basis. However, the quantities of each of the amino acids found in the first layer, except questionable β-alanine, decreased rapidly when we drilled further. But molar proportions of glycine, alanine, and β-alanine increased, and the amount and the composition of amino acids from the fourth and the fifth layers became very similar to each other. The relatively high content of serine (14.6%) in the first layer, which is characteristic of fingerprint contamination, decreased to 3–4% in the fourth and fifth layers. The increase of glycine and alanine in the third, fourth, and fifth layers is interesting because high content of glycine and alanine is a characteristic of abiotic syntheses of amino acids. From the results shown in Figure 2 and Table 1, it was possible to infer that amino acids found from the interior of the sample were derived (at least mostly) from indigenous precursors of amino acids in the meteorite.

Table 2 and Table 3 show amounts of total amino acids and molar compositions of amino acids from the samples of the Allende meteorite supplied by the Smithsonian Institution and Arizona State University. In these cases also, the first layers were contaminated and contained almost all common amino acids. However, in the second and the third layers, the amount of each amino acid decreased considerably; it attained almost constant distribution in quantity and in species. Again, the molar contents of glycine and alanine from the interior of the meteorites were increased by compensation of the decrease of other amino acids. Therefore, it is also possible to infer that at least most of the amino acids from the second and the third layers of the Smithsonian and the Arizona samples are derived from indigenous precursors of amino acids in the meteorites.

The content of amino acids from the Allende carbonaceous chondrite (Type III) was found to be low (0.2–0.5 nanomole/g) as expected. The amino acid level is close to that of the amino acids from the Apollo lunar samples (0.06–0.5 nanomole/g). The Murray and Murchison carbonaceous chondrites (Type II) contain two to three orders more of total amino acids. The relationship of the amino acid contents from the Murchison,

**Table 1. Analyses of Amino Acids from the Allende Meteorite (Miami sample), First Extract nmole/g (mole %)**

| | 1st (1.75 g) | 2nd (1.60 g) | 3rd (1.90 g) | 4th (2.55 g) | 5th (6.30 g) | Overall blank, 20 ml $H_2O$[a] |
|---|---|---|---|---|---|---|
| Asp | 2.97 ( 7.6) | 0.62 (12.0) | 0.08 ( 5.2) | 0.016 ( 3.0) | 0.010 ( 2.1) | 0.002 |
| Thr | 1.60 ( 4.1) | 0.15 ( 2.9) | 0.02 ( 1.3) | 0.003 ( 0.5) | 0.005 ( 1.0) | 0.001 |
| Ser | 5.72 (14.6) | 0.33 ( 6.4) | 0.06 ( 3.9) | 0.024 ( 4.5) | 0.014 ( 3.0) | 0.006 |
| Glu | 7.55 (19.3) | 1.00 (19.4) | 0.21 (13.8) | 0.05 ( 9.5) | 0.060 (12.9) | 0.002 |
| Pro | 1.15 ( 2.9) | 0.13 ( 2.5) | — — | — — | — — | — |
| Gly | 11.45 (29.3) | 1.50 (29.2) | 0.63 (41.4) | 0.24 (45.7) | 0.190 (41.0) | 0.015 |
| Ala | 2.75 ( 7.0) | 0.40 ( 7.7) | 0.42 (27.6) | 0.16 (30.4) | 0.085 (18.3) | 0.003 |
| ($\alpha$-$NH_2$-But)? | — — | — — | — — | 0.016 ( 3.0) | — — | — |
| Val | 1.37 ( 3.5) | 0.75 (14.6) | — — | — — | 0.003 ( 0.6) | 0.003 |
| Met | 0.02 — | — — | — — | — — | — — | 0.003 |
| Alloiso | — — | — — | — — | — — | 0.010 ( 2.1) | — |
| Iso | 1.15 ( 2.9) | 0.10 ( 1.9) | 0.02 ( 1.3) | — — | 0.003 ( 0.6) | — |

| | | | | | | | | | | | |
|---|---|---|---|---|---|---|---|---|---|---|---|
| Leu | 1.15 | ( 2.9) | 0.10 | ( 1.9) | 0.02 | ( 1.3) | — | — | 0.003 | ( 0.6) | — |
| Tyr | 0.12 | ( 0.3) | — | — | — | — | — | — | — | — | 0.003 |
| Phe | 0.46 | ( 1.1) | — | — | — | — | — | — | — | — | 0.003 |
| (β-Ala)? | 0.05 | ( 0.1) | 0.05 | ( 0.9) | 0.06 | ( 3.9) | 0.16 | (30.4) | 0.08 | (17.2) | — |
| His | — | — | — | — | — | — | — | — | — | — | — |
| unknown | — | — | 0.25 | — | 0.21 | — | 0.47 | — | — | — | — |
| Lys (Orn) | 1.37 | ( 3.5) | — | — | — | — | — | — | — | — | — |
| Amm | 20.6 | — | 11.5 | — | 11.1 | — | 10.1 | — | 15.0 | — | 50.0 |
| Arg | 0.12 | ( 0.3) | — | — | — | — | — | — | — | — | — |
| (Me-Amine)? | 0.12 | — | — | — | — | — | — | — | — | — | — |
| Total except $NH_3$ and unknown | 39.00 | | 5.13 | | 1.52 | | 0.525 | | 0.463 | | 0.041 |

[a] In each extraction, 20 ml of water was used. Therefore, contamination from the overall blank is, for example in the fifth layer, about 1/6 of the values listed in the column.

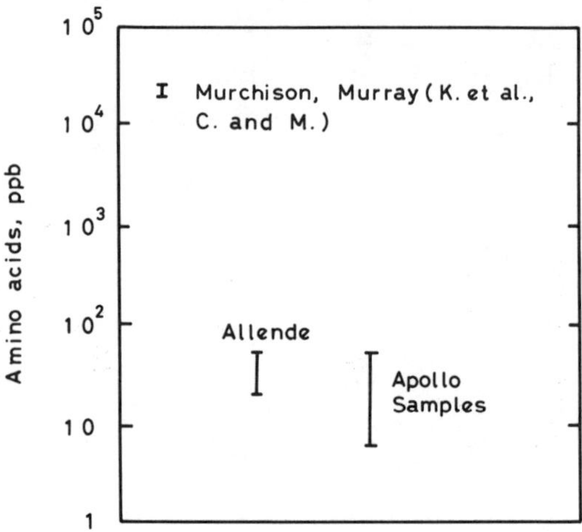


**Figure 3.** Comparison of total amount of amino acids found in the hydrolyzates of Murchison, Murray, and Allende meteorites, and Apollo samples.

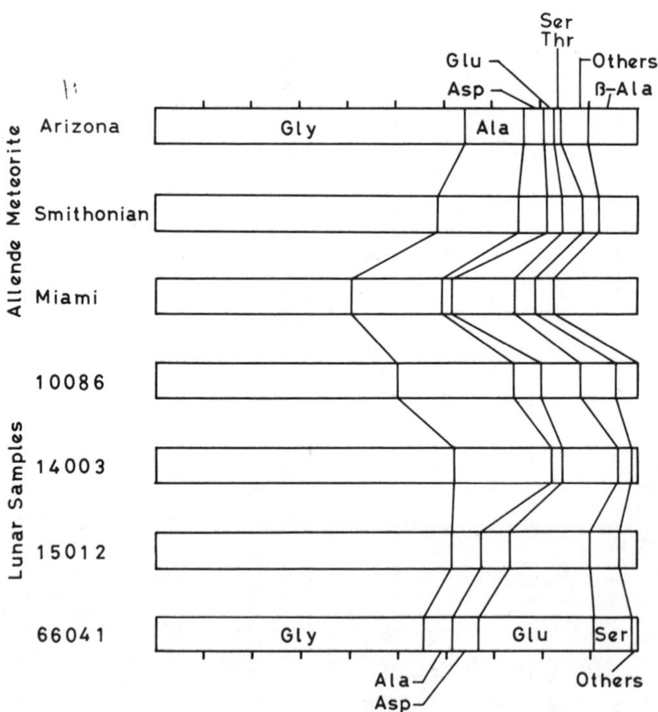


**Figure 4.** Comparison of molar composition of amino acids from samples of the Allende meteorite and the lunar samples.

**Table 2. Analyses of Amino Acids from the Allende Meteorite, First Extract (Smithsonian sample)**

| | 1st (3.90 g) | | 2nd (3.75 g) | | 3rd (6.10 g) | |
|---|---|---|---|---|---|---|
| | nmole/g | mole % | nmole/g | mole % | nmole/g | mole % |
| Asp | 0.42 | ( 8.4) | 0.010 | ( 4.5) | 0.015 | ( 5.8) |
| Thr | 0.15 | ( 3.0) | 0.004 | ( 1.8) | 0.005 | ( 1.9) |
| Ser | 0.65 | (13.0) | 0.017 | ( 7.7) | 0.006 | ( 2.3) |
| Glu | 0.85 | (17.0) | 0.021 | ( 9.6) | 0.010 | ( 3.9) |
| Pro | — | — | — | — | — | — |
| Gly | 1.80 | (36.1) | 0.10 | (45.8) | 0.15 | (58.5) |
| Ala | 0.44 | ( 8.8) | 0.025 | (11.4) | 0.045 | (17.5) |
| Cys | — | — | — | — | — | — |
| Val | 0.20 | ( 4.0) | 0.003 | ( 1.3) | 0.005 | ( 1.9) |
| Met | — | — | — | — | — | — |
| (Alloiso)? | 0.01 | ( 0.2) | 0.004 | ( 1.8) | — | — |
| Iso | 0.05 | ( 1.0) | 0.002 | ( 0.9) | — | — |
| Leu | 0.09 | ( 1.8) | 0.002 | ( 0.9) | — | — |
| Tyr | 0.04 | ( 0.8) | — | — | — | — |
| Phe | 0.07 | ( 1.4) | — | — | — | — |
| (β-Ala)? | 0.10 | ( 2.0) | 0.03 | (13.7) | 0.02 | ( 7.8) |
| His | — | — | — | — | — | — |
| Orn | 0.08 | ( 1.6) | — | — | — | — |
| Lys | 0.01 | ( 0.2) | — | — | — | — |
| Amm | 12.0 | — | 7.0 | | 9.0 | |
| Arg | 0.02 | ( 0.4) | — | — | — | — |
| Total except $NH_3$ | 4.980 | | 0.218 | | 0.256 | |

Murray, Allende meteorites and the Apollo lunar samples is shown in Figure 3.

Figure 4 shows molar compositions of amino acids from the Allende meteorite and from lunar samples. These celestial samples from different sources have unique similarity of amino acid distribution in quantity and species. In all cases, glycine, alanine, aspartic acid, glutamic acid, and serine are the major components.

These results strongly indicate that precursors of amino acids were indigenous in the meteoritic and lunar samples. The results also support the concept of cosmochemical unity in chemical evolution of matter in the Universe (Harada 1969; Fox and Dose 1972).

**Table 3. Analyses of Amino Acids from the Allende Meteorite, First Extract (Arizona sample)**

| | 1st (4.6 g) | | 2nd (4.3 g) | | 3rd (7.6 g) | |
|---|---|---|---|---|---|---|
| | nmole/g | mole % | nmole/g | mole % | nmole/g | mole % |
| Asp | 0.68 | (16.9) | 0.017 | ( 5.2) | 0.008 | ( 4.0) |
| Thr | 0.17 | ( 4.0) | 0.005 | ( 1.5) | 0.001 | ( 0.5) |
| Ser | 0.19 | ( 4.5) | 0.011 | ( 3.3) | 0.002 | ( 1.0) |
| Glu | 0.61 | (15.1) | 0.023 | ( 7.0) | 0.005 | ( 2.5) |
| Pro | — | — | — | — | — | — |
| Gly | 1.25 | (29.7) | 0.18 | (55.5) | 0.13 | (65.0) |
| Ala | 0.51 | (12.6) | 0.018 | ( 5.5) | 0.024 | (12.0) |
| Cys | — | — | — | — | — | — |
| Val | 0.18 | ( 4.4) | — | — | — | — |
| Met | — | — | — | — | — | — |
| (Alloiso)? | 0.06 | ( 1.4) | 0.02 | ( 6.1) | 0.01 | ( 5.0) |
| Iso | 0.04 | ( 0.9) | — | — | — | — |
| Leu | 0.08 | ( 1.9) | — | — | — | — |
| Tyr | — | — | — | — | — | — |
| Phe | — | — | — | — | — | — |
| (β-Ala)? | 0.25 | ( 6.2) | 0.05 | (15.4) | 0.02 | (10.0) |
| His | — | — | — | — | — | — |
| Orn | — | — | — | — | — | — |
| Lys | — | — | — | — | — | — |
| Amm | 14. | | 17. | | 8.5 | |
| Arg | — | — | — | — | — | — |
| Total except $NH_3$ | 4.020 | | 0.324 | | 0.200 | |

## *ACKNOWLEDGMENTS*

This study was supported by NASA grants NGR 10-007-088 and NGR 10-007-052. We thank Dr. S. W. Fox for his discussion and help, Dr. I. A. Breger for valuable suggestions, Dr. R. S. Clark, Jr., and Dr. C. B. Moore for arrangement of meteoritic samples.

## *REFERENCES*

Anders, E., E. R. Dufresne, R. Hayatsu, A. Dufresne, A. Cavaille, and F. W. Fitch, Contaminated meteorite, *Science,* **146,** 1157–1161, 1964.

Breger, I. A., P. Zuboviv, and J. C. Chandler, Occurrence and significance of formaldehyde in the Allende carbonaceous chondrite, *Nature,* **236,** 155, 1972.

Clarke, Jr., R. S., E. Jarosewich, B. Mason, J. Nelen, M. Gómez, and J. R. Hyde, The

Allende, Mexico, meteorite shower, *Smithsonian Institution Press,* Washington, D.C., 1970.

Cronin, J. R., and C. B. Moore, Amino acid analyses of the Murchison, Murray, and Allende carbonaceous chondrites, *Science,* **172,** 1327, 1971,

Degens, E. T., and M. Bajor, Amino acids and sugars in the Bruderheim and Murray meteorites, *Naturwiss.,* **49,** 605, 1962.

Fox, S. W., K. Harada, P. E. Hare, G. Hinsch, and G. Mueller, Bio-organic compounds and glassy microparticles in lunar fines and other material, *Science,* **167,** 767, 1970.

Fox, S. W., K. Harada, and P. E. Hare, Amino acid precursors in lunar fines from apollo 14 and earlier missions, Supplement 3, *Geochim. Cosmochim. Acta,* **2,** 2109, MIT, Cambridge, 1972a.

Fox, S. W., K. Harada, and P. E. Hare, Amino acid precursors in lunar samples, *Space Life Science,* **3,** 425–531, 1972b.

Fox, S. W., and K. Dose, *Molecular Evolution and the Origin of Life,* Freeman, San Francisco, 1972.

Fox, S. W., K. Harada, and P. E. Hare, Accumulated analyses of amino acid precursors in returned lunar samples, proceedings of the fourth Lunar Science Conference, Supplement 4, *Geochim. Cosmochim. Acta,* **2,** 2241–2248, 1973.

Fox, S. W., K. Harada, and P. E. Hare, Amino acid precursors in lunar fines: Limits to the contribution of jet exhaust, *Geochim. Cosmochim. Acta,* **40,** 1069–1071, 1976.

Hamilton, P. E., Amino acids on hands, *Nature,* **205,** 284–285, 1965.

Harada, K., The origin of organic compounds in meteorites, *Protein, Nucleic Acid, Enzyme,* **14,** 1197–1201, 1969.

Harada, K., P. E. Hare, C. R. Windsor, and S. W. Fox, Evidence for compounds hydrolyzable to amino acids in aqueous extracts of lunar fines from the Apollo 11 and Apollo 12 lunar fines, *Science,* **173,** 433, 1971.

Hare, P. E., Automatic multiple column amino acid analysis, the use of pressure elution in small bore ion-exchange columns, *Fed. Proc.,* **25,** 709, 1966.

Hayes, J. M., Organic constituents of meteorites, *Geochim. Cosmochim. Acta,* **31,** 1395, 1967.

Kaplan, I. R., E. T. Degens, and J. H. Reuter, Organic compounds in stony meteorites, *Geochim. Cosmochim. Acta,* **27,** 805–834, 1963.

Kvenvolden, K., J. Lawless, K. Pering, E. Peterson, J. Flores, C. Ponnamperuma, I. R. Kaplan, and C. Moore, Evidence for extraterrestrial amino-acids and hydrocarbons in the Murchison meteorite, *Nature,* **228,** 923–926, 1970.

Kvenvolden, K. A., J. G. Lawless, and C. Ponnamperuma, Non-protein amino acids in the Murchison meteorite, *Proc. Nat. Acad. Sci.,* **68,** 486–490, 1971.

Mueller, G., The properties and theory of genesis of the carbonaceous complex within the Cold Bokkeveld meteorite, *Geochim. Cosmochim. Acta,* **4,** 1–10, 1953.

Oró, J., and H. B. Skewes, Free amino-acids on human fingers: The question of contamination in microanalysis, *Nature,* **207,** 1042–1045, 1965.

Oró, J., and E. Gelpi, Organic analysis of the Pueblito de Allende meteorite, *Meteoritics,* **4,** 287, 1969 (abstract).

Ponnamperuma, C., et al., Search for organic compounds in the lunar dust from the Sea of Tranquillity, *Science,* **167,** 760–762, 1970.

Vallentyne, J. R., Two aspects of the geochemistry of amino acids, in *The Origin of Prebiological Systems and of their Molecular Matrices,* S. W. Fox, ed., Academic Press, New York, 1965.

# Enantiomeric Ratios of Amino Acids in Southern-Ocean Siliceous Oozes

**DETLEF A. WARNKE**
**DAVID J. BLUNT**
*Department of Geological Sciences, California State University, Hayward*
**GLENN E. POLLOCK**
*NASA-Ames Research Center, Moffett Field, California*

---

Amino acids contained in the tests of planktonic (and benthonic) microorganisms undergo complex racemization reactions after the death of the organisms. Unless the tests are dissolved after settling on the sea floor, they become part of the sediment, either as an admixture to other sedimentary components or exlusively as biogenic sediment (ooze). Such sediments therefore should show increasing enantiomeric ratios (D/L) of a given amino acid with time. These relationships and attendant racemization reactions in marine sediments have been studied by Kvenvolden et al. (1970, 1973), Bada et al. (1970), Bada and Man (1973), Wehmiller and Hare (1971), and Bada and Schroeder (1972). The kinetics of amino acid racemization have been studied in several ways and the subject has been reviewed by Bada and Schroeder (1972), Schroeder and Bada (1976), Kvenvolden (1975), Kvenvolden et al. (1973), Dungworth (1976), and Wehmiller et al. (1976). Williams and Smith (1977), urged great caution in the application of amino-acid racemization research to geologic problems.

Although much of the earlier work was based on the assumption of first-order reversible kinetics, it became clear that such models could not completely explain racemization reactions in marine sediments, specifically in calcareous foraminifera. The reason for this fact lies in the complexity of pathways which protein follows during diagenesis. Also, there are species effects as shown by King and Hare (1972) and King and Neville (1977). Therefore, racemization reactions can only be used as paleothermometric devices or indicators of absolute age if the kinetics are well understood, or if racemization rates have been established through calibration with independently available age and temperature data.

The composition of preserved amino acids in fossil Radiolaria and in other siliceous organisms has been investigated (King 1974, 1975, 1977). The present report deals with the extent of amino acid racemization in siliceous oozes from cores taken in the Southern Ocean east of the Scotia Arc, particularly core I0 07-75-25.

## EXPERIMENTAL

Core I0 07-75-25 was obtained from a depth of 5014 m at lat. 56°34.7′S, long. 20°17.2′W during *Islas Orcadas* Cruise 07-75 (Warnke et al., 1976). The temperature in the overlying water column is −0.2°C. The entire core is normally magnetized (A. Federman, Oregon State University, personal communication), and belongs to the *Coscinodiscus lentiginosus* Zone (P. Ciesielski, Ohio State University personal communication). It follows that the bottom of the core is less than 690,000 years old.Seven intervals in the core were sampled.

Analytical Procedure:

Approximately 20 g of wet sediment were placed in a polypropylene tube into which 40 ml of distilled water were added. This mixture was shaken on a reciprocating shaker for 1 hour followed by centrifugation. The distilled-water wash was discarded. This procedure was repeated twice to wash interstitial water from the sediment. After centrifugation and removal of the water, the weight was reduced to 10–15 g. To break down the silicate matrix, 40 ml 6 *N* HF were added, and the tube was shaken on a reciprocating shaker for 20 hours. The HF was removed on a rotary evaporator and 40 ml of 6 *N* HCl were added. The tube was placed in an oil bath at 110°C, capped, and allowed to hydrolyze for 20 hours. Following hydrolysis the hydrolysate was isolated from the sediment by centrifugation and removed. Two additional rinses of the sediment with 6 *N* HCl were added to the hydrolysate, and the mixture was evaporated to dryness.

Removal of cations was accomplished on a 20-ml column of Biorad AG 50W-X8 ($H^+$) cation-exchange resin. Amino acids were eluted with 2 *N* $NH_4OH$ and evaporated to dryness.

Amino acid extracts were derivatized as N-pentafluoropropionyl-(+)-(2)-butyl esters and were resolved on stationary phases of Carbowax 20 M and UCON 75H-90,000 by gas chromatography. Amino acid D-L ratios were calculated by peak height and were corrected for possible hydrolysis-induced racemization based on the ribonuclease control of Pollock et al. (1977). These data reflect the average of measurements between the two stationary phases. Where interfering substances did not permit such a measurement, a well-resolved amino acid from a single column is reported.

## RESULTS AND DISCUSSION

The results of these analyses are plotted in Figure 1. Five amino acids could be measured. The D-L ratios for leucine, proline, glutamic acid, and aspartic acid are low in the near-surface sediment, whereas alanine is significantly higher. These five amino acids show increasing extent of racemization to the bottom of the core. These data describe remarkably linear trends with depth, perhaps reflecting rather uniform proportions

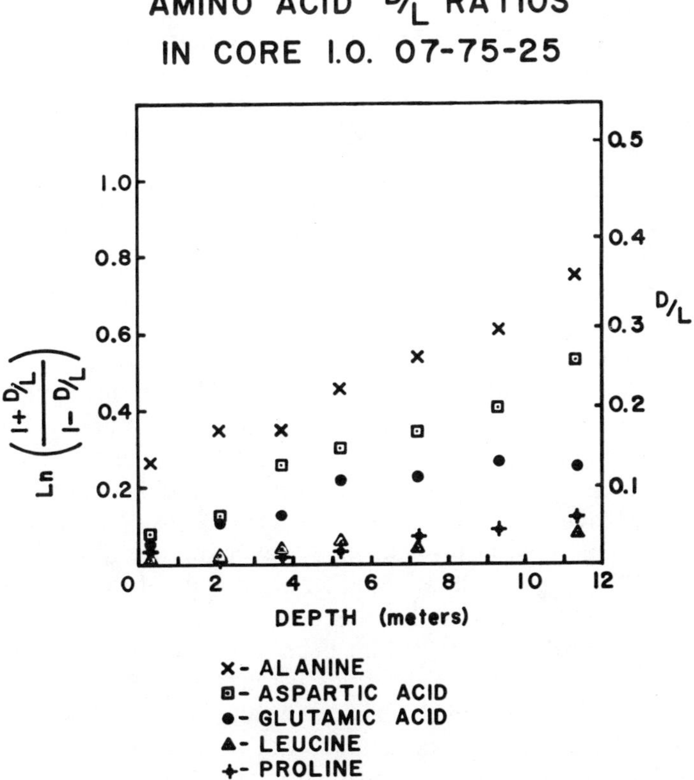


**Figure 1.** D-L ratios of five amino acids in Core I0 07-75-25, raised east of the Scotia Arc.

of the various siliceous components of this core. The scatter reflects analytical uncertainty rather than changes in rates of sedimentation (which are indicated by layers of micro manganese nodules) because such changes would affect all amino acids in the same sense. Instead, within the limits imposed by the number of data points, the evidence points to *overall* uniform sedimentation over the time span represented in this core. Linearity of the extent of racemization with depth (i.e. time) suggests that reversible first-order kinetics of amino acid racemization in a siliceous matrix may be operative within the time-span represented in this core(<0.69 m.y.), and in the prevailing effective temperature range (~0°C). It is noteworthy that the intercepts on the y-axis for alanine, aspartic acid, and glutamic acid are higher than zero, despite the application of all available correction factors connected with the methodology. These relationships are not exceptional, because the surface or near-surface sediments of cores I0 07-75-27 and 29, also raised east of the Scotia Arc, show the same pattern (Figure 2). It seems that some reworking by benthic organisms or erosion of surface material has also occurred (see particularly core I0 07-75-29). Still, the general pattern is the same.

Dungworth et al. (1977) described similar relationships for a Lake Ontario core. They explained the presence of substantial abundances of D-glutamic acid and D-alanine in surface sediments as caused by bacterial degradation. Pollock et al. (1977) found very significant amounts of D-glutamic acid, D-alanine, and D-aspartic acid in contemporary soils and stated that these D-amino acids are probably contributed to the soils primarily via microorganisms. Our data show alanine to have the highest percentage of D-isomers in contrast to Pollock et al., (1977), who report aspartic acid as highest. In a study of dissolved amino acids in the marine water column, Lee and Bada (1977) determined that D-alanine was the relatively highest D-enantiomer (of the amino acids with high percentages of D-enantiomers: alanine, glutamic acid, and aspartic acid). These authors also attributed the "high" D-L ratios of those amino acids to microorganisms (bacteria). These relationships indicate that marine bacteria may play a decisive role as a source for high D-L ratios of alanine, aspartic acid, and glutamic acid in the "surface" layer of core I0 07-75-25. Lee and Bada (1977) suggested that significant amounts of the D-enantiomer of aspartic acid may occur in diatoms and other algae. This finding certainly warrants further investigation because it would at least partially explain "high" D-L aspartic acid ratios in surface or near-surface materials. In any event, the high D-L ratios for alanine and glutamic acid in the near-surface sediment point to bacterial contributions. Other explanations, such as the decomposition reaction of serine to alanine (see Val-

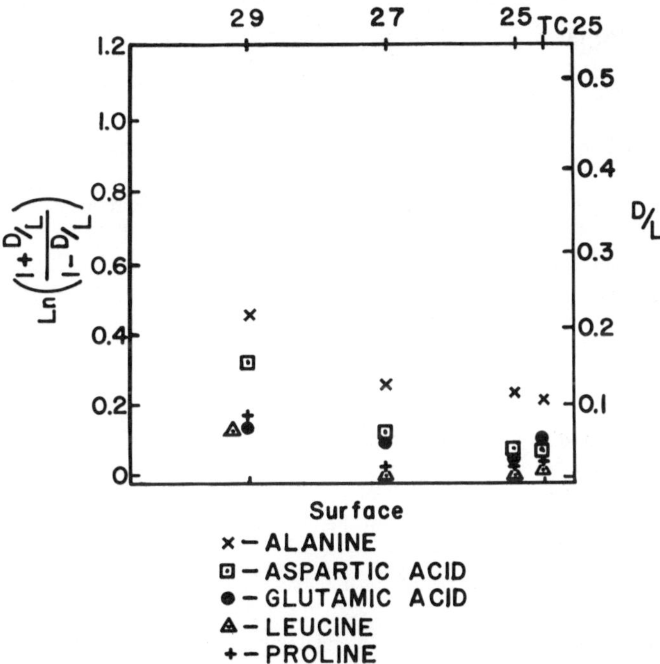


**Figure 2.** D-L ratios of five amino acids in surface or near-surface samples from piston cores raised east of the Scotia Arc (10-07-75-25, 27, 29) The arrangement in this diagram indicates schematically the increasing distance from the Scotia Arc. The sample labeled *TC* is from the 18 cm level of the trigger core of Core 25.

lentyne, 1964) are possible but seem to be less plausible because of the low temperatures and the short time span involved.

Finally, the relative order of rates of racemization for the five amino acids involved is not the same as the relative order reported for foraminiferal materials, as described for instance by Kvenvolden et al. (1973). In particular, proline has the second highest extent of racemization in core V12-122, and it has the highest or second highest extent in core V23-110, of the eight amino acids analyzed by Kvenvolden et al. (1973). In contrast, this study shows proline to have the lowest or second lowest extent of racemization. This contrast may stem from differences in protein composition between siliceous and calcareous organisms.

## *ACKNOWLEDGMENTS*

We gratefully acknowledge assistance through NASA-Ames Joint Research Interchanges Nr. NCA2-OR290-601 and NCA2-OR290-704, and through National Science Foundation Grant Nr. DPP76-01005-A01.

## *REFERENCES*

Bada, J. L. and R. A. Schroeder, Racemization of isoleucine in calcareous marine sediments: Kinetics and mechanism, *Earth Planet. Sci. Lett.*, **15,** 1–11, 1972.

Bada, J. L. and E. H. Mann, Racemization of isoleucine in cores from Leg 15, Site 148, in *Initial Reports of the Deep Sea Drilling Project,* Vol. 20, pp. 947–951, 1973.

Bada, J. L. B. P. Luyendyk, and J. B. Maynard, Marine Sediments: Dating by the racemization of amino acids, *Science,* **170,** 730–732, 1970.

Dungworth, Graham, Optical configuration and the racemization of amino acids in sediments and in fossils—a review, *Chem. Geol.*, **17,** 135–153, 1976.

Dungworth, G., M. Thijssen, J. Zuurveld, W. Van Der Velder, and A. W. Schwartz, Distribution of amino acids, amino sugars, purines and pyrimidines in a Lake Ontario sediment core, *Chem. Geol.*, **19,** 295–308, 1977.

King, K., Jr., Preserved amino acids from silicified protein in fossil Radiolaria, *Nature,* **252,** 690–692, 1974.

King, K., Jr., Amino acid composition of the silicified organic matrix in fossil polycystine Radiolaria, *Micropaleontol.*, **21,** 215–226, 1975.

King, K., Jr., Amino acid survey of recent calcareous and siliceous deep-sea microfossils, *Micropaleontol.*, **23,** 180–193, 1977.

King, K., Jr. and P. E. Hare, Species effect in the epimerization of L-isoleucine in fossil planktonic foraminifera, *Carnegie Inst. Wash. Year Book 71,* 596–598, 1972.

King, K., Jr. and C. Neville, Isoleucine epimerization for dating marine sediments: Importance of analyzing monospecific foraminiferal samples, *Science,* **195,** 1333–1335, 1977.

Kvenvolden, K. A., Advances in the geochemistry of amino acids, *Ann. Rev. Earth Planet. Sci.*, **3,** 183–212, 1975.

Kvenvolden, K. A., E. Peterson, and F. S. Brown, Racemization of amino acids in sediments from Saanich inlet, British Columbia, *Science,* **169,** 1079–1082, 1970.

Kvenvolden, K. A., E. Peterson, J. Wehmiller, and P. E. Hare, Racemization of amino acids in marine sediments determined by gas chromatography, *Geochim. Cosmochim. Acta,* **37,** 2215–2225, 1973.

Lee, C. and J. L. Bada, Dissolved amino acids in the equatorial Pacific, the Sargasso Sea, and Biscayne Bay, *Limol. Oceanog.*, **22,** 502–510, 1977.

Pollock, G. E., C-N Cheng, and S. E. Cronin, Determination of the D and L isomers of some protein amino acids present in soils, *Anal. Chem.*, **49,** 2–7, 1977.

Schroeder, R. A. and J. L. Bada, A review of the geochemical application of the amino acid racemization reaction, *Earth Sci. Rev.*, **12,** 347–391, 1976.

Vallentyne, J. R., Biogeochemistry of organic matter—II. Thermal reaction kinetics and transformation products of amino compounds, *Geochim. Cosmochim. Acta,* **28,** 157–188, 1964.

Warnke, D. A., P. Bruchhausen, J. LaBreque, P. F. Ciesielski, and A. Federman, ARA *Islas Orcadas* cruise 7. *Antarctic J.U.S.,* **11,** 70–73, 1976.

Wehmiller, J. F. and P. E. Hare, Racemization of amino acids in marine sediments, *Science,* **173,** 907–911, 1971.

Wehmiller, J. F., P. E. Hare, and G. A. Kujala, Amino acid in fossil corals: Racemization (epimerization) reactions and their implications for diagenetic models and geochronological studies, *Geochim. Cosmochim. Acta,* **40,** 763–776, 1976.

Williams, K. M. and G. G. Smith, A Critical evaluation of the application of amino acid racemization to geochronology and geothermometry, *Origin of Life,* **8,** 91–144, 1977.

# II

# TECHNIQUES AND METHODS

Biogeochemistry is completely dependent upon developments in chromatography and instrumental methods of analysis. The concentration of amino acids and peptides in geological materials is low, and sensitive analytical methods are needed. The earliest experiments in 1953 were made by paper chromatography, a technique that was then fairly new. Quantitative ion exchange chromatography soon became the preferred analytical method.

There has been a pattern in the development of techniques and methods. Procedures have been taken from the organic chemical and biochemical literature, adapted to the special needs of biogeochemistry, and then the sensitivity of the technique pushed to its limit. The analysis of amino acids by liquid chromatography is an excellent example. We are now witnessing a cycling of this ultrasensitive technique back into biochemistry.

Similarly, the gas chromatographic technique for the analysis of amino acid optical isomers is being developed rapidly by biogeochemists. Their methods are already being adopted by biochemists.

Research on isolating protein and peptides from fossils, characterizing them by gel electrophoresis and determining their amino acid sequence is just beginning. There are great opportunities to increase our knowledge of paleobiochemistry. The techniques are already available.

One indication that the field of amino acid biogeochemistry is maturing is the concern with comparability of results from different laboratories. A suite of calibration samples is being circulated for analysis. Systematic differences in analytical results for amino acid content and optical isomer content have been noted. Eventually the causes of these biases will be discovered and the precision of measurement will be increased.

# The Organic Constituents of Fossil Mollusc Shells

**THOMAS C. HOERING**
*Geophysical Laboratory, Carnegie Institution of Washington, Washington, D. C.*

***ABSTRACT***

The nature of organic substances in fossil mollusc shells has been examined. A suite of low-molecular-weight acids, corresponding to deamination products of amino acids, has been detected in Miocene-aged shells. Humic acid indigenous to the shells has been isolated. Carbon isotope analysis, amino acid compositions and optical isomer analysis show that the humic material has not migrated into the shell. Attempts to identify peptides from the partial hydrolysate of the insoluble residue of Pleistocene-aged molluscs have been unsuccessful.

The results indicate that diagenetic pathways of amino acids in fossils are complex. Quantitative treatment of amino acid compositions for geochronological purposes must take into account multiple pathways.

A number of studies have been carried out on the amino acid composition of fossil mollusc shells. Identifiable amino acids contribute only a minor fraction of the organic matter in a fossil, and the rest has not been well characterized. In the work described here, a wider range of organic substances has been studied. Molecules have been discovered that have a bearing on the understanding of organic diagenesis.

Fossil mollusc shells were chosen for this work because they are a conspicuous feature in the fossil record and their dense calcite or aragonite matrix provides a tight, inert medium for preserving organic matter. In addition, the biochemistry of the living species gives a base line for extrapolating back in time.

Figure 1 shows a simplified outline of possible pathways for the diagenesis of organic matter in mollusc shells that have been preserved at low temperatures. All living matter contains varying proportions of three major classes of biochemicals: lipids, proteins, and carbohydrates. Mollusc shells have a high concentration of the proteins. The three classes have a common property—in the presence of water, they undergo hydrolysis (reaction I) to simpler "building blocks." The hydrolysis products are chemically reactive and unstable, and over geological lengths of time they are transformed into more stable entities.

All proteins are composed predominantly of 20 amino acids whose stabilities vary greatly. Some disappear rapidly; some can persist for great lengths of time. The work described here has disclosed one pathway by which amino acids disappear, the deamination (reaction II) to aliphatic acids. Exposure to higher temperatures for longer periods of time apparently causes the decarboxylation of the aliphatic acids (reaction III) to yield the low-molecular-weight hydrocarbons found in fossils (Thompson and Creath 1966).

There are other possible pathways for the transformation of amino acids. The organic molecules in a living cell are reactive and can interact with each other upon the death of the cell. Sugars are particularly reactive, although they appear to be a minor component in molluscan shells. Previous work in the laboratory (Hoering 1973) showed that amino acids and sugars react rapidly (reaction IV) to form a substance called melanoidin that has many of the properties of natural humic acid found in soils and

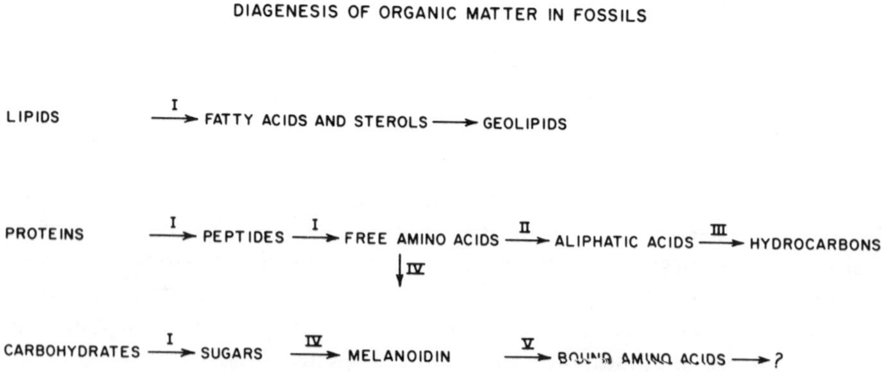

DIAGENESIS OF ORGANIC MATTER IN FOSSILS

LIPIDS —I→ FATTY ACIDS AND STEROLS → GEOLIPIDS

PROTEINS —I→ PEPTIDES —I→ FREE AMINO ACIDS —II→ ALIPHATIC ACIDS —III→ HYDROCARBONS

↓IV

CARBOHYDRATES —I→ SUGARS —IV→ MELANOIDIN —V→ BOUND AMINO ACIDS → ?

**Figure 1.** Diagenetic pathways for organic matter in fossil mollusc shells.

sediments. Indigenous humic acid has been isolated from fossil mollusc shells and partially characterized. This material adds a complication to amino acid studies because it liberates free amino acids on hydrolysis (reaction V) at different rates than does the hydrolysis of peptides. The stability of amino acid residues bound in humic acid is not known.

The following molluscan shells were studied in this work: (1) Live specimens of *Mercenaria mercenaria, Mytilus edulis, Ostrea virginica,* and *Mya arenacia* were obtained locally from commercial suppliers. (2) Fossil *Mercenaria mercenaria* was collected from the Pleistocene Pamlico formation at Wailes Bluff, Maryland. (3) *Mercenaria mercenaria, Isognomon maxillata,* and *Chlamys madisoniensis* were collected from the Miocene Choptank formation at Drum Cliffs, Jones Wharf, Maryland. The fossils are found in marine terrace deposits, and they are notable for their abundance and excellent state of preservation.

Contamination by younger organic matter is an ever-present problem, and great care must be taken in the selection and cleaning of fossils. The shells were given a preliminary cleaning and then crushed into coarse fragments. The material fractures along cracks and weathered surfaces or where the shell has been penetrated by boring organisms. The surfaces of the fragments were abraded by tumbling in a porcelain ball mill for several hours. Loosely cemented material was removed, and solid, dense particles were selected. They were etched in cold, dilute hydrochloric acid until 10–15% of their weight was removed. The etched fragments were cleaned further by extraction with distilled water in an ultrasonic tank.

About 200 g of cleaned shell fragments was dissolved by slowly dripping redistilled, constant boiling hydrochloric acid on them in an ice bath. About 24 hr is required for dissolution. A strong, fetid odor of organic acids was easily detectable above the solution of the fossils. The solution and the insoluble residue were extraced for 16 hr with diethyl ether in a continuous liquid-liquid extractor. The ether extract was dried over anhydrous sodium sulfate, and its volume was reduced by vacuum distillation at low temperature. The extract was analyzed by gas chromatography and mass spectrometry.

The mixture remaining in the extractor was centrifuged. The solution was separated, demineralized, and evaporated. This fraction was examined for partially hydrolyzed peptides.

The insoluble residue was extracted with a dilute solution of sodium pyrophosphate and sodium hydroxide. The resulting dark-brown extract yielded a flocculent precipitate of humic acid when acidified with hydrochloric acid. The humic acid was purified and freeze-dried.

Several classes of compounds were detected in the ether extract of the fossils. The most striking one is illustrated by the gas chromatogram in Figure 2. The low-molecular-weight acids in fossils are exactly those expected from the deamination of simple amino acids. Molecular structures of the compounds involved in the deamination reaction are illustrated in Figure 3. The quantitative analysis for some of the acids is shown in Table 1. Such low-molecular-weight acids are not present in live shells.

The fatty acid content of live shell is low; only about 10–50 ppm of the shell exists, mainly as myristic, palmitic, oleic, and stearic acid. In fossil shell, the unsaturated fatty acids have completely disappeared, and the concentration of the saturated acids is about a factor of 20 lower.

An unsuccessful attempt was made to isolate peptides from the insoluble residue of Pleistocene-aged fossils and to determine amino acid sequences in their partial hydrolysates. A gas chromatographic–mass spectrometric procedure was used (Nau 1974; Nau and Biemann 1976). The results, so far, have been inconclusive because there are serious analytical difficulties when the biochemical procedure was applied to a geological material. A dark, polymeric organic material, probably related to the humic acids, interfered. Exclusion chromatography on Sephadex G-25

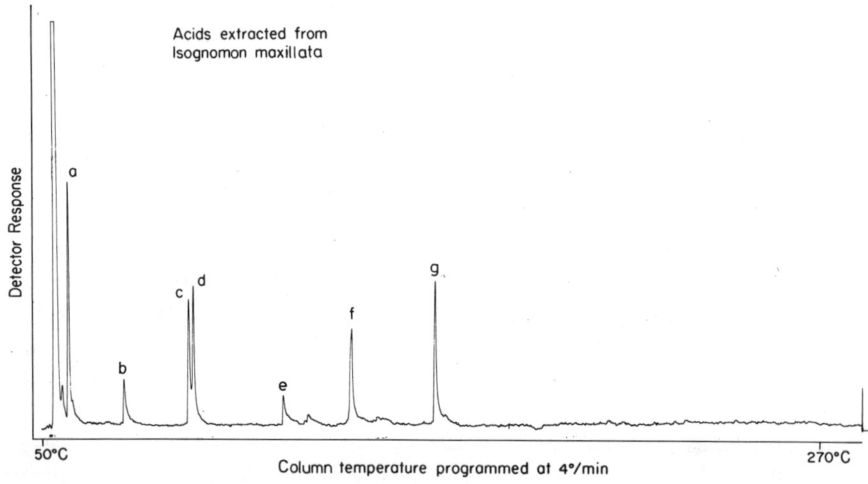


**Figure 2.** Gas chromatogram of the methyl esters of low-molecular-weight acids extracted from the shell of Miocene *Isognomon maxillata*. The chromatogram was obtained using a 150 ft × 0.020 inch i.d., porous-layer, open-tubing column coated with OV-101. The column temperature was programmed linearly from 50° to 250°C at a rate of 4°/min. Identification of the compounds in the lettered peaks was made by gas chromatography–mass spectrometry using the same column. They are as follows: (*a*) butyric acid methyl ester, (*b*) 3-methyl butyric acid methyl ester, (*c*) 3-methyl valeric acid methyl ester, (*d*) 4-methyl valeric acid methyl ester, (*e*) succinic acid dimethyl ester, (*f*) glutaric acid dimethyl ester, (*g*) phenyl propionic acid methyl ester.

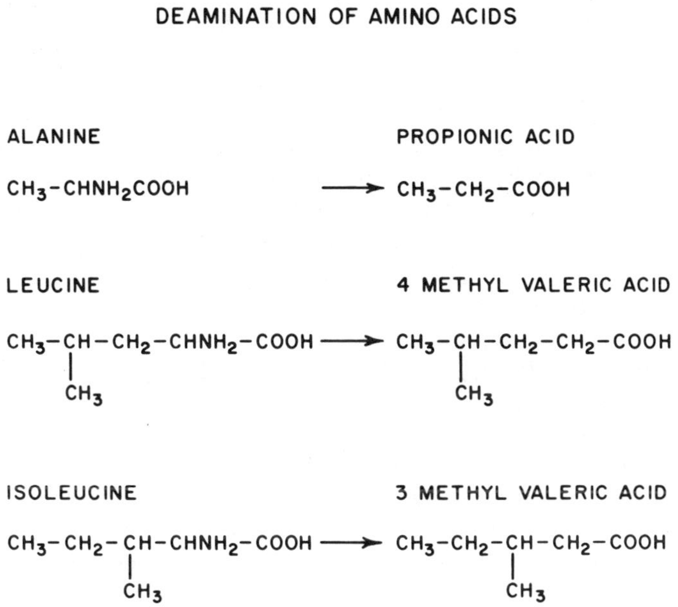

DEAMINATION OF AMINO ACIDS

| | | |
|---|---|---|
| ALANINE | | PROPIONIC ACID |
| $CH_3-CHNH_2COOH$ | $\longrightarrow$ | $CH_3-CH_2-COOH$ |
| LEUCINE | | 4 METHYL VALERIC ACID |
| $CH_3-CH(CH_3)-CH_2-CHNH_2-COOH$ | $\longrightarrow$ | $CH_3-CH(CH_3)-CH_2-CH_2-COOH$ |
| ISOLEUCINE | | 3 METHYL VALERIC ACID |
| $CH_3-CH_2-CH(CH_3)-CHNH_2-COOH$ | $\longrightarrow$ | $CH_3-CH_2-CH(CH_3)-CH_2-COOH$ |

**Figure 3.** Products of the deamination of common amino acids. Other amino acid–product pairs are as follows: glycine → acetic acid; valine → 3-methyl butyric acid; glutamic acid → succinic acid; aspartic acid → glutaric acid; phenylalanine → phenyl propionic acid.

medium effectively separates the humic materials from peptides, but relatively little of the insoluble residue from fossils was hydrolyzed to peptides under conditions applicable to proteins. Experiments are under way to isolate soluble peptides from the aqueous solution after dissolution of the fossils. The ability to determine the amino acid sequence in partially

**Table 1. Concentration of Some Low-Molecular-Weight Acids in Miocene Fossils**[a]

| | *Isognomen maxillata* | *Mercenaria mercenaria* |
|---|---|---|
| Propionic acid | 15.0 | 11.2 |
| Isobutyric acid | 0.5 | 0.7 |
| Butyric acid | 2.8 | 2.5 |
| 3-Methyl butyric acid | 7.5 | 6.7 |
| 3-Methyl valeric acid + 4-methyl valeric acid[b] | 17.8 | 9.8 |

[a] The units of concentration are in parts per million by weight of the fossil shell.
[b] The gas chromatographic analyses were made with a 10 ft × 1/8 inch column packed with Poropak QS (Waters Associates, Framington, Mass.). The 3-methyl and 4-methyl valeric acids are not resolved and are summed together.

hydrolyzed fossil peptides could add a new dimension to amino acid biogeochemistry.

The presence of humic material as a major component in fossil shells is unexpected. The possibility that this material represents contamination by organic matter in percolating ground waters was examined by measuring the stable carbon isotope ratios in solvent-extracted, insoluble organic matter from shells. Molluscs are marine organisms that derive their organic carbon from other marine organisms, whereas the organic matter in ground waters has been derived primarily from terrestrial plants. They are systematic differences in the $^{13}C/^{12}C$ of marine and terrestrial organic matter that are reflected in the humic acids from either source (Parker and Hedges 1976; Hoering 1974). The data in Table 2 show that the $^{13}C$ content of fossil humic acid is comparable with that in the organic matter of live shells and falls in the range expected for marine humic acid. On this scale, terrestrially derived organic matter would fall in the range $-22$ to $-26$. The humic acid is therefore probably indigenous and has not migrated into the shells. In addition, results from the acids described above show that there has not been a large-scale migration of organic

**Table 2. The $^{13}C$ Content of Solvent-Extracted, Insoluble Organic Matter in Live Mollusc Shells and of Humic Acid from Miocene Fossils**

| Sample | $\delta^{13}C$ |
|---|---|
| Living specimens: | |
| *Mercenaria mercenaria* | −16.20[a] |
| *Mya arenacia* | −20.50 |
| *Ostrea virginica* | −18.09 |
| *Mytilus* | −15.16 |
| Fossil humic acid: | |
| *Mercenaria mercenaria* | −19.58 |
| *Isognomen maxillata* | −18.83 |

[a] The units of $^{13}C$ are defined as follows:

$$\delta^{13}C = \frac{(^{13}C/^{12}C)_x - (^{13}C/^{12}C)_s}{(^{13}C/^{12}C)_s} \times 1000,$$

where the subscript $x$ refers to the unknown and the subscript $s$ refers to the standard. The standard used in this work is NBS-20. On this scale, the PDB standard is $+1.08$ and the NBS-22 standard is $-28.33$.

**Table 3. Analyses of Humic Acid from Miocene Shells[a]**

| | Humic Acid Recovered (wt %) | C/N | Nitrogen from Humic Acid (ppm shell) | Percentage of Total N in Shell |
|---|---|---|---|---|
| *Mercenaria* | 0.02 | 10 | 2 | 10 |
| *Isognomon* | 0.09 | 9 | 10 | 25 |
| *Chlamys* | 0.006 | 8 | 0.5 | 1.5 |
| Soil surrounding fossils | 0.006 | — | — | — |

[a] The nitrogen analyses were made by two methods: (1) analysis using a carbon, nitrogen, hydrogen analyzer, Hewlett-Packard model 185; (2) Micro-Kjeldahl digestion followed by analysis of ammonia on a short ion-exchange column with detection by ortho-phthalaldehyde as used for amino acids.

matter out of the shells. Fossil mollusc shells probably form a closed system with regard to their organic matter.

The distribution and fate of nitrogen-containing organic matter were investigated by comparing the nitrogen and amino acid content of the total fossil shell and the humic acid derived from them. Earlier work (Hare, 1969) has shown that the amino acid content of *Mercenaria* shells drops from 0.15 weight percent for Recent shells, to 0.06 for Pleistocene shells, and to less than 0.01 for Miocene shells. The degree of racemization of the individual amino acids increases from near zero for the Recent, to about 25 percent for Pleistocene, and to nearly 100 percent for Miocene (see Table 4). In addition, two classes of amino acids can be recognized in fossils: (1) free molecules and (2) bound acids that are assumed to be in the form of peptides. The percentage of free amino acids increases from zero in the Recent shells to over 95 percent in the Miocene shells.

Table 3 presents the analyses for humic acid from Miocene fossils. There is more than an order of magnitude variation in the total amount recovered from the three species. The material averaged around one per-

**Table 4. Alloisoleucine-to-Isoleucine Ratios in Fractions from *Mercenaria***

| | Total Shell | Humic Acid |
|---|---|---|
| Modern live shell | 0.01 | — |
| Pleistocene, Wailes Bluff | 0.25 | 0.06 |
| Miocene, Drum Cliff | 1.25 | 0.45 |

cent by weight organic nitrogen with a carbon-to-nitrogen atomic ratio of 9. A significant proportion of the organic nitrogen is bound in the humic fraction.

Aliquots of the humic acid were hydrolyzed in 6N. hydrochloric acid for 25 minutes at 150°C. and the amino acids released were analyzed by ion-exchange chromatography. Suitable derivatives of the hydrolysates were made, and the ratio of the D and L amino acid isomers were determined by gas chromatography (Hare and Hoering, 1973).

The suites of amino acids released from the humic material in Miocene shells included several of the less stable ones (threonine, serine and arginine) and were significantly different from those released when the intact shell (calcium carbonate plus organic matter) was hydrolyzed. In each of the three species, the amino acids released from the humic material showed significantly less racemization than that obtained from the total shell material.

A small amount of humic acid was isolated from the fine fraction in the sands that surrounded the Miocene fossils. The amount was insufficient for measurement of the carbon isotope ratio but was adequate for an amino acid isomer analysis. The amino acids released from the hydrolysis of this material were exclusively in their original L configuration, a characteristic of recent humic acid. Also, the composition of the suite ofamino acids was completely different from that obtained from the fossil humic acid. These observations are further evidence that the humic acid in fossil shells is indigenous.

New perspectives on the geochemistry of amino acids in fossils result from the research described in this report. It is apparent that several pathways are operative for the preservation and degradation of these molecules. The interaction of reactive biochemicals in a shell produces a high-molecular-weight organic polymer (humic acid). The reaction between sugars and amino acids is a possible pathway for its formation, but it is not known if sugars are quantitatively important. It has been demonstrated (Hoering 1973) that the sugar–amino acid reaction leads to deaminated compounds similar to those found in fossils. Additional experimentation with the melanoidin reaction and measurements of the sugar content of live shells may lead to an evaluation of this process.

The discovery of multiple pathways has implications for the application of amino acid geochemistry to geochronology and paleothermometry. The relative importance of the different modes of amino acid degradation and the factors that govern them must be evaluated before the method can be used as a quantitative tool.

The role of water in initiating the complex set of reactions is important. A quantitative measure of the effective water concentration in a fossil would be useful in judging the relative stabilities of organic assemblages. Reports of intact proteins found in fossils are surprising in view of the extreme reactivity that is implied from the results of this work. Perhaps the concentration of water in a fossil is the controlling factor. The development of a successful amino acid sequencing technique for fossil organic matter will be crucial for evaluating the preservation of proteinaceous material.

## *REFERENCES*

Hare, P. E., Geochemistry of proteins, peptides and amino acids, in *Organic Geochemistry, Methods and Results,* G. Eglinton and M. T. J. Murphy, eds., Springer-Verlag, New York, pp.438–463, 1969.

Hare, P. E. and T. C. Hoering, Separation of amino acid optical isomers by gas chromatography, *Carnegie Inst. Wash. Year Book . 72,* 690–694, 1973.

Hoering, T. C., A comparison of melanoidin and humic acid, *Carnegie Inst. Wash. Year Book 72,* 682–690, 1973.

Hoering, T. C., The isotopic composition of the carbon and hydrogen in organic matter of Recent sediments, *Carnegie Inst. Wash. Year Book. 73,* 590–595, 1974.

Nau, H., New dideutero-perfluoroalkylated oligopeptide derivatives for protein sequencing by gas chromatography–mass spectrometry, *Biochem. Biophys. Res. Comm.,* **59,** 1088–1096, 1974.

Nau, H. and K. Biemann, Amino acid sequencing by gas chromatography–mass spectrometry using perfluoro-dideuteroalkylated peptide derivatives, *Anal. Biochem.,* **73,** 139–153, 1976.

Parker, P. L. and J. I. Hedges, Land-derived organic matter in surface sediments of the Gulf of Mexico, *Geochim. Cosmochim. Acta,* **40,** 1019–1029, 1976.

Thompson, R. B. and W. B. Creath, Low molecular weight hydrocarbons in Recent and fossil shells, *Geochim. Cosmochim. Acta,* **30,** 1137–1152, 1966.

# The Resolution of Enantiomeric Amino Acids by Gas Chromatogaphy

**GRANT GILL SMITH**

**DAVID M. WONNACOTT**

***Department of Chemistry and Biochemistry, Utah State University***

*ABSTRACT*

A 150-foot stainless steel capillary column loaded with a mixed phase of N-docosanoyl-L-val *t*-butylamide and N-octadecanoyl-L-val-L-val CHE has satisfactorily separated N-TFA isopropyl ester derivatives of neutral and acidic amino acid enantiomers in less than 200 minutes. Aspartic acid was separated better by this gc column than by any other reported so far. Metal columns suitably cleaned give adequate resolution and are considerably less troublesome to handle than glass columns.

Diastereomers, such as N-TFA (+)-2-butyl esters of amino acids, are very well resolved on commercially available achiral phases. Several of the optically active amino acids are better resolved by a diastereometric method than by the chiral phase method. The problems with this method are discussed.

Reported in part at the American Chemical Society Northwest Regional Meeting in Portland, Oregon, June 1977, and at the Carnegie Institution of Washington Conference: Advances in the Biogeochemistry of Amino Acids, Arlie House, Warrenton, VA, October 29–November 1, 1978. This material was taken in part from David M. Wonnacott's Ph.D. Thesis, Utah State University, 1979.

The resolution of enantiomeric amino acids by gas chromatography is of wide interest to geochemists. This paper presents a reliable method that separates most of the enantiomeric amino acids found in racemized collagen hydrolysates and in fossil bone and shell. It employs temperature programming of a metal capillary column coated with chiral phases having good temperature stability. It is suitable for alanine, valine, isoleucine,

leucine, serine, threonine, aspartic glutamic, and phenylalanine. It is not suitable for proline, hydroxyproline, and the basic amino acids.

## THE DIASTEREOMERIC METHOD

Before amino acid enantiomers can be resolved by gas chromatography, the carboxyl and amino fuction groups must be derivatized. One method is the derivatization of the amino acids with an optically active reagent (reviewed by Gil-Av and Norok, 1974). The introduction of a second asymmetric center to the D- and L-amino acid enantiomers gives rise to diastereomeric derivatives, which are resolved on capillary columns coated with commercially available achiral polar stationary phases. Aspartic acid, (one of the most important amino acids in a biogeochemistry study) unfortunately is often not satisfactorily resolved by this column, but D,L-proline diastereomers are completely resolved.

(+)-2-Butanol has been the most commonly used resolving agent, but its cost is high. Often, it is not optically pure and a correction is necessary for each analysis. With optically active alcohols, there is the potential problem of diastereometric fractionation taking place during the derivatization process, that is, the formation of one diastereomer being kinetically or thermodynamically favored over the other. Although some have reported that this is not a problem, it is clearly shown (Table 1) that fractionation occurred when L-valine was derivatized with racemic alcohols, resulting in unequal amounts of the diastereometric products. These data show that the more highly substituted alcohols (e.g. 3,3-dimethyl-2-butanol) provided increased resolution ($RT_L/RT_D = 1.101$) but

**Table 1. Diastereomeric Fractionation in the Derivatization of L-Valine with Racemic Alcohols**

| Percent N-TFA-Amino Acid (+) Alcohol Derivative Determined by GC[a] | | |
|---|---|---|
| *Alcohol* | *Percent* | $RT_L/RT_D$ |
| 2-Butanol[b] | 51.4 ± 0.1 | 1.046 |
| 2-Octanol[c] | 51.7 ± 0.1 | 1.044 |
| 3-Methyl-2-Butanol[c] | 54.6 ± 0.1 | 1.077 |
| Menthol[d] | 54.3 ± 0.1 | 1.095 |
| 3,3-Dimethyl-2-Butanol[c] | 62.6 ± 0.1 | 1.101 |

[a] Duplicate runs and duplicate gc analyses on 150-ft s.s. capillary column, UCON 75-H-90,000.
[b] Pollock et al., 1965.
[c] Ayers et al., 1971; König et al., 1977.
[d] Hasegawa and Matsubara 1975.

also caused considerably greater fractionation. A 62.6% yield of one diastereomer was obtained using 3,3-dimethyl-2-butanol; a value of 50.0% should have resulted from the racemic mixture had no fractionation occurred. Selective fractionation in esterification can cause an appreciable error; this should be kept in mind when attempting quantitative analyses with methods dependent on these alcohols.

There are other disadvantages in using some of these alcohols for esterification. When menthol was used, it was difficult to solubilize the amino acid during the derivatization process. Sonication assisted but still the yields were low. Menthol, like other higher molecular weight alcohols, has a low volatility which makes removing the excess alcohol after esterification more time consuming. Passing a stream of nitrogen over the surface while heating to 110–120° in an oil bath assisted the removal of the menthol·HCl. It was almost impossible to remove all the high boiling alcohol which often eluted with the less volatile amino acids on polar stationary phases. Another problem with L-menthol eliminates it as a useful chiral reagent. Normally, the D-amino acid derivative elutes first, which is desirable because a more accurate quantitation can be made with the small peak preceding the larger one. With L-menthyl ester, however, the larger L-isomer eluted first. D-Menthol is not readily available. The authors have taken advantage of this reverse elution in studying the kinetics of racemization of D-phenylglycine, the readily available commercial isomer. N-TFA-D-phenylglycine (+)-2-butyl ester were poorly resolved on polar stationary phases. Baseline resolution ($RT_D/RT_L = 1.06$) was obtained, however, with the L-menthyl ester on Apiezon-L.

It has been argued that when an excess of an optically pure alcohol is used in derivatization of a D,L-mixture of amino acids and refluxed to increase esterification, the fractionation is eliminated, and nearly quantitative esterification can be obtained. This would be true if another complicating reaction were not present. The conditions for esterification are also those necessary for conversion of alcohols to alkyl chlorides and water (Lucas test for 1°, 2°, and 3° alcohols). When a large excess of alcohol was used, esterification was not complete because the water from the Lucas reaction prevented complete esterification (Table 2).

2-Butanol and 2-octanol appeared to be the preferred reagents for quantitation because of their smaller tendency to show diastereometric fractionation. (+)-2-Butanol has been the most popular reagent used in this method. The diastereometric method using (+)-2-butanol has been extremely successful in small-sample studies of fossils because of the excellent resolution in a stainless steel column and the thermal stability and availability of the achiral phases (Gil-Av, 1975). However, when a large number of amino acid samples require derivatization as in a kinetic study,

**Table 2. Formation of 2-Butyl Chloride During the Derivatization of Amino Acids in 2-Butanol**

| Heating Time (hr)[a] | Anhydrous Hydrogen Chloride Concentration | Percent 2-Butyl Chloride[b,c] |
|---|---|---|
| 1 | 2.5 *M* | 16 |
| 1 | saturated | 26 |
| 24 | 2.5 *M* | 40 |

[a] Heating of 2-butanol/HCl solution was at 105° in a sealed tube.
[b] Percent 2-butyl chloride was determined from the area of 2-butyl chloride and 2-butanol peaks. No correction was made for the molar response factor. The analyses were carried out isothermally (50°) on a 150-foot UCON-75H-90,000 capillary column. The hydrogen chloride was neutralized with sodium bicarbonate solution before gc analysis.
[c] An equal amount of water is formed, preventing complete esterification.

the cost of (+)-2-butanol becomes a limiting factor and chiral phase methods are advised.

## THE CHIRAL PHASE METHOD

The second approach in resolving amino acid enantiomers by gas chromatography involves the derivatization of the amino acid with achiral reagents followed by resolution on capillary columns coated with chiral stationary phases. The development of optically active stationary phases capable of exhibiting enantiomeric selectivity for these derivatives has been pioneered by Gil-Av and co-workers (1975). In this method, the amino acid enantiomers are usually derivatized with isopropyl alcohol and trifluoroacetic anhydride or pentafluoropropionic anhydride. The stereoselective partitioning observed with this technique appears to result from a diastereomeric interaction between the chiral stationary phase (solvent) and the derivatized D- and L-amino acid solutes (Parr and Howard 1972; Lochmüller and Souter 1975).

In choosing an optically active stationary phase several criteria are to be considered. For example, the phase should have good thermal stability and high-coating efficiency, a low melting point and good selectivity for amino acids and their corresponding enantiomers.

Most optically active stationary phases reported for resolving amino acid enantiomers have been N-trifluoroacetyl-dipeptide cyclohexyl esters

(N-TFA-dipeptidide-CHE). N-TFA-L-valyl-L-valine CHE was developed in conjunction with lunar analyses and has been one of the most popular of these types of phases (Nakaparkin et al., 1970). Some N-TFA dipeptide derivatives are available commercially and König et al., (1970) have described their synthesis. Although they give a fairly high separation factor for enantiomers and show good selectivity for the separate amino acid components, they lack temperature stability.

Attempts have been made to prepare more stable stationary phases by using higher molecular weight, less volatile amino acids in the dipeptides, such as L-phenylalanyl-L-leucine (König et al., 1970) di-L-phenylalanine (Parr et al., 1971), L-phenylalanyl-L-aspartic acid (König and Nicholson, 1975) and di-L-methionine (Andrawes et al., 1975). Although these phases demonstrated improved temperature stability and can separate enantiomers, they show poor selectivity for certain important pairs of amino acid derivatives, which results in peak overlapping. Feibush (1971) has shown that *t*-butylamide derivatives are effective in making stationary phases more stable.

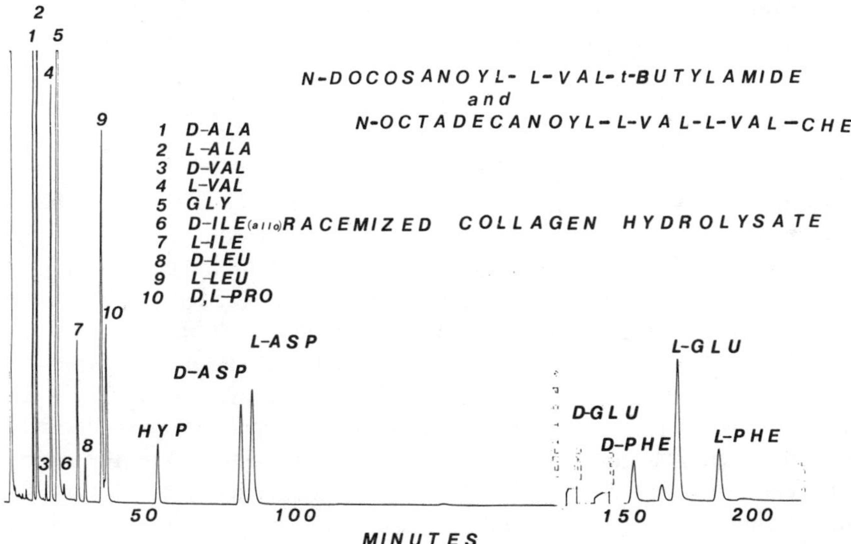


**Figure 1.** GC separation of N-TFA *i*-propyl esters of amino acids from racemized collagen hydrolysate using a 150 ft × 0.026 in. stainless steel capillary column coated with a mixed phase of N-docosanoyl-L-val *t*-butylamide and N-octadecanoyl-L-val-L-val CHE. Temperature and time programming: $T_1$ = 135°C for 110 min. $T_2$ = 150°C for 100 min. Nitrogen flow rate, regulated to elute the solvent, $CH_2Cl_2$, in 5.5 min. ala (1, 2), val (3, 4), gly (5), allo-ile (6), ile (7), leu (8, 9), pro (10).

## A MIXED PHASE COLUMN

An excellent column has been prepared by loading a mixture of two phases, N-octadecanoyl-L-val-L-val-cyclohexyl ester and N-docosanoyl-L-val *t*-butylamide, on a 150-foot stainless steel column. With the exception of the basic amino acids proline and hydroxyproline, satisfactory resolution is attained. The quality of the results is shown in Figure 1 for the resolution of racemized collagen hydrolysate. Although this column separated D,L-serine, theonine and methionine, the mode of derivatization and work-up were such that they did not appear as volatile components. Precise quantitation was possible for ala, val, ile, leu, asp, glu, ser, thr, and phe by using this mixed phase stainless steel capillary column. The details of its preparation are given by Wonnacott (Ph.D. thesis, Utah State University, 1979).

## GAS CHROMATOGRAPHIC–MASS SPECTROMETRIC ANALYSIS

In the quantitative analysis of racemic amino acids from fossils, it is essential that a careful examination is made of each peak by gas chromotography–mass spectrometry. Mass spectral confirmation is readily obtained. Further characterization of each peak is accomplished by single-ion monitoring. Each N-TFA amino acid isopropyl ester has a characteristic peak, often $M^+$-87. By drawing the chromatograms using single ion monitoring, each peak can be identified, and when more than one substance elutes at the same time, this is revealed. This is clearly shown by the gas chromatograms obtained on a mixture of racemic amino acids from a butter clam fossil shell. Figure 2 shows the total ion current. Figure 3 gives the total ion current and the current for selected single ions characteristic of selected amino acids. For example, *m/e* 140 for ala, *m/e* 55 for val, *m/e* 166 for asp, *m/e* 180 for glu, and *m/e* 91 for phe. Note that with single ion detection greater peak intensity is attained (e.g., D-val). Furthermore, with single ion detection, this specially prepared column did not separate L-glu from D-phe when helium was used as the carrier gas. When nitrogen was used as the carrier gas, baseline resolution was attained between L-glu and D-phe (Figure 1). A thorough GC-MS analysis on a fossil can give confidence in D-L ratios.

Occasionally, it is desirable to elute the L-isomer before the D-isomer, for example, where the D-isomer is in greater concentration. Chiral phases can be readily adapted to accommodate this need. In a kinetic study of the racemization of D-phenylglycine the rate was monitored on a specially prepared stainless steel capillary column coated with N-docosanoyl-D-

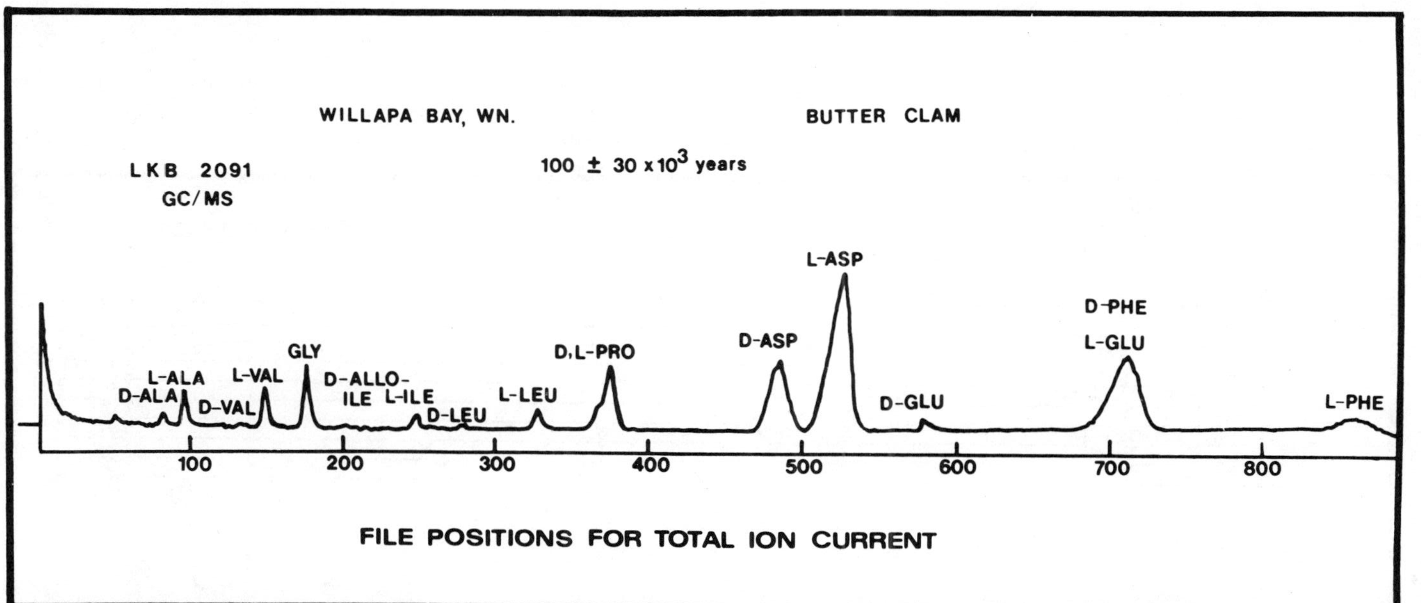


**Figure 2.** GC trace from the total ion current obtained on an LKB 2091 GC/MS/DS of N-TFA *i*-propyl esters of amino acids isolated from a fossil butter clam shell reported to be approximately 100,000 years old. The separation was made on a 150 ft × 0.026 in. stainless steel capillary column coated with a mixed phase of N-docosanoyl-L-val *t*-butylamide and N-octadecanoyl-L-val-L-val CHE. Column temp., 135° isothermal. D-phe and L-glu were not resolved using helium as the carrier gas (see Figure 1 and discussion of Figure 3).

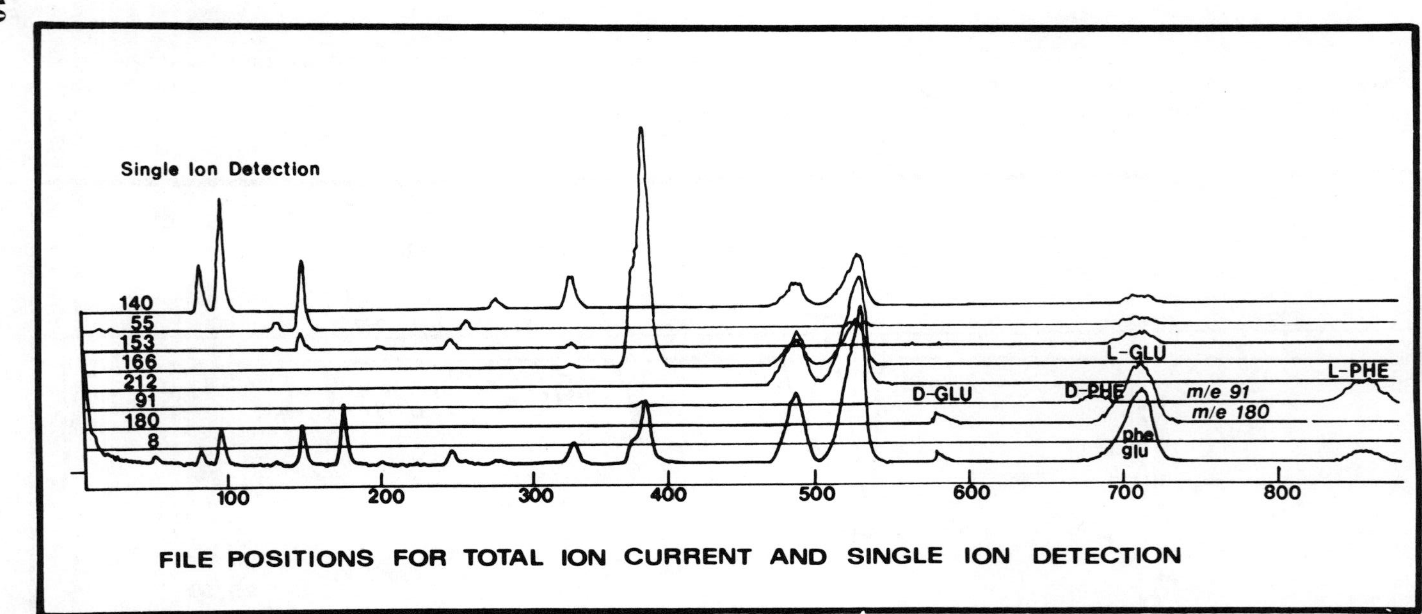


**Figure 3.** GC trace from both the total ion current and from single ion detection obtained on an LKB 2091 GC/MS/DS of N-TFA *i*-propyl esters of amino acids isolated from a fossil butter clam shell. For column details see Figure 2. Note the increased intensity for D-val (file position 132) by monitoring *m/e* 55 as well as other amino acids and detection of both D-phe and L-glu at file position approximately 700.

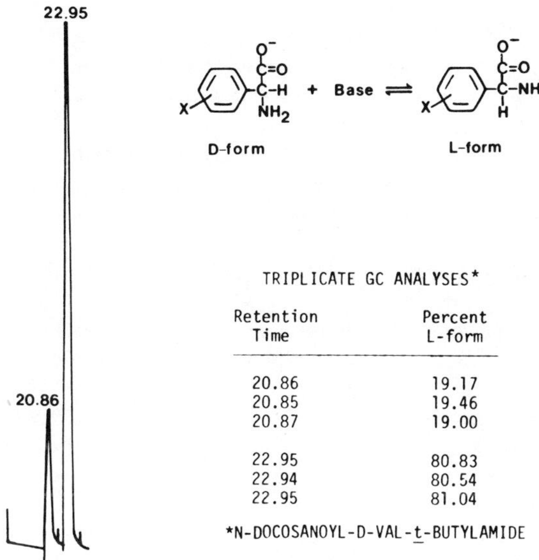


TRIPLICATE GC ANALYSES*

| Retention Time | Percent L-form |
|---|---|
| 20.86 | 19.17 |
| 20.85 | 19.46 |
| 20.87 | 19.00 |
| 22.95 | 80.83 |
| 22.94 | 80.54 |
| 22.95 | 81.04 |

*N-DOCOSANOYL-D-VAL-t-BUTYLAMIDE

**Figure 4.** Quantitative gc analysis for three separate but comparable experiments of the racemization of D-phenylglycine at 110.3°C for 4 hours. Measurements were made by gc resoltuion of the N-TFA-phenylglycine methyl ester on a 160 ft stainless steel capillary column coated with N-docosanoyl-D-valine *t*-butylamide at 130°C. The precision of the three experiments was within 0.2%. The L-isomer eluted first.

valine *t*-butylamide in which the smaller L-isomer eluted first. By changing the L-val to D-val on the chiral phase, this simple reversal is attained.

Figure 4 illustrates the resolution and precision in determining the percentage of L-and D-phenylglycine from racemization experiments of D-phenylglycine. The data were taken from triplicate racemization experiments (Sivakua 1978). The N-TFA-D,L-phenylglycine methyl ester was resolved on a 160-foot stainless steel capillary column coated with N-docosanoyl-D-val *t*-butylamide at 130°. The precision of the three racemization experiments is within 0.2%.

## *ACKNOWLEDGMENT*

Grateful acknowledgment is given to the *National Aeronautics and Space Administration, NSG-7038*, who supported some of the research reported in this chapter.

## *REFERENCES*

Andrawes, F., R. Brazell, W. Parr, and A. Zlatkis, Methionine dipeptide stationary phases for the resolution of enantiomers, *J. Chromatogr.,* **112,** 197–202, 1975.

Ayers, G. S., R. E. Moore, and J. H. Mossholder, Resolution of amino acid diastereomers by means of packed column gas chromatography, *J. Chromatogr.,* **63,** 259–265, 1971.

Bayer, E., E. Gil-Av, W. A. König, S. Nakaparksin, J. Oró, and W. Parr, Retention of configuration in the solid phase synthesis of peptides, *J. Amer. Chem. Soc.,* **92,** 1738–40, 1970.

Bodansky, M. and M. A. Ondetti, *Peptide Synthesis,* G. A. Olah, ed., Wiley, New York, 1966.

Bonner, W. A., The application of diastereomeric S-prolyl dipeptide derivatives to the quantitative estimation of R- and S-leucine enantiomers, *J. Chromatogr. Sci.,* **10,** 159–164, 1972.

Carrigan, J. J., D-Amino acids in animals, *Science,* **164,** 142–149, 1969.

Charles, R., U. Beitler, B. Feibush, and E. Gil-Av, Separation of enantiomers on packed columns containing optically active diamide phases, *J. Chromatogr.,* **112,** 121–133, 1975.

Corbin, J. A., J. E. Rhoad, and L. B. Rogers, Effects of structure of peptide stationary phases on gas chromatographic separation of amino acid enantiomers, *Anal. Chem.,* **43,** 327–331, 1971.

Dungworth, G., N. J. Uincken, and A. W. Schwartz, Compositions of fossil collagens: Analysis by gas-liquid chromatography, *Comp. Biochem. Physiol.,* **47B,** 391–399, 1974.

Feibush, B. and E. Gil-Av, Interaction between asymmetric solutes and solvents. Peptide derivatives as stationary phases in gas liquid partition chromatography, *Tetrahedron,* **26,** 1361–1368, 1970.

Feibush, B., Interaction between asymmetric solutes and solvents. N-Lauroyl-L-valyl *t*-butylamide as a stationary phase in gas liquid partition chromatography, *Chem. Commun.,* 544–545, 1971.

Fenimore, D. C., J. H. Whitford, and C. M. David, Nickel gas chromatographic columns: An alternative to glass for biological samples, *J. Chromatogr.,* **140,** 9–15, 1977.

Frank, H., G. J. Nicholson, and E. Bayer, Rapid gas chromatographic separation of amino acid enantiomers with a novel chiral stationary phase, *J. Chromatogr. Sci.,* **15,** 184–186, 1977.

Gil-Av, E., B. Feibush, and R. Charles-Segler, Separation of enantiomers by gas liquid chromatography with an optically active stationary phase, *Tetrahedron Lett.,* **10,** 1009–1015, 1966.

Gil-Av, E. and D. Norok, *Advances in Chromatography,* Vol. 10, J. C. Giddings and R. A. Keller, eds., Marcel Dekker, New York, pp. 99–172, 1974.

Gil-Av, E., Present status of enantiomeric analysis by gas chromatography, *J. Mol. Evol.,* **6,** 131–144, 1975.

Greenstein, J. P. and M. Winitz, *Chemistry of Amino Acids,* Vol. 2, Wiley, New York, p. 945–949, 1961.

Hasegawa, M. and J. Matsubara, Gas chromatography determination of the optical purities of amino acids using N-trifluoroacetyl menthyl esters, *Anal. Biochem.,* **63,** 308–320, 1975.

Hoopes, E. A., E. T. Peltzer, and J. L. Bada, Determination of amino-acid enantiomeric ratios by gas-liquid chromatography of N-trifluoroacetyl-L-peptide methyl ester, *J. Chrom. Sci.*, **16,** 556–560, 1978.

Iwase, H. and H. Murai, Resolution of racemic amino acids by gas chromatography, I. N-trifluoroacetyl-L-prolyl derivatives, *Chem. Pharm. Bull.*, **22,** 8-14, 1974; *C A 80*, 96325K, 1974.

König, W. A., W. Parr, H. A. Lichtenstein, E. Bayer, and J. Oró, Gas chromatographic separation of amino acids and their enantiomers: Non-polar stationary phases and a new optically active phase, *J. Chromatogr.*, **8,** 183–186, 1970.

König, W. and R. Geiger, Eine neue methode zur synthese von peptides: Aktivierung der carboxylgruppe mit dicyclohexylcarbodiimide unter zusatz von 1-hydroxy-benzotriazolen, *Chem. Ber.* **103,** 778–798, 1970.

König, W. A. and G. J. Nicholson, Glass capillaries for fast gas chromatographic separation of amino acid enantiomers, *Anal. Chem.*, **47,** 951–952, 1975.

König, W. A., W. Rahn, and J. Eyen, Gas chromatographic separation of diastereoisomeric amino acid derivatives on glass capillaries. The use of pentafluoropropionyl-amino acid (+)-3-methyl-2-butyl esters, *J. Chromatogr.*, **133,** 141–146, 1977.

Kvenvolden, K. A., E. Peterson, and G. E. Pollock, *Advan. Org. Geochem. Proc. 5th Int. Meeting*, Hanover, 1971, Pergamon, Oxford, p. 387, 1972.

Kvenvolden, K. M., Criteria for distinguishing biogenic and abiogenic amino acids—preliminary considerations, *Space Life Sciences*, **4,** 60–68, 1973.

Lapivot, L., S. Rappoport and L. Wolman, *J. Lipid Res.*, **8,** 142–145, 1967.

Mengelberg, M., Notizüber aminosäure—cyclohexyl ester, *Chem. Ber.* **90,** 451–452, 1957.

Lochmüller, C. H. and R. W. Souter, Chromatographic resolution of enantiomers selective review, *J. of Chromotogr.*, **113,** 283–302, 1975.

Mengelber, M., Natizüber aminosäure—cyclohexyl ester, *Chem. Ber.*, **90,** 451–452, 1957.

Nakaparksin, S., P. Birrell, E. Gil-Av, and J. Oró, Gas chromatography with optically active stationary phases: Resolution of amino acids, *J. Chromatogr. Sci.*, **8,** 177–82, 1970.

Nakaparksin, S., E. Gil-Av, and J. Oró, Study of the racemization of some neutral α-amino acids in acid solution using gas chromatographic techniques, *Anal. Biochem.*, **33,** 374–82, 1970.

Parr, W., J. Pleterski, C. Yang, and E. Bayer, Resolution of racemic amino acids by gas chromatography on optically active stationary phases, *J. Chromatogr. Sci.*, **9,** 141–147, 1971.

Parr, W. and P. Y. Howard, Molecular interactions in a unique solvent-solute system, *J. Chromatogr.*, **71,** 193–201, 1972.

Parr, W. and P. Y. Howard, Structural effects of selected dipeptides as stationary phases for the chromatographic separation of enantiomeric amino acids, *Anal. Chem.*, **45,** 711–724, 1973.

Pollock, G. E. and A. H. Kawauchi, Resolution of racemic aspartic acid, tryptophan, hydroxy and sulfhydryl amino acids by gas chromatography, *Anal. Chem.*, **40,** 1356–1358, 1968.

Pollock, G. E., U. I. Oyama, and R. D. Johnson, Resolution of racemic amino acids by gas chromatography, *J. Gas Chromatogr.*, **3,** 174–6, 1965.

Pollock, G. E. and A. H. Kawauchi, Resolution of racemic aspartic acid, tryptophan, hydroxy and sulfhydryl amino acids by gas chromatography, *Anal. Chem.*, **40,** 1356–1358, 1968.

Raulin, F. and B. N. Khare, Gas liquid chromatographic resolution of several protein amino acid enantiomers on a packed column, *J. Chromatogr.*, **75,** 13–18, 1973.

Robinson, T., D-Amino acids in higher plants, *Life Science,* **19,** 1097–1102, 1976.

Schopf, J. W., K. M. Kvenvolden, and E. S. Barghoorn, Amino acids in Precambrian sediments: An assay, *Proc. Nat. Acad. Sci. U.S.*, **59,** 639–648, 1968.

Sivakua, T., Structural effects on the kinetics and mechanism of the racemization of arylglycines, Ph.D. Thesis, Utah State University, Logan, UT, 1978.

Synge, R., The synthesis of some dipeptides related to gramicidins, *Biochem. J.*, **42,** 99–104, 1948.

Tapidot, Y., S. Rappoport, and Y. Wolman, Use of esters of N-hydroxysuccinimide in the synthesis of N-acylamino acids, *J. Tipid Res.*, **8,** 142–145, 1967.

Van Dort, M. A. and W. A. Bonner, The quantitative gas chromatographic resolution of amino ester diastereomers, *J. Chromatogr.*, **133,** 210–213, 1977.

Westley, J. W., B. Halpern, and B. L. Karger, Effect of solute structure on separation of diasteroisomeric esters and amides by gas chromatography, *Anal. Chem.*, **40,** 2046–2049, 1968.

Weygand, F., D. Hoffmann, and A. Prox, Ein einfacher gaschromatographischer racemisierungstest für die peptidesynthese, *Z. Naturforsch.*, Teil B, **23,** 279–281, 1968.

Williams, K. M. and G. G. Smith, A critical evaluation of the application of amino acids racemization to geochronology and geothermometry, *Origins of Life,* **8,** 91–144, 1977.

Windridge, G. and E. C. Jorgensen, 1-Hydroxybenzotriazole as a racemization-suppressing reagent for the incorporation of *im*-benzyl-*L*-hisidine into peptides, *J. Amer. Chem. Soc.*, **93,** 6318–6319, 1941.

Wonnacott, D. M. and G. G. Smith, The application of gas chromatography to the resolution of enantiometric mixtures of amino acids, *Abstract of the Northwest Regional-American Chemical Society Meeting,* Portland, OR, June 16, 1977.

Wonnacott, D. The application of gas chromatography to the resolution of enantiomeric mixtures of amino acids. Factors effecting the racemization of amino acids, Ph.D. Thesis, Utah State University, Logan, UT, 1979.

# The Effects and Implications of Ultrasonic Cleaning on the Amino Acid Geochemistry of Foraminifera

**BARRY J. KATZ***

*Division of Marine Geology and Geophysics, Rosenstiel School of Marine and Atmospheric Science, University of Miami*

**EUGENE H. MAN**

*Department of Chemistry and Division of Marine and Atmospheric Chemistry, University of Miami*

## *ABSTRACT*

Procedural investigations have been directed to the effects of hydrolysis, desalting, and instrumentation. But there has been no attempt to determine what effects ultrasonic cleaning may have on the observed amino acid geochemistry of foraminifera. This work represents a study to determine the effects of cleaning on 11 amino acids typically found in foraminiferal matrices. Although all the amino acids undergo an exponential decrease in concentration with time, the rate of the decrease varies within the initial 45 minutes of treatment. This variation has permitted three loosely defined groups to be established: Group I, stable; Group II, decreasing; Group III, slight increases prior to the ultimate decrease. When the relative concentrations of the amino acids are compared, it appears that racemization studies are less affected than other studies involving diagenetic changes or paleobiochemical studies by variations introduced by ultrasonification.

* Present address: Bellaire Research Laboratories, Texaco Inc. P.O. Box 425, Bellaire, Texas 77401

The study of amino acids in geologic materials, both fossils and sediments, has led to several investigations of the common laboratory procedures and their implications on the observed results. Procedures studied included desalting (Cheng 1975), hydrolysis (Hare 1971), and instrumentation (Kvenvolden 1978). However, specific materials require unique procedures such as the ultrasonic cleaning of foraminifera to remove adhering clay particles. Although several laboratories are studying the amino acid geochemistry of foraminifera, there has been no attempt to determine the effects of the procedure or whether the lack of standardization among laboratories is important.

The question of when a foraminifera test was organically clean was addressed by Schroeder (1975). He examined the concentration of two amino acids he considered unique to the clay fraction, β-alanine and γ-amino-n-butyric acid. The test was "clean" when these two amino acids had been removed. Although Schroeder concluded that the amino acid pattern produced was unique to the foraminifera test, he made no attempt to determine if during the cleaning, test amino acids were altered by the procedure. Furthermore, there was no time limit placed on this treatment. An examination of the literature (King 1976; Staas and Spurlock 1975) coupled with some preliminary work in our laboratory suggested that there may be several problems with this open-ended treatment.

We considered it important to determine what effects the cleaning procedure has on both the absolute and relative concentrations of the amino acids and whether the observed effects would interfere with the interpretation of the data. In addition we attempted to determine whether an optimum period of effective cleaning could be established when a maximum amount of clay can be removed with only minimum alteration of the amino acid pattern.

To accomplish these objectives, foraminifera were isolated from approximately 200 $cm^3$ of University of Miami core P6609-13 13/18 by sieving through a 63-micron mesh sieve. The collected foraminifera were dried and mechanically split into aliquots of approximately one gram. The samples were transferred to clean test tubes and then ultrasonically cleaned. The length of cleaning was variable and consisted of repetitive five-minute cycles. Each cycle consisted of 5 minutes of ultrasonification in a Cole-Palmer 8845-50 ultrasonic cleaner, after which the supernatant was decanted, and doubly distilled water was added.

The "cleaned" samples were then dissolved in 6 *M* HCl and dried under reduced pressure. The residue was hydrolyzed in 6 *M* HCl for 24 hours at 100°C under vacuum. Following the acid hydrolysis, desalting was accomplished on Dowex AG 50W-X8 $H^+$ as described by Bada and Man (1973) with 1.6 *M* $NH_4OH$ as the elutant. The $NH_4OH$ was evaporated and the residue was analyzed on a Beckman-Spinco Model 118 Amino Acid Analyzer in the laboratory of Dr. J.L. Bada at Scripps Institution of Oceanography.

The raw data were corrected for recovery by the use of a norleucine internal standard, which was added prior to hydrolysis. The corrected values are summarized in Table 1. A normalization of each of the 11 amino acids to the concentration of the amino acid content of the untreated material shows that all the amino acids undergo a decrease in concentration (peak area) with time.

A closer examination of each of the amino acids suggests that three loosely defined groups may be established based on the changes in concentration that occur during the initital 45 minutes of treatment. Group I includes serine threonine, proline, and valine and shows no significant change during the initial period of treatment. Group II, which contains aspartic acid, alanine, and alloisoleucine, exhibits an exponential decrease throughout the study. Group III, which includes glycine, glutamic acid leucine, and tentatively, isoleucine, exhibits increases in absolute concentrations within the early stages of treatment prior to decreasing in a manner similiar to that of Groups I and II.

It appears probable that there are two modes that account for these observed changes. One is the mechanical removal of the amino acids associated with either the foraminifera test or the adhering clay fraction; the second is the chemical alteration of the associated amino acids. The Group I amino acids, which exhibit no quantitative change during the initial treatment, appear to be associated with the tests rather than the clay fraction. Clay removal is particularly efficient during this initial period, and no changes are observed in the amino acid concentrations. The exponential decrease in concentration that follows appears to be the result of the stripping away of the outer layers of the foraminifera tests which are rich in amino acids (King 1976). The subsequent reduced rate of reduction in concentration is thus probably due to the lower concentrations of amino acids in the inner test layers.

The Group II amino acids, unlike those of Group I, behave in a manner suggestive of an association with both the clay fraction and the test. The initial rapid decrease in concentration parallels the clay loss during this period. An examination of both the test and the supernatant indicates that up to 90% of the adhering clay particles may be removed during this period of treatment, depending upon shell structure and extent of diagenesis. Although it appears that most of the decrease is simply mechanical, some is assumed to be chemical, since Group III amino acids increase during this period partially by the recombination of fragments from Group II compounds. Following this period (30–45 minutes), the reduction of

**Table 1. Corrected Peak Areas**

| Time of Ultrasonification (min) | Asp | Thr | Ser | Glu | Pro | Gly | Ala | Val | Allo | Iso | Leu |
|---|---|---|---|---|---|---|---|---|---|---|---|
| 0 | 206.0 | 68.5 | 77.0 | 97.0 | 8.0 | 195.0 | 156.1 | 67.0 | 3.2 | 46.0 | 60.0 |
| 15 | 164.6 | 70.5 | 77.0 | 111.1 | 8.7 | 142.4 | 111.9 | 68.3 | 3.3 | 34.9 | 54.4 |
| 30 | 107.6 | 65.8 | 77.8 | 83.7 | 8.9 | 95.7 | 95.7 | 67.2 | 1.4 | 37.7 | 69.5 |
| 45 | 73.7 | 59.7 | 65.6 | 81.2 | 7.2 | 107.4 | 88.4 | 57.2 | 1.0 | 30.6 | 48.7 |
| 105 | 41.6 | 19.1 | 24.2 | 22.1 | 2.9 | 38.6 | 29.3 | 20.5 | 0.6 | 11.3 | 15.5 |
| 135 | 33.7 | 20.2 | 25.2 | 21.8 | 3.5 | 43.7 | 33.3 | 23.2 | 0.7 | 12.4 | 16.7 |
| 240 | 0.6 | 1.9 | 1.2 | 0.6 | 0.4 | 2.1 | 0.6 | 0.4 | 0.3 | 0.2 | |

the amino acids associated with the tests as in Group I begins to dominate, resulting in the nearly identical behavior of these two groups.

Group III amino acids, which exhibit an increase in chemical concentration, indicate that ultrasonic cleaning must be inducing diagenesis within the foraminifera matrix. Both the apparent independent behavior of each of the amino acids of the group and the failure to detect any significant increase in serine, threonine, and aspartic acid indicates that the concentration increases are real and not the result of external contamination during handling (Hare 1969).

The alteration of amino acids by ultrasound had been reported by Staas and Spurlock (1975) for aqueous solutions, with deamination and decarboxylation occurring as well as interconversion among some of the protein amino acids. Thermally induced changes have been observed within calcareous matrices. The pyrolysis of *Mercenaria* shells resulted in changes in both the absolute and the relative amino acid concentrations (Abelson 1954; Jones and Vallentyne 1960; Mitterer 1966; Vallentyne 1969). Glycine was produced at the expense of threonine, serine, and valine. Similar diagenetic changes have been reported by Bada et al. (1977 and 1978) in foraminifera at reduced temperatures typical of the deep sea sediment water interface (~5°C). Serine and threonine undergo a reversible aldol cleavage to yield glycine. In addition, through a dehydration reaction they suggest alanine and α-amino–n-butyric acid may be produced from serine and threonine, respectively. Amino acids associated with clay minerals have also exhibited changes similar to those of Group III while undergoing heating in the presence of water (Sellers 1966). However, Sellers was unable to propose a mechanism and noted that several of the amino acids that increased (isoleucine, leucine, and glutamic acid) were not normally products of pyrolysis.

If one examines the reported diagenetic pathways (e.g., aldol cleavage, the breaking of interior carbon-carbon bonds) (Bada et al., 1977, 1978; Staas and Spurlock 1975; Mitterer 1966), one concludes that glycine may be produced from several amino acids if sufficient energy is available. We have concluded that ultrasonic cleaning could provide sufficient energy, particularly at sides of cavitation, for such reactions to occur. As a result of the stability of the Group I amino acids one must conclude that the conversions involved Group II and/or III or that glycine was synthesized from other components available within the test, thus explaining the observed increase in glycine.

The three other members of Group III rely upon the recently reported catalytic conversion of glycine into the larger straight-chain amino acids (Ivanov and Slavcheva 1977). Though the circumstances are dramatically different, both the deamination and decarboxylation reactions required for this conversion have been reported by Staas and Spurlock (1975) during ultrasonification of amino acids. The mechanism proposed is the elimination of $NH_3$ and $CO_2$ and may be expressed as:

$$n\mathrm{NH_2CH_2COOH} \longrightarrow \underset{\substack{| \\ (\mathrm{CH_2})_{n-2} \\ | \\ \mathrm{CH_3}}}{\mathrm{NH_2CHCOOH}} + (n-1)\mathrm{NH_3} + (n-1)\mathrm{CO_2}$$

The stability of the allo/iso value is significant, since it represents a racemization (epimerization) reaction. Racemization reactions involving amino acids in foraminifera are being used for both geochronology and geothermometry (Schroeder and Bada 1976). This relative stability indicates that even slightly excessive ultrasonic cleaning is probably not introducing a significant procedural error into such studies. Though only one of these reactions was studied, it is probable that the other racemization studies are also unaffected. This conclusion is based on the nearly identical activation energies and mechanisms for the racemization reactions (Dungworth 1976).

However, studies like the one reported by Bada et al. (1978) using other diagenetic reactions are much more sensitive to such procedures. The sensitivity of the Thr/Leu, Ser/Leu and ABA (α-amino–n-butyric acid)/(Thr) ratios to the cleaning procedure may result in unacceptable errors within calculated ages. Through the use of equations cited by Bada et al. (1978):

$$\mathrm{Ln(Thr/Leu)} = 0.62 - 1.8 \times 10^{-6} t \tag{4}$$

$$\mathrm{Ln(Ser/Leu)} = 1.1 - 3.1 \times 10^{-6} t \tag{5}$$

and the maximums and minimums of the two ratios as reported above, estimates on the magnitude of these errors may be made. Thus the calculated ages for Thr/Leu range from $2.0 \times 10^5$ to $3.7 \times 10^5$ yr. Similarly, the Ser/Leu ages range from $2.1 \times 10^5$ to $3.05 \times 10^5$ yr. Clearly, if such uncertainty results from experimental methodology, the procedure is unacceptable.

Paleobiochemical investigations have been aimed at developing evolutionary pathways by factor analysis of eight protein amino acids (King and Hare 1972a, b). This type of analysis is highly sensitive to relative changes in amino acid concentrations. In fact, some of the observed variations between species which have been used to establish a "natural" classification are less than those observed within our study. Though this does not invalidate such work, it should introduce a degree of caution into such studies.

One other interesting point established by King and Hare (1972a, b) was that spinose forms of foraminifera were more variable than nonspinose forms. This may be the result of the greater surface area of the spinose forms, which allows a larger amount of clay (with associated amino acids) to adhere to the surface, thus producing some of the observed variability.

## CONCLUSIONS

From the evidence presented we have shown that ultrasonic cleaning of foraminifera alters their observed amino acid content. The stability observed in the allo/iso ratios suggests that there probably has been little procedural error resulting from ultrasonification in racemization studies. Observed variations, however, are of sufficient magnitude to cast doubt on some geochronological conclusions based on amino acid ratios (i.e., Ser/Leu; Thr/Leu) and also suggest a more cautious approach to paleobiochemical investigations. It also appears that a maximum of 30 minutes of ultrasonification should be established, based upon the extent of clay removal and the limited alteration of the amino acids associated with the test (Group I) during this period. For studies involving reactions other than racemization, it is suggested that ultrasonic cleaning be kept to a minimum (less than 30 min.) and that a clay blank be determined based upon the weight percent of alumina found with the cleaned tests.

### *ACKNOWLEDGMENTS*

The authors wish to thank Dr. J.L. Bada of Scripps Institution of Oceanography and D. Roy for their assistance. Sample material was provided by the Division of Marine Geology and Geophysics, University of Miami. This research was supported by National Science Foundation Grants DES 7520937 and OCE 7719573. The authors also acknowledge contribution from the University of Miami, Rosenstiel School of Marine and Atmospheric Science.

### *REFERENCES*

Abelson, P. H., Organic constituents of fossils, *Carnegie Inst. Wash. Year Book 53*, 97–101, 1954.

Bada, J. L. and E. H. Man, Racemization of isoleucine in cores from Leg 15 Site 148, in *Initial Reports of the Deep Sea Drilling Project*, Vol. 20, B. C. Heezen et al., eds., U.S. Govt. Printing Office, Washington, D. C. pp 197–201, 1973.

Bada, J. L., M. Y. Shou, and E. H. Man, Decomposition reactions of hydroxy-amino acids; applications in geochronology and in paleotemperature determinations, *Abstr. w. Progr. Geol. Soc. Amer.*, **9,** 885–886, 1977.

Bada, J. L., M. Y. Shou, E. H. Man and R. A. Schroeder, Decomposition of hydroxy amino acids in foraminiferal tests; kinetics, mechanism and geochronological implications. *Earth Planet. Sci. Lett.*, **41,** 67–76, 1978.

Cheng, C. N., Extracting and desalting amino acids from soils and sediments: Evaluation of methods, *Soil Biol. Biochem.*, **7,** 319–322, 1975.

Dungworth, G., Optical configuration and the racemization of amino acids in sediments and in fossils—a review. *Chem Geol*, **17,** 135–153, 1976.

Hare, P. E., Geochemistry of proteins, peptides and amino acids, in *Organic Geochemistry*, Eglinton and Murphy, eds., Springer-Verlag, New York, pp. 438–463, 1969.

Hare, P. E., Effect of hydrolysis on the racemization rates of amino acids *Carnegie Inst. Wash. Year Book 70*, 256–258, 1971.

Ivanov, C. P. and N. N. Slavcheva, Formation of amino acids on heating glycine and alumina, *Origins of Life*, **8,** 13–19, 1977.

Jones, J. D. and J. R. Vallentyne, Biogeochemistry of organic matter, I. Polypeptides and amino acids in fossils and sediments in relation to geothermometry, *Geochim. Cosmochim. Acta*, **21,** 1–34, 1960.

King, K., Jr. and P. E. Hare, Amino acid composition of planktonic foraminifera: A paleobiochemical approach to evolution, *Science*, **175,** 1461–1463, 1972a.

King, K., Jr. and P. E. Hare, Amino acid composition of the test as a taxonomic character for living and fossil planktonic foraminifera, *Micropaleontology*, **18,** 285–293, 1972b.

King, K., Jr., Amino acid concentration in planktonic foraminifera: A new approach for studying the dissolution of deep sea carbonates, *Abst. w. Progr. Geolog. Soc. Amer.*, **8,** 955.

Kvenvolden, K., Interlaboratory comparison of amino acid racemization in a Pleistocene mollusk, *Saxidomus giganteus*, this volume. Presented at the *Carnegie Institution of Washington Conference* on the Advances in the Biogeochemistry of Amino Acids. Airlie House, Warrenton, Virginia, Oct 29–Nov 1, 1978.

Mitterer, R. M., Amino acid and protein geochemistry in mollusk shells, Ph.D. thesis, Florida State University, Tallahassee, FL, pp. 151, 1966.

Rodriquez, J. L. P., A. Weiss, and G. Lagaly, A natural clay organic complex from andalusian black earth, *Clay and Clay Minerals*, **25,** 243–251, 1977.

Schroeder, R. A., Absence of β-alanine and γ-aminobutyric acid in cleaned foraminiferal shells: Implications for use as a chemical criterion to indicate removal of nonindigenous amino acid contaminants, *Earth Planet. Sci. Lett.*, **25,** 274–278, 1975.

Schroeder, R. A. and J. L. Bada, A review of the geochemical applications of the amino acid racemization reaction, *Earth Sci. Rev.*, **12,** 347–391, 1976.

Sellers, G. A., Hydrothermal experiments on the thermal stability of amino substances in sediments, Ph.D. thesis, California Institute of Technology, Pasadena, CA. pp. 142, 1966.

Staas, W. H. and L. A. Spurlock, Chemistry of Ultrasound: Pt. IV, Effects of ultrasound on some amino acids, *J.C.S. Perkin Transactions*, 1675–1679, 1975.

Vallentyne, J. R., Pyrolysis of amino acids in Pleistocene mercenaria shells, *Geochim. Cosmochim. Acta*, **33,** 1453–1458, 1969.

# Interlaboratory Comparison of Amino Acid Racemization in a Pleistocene Mollusk, *Saxidomus giganteus*

**KEITH A. KVENVOLDEN**
*U.S. Geological Survey, Menlo Park, California*

*ABSTRACT*

Thirteen laboratories have participated in an exercise to compare automated ion exchange and gas chromatographic measurements of D/L ratios of amino acids from an homogenized sample of aragonitic shells of the late Pleistocene clam *Saxidomus giganteus*. The mean and relative standard deviation determined by both analytical techniques for 15 measurements of D-alloisoleucine/L-isoleucine are 0.23 ± 19%. Mean D/L ratios and relative standard deviations for seven amino acids measured by gas chromatography are: valine, 0.19 ± 18%; alanine, 0.45 ± 5%; leucine, 0.27 ± 9%; proline, 0.45 ± 15%; aspartic acid, 0.47 ± 5%; phenylalanine, 0.32 ± 15%; and glutamic acid, 0.21 ± 10%. Aspartic acid and alanine D/L ratios have the lowest relative standard deviations; D/L ratios of valine and the ratios of D-alloisoleucine/L-isoleucine have the highest relative standard deviations.

Although it has been recognized since 1954 that amino acids commonly occur in fossil shells (Abelson 1954), it was not until thirteen years later that the stereochemical properties of these fossil molecules began to be appreciated (Hare and Mitterer 1967; Hare and Abelson 1968). In this early work, measurements were made of the relative abundances of isoleucine and alloisoleucine, and consideration of the ratios of these amino acids in shells led to the realization that L-isoleucine interconverts to D-alloisoleucine over geological spans of time (possibly $10^7$ years). From this observation it was inferred that the extent of racemization (epimerization in the case of diastereomers such as isoleucine) of amino acids might be used to determine the age of shells. The idea of employing amino acids as geochronometers was later expanded to include potential applications to stratigraphic correlations and paleothermometry (Hare and Mitterer 1969). One reason that attention was first focused on isoleucine-alloisoleucine was the convenience of their quantitative measurement by conventional automated ion-exchange chromatography. With the development of specialized gas-chromatographic techniques it became possible to measure the enantiomeric (D/L) ratios of many chiral amino acids in fossils (Kvenvolden et al., 1973). Since these early investigations, numerous papers have been written concerning racemization of amino acids in fossils, and a number of review articles summarize the field (Bada and Schroeder 1975; Kvenvolden 1975; Dungworth 1976; Schroeder and Bada 1976; Williams and Smith 1977; Masters and Bada 1979).

In this work on racemization of amino acids in fossils, the basic measurements are the relative abundances of D- and L-isomers of amino acids. Because of the number of laboratories and individuals contributing to the field of amino acid biogeochemistry and because of the broad applications to geochemistry of amino acid stereochemistry, it is of interest to ascertain the reproducibility of these basic measurements by several laboratories. Previously an interlaboratory comparison was made of amino racemization in bone done by three laboratories (Bada et al., 1979). I report here

**Table 1. Participants in Interlaboratory Comparison**

| *Analyst* | *Laboratory* |
|---|---|
| J. L. Bada | Scripps Institution of Oceanography |
| D. J. Blunt | U.S. Geological Survey, Menlo Park |
| P. E. Hare | Carnegie Institution of Washington |
| K. L. King, Jr. | Lamont-Doherty Geological Observatory |
| G. H. Miller | University of Colorado |
| R. M. Mitterer | University of Texas at Dallas |
| B. Nagy | University of Arizona |
| E. Peterson | NASA—Ames Research Center |
| G. E. Pollock | NASA—Ames Research Center |
| N. W. Rutter | University of Alberta, Canada |
| R. A. Schroeder | University of Utah |
| G. G. Smith | Utah State University |
| J. F. Wehmiller | University of Delaware |

on a comparison of thirteen laboratories' measurements of the racemization of amino acids in a late Pleistocene butter clam *Saxidomus giganteus* (Deshayes). The participants in this study are listed in Table 1; I am grateful for their cooperation.

## SAMPLE PREPARATION

For this comparison, specimens of the pelecypod *Saxidomus giganteus* were collected from USGS Cenozoic Locality M 1497 at Willapa Bay, Pacific County, Washington. The aragonitic shells came from a well-exposed deposit in a cliff about one meter above sea level on the east side of the bay. Amino acids in shells at this locality have been extensively investigated by Kvenvolden et al. (1979), who have estimated that the shells used in this interlaboratory comparison are about 120,000 ± 40,000 years old and were deposited during the Sangamon Interglaciation.

Seven valves of *Saxidomus* were cleansed of all adhering sediment and immersed briefly (about 3 minutes) in 1 *N* HCl, then washed with distilled water and dried in a vacuum oven at 40°C. The cleaned shells were pulverized in an agate mortar until the material passed a 60-mesh screen. About 60 g of pulverized shell were homogenized by tumbling in a round-bottom flask. After thorough mixing, 2-g aliquots were placed in 1-dram vials. From a collection of 30 vials, samples were withdrawn randomly for distribution to participating laboratories.

## RESULTS AND DISCUSSION

The primary goal of this study is to compare D/L ratios of amino acids. In making these comparisons all laboratories were asked to measure amino acids from the same fraction, the total hydrolysate, and to use the same hydrolysis conditions; that is, the D/L ratios of the amino acids were to be determined in the total hydrolysate obtained by 6 *N* HCl hydrolysis for 20 hours at 110°C under nitrogen in a closed container.

Many laboratories reported several sets of analyses, while others listed averages. To simplify the data, only one set of mean values from each laboratory is given. The results reported come from automated ion-exchange chromatographic (IEC) and from gas chromatographic (GC) analyses. Where gas chromatography was used, two different approaches were taken: (1) Diastereomeric derivatives, either (+)-2-butyl esters (Kvenvolden et al., 1972) or L-prolyl peptide methyl esters (Hoopes et al., 1978) were chromatographed on nonchiral stationary phases. (2) Non-

**Table 2. Gas Chromatographic Parameters**

| Laboratory | Derivative | Stationary Phase |
|---|---|---|
| A | N-PFP-(+)-2-butyl esters | UCON 75H 90,000 and Carbowax 20 M |
| B | N-PFP-(+)-2-butyl esters | Carbowax 20 M |
| C | N-PFP-(+)-2-butyl esters | Carbowax 20 M |
| D | N-TFA-(+)-2-butyl esters and N-PFP-(+)-2-butyl esters | OV-225 and Carbowax 20 M |
| D* | N-TFA-2-propyl esters | N-propionyl-L-valine-t-butylamide polysiloxane (Chirasil-Val) |
| E | N-PFP-2-propyl esters | N-propionyl-L-valine-t-butylamide polysiloxane (Chirasil-Val) |
| F | N-TFA-2-propyl esters | 50% *n*-docosanoyl-L-valine-t-butylamide ester<br>50% *n*-octadecanoyl-L-valine-cyclohexyl ester |
| G | N-TFA-L-prolyl-peptidemethyl ester | SP-2250 (methylphenyl silicone) |
| H | N-TFA-(+)-2-butyl esters | EGA (ethylene glycol adipate) |
| I | N-TFA-(+)-2-butyl esters | Carbowax 20 M, OV-225, Dexsil 400 |

TFA = Trifluoroacetyl
PFP = Pentafluoropropionyl

chiral ester derivatives were chromatographed on chiral (optically active) stationary phases (e.g., Dungworth et al., 1976).

Table 2 lists the derivatives and gas chromatographic stationary phases used by the contributing laboratories, which are coded by letter. Nine laboratories reported the ratios of D-alloisoleucine/L-isoleucine based on analyses from automated IEC, and six laboratories listed these same ratios determined by GC (Table 3, Fig. 1). The range of ratios obtained by IEC (0.19–0.30) is slightly less than that derived from GC (0.18–0.31). Eleven of the analyses, however, are between 0.18 and 0.23, and it is likely that the true D-alloisoleucine/L-isoleucine ratio lies in this region. The mean of all IEC derived ratios is 0.22 ± 17%, while the mean of all GC determined ratios is 0.25 ± 21%. Of the two analytical methods, IEC and GC, used to ascertain the ratios of D-alloisoleucine/L-isoleucine, IEC is preferred because interfering peaks often obscure determinations of the true ratios in the GC analyses of these compounds.

D/L ratios of valine, alanine, leucine, proline, aspartic acid, phenylalanine, and glutamic acid were determined by GC techniques and the

Table 3. D-alloisoleucine/L-isoleucine

| Laboratory | Ratio | Method[a] |
|---|---|---|
| A | 0.31 | GC |
| A | 0.30 | IEC |
| B | 0.21 | GC |
| B | 0.20 | IEC |
| C | 0.23 | GC |
| D | 0.31 | GC |
| E | 0.20 | IEC |
| F | 0.18 | GC |
| G | 0.27 | IEC |
| I | 0.20 | IEC |
| I | 0.26 | GC |
| J | 0.23 | IEC |
| K | 0.21 | IEC |
| L | 0.19 | IEC |
| M | 0.22 | IEC |

[a] GC = Gas chromatography
IEC = Ion exchange chromatography

results are listed in Table 4 along with means, standard deviations, and relative standard deviations for each amino acid. The overall variation, as indicated by the relative standard deviations, is largest for valine and least for aspartic acid and alanine.

A complication occurs, however, in comparing the results of Table 4 because some of the reported data have been corrected by various pro-

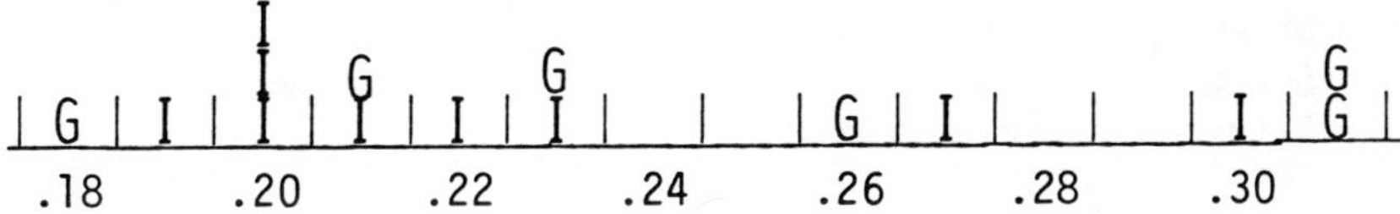


| | I | G | All ratios |
|---|---|---|---|
| Mean | 0.22 | 0.25 | 0.23 |
| Std. Deviation | .04 | .05 | .04 |
| Relative Std. Dev.(%) | 17 | 21 | 19 |

**Figure 1.** Histogram showing the distribution of ratios of D-alloisoleucine/L-isoleucine. I = ratios determined by IEC, G = ratios determined by GC.

**Table 4. D/L Ratios of Amino Acids by Gas Chromatography**

| Laboratory | Val | Ala | Leu | Pro | Asp | Phe | Glu |
|---|---|---|---|---|---|---|---|
| (+)-2-butyl derivatives[1] | | | | | | | |
| A | 0.21 | 0.44 | 0.32 | 0.50 | 0.46 | 0.38 | 0.22 |
| B | .14 | .46 | .24 | .41 | .49 | .29 | .18 |
| C | .17 | .42 | .27 | .45 | .47 | .30 | .21 |
| D | .20 | .46 | .30 | .46 | — | .38 | .22 |
| H | .24 | — | .26 | .40 | .47 | .31 | .21 |
| I | .22 | .50 | .28 | .58 | .48 | .40 | .24 |
| X̄ of [1] | .20 | .46 | .28 | .47 | .47 | .34 | .21 |
| SD " | .04 | .03 | .03 | .07 | .01 | .05 | .02 |
| RSD" | 18 | 7 | 10 | 14 | 2 | 14 | 9 |
| Optically active phase[2] | | | | | | | |
| D* | .16 | .45 | .26 | — | .50 | .26 | .21 |
| E | .20 | .43 | .26 | .40 | .43 | .29 | .19 |
| F | .14 | .43 | .25 | — | .48 | — | .17 |
| X̄ of [2] | .17 | .44 | .26 | .40 | .47 | .28 | .19 |
| SD " | .03 | .01 | .01 | — | .04 | .02 | .02 |
| RSD" | 18 | 3 | 2 | — | 8 | 8 | 11 |
| L-prolyl peptide derivatives | | | | | | | |
| G | .19 | .46 | .29 | .39 | — | .31 | .22 |
| X̄ of all above | .19 | .45 | .27 | .45 | .47 | .32 | .21 |
| SD " | .03 | .02 | .03 | .07 | .02 | .05 | .02 |
| RSD" | 18 | 5 | 9 | 15 | 5 | 15 | 10 |

X̄ = Mean
SD = Standard deviation
RSD = Relative standard deviation (%)

cedures and other data have not. To provide a basis for comparison, the applied corrections are listed in Table 5. Many of the corrections involve calculations to remove the effect of the optical impurity ((−)-2-butanol) which is always present in the chiral derivatizing reagent ((+)-2-butanol). Where the amount of the impurity is small (1%), the correction is negligible, equaling a reduction in D/L ratios of about 0.01. When the concentrations of the impurity are greater than 1%, the correction becomes significant.

The information listed in Table 4 and illustrated in Figure 2 provides

**Table 5. Corrections for Gas Chromatography**

| Laboratory | Correction |
|---|---|
| A | No correction. All D/L values would be reduced about 0.01 if a correction for the 1% (−)-2-butanol impurity in the (+)-2-butanol were made. |
| B | Correction made for 1% (−)-2-butanol impurity as follows: %D = 50(A − X)/(50 − X) Where A = apparent %D from chromatographic peak areas, and X = % impurity in (+)-2-butanol |
| C | Correction applied based on the extent of racemization of an L-amino acid standard carried through the same procedure as the sample. The following equation, solved for corrected D/L. was used: $\ln\left[\frac{1 + D/L}{1 - D/L}\right]_{\text{corrected}} = \ln\left[\frac{1 + D/L}{1 - D/L}\right]_{\text{sample}} - \ln\left[\frac{1 + D/L}{1 - D/L}\right]_{\text{standard}}$ |
| D | No correction. See A above. |
| D* | No correction |
| E | No correction |
| F | No correction |
| G | Correction applied based on chromatographic responses of standard mixtures of various proportions. |
| H | Correction made for 4%(−)-2-butanol impurity by means of the following equation: $D/L = (1 - p)(1 + p) = ((1 + q)t - (1 - q))/((1 + q) - (1 - q)t)$ where p = purity of amino acids, q = purity of alcohol, t = uncorrected D/L |
| I | Correction factors determined from the chromatographic response of racemic mixtures and applied to all ratios. |

a comparison of D/L ratios of amino acids obtained by the use of the two different GC approaches. The results from the two approaches fall in approximately the same ranges, with the spread of values usually less for ratios determined from chromatography on optically active phases; however, this observation may be due to the fact that only three laboratories reported ratios from this chromatographic approach. If the results from laboratories A and D were corrected for the 1% impurity in the derivatizing reagent, the difference in the mean ratios obtained by optically active derivatives and by optically active phases would be slightly smaller but would not be completely accounted for. Of the seven amino acids,

D/L ratios of aspartic acid and glutamic acid have the most narrow distributional ranges, as indicated on Figure 2. D/L ratio measurements for proline and phenylalanine have the widest spread of ratios. The relative standard deviations of D/L ratios of aspartic acid and alanine are the lowest. The D/L ratios of valine and for D-alloisoleucine/L-isoleucine show the highest deviations.

## CONCLUSIONS

The results of this interlaboratory comparison indicate a high variability in the determinations of D/L ratios of amino acids. Such high variability

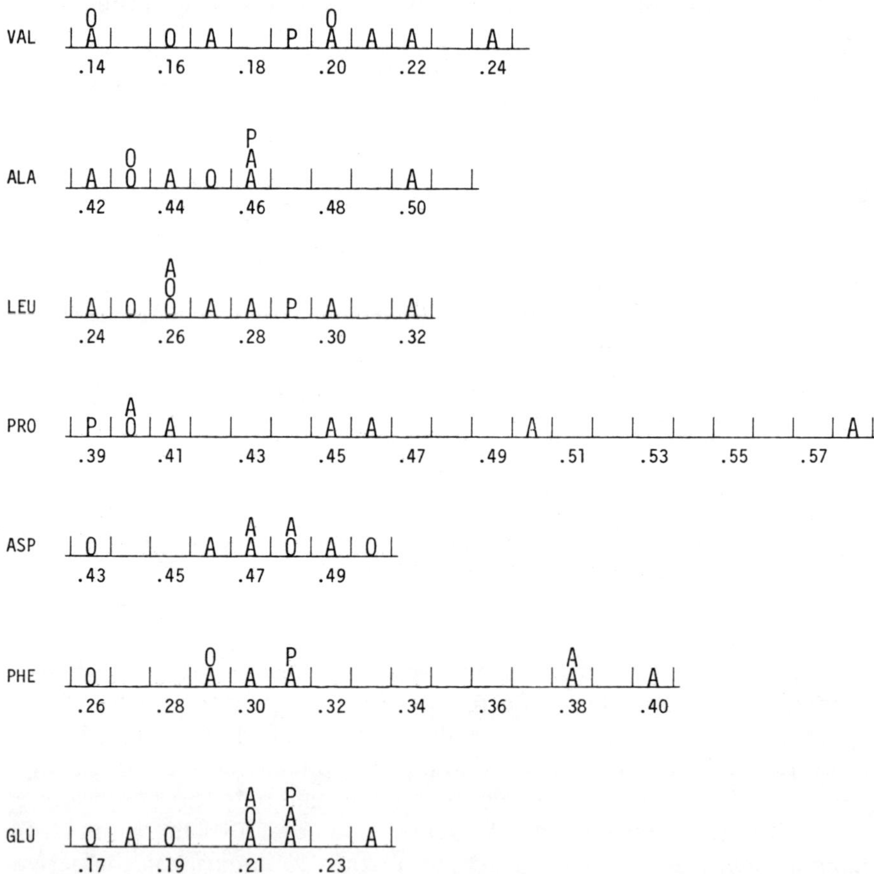


**Figure 2.** Histogram showing the distribution of D/L ratios of amino acids determined by chromatographic techniques. A = ratios obtained by use of (+)-2-butyl derivatives, P = ratios obtained by use of L-prolyl peptide derivations, O = ratios obtained by use of optically active stationary phases.

is somewhat disappointing but not totally unexpected. Different methods, techniques, and correction procedures undoubtedly have affected the reported ratios. Certainly the results of this study should be taken into account whenever comparisons are made between D/L amino acid measurements that have been generated in different laboratories. In the future, development of the methodology should continue in order to reduce the variability currently inherent in this work, and an exchange of information and samples should be encouraged in order to assess the uncertainties of the data.

## *REFERENCES*

Abelson, P. H., Organic constituents of fossils, *Carnegie Inst. Wash. Year Book 53,* 97–101, 1954.

Bada, J. L., E. Hoopes, D. Darling, G. Dungworth, H. J. Kessels, K. A. Kvenvolden, and D. J. Blunt, Amino acid racemization dating of fossil bones—I: Interlaboratory comparison of racemization measurements, *Earth Planet. Sci. Lett.*, **43,** 265–268, 1979.

Bada, J. L. and R. A. Schroeder, Amino acid racemization reactions and their geochemical implications, *Naturwissenschaften,* **62,** 71–79, 1975.

Dungworth, G., Optical configuration and the racemization of amino acids in sediments and in fossils—a review, *Chem. Geol.*, **17,** 135–153, 1976.

Dungworth, G. A. W. Schwartz, and L. van de Leemput, Composition and racemization of amino acids in mammoth collagen determined by gas and liquid chromatography, *Comp. Biochem. Physiol.*, **53B,** 475–480, 1976.

Hare, P. E. and P. H. Abelson, Racemization of amino acids in fossil shells, *Carnegie Inst. Wash. Year Book 66,* 526–528, 1968.

Hare, P. E. and R. M. Mitterer, Nonprotein amino acids in fossil shells, *Carnegie Inst. Wash. Year Book 65,* 362–364, 1967.

Hare, P. E. and R. M. Mitterer, Laboratory simulation of amino acid diagenesis in fossils, *Carnegie Inst. Wash. Year Book 67,* 205–210, 1969.

Hoopes, E. H., E. T. Peltzer, and J. L. Bada, Determination of amino acid enantiomeric ratios by gas liquid chromatography of the N-trifluoroacetyl L-prolylpeptide methyl esters, *J. Chromatog. Sci.*, **16,** 556–560, 1978.

Kvenvolden, K. A., Advances in the geochemistry of amino acids, in *Ann. Rev. Earth Planet. Sci.,* Vol. 3, Annual Reviews, Palo Alto, CA, pp. 183–212, 1975.

Kvenvolden, K. A., D. J. Blunt, and H. E. Clifton, Amino acid racemization in Quaternary shell deposits at Willapa Bay, Washington, *Geochim. Cosmochim. Acta,* **43,** 1505–1520, 1979.

Kvenvolden, K. A., E. Peterson, and G. E. Pollock, Geochemistry of amino acid enantiomers: Gas chromatography of their diastereomeric derivatives, in *Advances in Organic Geochemistry* 1971, H. R. von Gaertner and H. Wehner, eds., Pergamon, Braunschweig, pp. 387–401, 1972.

Kvenvolden, K. A., E. Peterson, J. F. Wehmiller, and P. E. Hare, Racemization of amino acids in marine sediments determined by gas chromatography, *Geochim. Cosmochim. Acta,* **37,** 2215–2225, 1973.

Masters, P. M. and J. L. Bada, Advances in amino acid racemization dating of bone and shell, in *Advances in Chemistry Series,* American Chemical Society, No. 171, 117–138, 1979.

Schroeder, R. A. and J. L. Bada, A review of the geochemical application of the amino acid racemization reaction, *Earth Sci. Rev.,* **12,** 347–349, 1976.

Williams, K. M. and G. G. Smith, A critical evaluation of the application of amino acid racemization to geochronology and geothermometry, *Origins of Life,* **8,** 91–144, 1977.

# KINETICS AND MECHANISMS

The application of amino acid biogeochemistry to geology and archeology depends on a knowledge of reaction rates. The problem is one of chemical kinetics and reaction mechanisms. It has been approached from several directions. One is an application of the wealth of information from physical-organic chemistry that has been obtained from studies of pure compounds under carefully controlled conditions. Another is a laboratory simulation of the processes in the complex environment of fossils where elevated temperatures substitute for long periods of time. A third direction uses nature as a laboratory in geological situations that are well known and where independent estimates of time and temperature can be made.

The first method has the advantage of precision and close control of the variables. But it carries the danger of over-simplifying the natural situation. Several of the papers using this approach, indicate the large number of factors governing the rates of reactions and rate of racemization of amino acids. Some of the important factors, such as the effective concentration of water in geological conditions are difficult to assess.

The simulation method is more realistic and is widely used. It combines many variables to yield empirical rate constants. However, the degree to which the laboratory reactions reflect the natural ones is subject to debate, and each case must be evaluated carefully. The sensitivity of new analytical methods permits the simulation experiments to be carried out at temperatures only slightly greater than the geological ones and should give a closer correspondence to reality.

The third method is powerful in the proper geological setting where the time, temperature, and chemical and geological history of the sediment is well known. However, such settings are rare.

The complexity of the problem is illustrated in papers that show how reaction rates are dependent on the matrices of individual species of fossils. There are variables in the chemistry that are not well known at present.

Continued progress depends on theoretical guidance from reaction kinetics and mechanisms. Several of the controversies, which arose at the meeting, can in principle be resolved by combined field and laboratory studies.

# Kinetics and Mechanism of Amino Acid Racemization in Aqueous Solution and in Bones

JEFFREY L. BADA

M.-Y. SHOU*

*Amino Acid Dating Laboratory, Scripps Institution of Oceanography, and Institute of Marine Resources, University of California, 92093*

## *ABSTRACT*

Amino acid racemization is an important new method for dating fossil materials. To be confident that this method is reliable, it is imperative that the kinetics and mechanism of the reaction be completely understood. In this paper we first discuss the kinetics and mechanism of free amino acid racemization. From studies at different pH values, it is possible to calculate the racemization rates for the various ionic forms of amino acids. These results are consistent with a carbanion mechanism. The pH studies also indicate that the rate of racemization for the various amino acids would be independent of pH in the range of approximately 3 to 9 at the temperatures found in nature.

The kinetics and mechanism of amino acid racemization in bones are also discussed. In historically dated fossil bones from localities that have similar present-day temperatures but whose other environmental characteristics differ drastically, the racemization reaction of aspartic acid was found to obey reversible first-order kinetics. The rates of racemization determined for these samples is consistent with expected rates, based on the temperature of the sites.

* Present address: Virginia Institute of Marine Science, Gloucester Point, Virginia, 23062.

Studies of the effects of pH on aspartic acid racemization rates in bones at 96°C indicate that the rates are unaffected by pH over the pH range approximately 3 to 9. This is probably due to buffering of the reaction by the bone matrix itself. Metal ions were also found to have no apparent effect on racemization rate in bones. These results strongly support the conclusion that temperature is the major factor controlling racemization rates in bones, and that other factors, such as leaching, pH, humidity, and metal-ion catalysis are unimportant. A comparison of the relative racemization rates for amino acids in bones with the rate in other systems indicates that racemization takes place in the proteinaceous fraction of bones.

## INTRODUCTION

Investigations on racemization reactions of amino acids and their geochemical and biological implications have been an area of research for over ten years (Bada and Schroeder 1975a; Schroeder and Bada 1976; Williams and Smith 1977; Vicar 1977). During this period there has been increasing use of amino acid racemization reactions for dating fossil materials as well as for age determinations of living mammals. There have been studies on the use of the reaction in estimating the temperature history of various environments. The possible role played by amino acid racemization in the aging process is also being investigated (Helfman et al., 1977).

These studies are based on the fact that all the amino acids commonly found in proteins, with the exception of glycine, can exist in two isomeric forms called the D- and L-enantiomers. The physical and chemical properties of these enantiomers are identical, except that they rotate plane polarized light in equal but opposite directions. Proteins of living organisms consist solely of the L-enantiomers; only a few microbial peptides contain D-amino acids. Since under conditions of chemical equilibrium both enantiomers would be present in equal quantities, a living organism maintains a disequilibrium state through the use of enzymes which stereoselectively utilize only the L-enantiomers. Thus, after an organism dies or a metabolically stable protein has been synthesized, the metabolic reactions that maintain the disequilibrium state cease, and a process called racemization begins. In racemization, the L-amino acids are reversibly converted into the corresponding D-amino acids, and this reaction continues until there are equal amounts of both enantiomers present (i.e., D/L ratio = 1.0). The racemization reaction can be written as

$$\text{L-amino acid} \underset{k}{\overset{k}{\rightleftharpoons}} \text{D-amino acid} \tag{1}$$

where $k$ is the first-order rate constant for interconversion of the enantiomers. The kinetic equation for this reaction is

$$\ln\left\{\frac{1 + D/L}{1 - D/L}\right\} - \ln\left\{\frac{1 + D/L}{1 - D/L}\right\}_{t=0} = 2\cdot k\cdot t \qquad (2)$$

where D/L is the amino acid enantiomeric ratio at any particular time, $t$. The $t = 0$ term is needed to account for the fact that the initial D/L ratio may not be exactly zero because some slight racemization occurs during sample preparation. For diastereomenic amino acids (i.e., isoleucine, threonine, hydroxyproline) equation 2 must be modified, since the equilibrium ratios of the diastereomers is not 1.0 (for the kinetic equation for diastereomeric amino acids see Bada and Schroeder 1975; Schroeder and Bada 1976).

Although the racemization reaction was found to occur in strongly acidic and basic solutions nearly a century ago, it was not until about ten years ago that it was recognized that this reaction also takes place at neutral pH at comparable rates (6).

The rate at which racemization occurs depends upon the particular amino acid and the temperature. In Table 1 we have listed the racemization half-lives (i.e., the time required to obtain a D/L ratio of 0.33) for several free amino acids at pH 7.6 and 0° and 25°C. The half-lives range from a few thousand years for aspartic acid at 25°C to millions of years for isoleucine at 0°.

Fossil materials have been found to contain both D- and L-amino acids, and the D/L ratio of a particular amino acid increases with increasing geological age of a fossil. From equation 2 it is apparent that there are two variables that can affect the extent of racemization of an amino acid in a fossil: time ($t$) and the rate ($k$) at which the racemization takes place, which is primarily a function of temperature. To use racemization to cal-

**Table 1. Racemization Half-Lives (years) for Several Free Amino Acids at 0° and 25°C and pH 7.6[a]**

| Amino | 0°C | 25°C |
|---|---|---|
| Aspartic Acid | 430,000 | 3,500 |
| Alanine | $1.4 \times 10^6$ | 12,000 |
| Isoleucine | $\sim 6 \times 10^6$ | ~50,000 |

[a] Taken from Bada and Schroeder 1975a, and Bada 1971.

culate one of these variables, the other must be known. For example, to date a fossil from a particular site, a value of $k$ for that site must be available. The value of $k$ for a particular locality can be determined by a "calibration" procedure by which a fossil of known age from the area of interest is analyzed and its D/L enantiomeric ratio determined. The age and the enantiomeric ratio of the calibration sample are then substituted into the equation 2, and an in situ $k$ value is thus determined. This "calibrated" interconversion rate constant is an average value integrated over the age of calibration sample. Thus any variation in the temperature or other environmental parameters of the locality over the depositional time period of the "calibration" sample is evaluated. Once a $k$ value has been determined at a particular site, it can be used, with certain limitations, to date other samples from the general area. This "calibration" procedure was first used for fossil bones (Bada and Protsch 1973) but recently has been used to date other fossil materials (Mitterer 1975; Lee et al., 1975).

In order to validate the racemization dating method, it is important to have a complete understanding of the kinetics and molecular mechanism of the reaction. For example, in aqueous solutions, racemization rates are affected by variables such as temperature, pH, ionic strength, buffer concentrations and metal-ion catalysis (Bada 1975a; Schroeder and Bada 1976; Bada 1971). By understanding how these effects operate, it is possible to predict whether they will be important in effecting the degree of racemization in a fossil and to estimate what effects these factors might have on amino acid–derived ages.

In this paper we will first discuss the kinetics and mechanism of free amino acid racemization in aqueous solution. Next, we will discuss the kinetics and mechanism of amino acid racemization in bones. We will use these results to predict what effect various environmental factors might have on the racemization-deduced ages of fossil bone.

## KINETICS AND MECHANISTIC STUDIES OF FREE AMINO ACID RACEMIZATION IN AQUEOUS SOLUTION

The racemization reactions of free amino acids in aqueous solutions are the simplest to investigate. Conditions such as temperature, pH, ionic strength, and buffer concentrations can be controlled, and the ionic state of the amino acids can be ascertained. During the last several years, we have carried out detailed investigations of the kinetics of racemization for several amino acids in aqueous solution. The results of these investigations are summarized in Figure 1.

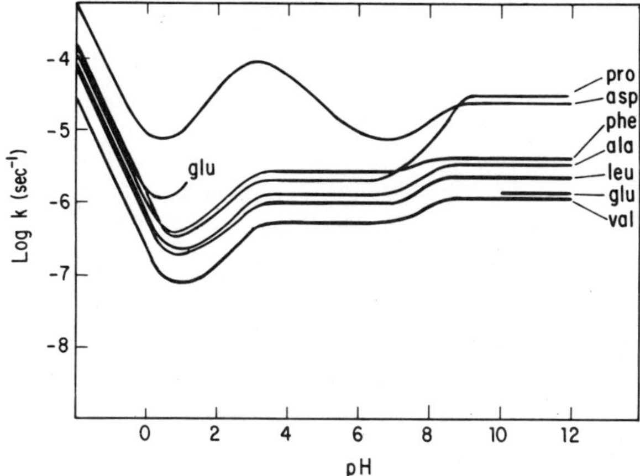


**Figure 1.** Rate of interconversion (plotted as log $k$) versus pH for several amino acids at 142°C. The data from which this figure was drawn are given in Bada 1972a, and Shou 1979.

From Figure 1 it can be seen that log $k$ from monocarboxylic amino acids shows three distinct regions: an acid-catalyzed region for pH less than 1; a region called Plateau I between pH 3 and approximately 6.5, where $k$ is independent of pH; and a region between pH 9 and 12 called Plateau II, where $k$ is again independent of pH. In addition to the pH regions described above for monocarboxylic amino acids, the kinetic curves for dicarboxylic amino acids (Bada 1972a) show a maximum pH of approximately 3.

The following equation (Bada 1972a; Shou 1979) can be written to describe the rate of interconversion of the D- and L-enantiomers for monocarboxylic amino acids between pH −1 and 12*:

$$k_{\mathrm{obs}}(\mathrm{AA})_{\mathrm{T}} = k^{\mathrm{H}^+}(\mathrm{AA}^{+0})(\mathrm{H}^+) + k^{+0}(\mathrm{AA}^{+0})(\mathrm{OH}^-) + k^{+-}(\mathrm{AA})^{+-})(\mathrm{OH}^-) \quad (3)$$

For dicarboxylic amino acids the corresponding equation is

$$k_{\mathrm{obs}}(\mathrm{AA})_{\mathrm{T}} = k^{\mathrm{H}^+}(\mathrm{AA}^{0+0})(\mathrm{H}^+) + k^{0+0}(\mathrm{A}^{0+0})(\mathrm{OH}^-) + k^{0+-}(\mathrm{AA}^{0+-})(\mathrm{OH}^-) + k^{-+-}(\mathrm{AA}^{-+-})(\mathrm{OH}^-) \quad (4)$$

The various $k$ values are the rate constants for the interconversion of the D- and L-enantiomers of the indicated ionic species of the amino acids.

* The abbreviations used here are the same as those used previously (Bada 1972a).

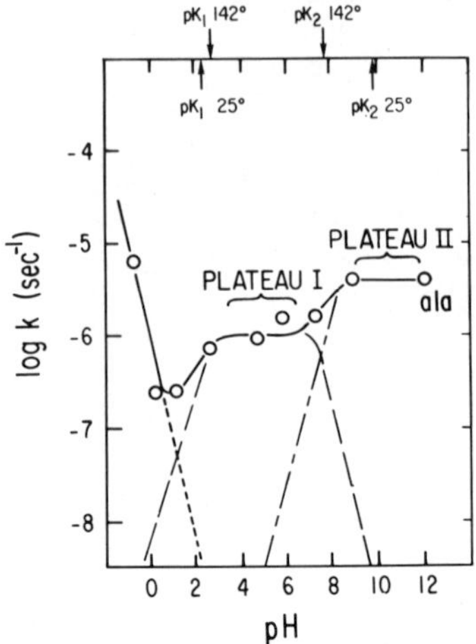


**Figure 2.** A plot of log $k$ versus pH for alanine at 142°C showing the various reactions given in equation 3. Short dashed lines indicate rate of interconversion of acid catalyzed reaction; long dashed lines, rate of interconversion of ala$^{+0}$; short and long dashes, rate of interconversion of ala$^{+-}$. Arrows indicate the values of the pK's at 25°C and 142°C. Data given in this figure are taken from Shou 1979.

Based on the extrapolated $pK$ values, the interconversion rates of various reactions of alanine at 142°C and aspartic acid at 100°C were calculated as the function of pH, as shown in Figures 2 and 3, respectively. As can be seen, eqs. (3) and (4) completely describe the pH dependence of $k_{obs}$ values of these amino acids. Using the above equations, the $k$ values at 142°C for the various ionic forms of the different amino acids were calculated. These values are listed in Table 2.

The results of Table 2 show that the relative order of racemization rates for all amino acids fall into the sequence $k^{+0} > k^{+-}$, which is in agreement with the carbanion mechanism for racemization (Bada and Schroeder 1975a; Bada 1972a; Neuberger 1948), since increasing the electron-withdrawing and resonance-stabilizing capacities of the substituents attached to the α-carbon causes a corresponding increase in the value of $k$.

The Hammett equation can be used to test whether the electron-withdrawing capacity of the R-substituents of the various amino acids is the principal factor that determines the relative order of racemization rates of a certain ionic species. The electron-withdrawing ability of a particular substituent is measured by the quantity $\sigma^*$, the polar substituent or Taft constant, which is an indicator of how well a substituent can stabilize a carbanion. Plots of log $k$ versus $\sigma^*$ for the ionic species $A^{+0}$ and $A^{+-}$

Table 2. The $k$ ($sec^{-1}$ $mole^{-1}$) Values of the Ionic Species of Several Amino Acids at 142°C and $I = 0.1$[a]

| | Ala | Val | Leu | Pro | Phe | Glu | Asp |
|---|---|---|---|---|---|---|---|
| $k_{int}^{H^+}$ | $9.84 \times 10^{-7}$ | $2.59 \times 10^{-7}$ | $9.62 \times 10^{-7}$ | $1.46 \times 10^{-6}$ | $8.97 \times 10^{-7}$ | $1.30 \times 10^{-6}$ | $4.12 \times 10^{-6}$ |
| $k_{int}^{+0}$ | $1.05 \times 10^{3}$ | $3.72 \times 10^{2}$ | $7.08 \times 10^{2}$ | $2.82 \times 10^{3}$ | $2.51 \times 10^{3}$ | — | $5.90 \times 10^{5}$ |
| $k_{int}^{+-}$ | $3.55 \times 10^{-2}$ | $1.32 \times 10^{-2}$ | $3.09 \times 10^{-2}$ | $6.92 \times 10^{-2}$ | $1.41 \times 10^{-1}$ | — | $1.78 \times 10^{2}$ (o+ −) |
| | | | | | | | $3.55 \times 10^{-2}$ (− + −) |

[a] Taken from Bada 1972a, and Shou 1979.

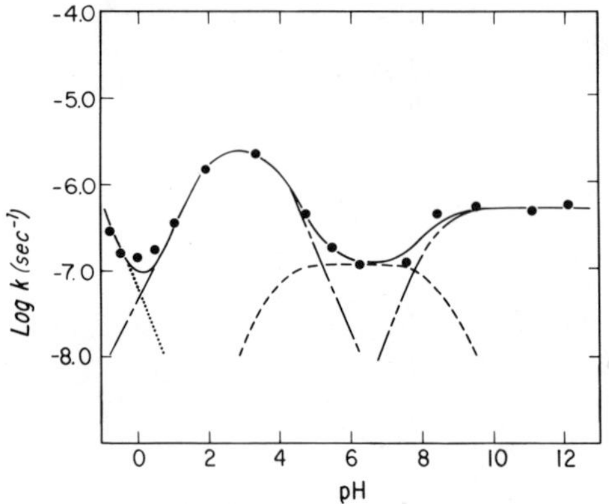


**Figure 3.** A plot of log $k$ versus pH for aspartic acid at 100°C showing the various reactions given in equation 4. Dots indicate rate of interconversion of acid catalyzed reaction; short and long dashes, rate of interconversion of $asp^{0+0}$; short dashes, rate of interconversion of $asp^{0+-}$; one long and two short dashes indicate rate of interconversion of $asp^{-+-}$. The data are taken from Bada 1972a.

are given in Figure 4. (The values used for $\sigma^*$ and the methods used to obtain them are described in Shou 1979). These results show that there is an excellent correlation between the $k$ values for the various ionic species and the $\sigma^*$ values. This implies that the electron-withdrawing capacity of the R-substituent of the various amino acids is the principal factor in determining relative rates of racemization for a given ionic species. It is important to note at this point that since at any particular pH, $k_{obs}$ observed is a function of both $k^{+0}$ (or $k^{+-}$) and $pK_1$ (or $pK_2$), any attempt to correlate the $k_{obs}$ with $\sigma^*$ is unjustified (for example, see Smith

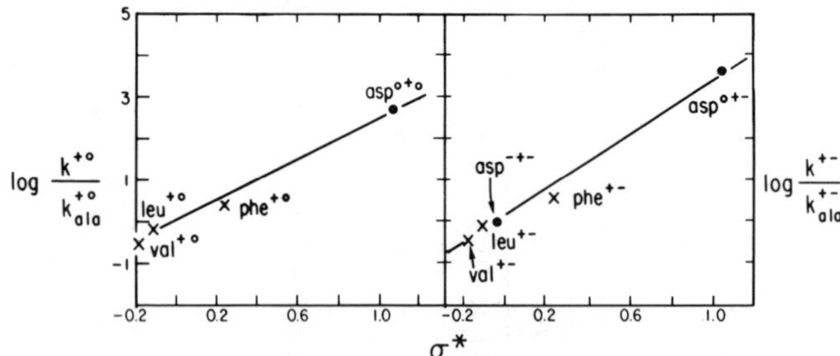


**Figure 4.** A plot of log $k/k_{ala}$ versus $\sigma^*$ for the ionic forms $A^{+0}$ and $A^{+-}$ at 142°C. The $k$ values relative to alanine are used, since the $\sigma^*$ for this amino acid is 0.0. The lines shown were obtained by a least-squares fit of the data. The correlation coefficient ($R$) is 0.996 for the $AA^{+0}$ results and 0.995 for $AA^{+-}$. The values of $\sigma^*$ were obtained as described in Shou 1979.

et al., 1978), especially at pH 7–9 where a mixture of the ionic reactions actually takes place. The values of $k$ for the particular ionic species (i.e., $k^{+0}$ or $k^{+-}$) should only be used in $\sigma^*$ correlations.

The relative $k$ values for the ionic species of the various amino acids are given in Table 3. These relative rates provide an important tool for ascertaining which ionic species is undergoing racemization at any particular pH. For example, a comparison of the relative $k_{obs}$ values at pH 7.6, which are also listed in Table 3, with those for the various ionic forms, indicate that at this pH, it is the form $AA^{+0}$ that is the principal reacting species for all the amino acids, with the exception of aspartic acid, proline, and phenylalanine. For phenylalanine, the principal species undergoing racemization at pH 7.6 is $AA^{+-}$ while for proline it is a mixture of $AA^{+0}$ and $AA^{+-}$. For aspartic acid $AA^{-+-}$ is the predominant reacting form with a small contribution from $AA^{0+-}$. Comparisons of the relative racemization rates, such as those presented here for free amino acids, should be useful indicators of the molecular species that undergoes racemization in fossils.

Our free amino acid results have important implications concerning various aspects of amino acid racemization at the lower temperatures commonly found in nature. To estimate racemization rates at low temperatures, the Arrhenius activation energy, $E_a$, is needed. Values of $E_a$ have been determined at pH 7 and pH 7.6 (Bada 1971; Smith et al., 1978; Dungworth et al., 1975). The values range from approximately 28 to 31

**Table 3. Relative Racemization Rates at 142°C of the Ionic Forms of Several Amino Acids Compared with the Observed Relative Racemization Rates at pH 7.6**

| Amino Acid | $k^{H+}$ | $k^{+0}$ | $k^{+-}$ | Observed $k_{int}$ at pH 7.6 |
|---|---|---|---|---|
| Aspartic Acid | | | | 7.5 |
| o+o form | 4.2 | $2.3 \times 10^3$ | — | — |
| o+− form | — | — | $5.0 \times 10^3$ | — |
| −+− form | — | — | 1.1 | — |
| Alanine | 1.0 | 1.5 | 1.1 | 1.4 |
| Glutamic acid | 1.4 | — | — | ~0.2[a] |
| Proline | 1.5 | 4.0 | 2.2 | 2.8 |
| Phenylalanine | 0.93 | 3.5 | 2.2 | 2.3 |
| Leucine | 1.0 | 1.0 | 1.0 | 1.0 |
| Isoleucine | — | — | — | 0.7 |
| Valine | 0.27 | 0.53 | 0.43 | 0.52 |

[a] Ratio relative to leucine at 161°C and pH 8.5 (14).

kcal mole$^{-1}$. In the past, there was no explanation for the different $E_a$ values determined by various workers other than variability in the precision of the experiments. However, if we look at Figure 1, it is apparent that the pH range 7 to 7.6, where measurements have been carried out in the past, is actually in the transition zone between Plateau I and Plateau II. As discussed above, the racemization reaction in this transition region may consist of a mixed reaction of different ionic species that have different activation energies (Bada 1972a). Hence slight differences in the pH of the buffers used by various workers could cause sizable differences in the $k_{obs}$ values and thus in the $E_a$ values themselves. This probably explains why the $E_a$ values determined by different workers show a large variability.

Actually, the use of $k_{obs}$ values in $E_a$ plots is not correct. The $k$ values for the ionic species should be plotted. If this is done, the $E_a$ values for the ionic species can be used to calculate the $k$ values of the ionic species at low temperatures, and from these values and equation 3 or 4, an estimate of the rate of racemization at pH 7 or pH 7.6 can be calculated.

Even though most of our measurements have been carried out at only 142°C, and thus $E_a$ parameters for each ionic species for the various amino acids are not available, except for aspartic acid (Bada 1972a), an important conclusion about the pH dependence of the amino acid racemization reaction at low temperatures can be made from our results. As can be seen in Figures 1–3, the racemization reaction is effectively independent of pH between approximately 3 and 6.5 pH. The width of this pH region, called Plateau I, depends mainly on the values of $pK\ 1$ and $pK\ 2$ for each amino acid. The width of the Plateau I region increases with decreasing temperature because, although $pK\ 1$ decreases only slightly with decreasing temperature, $pK\ 2$ increases greatly with decreasing temperature. For example, for alanine at 142°C, $pK\ 1 = 2.66$ and $pK\ 2 = 7.77$, while at 25°C $pK\ 1 = 2.35$ and $pK\ 2 = 9.87$; these values are summarized in Figure 2. Therefore, the width of Plateau I extends from pH 3 to 6.5 at 142°C but will increase to pH 3 to approximately 8.5–9 at 25°C. Since $pK\ 1$ does not change as much as $pK\ 2$, Plateau I will extend mainly toward higher pH at lower temperatures. This extension of Plateau I at low environmental temperatures has very important geochemical implications. Natural waters, such as lakes, rivers and the oceans, generally have temperatures on the order of $-2$ to 28°C and pH values (Stumm and Morgan 1970) in the range of 6–9. In this temperature and pH region it is apparent that the racemization rate of free amino acids would be independent of pH.

In summary, the results presented here further demonstrate that the amino acid racemization reaction follows the carbanion mechanism and hence the electron-withdrawing and resonance-stabilizing capacities of the substituents attached to the α-carbon are the principal factors that determine the relative order of racemization rates of the various amino acids. Furthermore, using the mechanism of racemization determined at elevated temperatures, it is possible to predict the pH dependence of the amino acid racemization reaction at low environmental temperatures. These predictions indicate that at low temperatures, the racemization reaction of free amino acids would be effectively independent of pH in the pH range commonly found in nature. Furthermore, these experiments illustrate that it is necessary to understand completely the molecular mechanism of the reaction at elevated temperatures in order to be sure that the extension of high temperature experiments to low environmental natural temperatures is indeed realistic. When this is done, then the elevated temperature can be used to make reliable predictions about the kinetics at low temperatures.

## KINETICS AND MECHANISM OF AMINO ACID RACEMIZATION IN BONES

During the last several years, there has been an increasing number of uses of the amino acid racemization reaction to determine the ages of fossil bones (Masters and Bada 1978; Bada et al., 1979). Fossil bones from virtually all over the world have been dated using this method. Some of the sites that have been dated include: Klasies River Mouth caves, South Africa; Olduvai Gorge, Tanzania; Old Kingdom sites in Egypt; several anthropologically significant caves on Mt. Carmel, Israel; the cave of Arago, French Pyrennes; Stránská skála and Külna, Czechoslovakia; and numerous sites in North America. In general, excellent agreement has been obtained when racemization ages have been compared with ages deduced from other independent evidence, including radiocarbon, historical records, or geological information (Masters and Bada 1978; Bada et al., 1979; Bada et al., 1974; Bada and Helfman 1975b).

However, some of the North American ages have generated intense controversy. In 1974 and 1975, the Amino Acid Dating Laboratory at Scripps determined that the ages of several human skeletons found in the coastal Southern California region were on the order of 40,000–50,000 years old (Bada et al., 1974; Bada and Helfman 1975b). These ages provided new evidence that human beings had migrated into the New World considerably earlier than was then generally accepted by North American archeologists. Primarily because of the racemization ages for these Southern California paleoindian skeletons, the validity of the racemization technique applied to bones has been questioned. Rather than continue to provide additional correlations between racemization-deduced ages and those ages deduced from other evidence, we would like to address some of the criticisms of the validity of the racemization dating method as applied to fossil bones.

Kinetic studies at elevated temperatures (Bada 1972b; Hare 1974a), using modern bone samples, show that racemization in bones follows the expected reversible first-order kinetics (i.e., equation 2). This is in contrast to carbonate fossils, in which reversible first-order kinetics are not obeyed (Bada and Schroeder 1975a). Because of this kinetic complication, it is more difficult to assign an absolute age to a carbonate fossil than it is to a fossil bone. It has been suggested, however, that certain effects, such as leaching by ground waters, might cause a deviation from reversible first-order kinetics for bones found in natural environments (Hare 1974a, 1974b, 1979). This deviation from first-order kinetics could then cause racemization-deduced age, using the "calibration" procedure outlined previously, to be in error. To further demonstrate that reversible first-order kinetics are followed in bones under natural conditions, we have analyzed several known-age bones from sites that have essentially identical present-day mean annual air temperatures but whose other environmental characteristics differ considerably. These bones, their measured D/L aspartic acid ratios, and a summary of the environmental characteristics of the sites, are given in Table 4. A plot of these results in the form of equation 2 is presented in Figure 5.

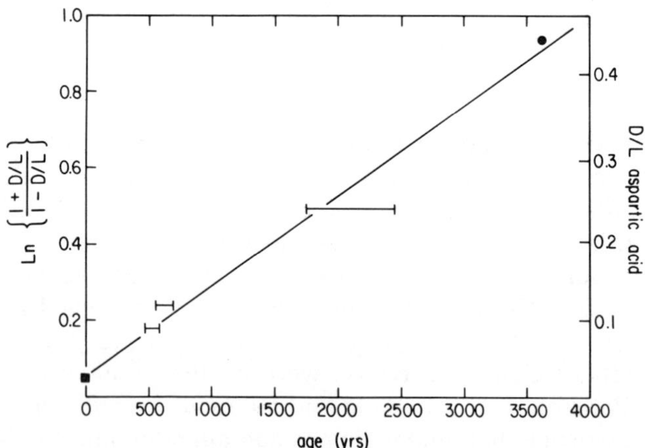


**Figure 5.** Racemization kinetics of aspartic acid in fossil bones from sites with nearly identical temperatures (data given in Table 4). A least-squares fit of the data ($R = 0.993$) yields the indicated straight line which has a slope ($2k_{asp}$) of $2.37 \times 10^{-4}$ yr$^{-1}$.

**Table 4. Site Description and Racemization Results for Fossil Bones from the Northern Sudan (Buhen Horse) and Mindanao, Philippines[a]**

| Site | Type of Environment | Present-day Mean Annual Air Temp. (°C) | Annual Precipitation (cm) | Age (yrs) | D/L Aspartic Acid |
|---|---|---|---|---|---|
| Butuan, Agusan del Norte, Mindanao, Philippines | estuarine swamp | 27°C | 230 | Ming Dynasty (15th–16th cent.) | 0.090 |
| | | | | Yuan Dynasty (1260–1368 A.D.) | 0.119 |
| Huluga, Cagayan de Oro City, Misamis Oriental Mindanao, Philippines | limestone burial crypt located ~30m high in a cliff | 27°C | 230 | 500 B.C.–200 A.D. | 0.241 |
| Buhan Horse, near Wadi Halfa, Nubia, Sudan | arid desert; skeleton sealed into brickwork of a New Kingdom fortress | 26°C | 0.4 | 1675 B.C. | 0.438 |

[a] The temperature and precipitation values were obtained from Wernstedt 1972. The ages were provided by Ms. Linda Burton, Xavier University, Cagayan de Oro City, Mindanao, Philippines. The samples were all hydrolyzed 4 hours.

As can be seen, reversible first-order kinetics are certainly observed for these bone samples even though they come from drastically different environments. We believe that these samples represent the optimum situation for detecting the "leaching effects" proposed by Hare (1974a, 1974b, 1979). For example, the Buhen Horse comes from the extremely arid environment of the Nubian desert of Northern Sudan. But the Mindanaoan samples come from, in one case, a waterlogged burial from an estuarine swamp, and in another case, from a dry limestone crypt. The precipitation in the Mindanaoan region exceeds that found in the Northern Sudan by over a factor of 500. If leaching could cause a deviation from reversible first-order kinetics, as Hare has claimed in his elevated temperature experiments (Hare 1974a, 1974b, 1979), it is certainly not detectable in these natural samples.

The slope of the line in Figure 5 should correspond to the value of $2k_{asp}$ at the temperature of the localities where the samples were obtained. Using the procedures described elsewhere (Bada et al., 1979), it is possible to calculate that this measured $k_{asp}$ value corresponds to a temperature of 26.4°C, which is in excellent agreement with the actual present-day mean temperature (see Table 4). This correlation provides additional evidence that temperature is the major factor affecting amino acid racemization in fossil bones and that other factors, such as leaching and moisure content, have little if any effect on racemization rates.

The pH of the environment is another factor that has been suggested (Smith et al., 1978; von Endt et al., 1975) as a variable which might affect racemization ages of bones. In the preceding section it was found that at low temperatures characteristic of those in nature, pH has no effect on the free amino acid racemization rates in the pH range 5–9. To test what effect pH might have on the racemization rate in bones, we conducted a series of experiments in which the racemization kinetics in modern bone fragments were studied as a function of pH. Small bone fragments (~1–1.5 g) were sealed in Pyrex test tubes containing the appropriate buffer (Shou 1979). The test tubes were heated at 96°C for various lengths of time. After heating, the samples were removed from the tube, cleaned and processed, using the standard procedures in amino acid racemization analyses of fossil bones (Masters and Bada 1978; Bada et al., 1979). The results of aspartic acid measurements are shown in Figure 6. These results demonstrate that the $k_{asp}$ value in bone is effectively independent of pH in the range pH 3–9 at 96°C. In contrast, the interconversion rate of free aspartic acid in aqueous solution shows an appreciable pH dependence in the pH range 3–9 (see Figs. 1 and 3). The reason for this pH independence of the interconversion rate in bone is probably that the bone matrix itself is an excellent buffer and controls the pH of the system.

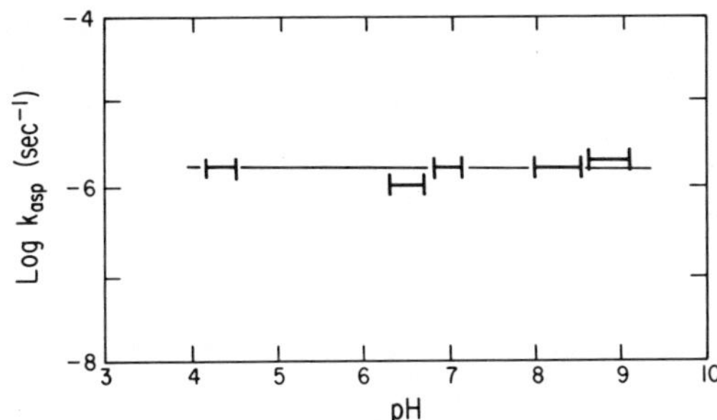


**Figure 6.** A plot of log $k_{asp}$ versus pH for bones at 96°C. The data from which this figure was drawn is given in Shou 1979.

Thus any fluctuations in environmental pH are not actually important. At extreme pH values, where the bone matrix itself begins to dissolve, pH could affect racemization rates in bone; but since bones are not preserved in these extreme pH environments, this effect obviously would have no direct bearing on the validity of racemization-deduced ages. We should emphasize that we have carried out our pH experiments at only one temperature; thus the extrapolation of these results to low temperatures must be viewed with some caution. However, if our interpretation is correct that the bone matrix determines the effective pH of the system, then it seems apparent that this effect would also be the primary controlling factor at low temperatures.

Catalysis by metal ions has also been suggested as a factor that could influence the validity of a racemization-derived age (Davis and Treloar, 1977). Metal ions are known to be effective catalysts of free amino acid racemization in aqueous solution (Bada and Schroeder 1975a; Bada 1971). To test whether metal ions could catalyze amino acid racemization in bones, we analyzed a bone from the southern Utah desert (near the town of Moab) which had been impregnated with copper. In contrast to normal fossil bone samples, this bone was dark green, indicating that copper from a surrounding ore deposit had been diffused into the sample. The radiocarbon age of the sample was determined to be 210 ± 70 years (UCLA 1942; tree-ring "calibrated" age is mid-17th century). Analysis of this bone yielded a D/L aspartic acid ratio that was only slightly greater than that of a modern bone sample carried through the same processing procedure. Since this bone is of very young age and comes from an environment that has relatively cool temperatures, there should be very little aspartic acid racemization in this sample, and indeed that is the case. These results suggest that, at least for this particular sample, metal ions do not catalyze the racemization reaction in bones.

In demonstrating the reliability of amino acid-derived ages, another important factor is the reproducibility of the "calibration" rate constant used to date fossils from a particular area. We have suggested (Bada and Protsch 1973; Masters and Bada 1978; Bada et al., 1974, 1979; Bada and Helfman 1975b) that once a "calibration" constant ($k$) has been determined for a particular site or region, this value can be used to date other samples from the immediate vicinity; the only condition is that the present-day mean annual temperature of the calibration site is effectively identical to that of the site to be dated. To demonstrate that this approach is valid, we have analyzed several samples from the Olduvai Gorge region in Tanzania. We have analyzed bones of Holocene and Upper Pleistocene ages from Olduvai Gorge itself and from the Nasera Rock Shelter, which is located about 30 kilometers northwest of the Olduvai Gorge (Mehlman 1977). These analyses are summarized in Table 5. As can be seen, the Holocene and the Upper Pleistocene rate constants of these two sites are effectively identical. We believe that these results provide convincing evidence that once a "calibration" constant is determined at a particular site, it can be used to date other bones nearby as long as the temperature of the site to be dated is the same as the calibration site. It is also important to emphasize that the physical characteristics of the two sites in Table 5 differ considerably. The Olduvai samples are from open-air localities exposed to the highly variable and often extreme physical conditions of the Serengeti Plain in East Africa. The Nasera Rock Shelter is a protected environment, which has had an extensive human occupation history, located in the overhang of a large rock outcroping in the Serengeti. Again, the concordance of the rate constants determined for the two sites suggests that the different physical characteristics of the environment have little effect on the amino acid racemization rates.

Finally, as we emphasized in the section on free amino acids, it is important to understand the molecular mechanism of racemization whether in fossils or in aqueous solution. Although there seems little doubt that a carbanion intermediate is always involved in racemization, there is a question concerning what is the major species which actually undergoes racemization in fossil materials. The relative rates of racemization of the various amino acids provide an important method for answering this question. In Table 6 we have listed the relative racemization rates of amino acids in various systems. Comparison of these relative values shows that fossil bones most closely match the relative ratios in various proteins. Thus it is the proteinaceous fraction that is the principal component undergoing racemization in fossil bones. Free amino acids comprise a small and negligible component in fossil bones (Schroeder and Bada 1976), and thus have little significance in determining the extent of

**Table 5. Comparison of $k_{asp}$ Values at Nasera Rock Shelter and Olduvai Gorge[a]**

| Site Location | | Radiocarbon Age (yrs) | D/L Aspartic Acid | $k_{asp}$ ($yr^{-1}$) |
|---|---|---|---|---|
| Nasera Rock Shelter, Ngorongoro Conservation Area, Eastern Serengeti Plain, Tanzania | Holocene-level 3A | 2180 ± 200 (USGS-438)[b] | 0.212 | $6.7 \times 10^{-5}$ |
| | Upper Pleistocene-level 5A | 21,600 ± 400 (ISGS-445)[b] | 0.428 | $1.8 \times 10^{-5}$ |
| Olduvai Gorge, eastern Serengeti Plain, Tanzania | Holocene-beneath Namord ash | 1360 ± 40 (LJ2979)[c]<br>1350 ± 100 (LJ3330)[d] | 0.165 | $7.4 \times 10^{-5}$ |
| | Upper Pleistocene-Naisuisui beds | 17,550 ± 1000 (UCLA 1695)[b] | 0.32 | $1.5 \times 10^{-5}$ |

[a] A description of Nasera can be found in Mehlman 1977. The Olduvai results were taken from Bada and Protsch 1973, and Bada and Helfman 1975b.

[b] Organic fraction ("collagen").

[c] Associated land snail.

[d] Amino acids isolated from same bone used for racemization analysis.

**Table 6. Relative Racemization Rates of Several Amino Acids in Various Systems and in Fossil Bones and Shells**

| Amino Acid | Free Amino Acids: pH 7.6; 142°C | Proteins[a] 0.1 *N* NaOH; 70°C | Proteins[b] pH 7.6; 135°C | Typical Uncontaminated Fossil Bones;[c] (total amino acids) | Fossil Shells (*chione*) Southern California;[d] (total amino acids) |
|---|---|---|---|---|---|
| Aspartic acid | 7.5 | 8–10 | 8–10 | 12–14 | ~2[e] |
| Alanine | 1.4 | ~2.5–4 | 2–3 | 2–3 | 2.0 |
| Glutamic acid | ~0.2 | ~3–4 | — | 3–4 | 1.0 |
| Proline | 3.9 | ~0.2–0.5 | — | ~0.2–0.8 | 1.3 |
| Phenylalanine | 2.3 | 5–7 | — | 3.3 | 1.2 |
| Leucine | 1.0 | 1.0 | — | 1.0 | 1.0 |
| Isoleucine | 0.7 | — | 1.0 | ~1 | ~1 |
| Valine | 0.5 | 0.4–0.9 | — | ~0.8–0.9 | 0.6 |

[a] Masters and Friedman 1979.
[b] Schroeder and Bada 1976.
[c] Kessels and Dungworth, this volume.
[d] Wehmiller 1977, and J. L. Bada and E. A. Hoopes, unpublished results.
[e] P. M. Masters and J. L. Bada, unpublished results.

racemization. In contrast, shells show a completely anomalous pattern of relative racemization rates compared to those found in any of the other systems. This is another indication of the complicated racemization kinetics in shells compared to the straightforward and easily understood racemization kinetics in bones. Any mechanism advanced to account for the observed kinetics in shells must also be able to explain the observed relative racemization rates.

In summary, it seems apparent from the results we have presented here and elsewhere that temperature is the major factor that affects the extent of racemization in a fossil bone. Other factors, such as leaching by ground waters, moisture content of the environment, pH and metal-ion catalysis appear, at least in the majority of situations, to have negligible effects on racemization rates in fossil bones. The "calibration" procedure used to date fossil bones effectively evaluates the temperature history of a particular region; once a "calibration" constant has been determined, it can be used to date other samples from the general region as long as the site to be dated has the same temperature of the "calibration" site. The relative racemization rates of the various amino acids indicate that in bones, the proteinaceous component is the principal fraction in which racemization takes place.

## *ACKNOWLEDGMENTS*

We thank S. Steinberg and P. M. Masters for reviewing the manuscript, D. Darling for technical assistance, and B. Fraser for editorial assistance and patience in typing the manuscript. This work was partially supported by grants (EAR 73-00320 and EAR 77-14490) from the National Science Foundation.

## *REFERENCES*

Bada, J. L. Kinetics of the non-biological decomposition and racemization of amino acids in natural waters, *Advances in Chemistry Series* (106), 309–331, 1971.

Bada, J. L. Kinetics of racemization of amino acids as a function of pH, *J. Amer. Chem. Soc.*, **95,** 1371–1373, 1972a.

Bada, J. L. The dating of fossil bones using the racemization reaction of isoleucine, *Earth Planet. Sci. Lett.*, **15,** 223–231, 1972b.

Bada, J. L, and R. Protsch. Racemization reaction of aspartic acid and its use in dating fossil bones, *Proc. Nat. Acad. Sci. (U.S.A.)*, **70,** 1331–1334, 1973.

Bada, J. L., R. A. Schroeder, and G. F. Carter. New evidence for the antiquity of man in North America deduced from aspartic acid racemization, *Science*, **184,** 791–793, 1974.

Bada, J. L., and R. A. Schroeder, Amino acid racemization reactions and their geochemical implications, *Naturwissenschaften*, **62**, 71–79, 1975a.

Bada, J. L., and P. M. Helfman, Amino acid racemization dating of fossil bones, *World Arch.*, **7**, 160–173, 1975b.

Bada, J. L., P. M. Masters, E. Hoopes, and D. Darling, The dating of fossil bones using amino acid racemization, *Proc. IX International Radiocarbon Conference*, R. Berger and H. Suess, eds., Univ. of California Press, in press, 1979.

Davies, W. D., and F. E. Treloar, The application of racemization dating in archaeology: A critical review, *The Artifact*, **2**, 63–94, 1977.

Dungworth, G., N. J. Vincken, and A. W. Schwartz, Racemization of aliphatic amino acids in fossil collagens of Pleistocene, Pliocene and Miocene ages, in *Advances in Organic Geochemistry*, B. Tissot and F. Bienner, eds., Editions Technip Pub., Paris, pp. 690–700, 1975.

Hare, P. E., Amino acid dating—a history and an evaluation, *MASCA Newsletter*, **10**, 4–7, 1974a.

Hare, P. E., Amino acid dating of bone—the influence of water, *Carnegie Inst. Wash. Year Book*, **73**, 576–581, 1974b.

Hare, P. E., Organic geochemistry of bone and its relation to the survival of bone in the natural environment, in *Taphonomy and Vertebrate Paleontology*, A. K. Behrensmeyer and A. Hill, eds., Univ. of Chicago Press, in press, 1979.

Hare, P. E., and T. C. Hoering, Separation of amino acid optical isomers by gas chromatography, *Carnegie Inst. Wash. Year Book 72*, 690–696, 1973.

Helfman, P. M., J. L. Bada, and M.-Y. Shou, Considerations on the role of aspartic acid racemization in the aging process, *Gerontology*, **23**, 419–25, 1977.

Kessels, H. J., and G. Dungworth, Necessity of reporting amino acid compositions of fossil bones where racemization analyses are used for geochronological applications. Inhomogeneities of D/L amino acids in fossil bones, this volume.

Lee, C., J. L. Bada, and E. Peterson, Amino acids in modern and fossil woods, *Nature*, **259**, 183–186, 1975.

Masters, P. M., and J. L. Bada, Amino acid racemization dating of bone and shell, *Advances in Chemistry Series* (171), 117–138, 1978.

Masters, P. M., and M. Friedman, Racemization of amino acids in alkali-treated food proteins, *J. Ag. Food Chem*, **27**, 507–511, 1979.

Mehlman, M. J., Excavations at Nasera Rock, Tanzania, *Azania*, **XII**, 111–118, 1977.

Mitterer, R. M., Ages and diagenetic temperatures of Pleistocene deposits of Florida based on isoleucine epimerization in *Mercenaria*, *Earth. Planet. Sci. Lett.*, **28**, 275–282, 1975.

Neuberger, A., Stereochemistry of amino acids, *Adv. Protein Chem.*, **4**, 298–383, 1948.

Schroeder, R. A., and J. L. Bada, A review of the geochemical applications of the amino acid racemization reaction, *Earth Sci. Rev.*, **12**, 347–391, 1976.

Shou, M. Y., Kinetics and mechanisms of several amino acid diagenetic reactions in aqueous solutions and in fossils, Ph.D. thesis, Scripps Institution of Oceanography, Univ. of California, San Diego, 1979.

Smith, G. G., K. M. Williams, and D. M. Wonnacott, Factors affecting the rate of racemization of amino acids and their significance to geochronology, *J. Org. Chem.*, **43**, 1–5, 1978.

Stumm, W., and J. J. Morgan, *Aquatic Chemistry*, Wiley-Interscience, New York, p. 408, 1970.

Vicar, J., Racemization of amino acids and its utilization for age determinations, *Chemicke Listy*, **71**, 160–172, 1977.

von Endt, D. W., P. E. Hare, D. J. Ortner, and A. I. Stix, Amino acid isomerization rates and their use in dating archaeological bone, *Soc. Amer. Arch. Abstracts Fortieth Annual Meeting*, Dallas, May 8–10, 1975, p. 66.

Wehmiller, J. F., Amino acid studies of the Del Mar, California, midden site: Apparent rate constants, ground temperature models and chronological implications, *Earth Planet. Sci. Lett.*, **37**, 186–196, 1977.

Wernstedt, F. L., *World Climatic Data*, Climatic Data Press, Lemont, PA, pp. 71, 344, 1972.

Williams, H. M., and G. G. Smith, A critical evaluation of the application of amino acid racemization to geochronology and geothermometry, *Origins of Life*, **8**, 91–144, 1977.

# The Effect of Structure and Conditions on the Rate of Racemization of Free and Bound Amino Acids

GRANT GILL SMITH

ROBERT C. EVANS

*Department of Chemistry and Biochemistry, Utah State University*

---

## *ABSTRACT*

The racemization of amino acids, both free and bound, is dependent on molecular structure. Steric effects, as well as electrostatic influences, alter racemization rates especially when they preclude intramolecular stabilization of the incipient α-carbanion. Where steric effects are minimal, C-terminal amino acids racemize faster than N-terminal residues in dipeptides. Neighboring group participation has been proposed to explain this surprising result.

The transition state leading up to the carbanion intermediate has only a modest charge separation along the C—H bond being broken and is strongly solvated either intermolecularly or intramolecularly. Racemization of amino acids in dipeptides is strongly affected by neighboring amino acid residues, a result of an intramolecular stabilization of the incipient α-carbanion. Environmental factors are also significant, especially changes in temperature, water concentration, pH, complexation with clay surfaces, and the presence of aldehydes (e.g., pyridoxyl associated with metal ions). The concentration of buffers, ionic strength, and metal ions are less important. The racemization of bound amino acids appears to be less influenced by environmental effects than are free amino acids.

Water plays a very significant role in the extent of racemization—

for example, through the hydrolysis of proteins and the leaching of fossils.

Reported in part at the Carnegie Institution of Washington Conference: Advances in the Biogeochemistry of Amino Acids, Arlie House, Warrenton, VA, October 29–November 1, 1978. This material was taken in part from Thipamon Sivakua's Ph.D. Thesis (1978), Beatriz Silva de Sol's M.S. Thesis (1978), and David M. Wonnacott's Ph.D. Thesis (1979), Utah State University, Logan, Utah.

## INTRODUCTION

This year marks the silver anniversary of the first report of the discovery of protein residues in various fossil bones and shells (Abelson 1954). Considering the numerous fates which befall proteinaceous material, it is amazing that any intact protein survives geological time. In addition to the usual degradations induced by microorganisms while in the biosphere, abiogenic environmental influences cause denaturations, for example, the hydrolyses of bone and shell proteins to smaller peptides and free amino acids. Peptides and amino acids may be transferred back and forth from the biosphere to the geosphere *via* leaching. This reversible transfer is thought to cause modification in the amino acid composition of geological deposits. Hence, it is with a certain amount of surprise to find that fossil proteins have been isolated and that significant data can be obtained for geochronology.

Abelson (1955) was the first to suggest the possibility of using the kinetics of degradation of amino acids as a dating method. Hare and Abelson (1967) reported the presence of increasing amounts of non-protein D-amino acids in fossil shells of increasing age. Thus, one change that the amino acids undergo is diagenetic racemization from the L-form to the D-form. Hare and Mitterer (1968) were the first to realize the importance of the laboratory simulation of diagenetic racemization and to apply the rate data to the dating of fossils. The factors influencing the rate of diagenetic racemization include amino acid structure and their sequence in peptides, pH, ionic strength and buffering effects, surfaces, metallic cations and the concentration of water.

## STRUCTURE-REACTIVITY RELATIONSHIPS OF AMINO ACIDS

From the relationships between the structures of amino acids and the rate at which they react, it is possible to discern transition state structures and to predict the reactivity of analogous compounds. In some cases, quantitative expressions (Hammett and Taft equations) can be used to describe structural effects on reactivity. In this section, these linear free energy relations have been used to relate rates of racemization to electronic (resonance, inductive or field effects) and steric effects of substituents.

## Free Amino Acids

The substituents attached to the α carbon in a free amino acid influence the rate at which an α hydrogen ionizes to yield a negatively charged carbanion intermediate (Scheme 1)

Scheme 1

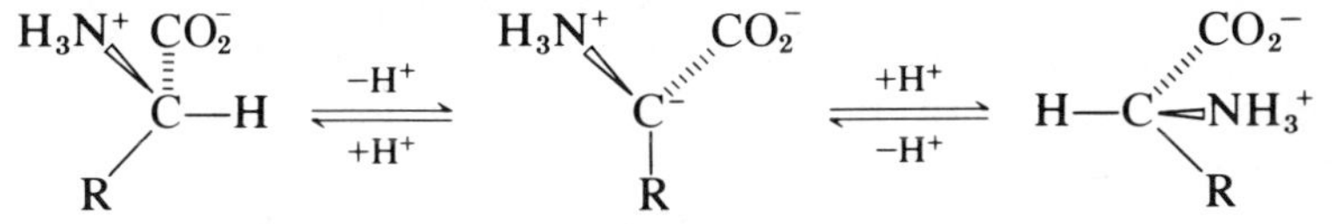

$$H_3N^+(CO_2^-)(R)C\text{—}H \underset{+H^+}{\overset{-H^+}{\rightleftharpoons}} H_3N^+(CO_2^-)(R)C^- \underset{-H^+}{\overset{+H^+}{\rightleftharpoons}} H\text{—}C(CO_2^-)(R)\text{—}NH_3^+$$

L-amino acid — carbanion intermediate — D-amino acid

Groups that stabilize or destabilize charge will accelerate or retard racemization. The effects of the amino and carboxyl groups are considered first, then the effect of the R group.

**The carboxyl and amine groups.** At the isoelectronic point (the midpoint between the pKa's of the acid and amino group) the amino acid exists predominantly in its zwitterionic form. It is in this form that the amino and carboxylate groups are first considered.

The carboxylate group, —$CO_2^-$ , would be expected to destabilize the incipient α-carbanion in the transition structure (Neuberger 1948). Despite the potential for resonance, the proximity of two negative charges is sufficiently destabilizing to raise the energy of the transition structure leading to a carbanion. In this case, the electrostatic inductive and field effects are more important than resonance.

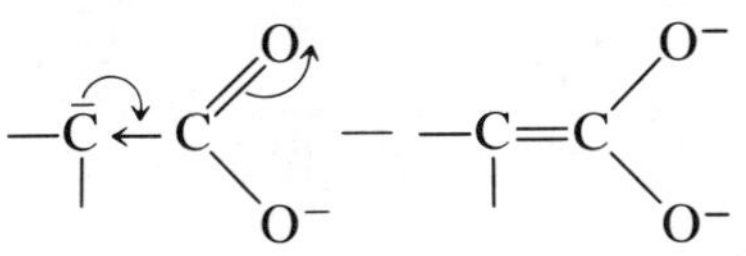

$$\text{—}\bar{C}\text{—}C(=O)O^- \longleftrightarrow \text{—}C\text{=}C(O^-)O^-$$

Substitution on the carboxylate removes the negative charge from the oxygen and enhances racemization. This may be accomplished by either protonation (by lowering the pH) or by transformation into an carboxylic acid derivative (I).

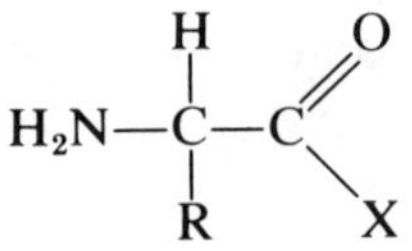

$$\mathrm{H_2N{-}\overset{\displaystyle H}{\underset{\displaystyle R}{C}}{-}C({=}O)X} \qquad X = OR;\quad X = NH_2, NR_2$$

I

Matsuo et al. (1967) studied the base-catalyzed deuteration and racemization of phenylalanine, alanine, and their ethyl esters using nmr. Their work revealed that an ester group had a small accelerating effect. Substitution by the carboxamide ($CONH_2$) group for the carboxylate group greatly enhanced the rate. The carboxylic acid group and its derivatives all are characterized by inductive and resonance effects which withdraw the electron density of the α-carbanion while not possessing an appreciable negative field effect.

A protonated amine, $—NH_3^+$ withdraws electrons inductively and stabilizes the incipient carbanion (Neuberger 1948). Also, the electrostatic field effect is favorable, as unlike charges are in close proximity. Thus a protonated amine will favor racemization. Furthermore, substitution by an electronegative group on the nitrogen further accelerates the rate of racemization. Matsuo et al. (1967) reported that an unprotonated amino group slightly retarded base catalyzed deuterations of phenylglycine (II) and alanine (III) when compared to the respective des-amino compounds.

*Relative rate of deuteration exchange at the α-position*

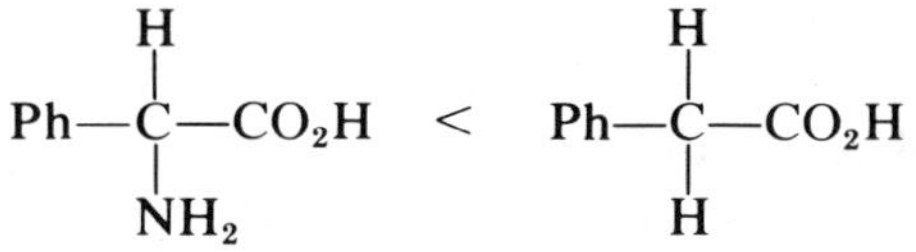

$$\mathrm{Ph{-}\overset{\displaystyle H}{\underset{\displaystyle NH_2}{C}}{-}CO_2H} < \mathrm{Ph{-}\overset{\displaystyle H}{\underset{\displaystyle H}{C}}{-}CO_2H}$$

II

$$\mathrm{CH_3{-}\overset{\displaystyle H}{\underset{\displaystyle NH_2}{C}}{-}CO_2H} < \mathrm{CH_3{-}\overset{\displaystyle H}{\underset{\displaystyle H}{C}}{-}CO_2H}$$

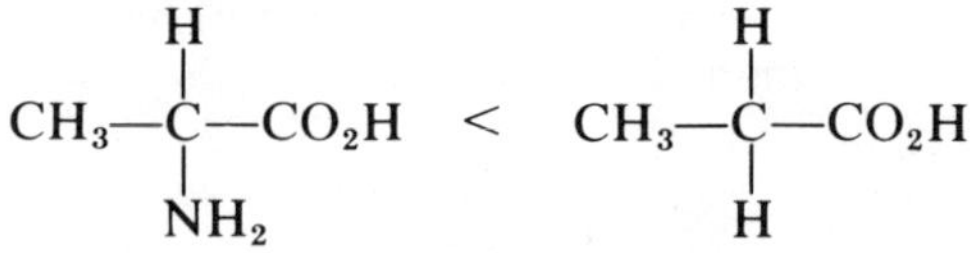

III

This is not unexpected, as the amino group is characterized by an electron-donating resonance capability that more than counterbalances any inductive electron-withdrawing effect. However, upon N-substitution by electron-withdrawing electronegative groups, racemization would be enhanced. This is what was observed by Matsuo and co-workers. N-Acyl derivatives of phenylglycine (IV) and alanine (V) both showed increased rates of deuteration.

$$\underset{\text{IV}}{\mathrm{RNH{-}\underset{\underset{\displaystyle C_6H_5}{|}}{C}HCO_2H}} \qquad \underset{\text{V}}{\mathrm{CH_3CONH{-}\underset{\underset{\displaystyle CH_3}{|}}{C}HCO_2H}}$$

$R = CH_3CO—$

$R = CH_3CH_2CO—$

Overall, the following sequence of reactivity of the α-hydrogen (shown in italics) was found by Matsuo et al. (1967).

$$\mathrm{R'{-}\overset{O}{\overset{\|}{C}}{-}NH{-}}C\mathit{H}\mathrm{(R){-}\overset{O}{\overset{\|}{C}}{-}NH_2} \gg \mathrm{R'{-}\overset{O}{\overset{\|}{C}}{-}NH{-}}C\mathit{H}\mathrm{(R){-}\overset{O}{\overset{\|}{C}}{-}OH} >$$

$$\mathrm{H_2N{-}}C\mathit{H}\mathrm{(R){-}\overset{O}{\overset{\|}{C}}{-}O{-}R'} \geq \mathrm{H_2N{-}}C\mathit{H}\mathrm{(R){-}\overset{O}{\overset{\|}{C}}{-}OH}$$

**The R group.** The effect of α-side chains on racemization apparently is a result of a combination of steric as well as electronic factors. Aryl groups stabilize the incipient carbanion by charge delocalization. Aralkyl groups also have a stabilizing effect, but through induction, not resonance. Alkyl groups have an inductive destabilizing effect since they are electron donors. Thus the general order of enhancing racemization reactivity would be:

$$\mathrm{Ar > ArCH_2 > H > R}$$

Studies (Satō et al., 1970; Bada 1972; Smith et al., 1978) have confirmed that this is indeed so.

Satō et al., investigated the kinetics of racemization of anilides of N-benzoyl-L-amino acids (VI) polarimetrically. The relative rates for several different amino acids in ethanol-DMSO displayed the following trend (R group given):

$$\underset{\text{VI}}{\mathrm{C_6H_5{-}CO{-}NH\underset{\underset{\displaystyle R}{|}}{C}HCO{-}NHC_6H_5}}$$

$$\mathrm{R = Ar \gg ArCH_2 > Me >} \mathit{i}\text{-Bu} > \mathrm{Et} \gg \mathit{i}\text{-Pr}$$

Smith et al. (1978) have similarly studied the effect of the R group on racemization. Their results, conducted on the free amines in aqueous solution, are presented in Table 1. As in the Satō study, the same general trend for the relative rates was followed.

In addition, Smith et al. (1978) investigated these racemizations at various temperatures. From transition state theory, the rate of racemization is related to the enthalpy of activation (ΔH†) and the entropy of activation (ΔS†) by the following equation:

$$\ln \frac{k}{T} = \ln \frac{k'}{h} + \frac{\Delta S^\dagger}{R} - \frac{\Delta H^\dagger}{RT}$$

where $k'$ is Boltzman's constant, h is Planck's constant and R is the gas constant. Typical plots of ln $k$/T versus 1/T for the amino acids are presented in Figure 1. The ΔH† and ΔS† were evaluated from linear regression analyses of the plots and are presented in Table 1.

Electronic effects are usually reflected by enthalpy changes. Steric inhibition to solvation and/or steric influences on loss of the α-hydrogen are reflected by changes in ΔS†.

Phenylglycine and phenylalanine have ΔH† values 5–7 kcal/mole lower than the aliphatic amino acids and entropy terms about 10 eu more negative. Smith et al. (1978) concluded from these data that electronic factors contributed to the increasing stability of the incipient carbanion where aryl substituents were present (lower ΔH†), while the bulky rings showed an increasing ordering of the amino acid and solvent at the transition state (more negative ΔS†). Apparently the electronic factors prevail here, as

**Table 1. The Relationship between Structure ΔH†, ΔS†, and Relative Rate of Amino Acid Racemization at pH 7.6[a]**

| Amino Acids | R | ΔH† (kcal $mol^{-1}$) | ΔS† (eu) | Rel. Rate[b] |
|---|---|---|---|---|
| Phenylglycine | $C_6H_5$ | 20.8 | −27.6 | 44.6 |
| Phenylalanine | $C_6H_5CH_2$ | 23.0 | −28.4 | 2.0 |
| Alanine | $CH_3$ | 28.6 | −16.3 | 1.0 |
| Leucine | $i$-$C_4H_9$ | 27.5 | −19.8 | 0.7 |
| Isoleucine | $s$-$C_4H_9$ | 27.1 | −21.6 | 0.4 |
| Valine | $i$-$C_3H_7$ | 28.0 | −19.8 | 0.3 |

[a] Sivakua (1978).

[b] All racemizations studied were carried out in sealed tubes, in phosphate-buffered solution, 0.05 *M*, ionic strength, 0.5, amino acid concentration 0.01 *M*. The studies for phenylglycine were taken in the temperature range of 104–138.9°C, all others between 124.6 and 175.4°C.

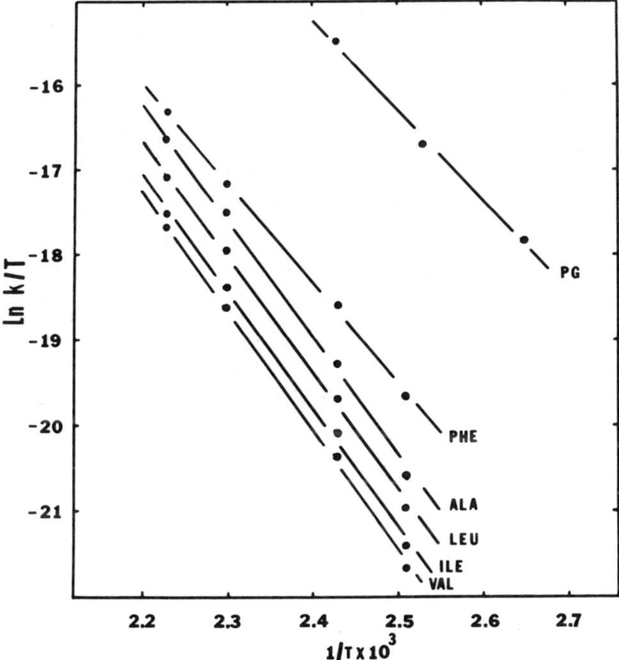


**Figure 1.** Arrhenius plots for the racemization of phenylglycine (PG), phenylalanine (PHE), alanine (ALA), leucine (LEU), isoleucine (ILE) and valine (VAL) (Sivakua 1978).

the phenyl-substituted amino acids display enhanced rates of racemization. Note that the ΔH† for phenylalanine is higher by 2 kcal/mole than that for phenylglycine. Placing a methylene group (—$CH_2$—) between the ring and the site of the developing carbanion (as in phe) has an attenuating effect on electronic stabilization. As a consequence, phenylalanine does not racemize nearly as fast as phenylglycine. Note also that the benzyl group ($C_6H_5CH_2$—) appears bulkier than the phenyl ($C_6H_5$—) group as reflected in the ΔS† term.

A comparison of the values of ΔH† for the four aliphatic groups finds only modest variations (1.5 kcal/mole maximum). Electronic factors evidently play a minimal role in racemization for these species. However, inspection of the entropy term reveals a difference of 5.3 eu between the smallest alkyl group (methyl of ala) and one of the largest and least symmetrical groups (*sec*-butyl of ile). Steric effects exhibited by these groups are apparently more critical than their electron-donating ability. The small difference between ile and val may best be explained as arising from a polarizability effect, which is larger for bulkier groups (Hine 1975).

A further attempt to correlate rates to structure was made by applying the Taft equation (Gilliom 1970):

$$\log k/k_{\text{ala}} = \rho^*\sigma^* + SE_s$$

where $k$ and $k_{\text{ala}}$ were set equal to the rate constants for any amino acid under study and ala, respectively. The electronic factors are presented by $\rho^*\sigma^*$ and the steric factors by $SE_s$. The $\sigma^*$ parameter measures the polar effect of the substituent, and $\rho^*$ measures the sensitivity of the reaction to changes in the polar effect (attenuation factor). Similarly, $E_s$ measures the steric effect introduced by the presence of the substituent, and S measures the sensitivity of the reaction to this steric effect. The plot (Figure 2) of $\log k/k_{\text{ala}}$ versus $\rho^*\sigma^* + SE_s$ (using $\rho^*$ and S as determined by a linear regression analysis of the kinetic data) is essentially linear for aliphatic amino acid racemization. This further serves to indicate the importance of steric as well as electronic factors in the role of amino acid racemization.

More recently, Wonnacott (1979) reported the kinetics of serine (R= $CH_2OH$), threonine (—$CH(CH_3)OH$), aspartic acid (—$CH_2CO_2H$), and glutamic acid (—$CH_2CH_2CO_2H$). These results, together with those of Smith et al. (1978), give the following selective order for the rate of racemization of free amino acids in aqueous solution:

$$\text{pg} > \text{ser} > \text{thr} > \text{asp} > \text{phe} > \text{ala} > \text{glu} > \text{leu} > \text{ile} > \text{val}$$

Apparently, a hetero atom (oxygen) has a further intramolecular stabilizing effect on this incipient carbanion and/or on solvation.

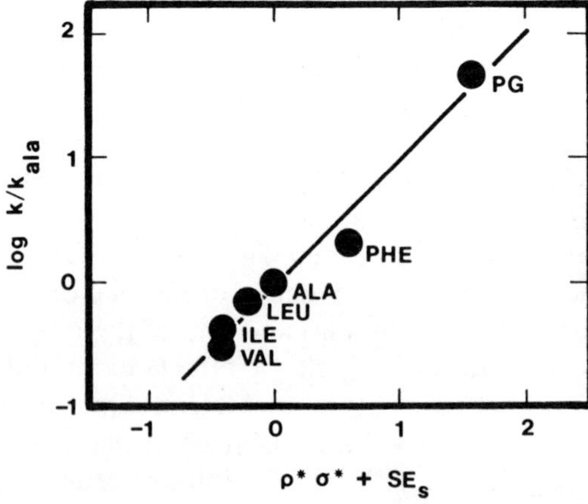


**Figure 2.** The Taft $\rho^*\sigma^* + SE_s$ plot for the racemization of phenylglycine (PG), phenylalanine, alanine, leucine, isoleucine, and valine (Sivakua 1978).

## Bound Amino Acids

The amino acids bound in peptide chains are the ones of greatest significance to biogeochemistry. Their racemization studies are the most interesting and also the most challenging to understand. There are a number of questions about the effect of primary structure, whether or not racemization is promoted by hydrolysis of the peptide chain and the influence of water and metal ions.

**Terminal versus interior.** Epand and Epand (1973) concluded that there is enhanced racemization of interior amino acids over terminal amino acids. Both Pickering and Li (1964) and Geschwind and Li (1964) reported extensive racemization of internal phenylalanine residues when certain peptide hormones were treated with hot alkali but no appreciable racemization of C-terminal phenylalanine. NMR studies of deuterium incorporation at the α-carbon of glycine peptides confirmed this and showed that there is more rapid exchange taking place with internal amino acids than with terminal amino acids (Fridkin et al., 1970). However, Kriausakul and Mitterer (1978) reported that terminal amino acids racemize faster than interior amino acids in peptides.

Further differences exist between C-terminal and N-terminal positions. Kriausakul and Mitterer (1978) studied the kinetics of epimerization of isoleucine at 152° and reported that N-terminal isoleucine racemized faster than isoleucine in the C-terminal position. Sol (1978) showed that just the opposite is true for most amino acids, although her results for isoleucine were the same as Kriausakul and Mitterer. It is apparent that bound amino acids are complex and interesting.

From the simple electrostatic theory discussed previously it would be predicted that an N-terminal amino acid residue of a peptide would racemize faster than internal ones and they would racemize faster than C-terminal residues at a pH near the isoelectric point of the amino acid residue.

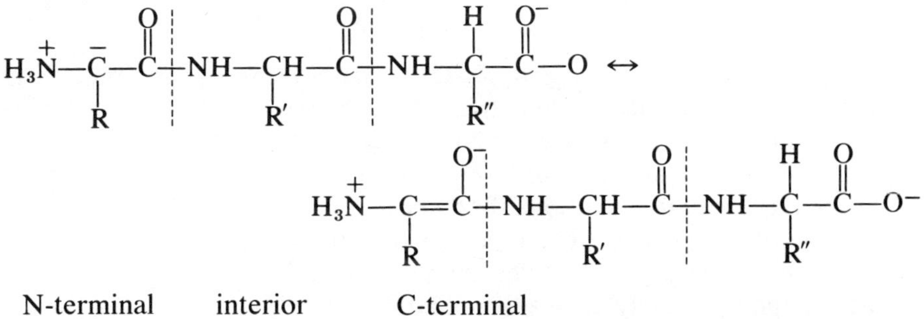

$$H_3\overset{+}{N}-\underset{R}{\overset{-}{C}}-\overset{O}{\overset{\|}{C}}\,\vdots\,NH-\underset{R'}{CH}-\overset{O}{\overset{\|}{C}}\,\vdots\,NH-\underset{R''}{\overset{H}{\overset{|}{C}}}-\overset{O^-}{\overset{\|}{C}}-O \leftrightarrow$$

$$H_3\overset{+}{N}-\underset{R}{C}=\overset{O^-}{\overset{|}{C}}\,\vdots\,NH-\underset{R'}{CH}-\overset{O}{\overset{\|}{C}}\,\vdots\,NH-\underset{R''}{\overset{H}{\overset{|}{C}}}-\overset{O}{\overset{\|}{C}}-O^-$$

N-terminal interior C-terminal

Scheme 2. Resonance stabilization in the carbanion of a tripeptide and illustration of the N-terminal, C-terminal and interior amino acid residues.

In peptides the N-terminal and interior residues no longer have the destabilizing carboxylate group. Hence with these residues the incipient carbanion formed during racemization benefits from resonance stabilization as illustrated in Scheme 2. (A similar resonance structure could be written from the interior residue.) Removal of the stabilizing quaternary amino group (as is the case with the interior and C-terminal amino acid residue) would decrease the incipient carbanion stability and result in slower racemization rates for these amino acids. Furthermore, the free carboxylate anion through inductive and field effects should destabilize an incipient carbanion when formed at the α-carbon of the C-terminal position. Hence, based on inductive and electrostatic effect, the C-terminal amino acid should racemize more slowly than interior amino acids. However, there is adequate evidence to show this is not always true. The simple electrostatic theory is not sufficient to explain all that is known about primary structural effects on racemization of bound amino acids.

Apparently, the activating effect of the $\text{—}\overset{+}{N}H_3$ group is essentially countered by the deactivating effect of the $\text{—}CO_2^-$ group, since interior amino acid residues racemize only slightly faster than free amino acids (Kriausakul and Mitterer 1978).

Schroeder (1974) was one of the first to systematically consider the effect of primary structure on racemization of amino acids. He studied the epimerization of isoleucine to alloisoleucine in various proteins. He reported considerable variation in the D-alloile-ile ratio in a study of oxytocin, insulin, ribonuclease, and collagen which may reflect variation in primary structure. A definite pattern did not develop. Certainly, secondary, tertiary, and quaternary structures of proteins are lost upon heating and cannot be considered as causes for the differences in the ratios.

Kriausakul and Mitterer (1978) studied the kinetics of epimerization of isoleucine at 152° in various dipeptide and polypeptide environments. The following relative order of epimerization was found:

$$\text{N-terminal} > \text{C-terminal} \gg \text{interior} \geq \text{free}$$

Isoleucine epimerized slightly faster at the N-terminal position than at the C-terminal position in dipeptides. Specifically, D-alloile/L-ile for ile-gly was approximately 0.30 after 8 hours, while for gly-ile it was approximately 0.25. The apparent difference in ratios was even smaller for ile-val and val-ile (~0.20 and ~0.18, respectively). The much slower racemization of the interior isoleucine was obtained from studies on various naturally occurring polypeptides and protein (i.e., oxytocin and insulin). The tripeptides, ile-gly-gly (N-terminal) and gly-gly-ile (C-terminal) were studied but not gly-ile-gly (interior). The bound isoleucine followed pseudo first-order kinetics to equilibrium.

Simultaneously but independently, Sol (1978) found that Kriausakul and Mitterer's observation that the N-terminal racemized faster than the C-terminal was the exception rather than the rule. By employing optically active phases on a gc capillary column, Sol studied not only isoleucine containing dipeptides but many others including ala, leu, val, asp, phe, met, ser dipeptides.

Unexpectedly, the peptides alanine, leucine, phenylalanine, aspartic acid, and methionine showed the C-terminal amino acid racemizing faster than the N-terminal for all peptide environments except proline dipeptides (Table 2).

This result was not anticipated from a knowledge of the electrostatic theory previously discussed. A reasonable explanation is given below. All proline dipeptides were unique in that none of the amino acids attached to proline showed a greater racemization at the C-terminal position (Table 3). This also is discussed in more detail subsequently.

Sol's work (1978) agrees with the work of Kriausakul and Mitterer (1978) concerning the racemization of isoleucine. Her work showed that ile racemized faster at the N-terminal position than at the C-terminal po-

**Table 2. Amounts of Racemization of Amino Acid Residues in Dipeptides, C-terminal Greater than N-terminal[a]**

| | Percent D-Amino Acid[b] | | |
|---|---|---|---|
| Dipeptides | N-Terminal | C-Terminal | C/N[c] |
| ala-gly, gly-ala | 27.5 (ala) | 34.5 (ala) | 1.25 |
| ala-val, val-ala | 20.2 (ala) | 32.5 (ala) | 1.61 |
| leu-gly, gly-leu | 20.3 (leu) | 25.6 (leu) | 1.26 |
| leu-val, val-leu | 11.1 (leu) | 21.0 (leu) | 1.89 |
| phe-gly, gly-phe | 22.7 (phe) | 34.1 (phe) | 1.50 |
| phe-val, val-phe | 11.8 (phe) | 19.5 (phe) | 1.65 |
| asp-gly, gly-asp | 39.4 (asp) | 42.3 (asp) | 1.07 |
| asp-val, val-asp | 28.1 (asp) | 40.9 (asp) | 1.46 |
| met-gly, gly-met | 29.4 (met) | 34.8 (met) | 1.18 |

[a] Sol (1978).

[b] Determined by gc on a stainless steel capillary column loaded with a chiral phase.

[c] All racemizations studied were carried out in sealed tubes, in phosphate-buffered solution, 0.05 *M*, ionic strength, 0.12, amino acid concentration, 0.02 *M*, pH 7.6, T, 122.5°C, time, 8 hours.

sition in a study of isoleucine dipeptides (Table 3). She also found that valine, another sterically hindered amino acid, and serine, a hydroxy amino acid, followed the same pattern (N> C) (Table 3). Although Sol's study confirmed Kriausakul and Mitterer's work qualitatively, there were significant quantitative differences. Where Kriausakul and Mitterer found only a slight difference in the rate of epimerization of ile between the N- and C-terminal positions in isoleucine dipeptides, Sol found a significant difference with these dipeptides (Table 4).

The variance in the C—N ratio in racemization might be explained by considering a balance among several factors: (−) inductive effect of $NH_3^+$, (+) inductive effect of $CO_2^-$, solvation of the incipient carbanion, intramolecular "solvation," and steric effects that can alter these influences.

Intramolecular stabilization of the transition structure through a five-membered ring involving the ammonium nitrogen, with its concomitant

**Table 3. Amounts of Racemization of Amino Acid Residues in Dipeptides, N-terminal Greater than C-terminal**[a]

| | Percent D-Amino Acid[b] | | |
|---|---|---|---|
| Dipeptides | N-Terminal | C-Terminal | N/C[c] |
| ile-pro, pro-ile | 52.8 (ile) | 0.9 (ile) | 58.67 |
| val-pro, pro-val | 34.6 (val) | 1.1 (val) | 31.45 |
| phe-pro, pro-phe | 48.7 (phe) | 12.2 (phe) | 3.99 |
| leu-pro, pro-leu | 44.1 (leu) | 16.8 (leu) | 2.63 |
| glu-pro, pro-glu | 33.0 (glu) | 15.8 (glu) | 2.09 |
| ala-pro, pro-ala | 43.2 (ala) | 26.5 (ala) | 1.63 |
| val-pro, pro-val | 34.6 (val) | 1.1 (val) | 31.45 |
| val-asp, asp-val | 29.5 (val) | 3.1 (val) | 9.52 |
| val-phe, phe-val | 16.7 (val) | 2.2 (val) | 7.59 |
| val-leu, leu-val | 18.1 (val) | 3.5 (val) | 5.17 |
| val-ala, ala-val | 26.0 (val) | 5.6 (val) | 4.64 |
| val-gly, gly-val | 13.0 (val) | 3.6 (val) | 3.61 |
| val-ile, ile-val | 5.4 (val) | 3.5 (val) | 1.54 |
| ile-gly, gly-ile | 20.6 (ile) | 5.6 (ile) | 3.68 |
| ile-val, val-ile | 11.8 (ile) | 4.0 (ile) | 2.95 |
| ser-gly, gly-ser | 46.5 (ser) | 39.1 (ser) | 1.19 |

[a] Sol (1978).

[b] Determined by gc on a stainless steel capillary column loaded with a chiral phase.

[c] For racemization conditions see Table 2.

**Table 4. Amount of Racemization of Isoleucine in Dipeptides as a Function of Position**

| | | D-alloile/L-ile | |
|---|---|---|---|
| Dipeptides | Ile Position | Kriausakul & Mitterer (1978)[a] | Sol (1978)[b] |
| ile-gly | N-terminal | ~0.30 | 0.259 |
| gly-ile | C-terminal | ~0.25 | 0.059 |
| ile-val | N-terminal | ~0.20 | 0.134 |
| val-ile | C-terminal | ~0.18 | 0.042 |

[a] Data obtained after 8 hours at 152°. Numbers were read from Kriausakul and Mitterer's (1978) graph.
[b] For racemization conditions see Table 2.

reduction of the entropy factor by reducing solvation requirements, would promote C-terminal faster than N-terminal.

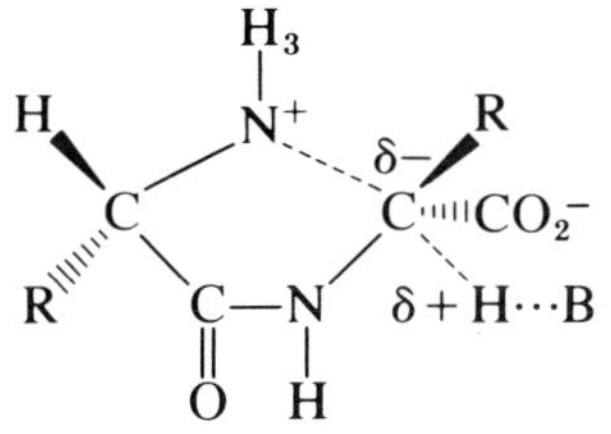

Less likely is the assistance of the amino group in the removal of the α-hydrogen, except with the amino acid species in which the amino group is in the free amino form.

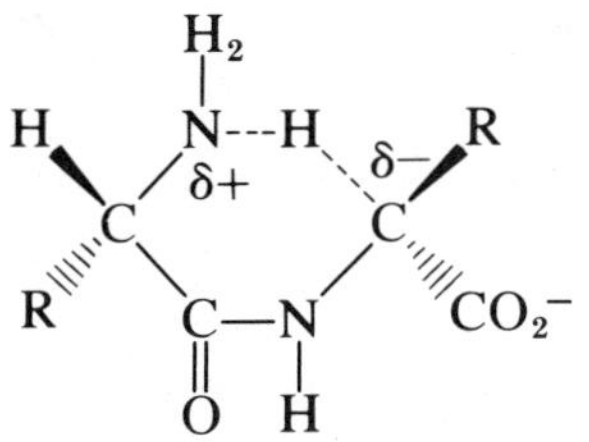

At certain pH values this could become a major contributor.

Steric requirements would alter the formation of the five- and six-membered rings. For proline to show intramolecular stabilization of the incipient carbanion would require the formation of fused five-membered rings, which is rarely observed.

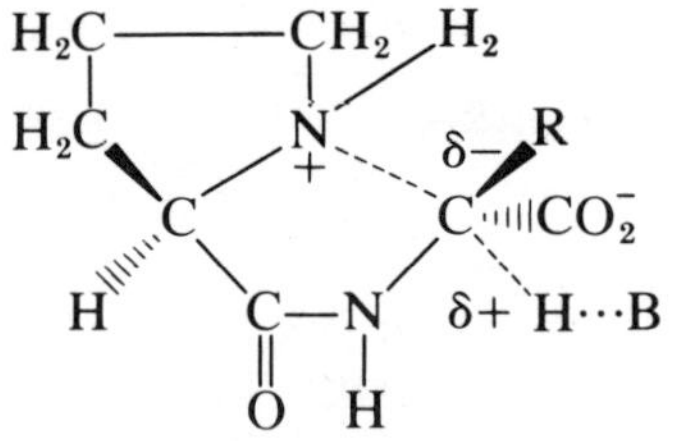

PRO–X

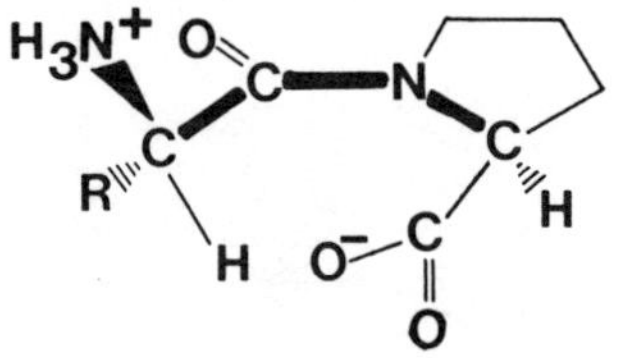

cis–X–PRO

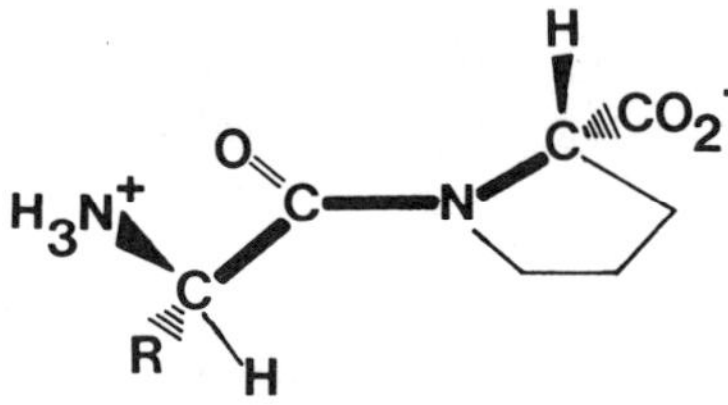

trans–X–PRO

Partial double bond character of the C-N bond places the four atoms of the peptide link plus the two adjacent α-carbon atoms in the same plane, making possible *cis* and *trans* isomers. For X-pro dipeptides the *cis*-conformation is favored (Wuthrich, *et al.*, 1972; Dorman and Bovey, 1973). This configuration makes removal of the α-proton easier and readily explains why the N-terminal amino acid residues in X-pro dipeptides racemize more readily than the C-terminal residues (Table 3).

Azlactone intermediate formation might be postulated, since such species are known to racemize with ease (Goodman and Levine 1964; Goodman and McGahren 1965).

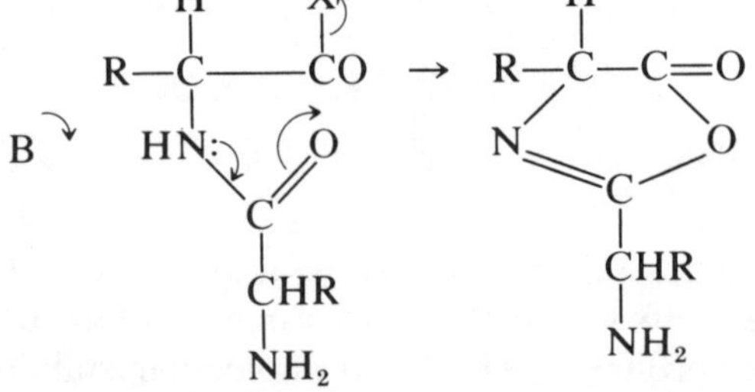

Azlactone formation is not favored in dipeptides, since the leaving group (negative oxygen in the carboxylate) is extremely poor. Thus the intermediacy of such a species is highly unlikely.

**Bound versus free amino acids.** Generally, bound amino acids racemized at a faster rate than the corresponding free amino acid. Serine showed no racemization when treated with excess sodium hydroxide solution at room temperature for 22 hours (Crawhill and Elliott 1951). Polyserine when treated with 0.6 *N* or 1.0 *N* sodium hydroxide solution at 30° for only 20 minutes showed significant racemization (Bohak and Katchalski 1963). Appreciable racemization was observed for albumin, casein, elastin, fibrin, and gelatin upon treatment with dilute alkali (Levene and Bass 1929; Davin and Dudley 1913). Studies by Sol (1978) on the racemization of free and collagen-bound amino acids also confirmed the more rapid racemization of the bound species (Table 5). A comparison of these data with Sol's data for dipeptides (see Tables 2 and 3) also reveals that bound amino acids racemize faster than free amino acids.

**Effect of hydrolysis.** It has been suggested that hydrolysis of bound amino acids produces an amino acid in an activated state such that the racemization process is facilitated (Wehmiller 1971; Wehmiller and Hare 1971; Hare 1971; Hare and Hoering 1973). This interesting hypothesis has

**Table 5. Amounts of Racemization of Free and Collagen-Bound Amino[a] Acids**

| | % D determined by GLC | |
|---|---|---|
| L-Amino Acid | Free[b] | Bound[c] |
| Val | 0.3 | n.d. |
| Ile | 0.5 | n.d. |
| Leu | 0.5 | 4.1 |
| Ala | 1.0 | 10.3 |
| Phe | 1.9 | n.d. |
| Met | 2.1 | n.d. |
| Asp | 2.5 | 42.3 |
| Ser | 21.9 | n.d. |

[a] Sol (1978).
[b] For racemization conditions see Table 2.
[c] pH, 8.0; 10-mg sample in 2 ml of 0.05 *M* phosphate buffer; ionic strength, 0.12; time, 8 hours; T, 122.5°C.

**Table 6. Racemization of Amino Acids in Various Dipeptides as Induced under Hydrolytic Conditions[a,b]**

| L-dipeptide | | L-dipeptide | |
|---|---|---|---|
| | *% D-leu* | | *% D-ala* |
| gly-leu | 1.6 | pro-ala | 0.0 |
| leu-gly | 1.6 | ala-pro | 0.0 |
| val-leu | 1.2 | | |
| leu-val | 0.8 | | *% D-phe* |
| free-leu | 1.5 | pro-phe | 0.1 |
| | | phe-pro | 0.0 |
| | *% D-val* | | |
| leu-val | 0.5 | | *% D-glu* |
| val-leu | 0.5 | pro-glu | 1.3 |
| pro-val | 0.0 | glu-pro | 2.0 |
| val-pro | 0.0 | | |

[a] 6 *N* HCl, 24 hours at 100–110°C.
[b] Sol (1978).

given rise to controversy. For example, Hare (1971) studied the alkaline racemization of the protein isolated from a natural sample. The extent of racemization for seven constituent amino acids was considerable under the conditions employed (6 *N* NaOH at 110° for 22 hours) when compared to the similarly treated free amino acids. From this work, Hare inferred the necessity of hydrolysis for the racemization reaction. It was argued by Bada and Schroeder (1972) that since hydrolysis and racemization involved different carbon centers, the two reactions were independent. They suggested that the enhanced rate of the protein-bound amino acids was the effect of metal-ion catalysis, the metal ions being a part of the *Mercenaria* matrix studied.

Hare (1971) and Sol (1978) studied amino acid racemizations induced by acid hydrolytic conditions of dipeptides; they both found no significant racemization. Table 6 contains Sol's data, which shows the minimal racemization that occurs under these conditions.

It has more recently been suggested (Kriausakul and Mitterer, 1978) that although hydrolysis is not a prerequisite for racemization, hydrolysis can markedly affect the racemization kinetics observed. The epimerization of isoleucine in protein was reported to follow nonlinear kinetics with a gradual decrease in the apparent first-order rate constant. The epimerization followed linear first-order kinetics to equilibrium and then changed to a second linear curve that approximated the kinetics of free isoleucine.

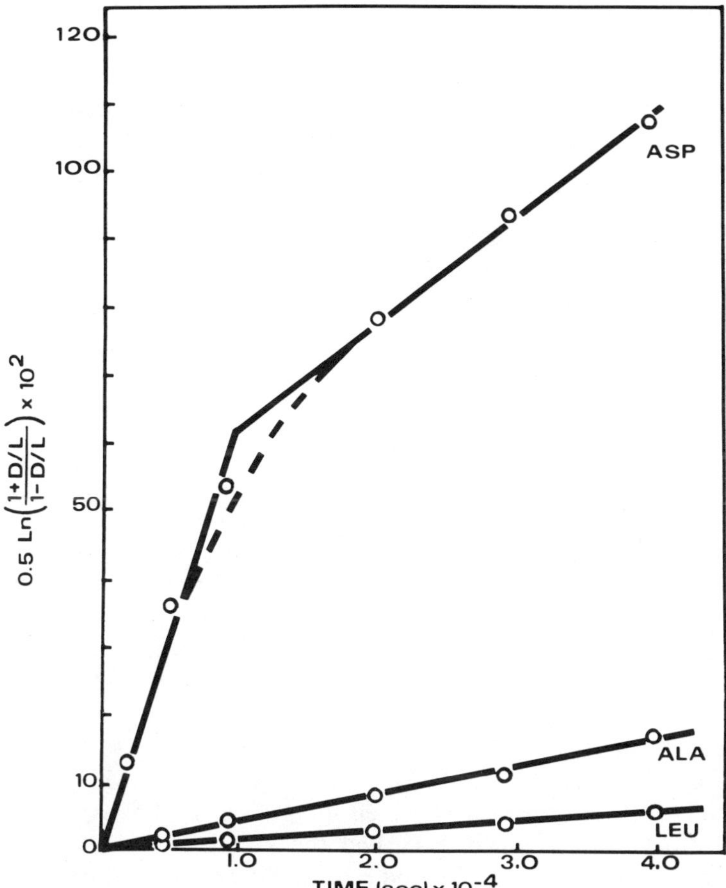


**Figure 3.** First order rate plots for the racemization of alanine, leucine and aspartic acid bound in collagen at 122.5° at pH 8.0 in phosphate-buffered solution (0.05M) (Sol 1978).

Similar results were obtained for isoleucine bound in dipeptides. Kriausakul and Mitterer rationalized these findings by suggesting that epimerization occurs before and after hydrolysis. The bound amino acids are hydrolyzed from internal to terminal to free. As a result, the kinetics of epimerization change to reflect the different epimerization rates of the amino acid species.

Sol (1978) studied the rates of racemization of collagen-bound leucine, alanine, and aspartic acid in the temperature range 95.5–140.5° and at pH 8.0 and 10.5. The rates of racemization of the leucine and alanine followed first-order kinetics throughout the entire range of temperatures at both

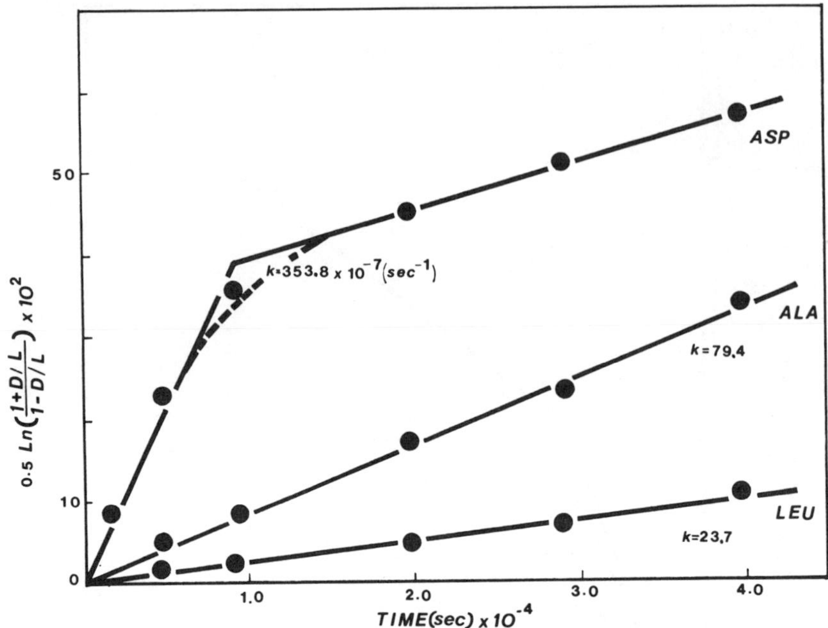


**Figure 4.** First order rate plots for the racemization of alanine, leucine and aspartic acid bound in collagen at 122.5° and pH 10.5 in phosphate-buffered solution (0.05M) (Sol 1978).

pH values (Figures 3 and 4). But with aspartic acid the first-order kinetic plots were not linear. As with the work of Kriausakul and Mitterer (1978), two distinct linear portions manifested themselves in the reversible first-order plots. The first curve described the kinetics of racemization of bound aspartic acid, and the second portion described the kinetics of the free and bound aspartic acid. Since the kinetics of the first-order plots were linear for leucine and alanine at both pH 8 and 10, no significant hydrolysis occurred at these residues to form free amino acids. However, at the higher pH, rates of racemization were doubled.

The plot for aspartic acid at pH 10 revealed the effect of increased hydrolysis over that at pH 8. The transition from the kinetics of bound aspartic acid to free aspartic acid occurred sooner and more extensively with a concomitant reduction in the apparent rate constant. Hence, when hydrolysis increases, racemization decreases.

## ENVIRONMENTAL FACTORS

Variations in other parameters may also affect the rates of racemization of amino acids. Among these factors are moisture, surface, pH, buffer, ionic strength, and metal cations.

### Presence of Water

Both fossil diagenesis and laboratory simulations show that water has the most significant influence on the rate of racemization of all the environmental factors studied. For example, the amount of racemization observed in fossil bones inside the LaBrea tar pits in California is small compared to that observed in samples found outside the tar pits (Kvenvolden and Peterson 1974).

Wonnacott (1979) has reported that serine showed little tendency toward racemization when heated under anhydrous conditions. However, when heated in aqueous solution, serine racemized faster than any other protein amino acid studied.

The laboratory simulation experiments of Hare (1974) on bones with varying amounts of water showed that under anhydrous conditions, recent bone samples showed little or no racemization in three days at 157°. Heating the same bone in the presence of water vapor resulted in a D-alloile/L-ile of 1.08 (the equilibrium constant for epimerization of isoleucine is 1.40). Increasing proportions of water resulted in successive *decreases* in the ratio of D to L. A 1:1 ratio yielded a D/L of 0.69, and a greater than 1000:1 ratio yielded only 0.19 for D/L. Hence water is involved, but it appears that excessive amounts decrease racemization.

When Hare examined the contents of the water used for the racemization of bone chips, he found that at the 1:1 ratio the D/L was almost racemic. At the increased ratios of water to bone, the D/L in water remained at approximately 1.0. Hare concluded that the water was leaching out the racemized collagen as evidenced by the D/L found in the water by the decreasing D/L in the bone chips. Furthermore, nitrogen analysis of the bone chips revealed that as more water is introduced, increasing amounts of organic matter are removed.

Hare suggested that leaching removed the collagen and left behind a noncollagenous material that represented an additional small protein component of the bone. This material was highly resistant to leaching and racemization.

### Surfaces

Research on the effect of clay surfaces upon racemization has revealed conflicting results. Hare and Hoering (1973) also found predominantly L-enantiomers in clay material of a core from which foraminiferal amino acids had almost completely racemized. Assuming that the amino acids associated with these surfaces were not of recent origin, retardation of racemization is indicated.

However, experiments on L-leucine and L-isoleucine in the presence of montmorillonite clay surfaces at elevated temperatures have reportedly led to an acceleration in racemization (Kroepelin and Georgaras 1968; Kroepelin 1969; Frenkel and Heller-Kallei 1977). These researchers concluded that montmorillonite clay surfaces act as a specific catalyst for racemization.

Most recently, Akiyama (1978) reported the epimerization of isoleucine that had been absorbed into the interlayer sites of sodium montmorillonite. His results are in direct contrast to those obtained from the work previously described. This study found that the rate constants for epimerization under these conditions were even slower than for epimerizations under anhydrous conditions. Akiyama confirmed the absorption of the amino acid by the clay, using X-ray diffraction experiments.

The effect of surfaces has not been totally resolved, although it is the opinion of the authors that clays do inhibit racemization. In all probability, the surface effect is a further manifestation of the importance of moisture to the racemization process. Amino acids while "locked" in interlayer sites of the surface are not exposed to moisture and thus suffer a retardation of racemization.

## Ionic Strength, pH, and Buffer

The effect of these three factors will be discussed together, as they are intimately associated with one another.

Bada and his associates have made some of the most extensive studies of the effect of pH on racemization over a pH range of −1 to 12. Bada (1972) reported a complex relationship between pH and the racemization rate of aspartic acid at various temperatures. His study showed that the rate of racemization of this dicarboxylic acid maximized at pH 3. Presumably this enhanced effect was due to the protonation of both carboxylate groups. Free carboxylate anions reduce the racemization rate at the α-carbon as discussed earlier. He observed that the rate decreased with an increase of pH from 3 to 5, and it leveled off in the range 5–8. Above 8 the rate increased until pH 10 where it became constant.

More recently Shou and Bada (1979) reported the rate of racemization of ala, pro, and leu. They observed two pH independent regions, one plateau occurring within the range 3–6.5 and the second between 8.5 and 12. They stressed that the measured rate of interconversion of the L to D amino acid enantiomers is a combination of several rates involving the several forms of the protonated and unprotonated amino acid.

Smith et al. (1978) reported an increase in racemization with increasing pH in the range of 6.5–8.5 for phosphate-buffered solutions of alanine. This is in agreement with the results of Shou and Bada (1979).

Wonnacott (1979) also studied the effect of pH changes on the rate of racemization of gelatin-bound alanine. His work revealed that pH increases led to enhanced racemization for bound alanine. Significantly, bound alanine racemized to a much greater extent than did the free alanine at the same pH and conditions. This is confusing because only one form of the bound protein would be expected to make up the rate expression, while for the free amino acid, various pH dependent species would all contribute to the rate equation:

$$\text{Rate} = [k_1[\text{H}_2\text{N—CHR—CO}_2\text{H}] + k_2[\text{H}_2\text{N—CHR—CO}_2^-] + k_3[\overset{+}{\text{H}_3\text{N}}\text{—CHR—CO}_2^-] + k_4[\overset{+}{\text{H}_3\text{N}}\text{—CHR—CO}_2\text{H}]]\,[\text{Base}]$$

Buffer effects were also studied by Smith, Williams, and Wonnacott (1978). They found that increasing concentration of phosphate buffer (at constant ionic strength and pH) results in a corresponding increase in the rate of racemization of alanine. Carbonate-bicarbonate buffer systems were explored in this study but were found to be unsuitable. The bicarbonate solutions lost carbon dioxide and resulted in higher pH's and higher rates. Unbuffered leucine and alanine were also found not to maintain their pH, presumably because of decarboxylation of the amino acids at higher temperatures. Unbuffered alanine racemized much less rapidly than phosphate-buffered samples. The role of the buffer, apart from providing a constant pH, is that of a base which assists in the abstraction of the α-hydrogen. Thus the racemization is seen to be an example of a general base catalyzed reaction.

The effect of changes in ionic strength on the rates of racemization of both free and gelatin-bound alanine was also studied by Smith and Wonnacott. At pH 7.6, changes in ionic strength had no marked effect on the rate of racemization of either species. However, at pH 10 (buffered) and pH 10.7 (unbuffered), Smith and co-workers (1978) reported increased racemization for free alanine with increasing ionic strength. These effects are predictable, since at the higher pH the amino acid is more negatively charged [$RCH(NH_2)CO_2^-$]. The reacting ionic species, the amino acid and the base ($B^-$), have the same charge, and their repulsions are more effectively shielded from one another by the increasingly polarized medium (increasing ionic strength).

Because pH, buffer concentration, and ionic strength are important to racemization, these factors need to be thoroughly understood and considered in a geochronological study.

## Metal Ions

The catalytic effect of metal ions on the racemization of both free and dipeptide bound amino acids has been reported in many studies. Olivard et al. (1952) reported that the racemization of alanine at pH 10 is catalyzed by $Cu^{2+}$, $Fe^{3+}$, and $Al^{3+}$ when in the presence of pyridoxal, VII

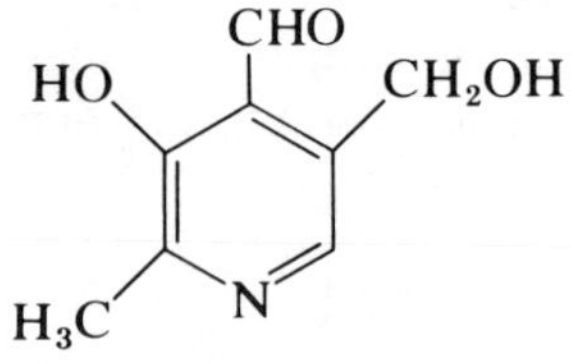

VII

In the absence of pyridoxal, extensive racemization does not occur. Olivard and co-workers suggested that a Schiff base was formed between the amino acid and pyridoxal which was stabilized by the metal ion, thus increasing racemization.

Toi (1963) has similarly investigated the influence of metal ions on the racemization of various free amino acids in the presence of resins with structures resembling pyridoxal. He found that the racemization catalyzed by resins required the presence of metal cations, with $Cu^{2+}$ being the most effective of those studied ($Cu^{2+}$, $Co^{2+}$, $Mg^{2+}$, $Ni^{2+}$, $Zn^{2+}$).

Gillard and Phipps (1970) investigated the effect of metal ions on the rates of racemization of amino acids bound in dipeptides. They showed that alanine in *bis*-glycyl-L-alaninatocobalt [III] complex (VIII) racemized faster than in the *bis*-L-alanyl-glycinatocobalt [III] complex (IX).

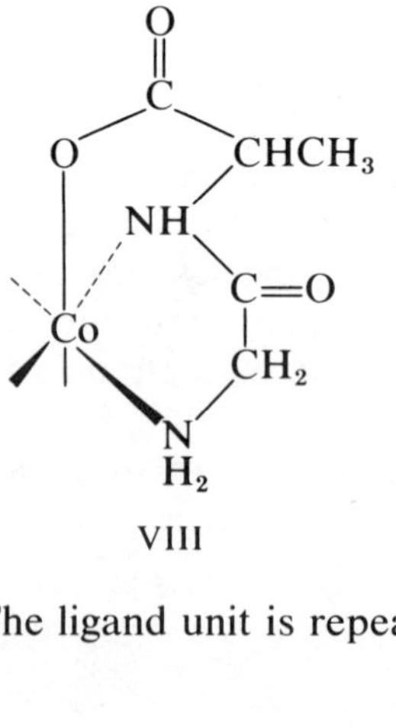

VIII

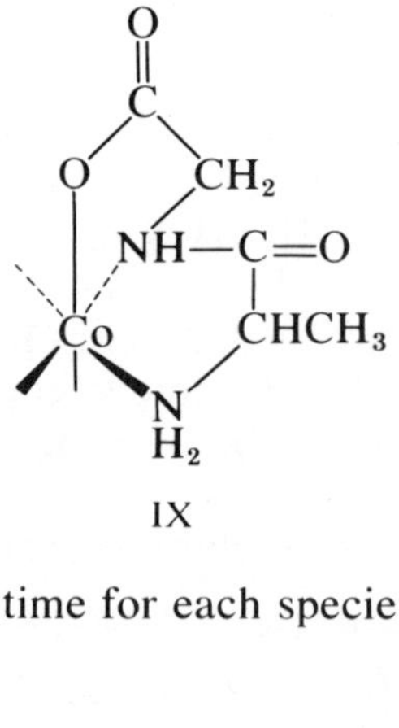

IX

The ligand unit is repeated a second time for each species.

Alanine occupies a "C-terminal" position in VIII and an "N-terminal" position in IX. There are insufficient data on the generality of C-terminal amino acids racemizing faster than N-terminal in such environments, and further investigation is needed, using varied dipeptides. It is interesting that in such systems the "terminal" amino group is unable to intramolecularly assist the ionization of the α-hydrogen at the C-terminus. It should be noted that Gillard and Phipps attempted unsuccessfully to synthesize cobalt tripeptide species for kinetic studies.

Buckingham et al. (1967) have studied the racemization of alanine and valine as ligands in ethylenediaminecobalt complexes at 34.3°C. Bada (1971) extrapolated their data to 0°C for pH 7.6 and reported that the chelated alanine racemized 3000 times faster than the nonchelated amino acid.

The enhancement of racemization in such complexes is understandable. The metal cation stabilizes the incipient carbanion via field effects and by removal of the destabilizing carboxylate moiety.

Wonnacott (1979) has studied the effect of $Cu^{2+}$ on the racemization of gelatin-bound and free alanine, leucine, and aspartic acid. The influence of the metal ion was negligible for the bound amino acids and only modest for the free species. The contrasting data seem to indicate that the free amino acids were not coordinated with the $Cu^{2+}$ as they were in the previously discussed studies.

## *ACKNOWLEDGEMENT*

Grateful acknowledgement is given to the National Aeronautic and Space Administration, NSG–7038, who supported some of the research reported in this chapter.

## *REFERENCES*

Abelson, P. H., Organic constituents of fossils, *Carnegie Inst. Wash. Year Book 53*, 97–101, 1954.

Abelson, P. H., Organic constituents of fossils, *Carnegie Inst. Wash. Year Book 54*, 107–109, 1955.

Abelson, P. H. and P. E. Hare, Recent amino acids in the Gunflint Chert, *Carnegie Inst. Wash. Year Book 67*, 208–210, 1969.

Akiyama, M., Epimerization of L-isoleucine on Na-montmorillonite and its implication to precambrian chemical fossils, *Origins of Life*, **9**, 541–545, 1978.

Bada, J. L., Kinetics of the nonbiological decomposition and racemization of amino acids in natural waters, in J. D. Hem (Symposium Chairman) and R. F. Gould (Series Editor),

Nonequilibrium Systems in Natural Water Chemistry, *Advan. Chem. Ser.* (106), American Chemical Society, Washington, D. C., pp. 309–331, 1971.

Bada, J. L., Kinetics of racemization of amino acid as a function of pH, *J. Amer. Chem. Soc.*, **95,** 1371–1373, 1972.

Bada, J. L. and R. A. Schroeder, Racemization of isoleucine in calcareous marine sediments: Kinetics and mechanism, *Earth Planet. Sci. Lett.*, **15,** 1–11, 1972.

Bohak, Z. and E. Katchalski, Synthesis, characterization, and racemization of poly-L-serine, *Biochemistry,* **2,** 228–237, 1963.

Bordwell, F. G. and F. G. Boyle, Jr., Kinetic isotope effects for nitroalkanes and their relationship to transition-state structure in proton-transfer reactions, *J. Amer. Chem. Soc.*, **97,** 3447–3452, 1975.

Buckingham, D. A., L. G. Marzilli, and A. M. Sargeson, Proton exchange and mutarotation of chelated amino acids via carbanion intermediates, *J. Amer. Chem. Soc.*, **89,** 5133–5138, 1967.

Crawhill, J. C. and D. F. Elliott, A note on the racemization of serine, *Biochem. J.*, **48,** 237–238, 1951.

Dakin, H. D., The catalytic racemization of optically active hydantoin derivatives and of related substances as the result of tautomeric change, *Amer. Chem. J.*, **44,** 48–60, 1910.

Davin, H. D. and H. W. Dudley, The racemization of proteins and their derivatives resulting from tautomeric change, Part II: The racemization of casein, *J. Biol. Chem.* **15,** 263–269, 1913.

Dorman, D. E. and F. A. Bovey, Carbon-13 magnetic resonance spectroscopy, the spectrum of proline in oligopeptides. *J. Org. Chem.*, **38,** 2379–2383, 1973.

Epand, R. M. and R. F. Epand, Studies on the alkaline denaturation of glucagon, *Can. J. Biochem.*, **51,** 140–147, 1973.

Frenkel, M. and L. Heller-Kallai, Racemization of organic compounds by montmorillonite. Implications for age determinations and stereocatalysis, *Chem. Geol.*, **19,** 161–166, 1977.

Fridkin, M., M. Wilchek, and M. Sheinblatt, NMR studies of H—D exchange of α-CH group of amino acid residues in peptides, *Biochem. Biophys. Res. Commun.*, **38,** 458–464, 1970.

Geschwind, I. I. and C. H. Li, Effects of alkali-heat treatment on β-melanocyte-stimulating hormone, *Arch. Biochem. Biophys.*, **106,** 200–206, 1964.

Gillard, R. D. and D. A. Phipps, The selective activation of amino acids and peptides, *Chem. Comm.*, 800–801, 1970.

Gilliom, R. D., *Introduction to Physical Organic Chemistry,* Addison-Wesley, Reading, MA, pp. 156–160, 1970.

Goodman, M. and L. Levine, Peptide synthesis via active esters, IV: Racemization and ring-opening reactions of optically active oxazolones, *J. Am. Chem. Soc.*, **86,** 2918–2922, 1964.

Goodman, M. and W. J. McGahren, Optically active peptide oxazolones. Preliminary racemization studies under peptide-coupling conditions, *J. Amer. Chem. Soc.*, **87,** 3028–3029, 1965.

Hare, P. E., Effect of hydrolysis on the racemization rate of amino acids, *Carnegie Inst. Wash. Year Book 70,* 256–258, 1971.

Hare, P. E., Amino acid dating of bone—the influence of water, *Carnegie Inst. Wash. Year Book 73,* 576–581, 1974.

Hare, P. E. and P. H. Abelson, Racemization of amino acids in fossil shells, *Carnegie Inst. Wash. Year Book 66,* 526–528, 1968.

Hare, P. E. and T. C. Hoering, Separation of amino acid optical isomers by gas chromatography, *Carnegie Inst. Wash. Year Book 72,* 690–694, 1973.

Hare, P. E. and R. M. Mitterer, Laboratory simulation of amino acid diagenesis in fossils, *Carnegie Inst. Wash. Year Book 67,* 205–208, 1969.

Helfman, P. M. and J. L. Bada, Aspartic acid racemization in tooth enamel from living humans, *Proc. Nat. Acad. Sci., U.S.A.,* **72,** 2891–2894, 1975.

Helfman, P. M. and J. L. Bada, Aspartic acid racemization in dentine as a measure of aging, *Nature,* **262,** 279–281, 1976.

Hine, J., *Structural Effects on Equilibria in Organic Chemistry,* Wiley, New York, 151 pp., 1975.

König, W. A., W. Rann, and J. Eyem, Gas chromatographic separation of diastereoisomeric amino acid derivatives on glass capillaries—use of pentafluoropropionyl amino acid (+)-3-methyl-2-butyl esters, *J. Chromatogr.,* **133,** 141–146, 1977.

Kosower, E. M., *An Introduction to Physical Organic Chemistry,* Wiley, New York, 49 pp., 1968.

Kriausakul, N. and R. M. Mitterer, Isoleucine epimerization in peptides and proteins: Kinetic factors and application to fossil proteins, *Science,* **201,** 1011–1014, 1978.

Kroepelin, H. and K. Georgaras, Racemisierung von aminosäuren an Silikaten, *Z. Natursforch. B.,* **23,** 1266, 1968.

Kroepelin, H., Racemization of amino acids on silicates, in *Advances in Organic Geochemistry,* P. A. Schenck and I. Havenaar, eds., Pergamon, New York, 1969.

Kvenvolden, K. A. and E. Peterson, Amino acids in late Pleistocene bone from Rancho La Brea, California, presented at 86th Annual Meeting of the Geological Society of America, November 12–14; *Abstracts with Programs,* **5,** 704, 1974.

Levene, P. A. and L. W. Bass, Studies on racemization, VIII: The action of alkali on proteins: Racemization and hydrolysis, *J. Biol. Chem.,* **82,** 171–190, 1929.

Manning, J. M., Determination of D- and L-amino acid residues in peptides. Use of tritiated hydrochloric acid to correct for racemization during acid hydrolysis, *J. Amer. Chem. Soc.,* **92,** 7449–7454, 1970.

Matsuo, H., Y. Hawazoe, M. Sato, M. Ohnishi, and T. Tatsuno, Studies on the racemization of amino acids and their derivatives, I: On the deuterium-hydrogen exchange reaction of amino acids derivatives in basic media, *Chem. Pharm. Bull.,* **15,** 391–398, 1967.

Neuberger, A., Stereochemistry of amino acids, in *Advances in Protein Chemistry,* Vol. 4, M. L. Anson and J. T. Edsall, eds., Academic Press, New York, pp. 297–383, 1948.

Olivard, J., D. E. Metzler, and E. E. Snell, Catalytic racemization of amino acids by pyridoxal and metal salts, *J. Biol. Chem.,* **199,** 669–674, 1952.

Petit, M. G., A rapid enzymatic technique for estimating the age of biological samples, *Anal. Lett.,* **7,** 215–221, 1974a.

Petit, M. G., The racemization rate constant for protein-based aspartic acid in woodrat middens, *Quat. Res.,* **4,** 340–345, 1974b.

Pfeffer, P. E., L. S. Gilbert, and J. M. Chirinko, Jr., α-Anions of carboxylic acids, II: The formation and alkylation of α-metalated aliphatic acids, *J. Org. Chem.,* **37,** 451–458, 1972.

Pickering, B. T. and C. H. Li, Adrenocorticotropins, XXIX: The action of sodium hydroxide on adrenocorticotropin, *Arch. Biochem. Biophys.,* **104,** 119–127, 1964.

Sato, M., T. Tatsuno, and H. Matsuo, Studies on the racemization of amino acids and their derivatives, III: The effect of alkyl-, aralkyl- and aryl-side chain at α-position of amino acids on their base-catalyzed racemization, *Chem. Pharm. Bull.*, **18,** 1794–1798, 1970.

Schroeder, R. A., Kinetics, mechanism and geochemical applications of amino acid racemization in various fossils, Ph.D. Thesis, Univ. of California, San Diego, 1974.

Schroeder, R. A. and J. L. Bada, A review of the geochemical applications of the amino acid racemization reaction, *Earth Sci. Rev.*, **12,** 347–391, 1976.

Shou, M.-Y. and J. L. Bada, The pK's of amino acids at elevated temperatures estimated from racemization data, personal communication, 1979.

Sivakua, T., Structural effects on the kinetics and mechanism of the racemization of arylglycines, Ph.D. Thesis, Utah State University, Logan, UT, 1978.

Smith, G. G., K. M. Williams and D. M. Wonnacott, Factors affecting the rate of racemization of amino acids and their significance to geochronology, *J. Org. Chem.*, **43,** 1–5, 1978.

Smith, G. G. and B. S. Sol, Racemization studies of carboxy and amino-terminal amino acid residues in dipeptides: C>N for nonsterically hindered amino acids, *Science*, in press, 1980.

Sol, B. S., Factors affecting the extent of racemization of peptide and protein bound amino acids: Positioning effect and kinetic studies, M.S. Thesis, Utah State University, Logan, UT, 1978.

Terent'ev, A. P. and E. I. Klabunovskii, The role of dissymmetry in the origin of living material, in *Aspects of the Origin of Life*, M. Florkin, ed., Pergamon, New York, pp. 74–84, 1960.

Toi, K., Synthetic resins catalyzing the racemization of amino acids, II: The activity of the resins, *Bull. Chem. Soc. Japan*, **36,** 739–743, 1963.

Wald, G., The origin of optical activity, *Ann. N. Y. Acad. of Sci.*, **69,** 352–368, 1958.

Wehmiller, J., Amino acid diagenesis in fossil calcareous organisms, Ph.D. Thesis, Columbia University, New York, 1971.

Wehmiller, J. and Hare, P. E., Racemization of amino acids in marine sediments, *Science*, **173,** 907–911, 1971.

Williams, K. M. and Smith, G. G., A critical evaluation of the application of amino acid racemization to geochronology and geothermometry, *Origins of Life*, **8,** 91–144, 1977.

Wonnacott, D. M., The application of gas chromatography to the resolution of enantiomeric mixtures of amino acids. Factors affecting the racemization of amino acids, Ph.D. Thesis, Utah State University, Logan UT, 1979.

Wüthrich, K., A. Tun–Kyi, and R. Schwyzer, Manifestation in the $^{13}C$—NMR spectra of two different molecular conformations of acyclic pentapeptides. *F.E.B.S Lett.*, **25,** 104–108 (1972).

# Some Factors Affecting the Epimerization of Isoleucine in Peptides and Proteins

**NIVAT KRIAUSAKUL**

**RICHARD M. MITTERER**

***Program in Geosciences, University of Texas at Dallas***

*ABSTRACT*

The major factors affecting the kinetics of isoleucine epimerization are the position of isoleucine in the peptide chain and the rate of hydrolysis of isoleucine to the free state. The rate-controlling steps in fossils are: (1) the rate of epimerization of terminal isoleucine; (2) the rate of hydrolysis of isoleucine; and (3) the rate of epimerization of free isoleucine.

In the presence of calcite the overall rate of epimerization of isoleucine in proteins and polypeptides increases, and in dipeptides decreases. Calcite has no direct effect on the epimerization of free isoleucine, but it does affect the rate of epimerization of bound isoleucine primarily by increasing the rate of hydrolysis. In the presence of calcite, interior isoleucine in proteins is partially hydrolyzed to the faster epimerizing terminal position, while terminal isoleucine in dipeptides is hydrolyzed to the slower epimerizing free state.

## INTRODUCTION

An understanding of the factors affecting the rate of amino acid racemization-epimerization reactions is essential if they are to be reliably utilized for dating purposes or other geological applications. In a previous study (Kriausakul and Mitterer 1978), a number of factors affecting the rate of epimerization of isoleucine in simple aqueous systems of dipeptides and polypeptides at elevated temperatures were elucidated. These factors were: (1) the position of isoleucine in the peptide chain, (2) the nature of the adjacent amino acids, and (3) the stability of the isoleucine peptide bonds. Isoleucine preferentially epimerizes at the terminal position of a peptide or protein chain, with a lower rate of epimerization occurring when it is in an interior position or exists as a free acid. The rate of hydrolysis, which is affected by the stability of the peptide bonds, will determine the rate of transfer of isoleucine from interior to terminal to free amino acid; consequently, the rate of epimerization of isoleucine will also be affected by the rate of hydrolysis of isoleucine.

The present study is a continuation of the previous work and provides additional data in support of the interpretation of isoleucine kinetics observed in fossils. Both the previous and present investigations utilize elevated temperatures to accelerate reaction rates. As with all elevated temperature experiments the inherent assumption is that the general phenomena observed at higher temperatures would also be observed at ambient temperature.

## EXPERIMENTAL PROCEDURE

### Heating Experiments

All dipeptides and tripeptides used in this study were purchased from ICN Pharamaceuticals, Inc. Oxytocin and insulin (bovine) were obtained from Calbiochem. Eledoisin and mellitin were purchased from Sigma Chemical Company. Stock solutions (0.1% wt/vol) of dipeptides, polypeptides, and proteins were adjusted to pH 8.0 by either NaOH or HC1.

For heating experiments, 1.0 ml. of the sample solution was sealed under nitrogen in a Pyrex tube and heated at 152°C in a sand bath maintained in a constant temperature oven. At the end of the heating period the sample was hydrolyzed in 6 *N* HCl at 110°C for 22 hours. The hydrolyzate was evaporated to dryness in vacuo at room temperature, the residue taken up in pH 2.2 HCl, and the alloisoleucine (aIle) to isoleucine (Ile) ratio determined with an automated amino acid analyzer.

Heating experiments in the presence of $CaCO_3$ were conducted by adding 1 ml of the sample solution to a Pyrex tube containing 100 mg $CaCO_3$. The suspension was mixed with a vortex mixer and left open to the atmosphere for 1 hour to allow the $CaCO_3$ system to reach equilibrium and to stabilize the pH. Random samples were subjected to pH measurement in order to check whether the pH had stabilized. The pH's of the slurries were in the range 8.05 to 8.15. The tubes were then sealed under nitrogen and heated at 152°C in a sand bath. After heating and acid hydrolysis, calcium was precipitated by the method described by Hare (1969).

### Molecular Fractionation and Partial Hydrolysis

Two 3.0-ml. albumin samples (0.1%) were sealed under $N_2$ and heated at 145°C for 66 hours. At the end of heating period, one tube was opened and five 0.25 ml. aliquots were taken. The first aliquot was totally hydrolyzed in 6 *N* HCl. The second and third aliquots were partially hydrolyzed in 0.5 *M* $CH_3COOH$ at 110°C for 22 hours and 48 hours, respectively. The fourth aliquot was heated without acid at the same temperature and time to serve as a control sample. The fifth aliquot was directly analyzed to determine the aIle-Ile ratio of the free fraction. The second albumin tube was opened and its contents subjected to molecular fractionation.

An ultrafiltration system was used to separate the high molecular weight peptide fraction (>500 MW) from the low molecular weight fraction (<500 MW) of heated albumin and an unheated fossil *Mercenaria* shell. An Amicon Model 402 Ultrafiltration system was employed using UM-05 ultrafilters. The fraction <500MW was directly analyzed for the aIle-Ile ratio. Two aliquots were taken from the >500 MW fraction which had previously been taken up in a small volume of acidic distilled water; one aliquot was totally hydrolyzed and the other was partially hydrolyzed as described above.

## RESULTS AND DISCUSSION

### Isoleucine Epimerization in Dipeptides and Polypeptides

Previous experiments utilizing dipeptides containing neutral amino acids have shown that the initial epimerization rate of isoleucine ($k_{epim}$) in simple bound form is about an order of magnitude greater than $k_{epim}$ of free isoleucine in aqueous solution (Kriausakul and Mitterer, 1978). In addition, different epimerization rates are found for isoleucine, depending on the nature of the adjacent amino acid in the dipeptide. Initial rates are faster for dipeptides containing glycine and slower for dipeptides containing valine. For each pair of dipeptides the $NH_2$-terminal isoleucine epimerizes slightly faster than the COOH-terminal form. This work has now been extended to include dipeptides containing aromatic and heterocyclic amino acids adjacent to isoleucine (Fig. 1).

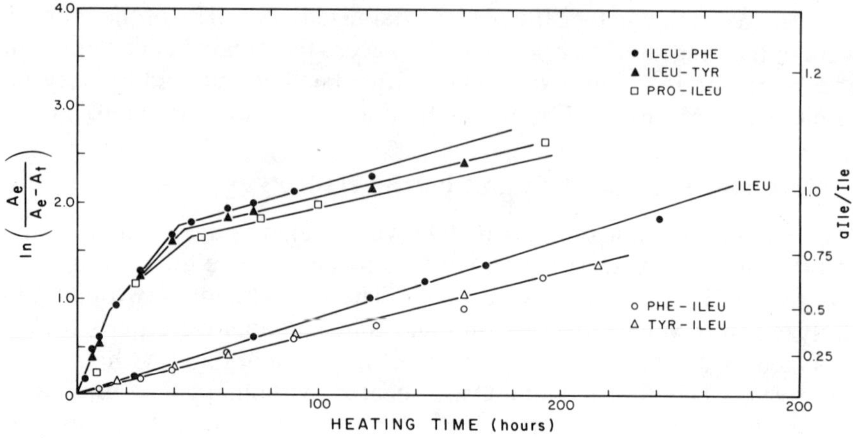


**Figure 1.** Kinetics of epimerization of isoleucine in dipeptides in aqueous solution (pH 8.0) at 152°C. Abbreviations : Ile, isoleucine; Phe, phenylalanine; Tyr, tyrosine.

The results for these dipeptides are distinctly different from those obtained for dipeptides containing neutral amino acids. Instead of different epimerization rates for isoleucine in each dipeptide, initial rates for both $NH_2$-terminal isoleucines of the aromatic dipeptides are almost identical, as are the rates for both COOH-terminal forms. Apparently, the aromatic groups of tyrosine and phenylalanine, which differ only in the presence of an OH group on the aromatic ring of tyrosine, have virtually identical effects on the isoleucine epimerization rate. In accordance with previous results, $k_{epim}$ for $NH_2$-terminal isoleucine is greater than that for the COOH-terminal form. However, for aromatic dipeptides the difference between the $NH_2$-terminal and COOH-terminal isoleucine epimerization rate constants is much greater than the difference for dipeptide pairs containing neutral amino acids. Indeed, $k_{epim}$ of COOH-terminal isoleucine in Phe-Ile and Tyr-Ile is even slightly less than the $k_{epim}$ of free isoleucine. One possible explanation for this result is that rapid hydrolysis of the COOH-terminal isoleucine would account for a smaller rate constant due to a greater concentration of the slower epimerizing free amino acid. The rate of hydrolysis of these aromatic dipeptides, however, is not significantly different from the hydrolysis rate for other dipeptides containing COOH-terminal isoleucine, so another explanation must be sought.

Compared to the calcified protein of mollusks and foraminifera, dipeptides are very simple systems and provide no information on the rate of epimerization of amino acids in interior positions of a polypeptide chain. However, the rate of epimerization of isoleucine located in interior positions of polypeptides and proteins is more difficult to deduce because isoleucine may be partially hydrolyzed to the terminal position during heating and because there is some uncertainty whether the $k_{epim}$ being measured is solely that of interior isoleucine. Despite this uncertainty, a maximum $k_{epim}$ for interior isoleucines can be estimated. As will be discussed later, it is likely that the initial $k_{epim}$ of isoleucine for some polypeptides and proteins is a good approximation for interior isoleucine.

Various polypeptides and proteins containing only one isoleucine residue (except for mellitin and albumin, which contain three and four, respectively) were heated in aqueous solution to investigate the kinetics of epimerization of isoleucine in interior positions. The rate curves obtained from these experiments vary considerably with the type of polypeptide or protein (Fig. 2). The epimerization rate constants are greater than that for free isoleucine and less than those for most of the dipeptides. As with most of the dipeptides, the initial $k_{epim}$ for isoleucine is different for each protein or polypeptide. The range of values for initial $k_{epim}$ is probably due partially to the effect of the adjacent amino acids on the epimerization rate and partially to the different degrees of stability of the isoleucine peptide bond. The larger epimerization rate constants for some of the polypeptides may be the result of partial hydrolysis of interior to terminal isoleucine during heating. In general, proteins having a greater rate of transfer of interior isoleucine to the terminal position because of a relatively weak peptide bond should have a larger initial $k_{epim}$. One possible exception to this generalization is when an isoleucine residue is adjacent to an aromatic amino acid such as that in eledoisin and oxytocin. In those

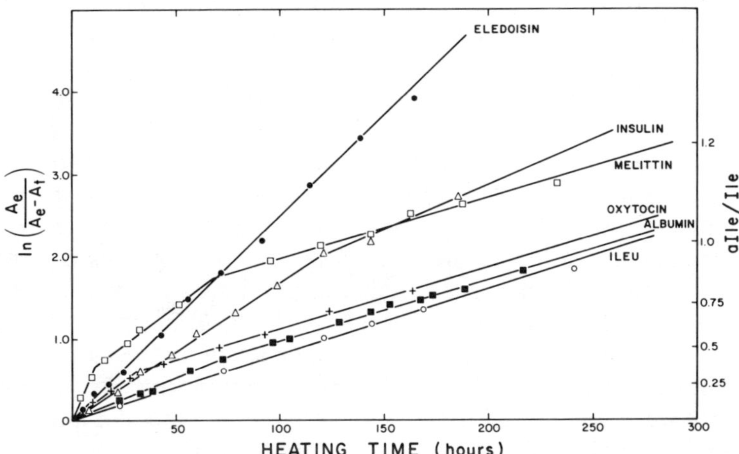


**Figure 2.** Kinetics of epimerization of isoleucine in polypeptides and proteins in aqueous solution (pH 8.0) at 152°C. The rate curve for free isoleucine is also shown.

cases, if the peptide bond binding isoleucine to the aromatic amino acids is stronger than the bond binding isoleucine to the other adjacent amino acid, partial hydrolysis would yield a COOH-terminal isoleucine adjacent to an aromatic acid. In this position $k_{epim}$ for terminal isoleucine would be even less than $k_{epim}$ for free isoleucine as shown in Figure 1. However, the observed rate of epimerization of isoleucine in these two polypeptides is greater than that of the free isoleucine. Apparently, some of the isoleucine is partially hydrolyzed to the $NH_2$-terminal form, which has a much greater rate of epimerization than the COOH-terminal form.

The results from the polypeptides and proteins are also similar to those from the dipeptides in that the overall rate curves, except for eledoisin, do not completely follow first-order kinetics. The slopes of the second portion of the rate curves are significantly less than the initial slopes, and, in some cases, approximately parallel to the slope of the free isoleucine curve. The rate of production of free isoleucine was monitored in these experiments. The isoleucine epimerization rate for dipeptides, polypeptides, and proteins decreases when the production of the slower epimerizing isoleucine by hydrolysis becomes significant. The rate of production of free isoleucine will depend on the stability of both adjacent peptide bonds. As previously noted, there is a very wide range in the rate of release of isoleucine from different peptides and proteins (Kriausakul and Mitterer 1978).

## Effect of $CaCO_3$ on Rate of Epimerization

The rate of isoleucine epimerization may differ considerably depending on the sedimentary environment. For example, Hare (1972) noted that the carbonate fraction of a sediment core has a greater aIle-Ile ratio than the clay fraction. Although the application of racemization-epimerization reactions as a geochronological tool for carbonate systems has received considerable attention, kinetic studies designed to elucidate the specific effect of calcium carbonate on the epimerization rate of isoleucine have not been reported. Indeed, relatively little experimental information is available concerning the possible effect of various inorganic minerals on the rate of epimerization of isoleucine in peptides and proteins. Recent studies on the effects of clay minerals on the rate of epimerization are limited to free isoleucine (Frenkel and Heller-Kallai 1977; Akiyama 1978). To investigate the role of $CaCO_3$ in epimerization kinetics, heating experiments were conducted using free isoleucine, peptides, and proteins in the presence of solid $CaCO_3$ (reagent grade calcite).

Results for free isoleucine, insulin, and albumin in aqueous solution and in the presence of $CaCO_3$ are given in Figure 3. The effect of $CaCO_3$

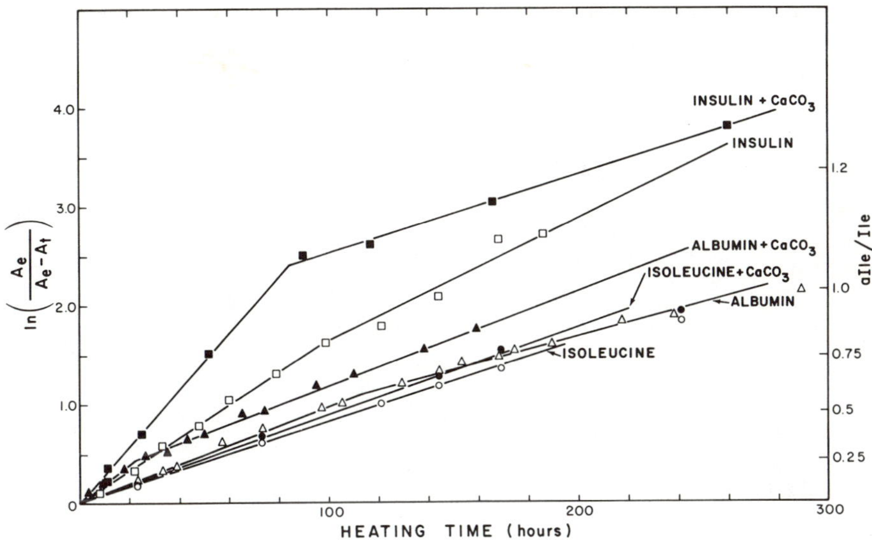


**Figure 3.** Kinetics of epimerization of free isoleucine, insulin and albumin in aqueous solution and in the presence of calcite at 152°C.

on the epimerization of free isoleucine is minimal, as the kinetic curve is only slightly affected in the presence of $CaCO_3$ compared to that obtained in aqueous solution. Apparently, $CaCO_3$ exhibits no catalytic effect on epimerization of free isoleucine. The rate curves of insulin and albumin, however, are drastically different with an enhanced initial rate of isoleucine epimerization when these proteins are heated in the presence of $CaCO_3$. In the latter portion, rate curves decrease by about an order of magnitude to a slope approximately parallel to that of free isoleucine. The rate curves obtained in the absence of $CaCO_3$ do not exhibit this order of magnitude decrease in rate constants.

The enhanced epimerization of bound isoleucine observed in the presence of $CaCO_3$ may be due to an increase in the rate of epimerization or to an increase in the rate of hydrolysis of the proteins. An increase in hydrolysis leads to enhanced epimerization by converting the slower epimerizing interior isoleucine to the faster epimerizing terminal isoleucine. In free isoleucine, hydrolysis has no affect on the state of the amino acid and $CaCO_3$ can only affect $k_{epim}$ directly; this influence, as noted above, is very slight.

To determine whether $CaCO_3$ enhances either $k_{epim}$ or $k_{hyd}$ of bound isoleucine, a pair of dipeptides was heated with $CaCO_3$. The results are presented in Figure 4, together with the rate curves of the dipeptides and free isoleucine obtained by heating in aqueous solution without $CaCO_3$.

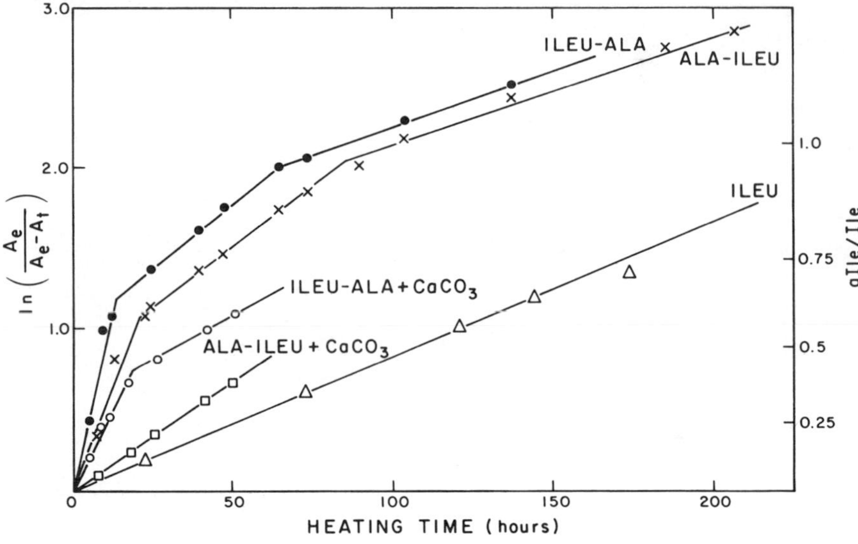


**Figure 4.** Kinetics of epimerization of isoleucine in the dipeptide pair Ile-Ala and Ala-Ile in aqueous solution and in the presence of calcite.

Unlike the proteins (Fig. 3) the rates of epimerization decrease significantly when the dipeptides are heated with $CaCO_3$ compared to the rate curves obtained in aqueous solution. If $Ca^{2+}$ or solid $CaCO_3$ is catalyzing the epimerization reaction of bound isoleucine in these dipeptides, it would be expected that the slopes of the rate curves should increase rather than decrease. The decrease in the slopes of the rate curves indicates that the preferentially epimerizing terminal isoleucine is being converted more rapidly to the slower epimerizing free isoleucine. Thus, with the dipeptides, $CaCO_3$ enhances the rate of hydrolysis to a greater extent than it enhances the rate of epimerization. It is likely that $CaCO_3$ also enhances the rate of hydrolysis of isoleucine in the proteins; however, with the proteins this increased hydrolysis brings about an increase in the slopes of the rate curves rather than a decrease because more of the isoleucine is converted to the rapidly epimerizing terminal position.

The influence of $CaCO_3$ on the hydrolysis rate may be either through free $Ca^{2+}$ present in solution in equilibrium with solid $CaCO_3$ or through the solid $CaCO_3$, or both. While it is not clear yet, which is important, the effect of the $CaCO_3$ (or $Ca^{2+}$) is to increase $k_{hyd}$; neither $Ca^{2+}$ nor solid $CaCO_3$ appear to have a direct effect on $k_{epim}$ as evidenced by the very slight increase in $k_{epim}$ for free isoleucine in the presence of $CaCO_3$. If metal ions are important in enhancing $k_{epim}$ of free isoleucine as suggested by Bada and Schroeder (1972), ions other than calcium must be responsible.

These results further support our interpretation that interior isoleucine epimerizes relatively slowly. Because of the lesser amount of protein hydrolysis in the absence of $CaCO_3$ the lower rates of epimerization of isoleucine in the polypeptides and proteins, noted previously (Fig. 2), are a good approximation of the epimerization rates for interior isoleucine. These rates are greater than that for free isoleucine but less than the rates for terminal isoleucine.

## Molecular Fractionation and Partial Hydrolysis of Fossil Protein and Heated Albumin

Intact proteins contain one or more interior isoleucines, and each of these may have different rates of epimerization. After a period of time the large protein molecule is converted through partial hydrolysis to a variety of peptides of different sizes. Isoleucine may be present then as interior, terminal, and free amino acid with each form having a characteristic rate of epimerization. Consequently, it is not easy to determine the kinetics of epimerization for isoleucine in the bound fraction, because of the variety of forms present and the difficulty in isolating each one. One approach to obtain additional information on protein systems is through molecular fractionation and partial hydrolysis (Table 1). For these analyses, an aqueous solution of albumin, which was heated at 145° for 66 hours, and a Pleistocene *Mercenaria* were utilized.

Analysis of the free isoleucine fraction produced by heating albumin shows that there is good agreement between the aIle-Ile ratios of the free fraction obtained by analysis of an unhydrolyzed sample (aIle/Ile = 0.751)

**Table 1. Extent of Isoleucine Epimerization in Different Fractions of Heated Albumin (145°C, 66 hrs)**

| | AIle/Ile | % |
|---|---|---|
| Total | 0.282 | 100 |
| Free | 0.751 | 11 |
| Bound (calculated) | 0.263 | 89 |
| <500 MW free | 0.767 | |
| >500 MW | | |
| Total hydrolysis | 0.215 | 100 |
| Partial hydrolysis | 0.958 | 10 |

and the free fraction obtained by molecular separation of the <500MW fraction (aIle/Ile = 0.767). The ratio in the total bound fraction obtained by calculation (aIle/Ile = 0.263) is, as expected, considerably less than the free ratio but is slightly greater than the ratio obtained by hydrolysis of the >500MW fraction (aIle/Ile = 0.215). The greater ratio in the calculated bound fraction possibly results from the inclusion of small molecular weight peptides such as dipeptides and tripeptides which would contain more isoleucine in the highly epimerized terminal form. The >500MW fraction excludes these small peptides. Partial hydrolysis of the >500MW fraction yields an aIle-Ile ratio of 0.958 for the new free fraction. Partial hydrolysis would tend to release terminal, rather than interior, amino acids because only one peptide bond must be broken. The high aIle-Ile ratio of new free isoleucine after partial hydrolysis of the >500MW fraction supports our suggestion that this fraction contains highly epimerized isoleucine in the terminal position.

High-temperature heating experiments with albumin may not exactly duplicate the processes occurring in fossils. Accordingly, it is desirable to investigate the aIle-Ile ratios in different molecular weight fractions of fossil proteins. Preliminary fractionations have been carried out for the protein and peptide constituents of a fossil *Mercenaria* specimen of Late Pleistocene age (Table 2). The >500MW fraction has the lowest aIle-Ile ratio, in agreement with the results from heated albumin. Bound isoleucine in the <500MW fraction, which contains a mixture of dipeptides, tripeptides, and possibly tetrapeptides, has an aIle-Ile ratio (0.234) that exceeds that of both the >500MW bound fraction (0.026) and the total fraction (free + bound) for the fossil (0.140). Evidently, in fossils, isoleucine in low molecular weight peptides epimerizes at a much greater rate than it does in the higher molecular weight fractions, where it is less likely to be in the terminal position. Low molecular weight tripeptides

**Table 2. Extent of Isoleucine Epimerization in Different Molecular Weight Fractions of Fossil Protein from a Late Pleistocene *Mercenaria***

| | AIle/Ile | % |
|---|---|---|
| Total | 0.140 | 100 |
| >500 MW | 0.026 | |
| <500 MW | | |
| Total | 0.328 | 100 |
| Free | 0.449 | 48 |
| Bound (calculated) | 0.234 | 52 |

and tetrapeptides may contain some isoleucine in the slower epimerizing interior position; hence, the ratio in the low molecular weight bound fraction is somewhat less than the ratio in the free fraction which is derived entirely from the terminal form.

It is noteworthy that the aIle-Ile ratios of the free fractions of both heated albumin and fossil *Mercenaria* are much greater than those of the total fractions. This is the usual relationship for carbonate fossils; the greater degree of epimerization of free isoleucine in fossils has been attributed to metal ion catalysis (Bada and Schroeder 1972). However, albumin was heated in aqueous solution with no metal ions present. The similar results for heated albumin and fossil *Mercenaria* provide additional support for our interpretation that the greater aIle-Ile ratio in the free fraction of fossil carbonates is due to hydrolysis of preferentially epimerized terminal isoleucine.

## Isoleucine Kinetics in *Mercenaria*

Preliminary kinetics for the epimerization of isoleucine in *Mercenaria* have been given previously (Mitterer 1975). The preliminary results, indicated that linear first-order kinetics were followed to an aIle-Ile ratio of almost 0.9. A detailed kinetic curve, based on more closely spaced heating times at 152°C, is presented in Figure 5. The more detailed results

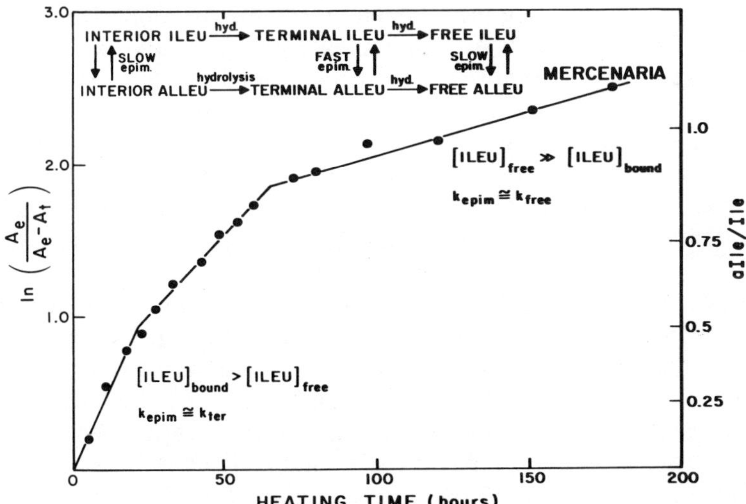


**Figure 5.** Kinetics of epimerization of isoleucine in heated *Mercenaria* at 152°C and schematic representation of possible reaction sequence of isoleucine epimerization in carbonate systems.

indicate that linear first-order kinetics are followed to an aIle-Ile ratio of about 0.6. Preliminary estimates of ages of *Mercenaria* from Plio-Pleistocene deposits of Florida based on the previous kinetic curve are too low. The aIle-Ile ratios of fossil *Mercenaria* from these deposits fall on the nonlinear portion of the revised kinetic curve, and new estimates of ages for the fossils cannot accurately be made. The previous estimates, however, represent minimum ages for these Plio-Pleistocene deposits.

The kinetic curve for *Mercenaria*, although differing in some details, is typical of kinetic curves for the epimerization of total isoleucine (free + bound) in carbonate tests such as mollusc shells and foraminifera. The curve is generally characterized by two linear segments of different slopes joined by a transition zone. The aIle-Ile ratio of the free fraction exceeds that of the bound fraction. Bada and Schroeder (1972) have interpreted this effect to be due to preferential epimerization of free isoleucine through metal ion catalysis. Based on experiments with dipeptides Kriausakul and Mitterer (1978) have postulated that the greater degree of epimerization of free isoleucine in fossils is due to hydrolysis of the preferentially epimerized terminal isoleucine. The results discussed here give added support to this hypothesis; a more detailed interpretation of the kinetics of isoleucine epimerization in carbonate systems is presented in Figure 5.

Initially, in modern shells almost all of the isoleucine residues are in interior positions, where they epimerize very slowly. Through partial hydrolysis some of the interior isoleucine residues will be hydrolyzed to terminal positions, where they will be preferentially epimerized. Because the rate of epimerization of terminal isoleucine greatly exceeds that of interior isoleucine, the epimerization of terminal isoleucine will be the rate controlling step as soon as significant hydrolysis has occurred. Partial hydrolysis probably occurs very early after burial, depending on the diagenetic temperature. For example, significant hydrolysis has already occurred in a 1000-year-old *Mercenaria* from Florida as evidenced by the relative concentrations of the various peptide fractions (Hare and Mitterer 1967).

With further hydrolysis the concentration of the free fraction becomes significant and approaches that of the bound fraction. Because the rate of epimerization of free isoleucine is much less than the rate for terminal isoleucine, the presence of a significant amount of free isoleucine results in a decrease in the rate constant and a departure from linear first-order kinetics (transition zone). When the bound isoleucine is almost entirely epimerized or the concentration of the bound fraction is low compared to the concentration of the free fraction, or both, the rate of epimerization is controlled solely by the rate of epimerization of free isoleucine (second linear portion). This portion of the rate curve is approximately parallel to the rate curve of free isoleucine in aqueous solution.

In summary, then, the kinetic curve is controlled by the rate at which isoleucine is transferred from the faster epimerizing terminal position to the very slowly epimerizing free fraction. The first portion of the rate curve is controlled by the rate of epimerization of terminal (and, to some extent, interior) isoleucine; the second portion is dominated by the rate of epimerization of free isoleucine.

## CONCLUSIONS

The results presented here confirm and extend the hypothesis that the major factors affecting the kinetics of isoleucine epimerization are the position of isoleucine in the peptide chain and the rate of hydrolysis of isoleucine to the free state. The rate-controlling steps in fossils are: (1) the rate of epimerization of terminal isoleucine; (2) the rate of hydrolysis and; (3) the rate of epimerization of free isoleucine.

The associated mineral plays an important role in the kinetics of isoleucine epimerization. In the case of $CaCO_3$, this role is to enhance the rate of hydrolysis. Other minerals may have similar or different effects so that the results obtained for one mineral system should not be applied to other mineral systems.

Finally, the detailed investigations of the kinetics of isoleucine epimerization in simple systems such as dipeptides and polypeptides of known sequence enable us to understand somewhat the kinetics within the much more complicated systems of fossil shells.

### *ACKNOWLEDGMENT*

This work was supported by NSF Grant, DES75-14868. Contribution No. 366 of the Geosciences Program, University of Texas at Dallas.

### *REFERENCES*

Akiyama, M., Epimerization of L-isoleucine on Na-Montmorillonite and its implication to Precambrian chemical fossils, *Origins of Life*, **9**, 541–545, 1978.

Bada, J. L. and R. A. Schroeder, Racemization of isoleucine in calcareous marine sediments: Kinetics and mechanism, *Earth Planet. Sci. Lett.*, **15**, 1–11, 1972.

Frenkel, M. and L. Heller-Kallai, Racemization of organic compounds by montmorillonite—implications for age determinations and stereocatalysis, *Chem. Geol.*, **19,** 161–166, 1977.

Hare, P. E., Organic Geochemistry of Proteins, Peptides and Amino Acids, in *Organic Geochemistry,* G. Eglinton and M. T. J. Murphy, eds., Springer-Verlag, New York, pp. 438–463, 1969.

Hare, P. E., Amino acid geochemistry of a sediment core from the Caraico Trench, *Carnegie Inst. Wash. Year Book 71,* 592–596, 1972.

Hare, P. E. and R. M. Mitterer, Nonprotein amino acids in fossil shells, *Carnegie Inst. Wash. Year Book 65,* 362–364, 1967.

Kriausakul, N. and R. M. Mitterer, Isoleucine epimerization in peptides and proteins: Kinetic factors and application to fossil proteins, *Science,* **201,** 1011–1014, 1978.

Mitterer, R. M., Ages and diagenetic temperatures of Pleistocene deposits of Florida based on isoleucine epimerization in *Mercenaria, Earth Planet, Sci. Lett.*, **28,** 275–282, 1975.

# Protein Hydrolysis and Amino Acid Racemization in Sized Bone

**DAVID W. VON ENDT**
*Department of Anthropology, Smithsonian Institution*

*ABSTRACT*

To determine the effect of bone volume on the degradation of bone protein through time, samples of cow bone were cut into 1, 2, and 4 mm cubes, and placed in a constant volume of water heated to 120 C for varying periods of time up to 4 hours. Amino acid analysis of the solution indicates that protein is lost from bone at a rate inversely proportional to bone size. But the major effect of size is seen during the initial, relatively slow loss of bone protein; the final rate of protein loss is approximately equal in bone cubes of different size. Gas chromatographic analysis of amino acid enantiomers from the solution indicates that racemization is proceeding at an approximately linear rate in aspartic acid already released from bone. Results are interpreted as the consequence of initial protein-mineral bond disruption accompanied by hydrolysis and diffusion. A diffusion gradient from the interior to the exterior of bone is postulated, the gradient consisting of a mixture of peptides undergoing hydrolysis and attendant amino acid racemization.

## INTRODUCTION

The ubiquity of amino acids in geological formations has stimulated a search for amino acid reactions of use in chronology (Hare 1969, 1974; Bada and Helfman 1975). The occurrence of amino acids and proteins in various geologic formations and some of the reactions these organic molecules undergo have been summarized in several reviews (e. g., Hare 1969, and Kvenvolden 1975).

Much of the work that stimulated current research in the geochemistry of amino acids consisted of a series of laboratory experiments having application to the fossil record (Abelson 1957, 1959, 1963). Recent experiments indicate that the deterioration of calcified tissue buried in the ground proceeds by a complex series of reactions. Of major importance are reactions affecting the protein-mineral bond (Ortner et al., 1972; Von Endt and Ortner 1979; Von Endt et al., 1975). The results of these experiments indicate that the role of hydrolytic reactions are maximized, and that other reactions proceed as a direct consequence of protein–mineral bond disruption, mineral dissolution, and protein hydrolysis. Similar reasoning has been invoked to explain the nonlinear kinetics of deep sea core isoleucine epimerization (Wehmiller and Hare 1971; Bada and Schroeder 1972).

While environmental factors such as water and acidity have been shown to have little effect on certain other bone protein alterations such as racemization of amino acids (Bada 1972a), environmental conditions dramatically alter the rate and direction of protein-mineral bond disruption and protein hydrolysis. In this paper I wish to present (1) the results of two experiments designed to provide information on the rate of hydrolysis of protein in bone of different size, (2) the rate at which the hydrolysis products diffuse through these bones to the exterior environment, and (3) some indicator of the steric configuration of the protein amino acids which have been removed.

The implication that hydrolysis and dissolution do not proceed at a linear rate through long periods of time is of consequence to geochemists and archeologists. A firm knowledge of the postmortem environmental conditions faced by calcified tissue is needed to predict time or temperature. Bone was chosen because, along with teeth, it is common in archeological finds and represents the only remaining physical evidence of prehistoric peoples, some evidence of their food, and some of their technology.

## MATERIALS AND METHODS

Heating experiments were conducted in a covered oil bath at 120 C. Bone samples were taken from a cleaned, air-dried bovine tibia obtained from a local meat packing plant. Bone sections were first cut by hand on a diamond saw, then mounted on a glass microscope slide and parallel sections removed with an Ingram petrographic thin-sectioning saw. Cubes were taken from these slices in the same manner. The final sections were 1 mm., 2 mm., and 4 mm. cubes and measured within 2% of their stated values using a reticule and microscope. The newly cut cubes were cleaned in non-ionic detergent in a sonic cleaner for 10 minutes, then rinsed in 100 ml volumes of distilled water three times in the same manner. Each cube was sealed in a clean test tube containing 5 ml of distilled water. Eight tubes containing cubes of each size were placed in the oil bath, and two of each were removed at intervals of 1 hour, 2 hours, 3 hours, and 4 hours. The bone samples were dried over silica gel in a vacuum, then hydrolyzed in distilled 6 *N* hydrochloric acid for 20 hours. Samples of the water in each tube were dried over silica gel and solid solium hydroxide in a dessicator at 2 mm Hg pressure overnight, and then hydrolyzed as before.

Amino acid analysis was conducted on a ultrasensitive automatic amino acid analyzer similar in design to that of Hare (1973) using the ninhydrin reaction (Moore 1968) to detect the presence of amino acids. These results were confirmed with an analyzer (kindly supplied by P. E. Hare) using the *o*-phthalaldehyde fluorescent reaction (Benson and Hare 1975). The steric configuration and relative enantiomeric concentrations of selected amino acids were determined by two gas chromatographic techniques. Derivatization was performed directly on the dried amino acid bone sample with no initial sample deionization. The pentafluoropropionyl-D,L-amino acid-(+)-3-methyl-2-butyl esters were formed according to the method of König et al. (1977). Separation took place on a 25 m glass capillary column coated with purified SE-30 (LKB Instrument Co.). The gas chromatograph was a Hewlett-Packard Model 5830A with flame ionization detector.

Samples were injected in 1 μl of glass distilled dichloromethane (Burdick and Jackson Co.) with the column oven temperature set initially at 45 C. After 5 minutes of isothermal operation, temperature was raised to 250 C at a rate of 2 per minute. Helium was used as the carrier gas and flowed through the column at a velocity of 25 cm/sec (optimized for linear velocity with butane).

Duplication of the results was achieved in the laboratory of P. E. Hare using a solventless injection system mounted on a Perkin-Elmer Model 9000 instrument equipped with a nitrogen detector and a 25 m Chirasil-Val column (Applied Science Labs.). Helium was used as the carrier gas. The N-pentaflouropropionyl-O-acetyl-(+)-2-butyl esters of amino acids were formed directly on the dried samples using Hare's modification of the method of Kvenvolden et al. (1972).

## RESULTS AND DISCUSSION

Protein-mineral bond disruption and consequent protein hydrolysis have been studied by heating samples of ground and sized bone in water and determining the amount of nitrogen released into solution (Ortner *et al.*, 1972). These reactions proceed in bone of a given size at a log-linear (pseudo–first order) rate that is temperature dependent. Using similar experimental conditions, protein-mineral bond disruption as indicated by nitrogen loss has also been shown to proceed at a rate inversely dependant on the size of the heated bone sample (Von Endt and Ortner 1979). The previous heating experiments were conducted on ground bone of small size. The results of heating cortical bone samples which approach archeological material in thickness is given in Figure 1. This figure represents the results of heating 1, 2, and 4 mm. cubes of bone in water at 120 C for periods of time up to 4 hours. If the appearance of leucine in the solution surrounding the bones is used as an indicator of protein-mineral bond disruption and protein hydrolysis, the results indicate that these reactions proceed at an initial slow rate that is inversely proportional to bone size. It may also be noticed that the length of time for which the

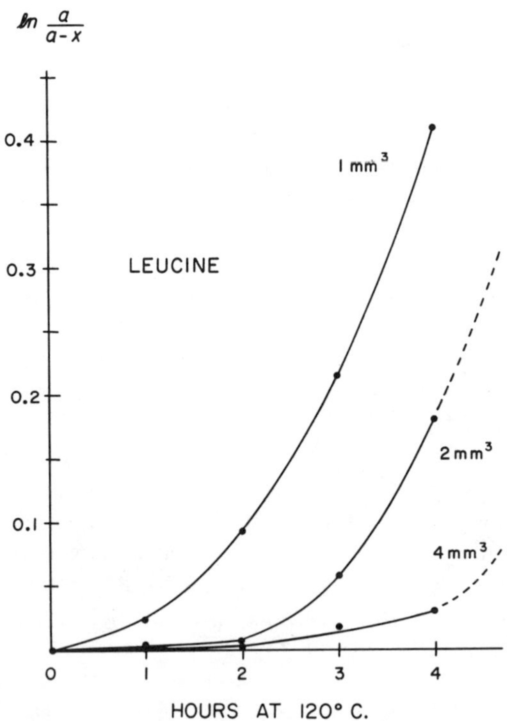


**Figure 1.** A plot of *ln a/a-x versus* time for heating 1, 2, and 4 mm cubes of bone in 5 ml of water heated at 120 C. Samples of water were removed and analyzed for leucine at 1, 2, 3, and 4 hour intervals.

reactions proceed slowly is proportional to the size of the sample. The rate then accelerates for each sample, so that the reactions are proceeding at approximately the same rate regardless of sample size. The data of Figure 1 are plotted in a log-linear fashion according to the integrated form of the first-order rate expression as $ln[a/a\text{-}x]$ versus time $t$, and indicate that leucine is lost to the environment surrounding bone at a nonlinear rate. The first-order rate expression does not adequately describe the degradation of bone as indicated by leucine loss.

I interpret these results to mean that protein-mineral bond disruption and protein hydrolysis proceed at an initial slow rate until alterations to the biological integrity of the organ have occurred which are sufficient to allow hydrolytic products to be removed more rapidly. The integrity of the organ is slowly altered, allowing more environmentally introduced water and dissolved bone ions to further disrupt the organ. The net velocity of these reactions increases until sufficient disruption and hydrolysis have occurred that peptides and amino acids are reaching the environment at an approximately constant rate. Overlying these reactions is the ability of protein fragments to diffuse from the interior to the exterior of bone. Diffusion is facilitated by the continuing bone disruption and the rapidly hydrolyzing, and thus shortened, bone proteins. It is postulated that a diffusion gradient exists from the not yet hydrolyzed interior which contains native protein and protein-mineral bonding, through partially altered large peptides, to peptides and amino acids small enough to diffuse through the degraded portion of bone and reach the surface. Racemization and epimerization then occurs at various position dependent rates (Kriausakul and Mitterer 1978) at or near the hydrolyzed ends of polypeptides.

The data of Figure 2 demonstrate that hydrolysis is a more facile reaction than racemization. The figure describes data on the racemization of L-aspartic acid which has appeared in the solution surrounding a 2 mm cube of bone heated at 120 C for 1, 2, 3, and 4 hours. It may be seen that the amino acids and small peptides that have entered the environment, as indicated by the appearance of aspartic acid in solution, are not completely racemized, and that racemization is continuing at an approximately linear rate. These observations indicate that peptides of unknown but varying length only partially hydrolyze within bone as they are migrating toward the surface. It is inferred that not only is there a diffusion gradient for hydrolytic products which decrease in chain length as they near the surface but also that a 'racemization gradient' exists, with less racemized material being found in the interior of bone. Thus samples of whole bone taken to obtain racemization data would contain an average of partially hydrolyzed and racemized peptides varying from relatively unreacted in the interior to much more reacted near the surface. If smaller

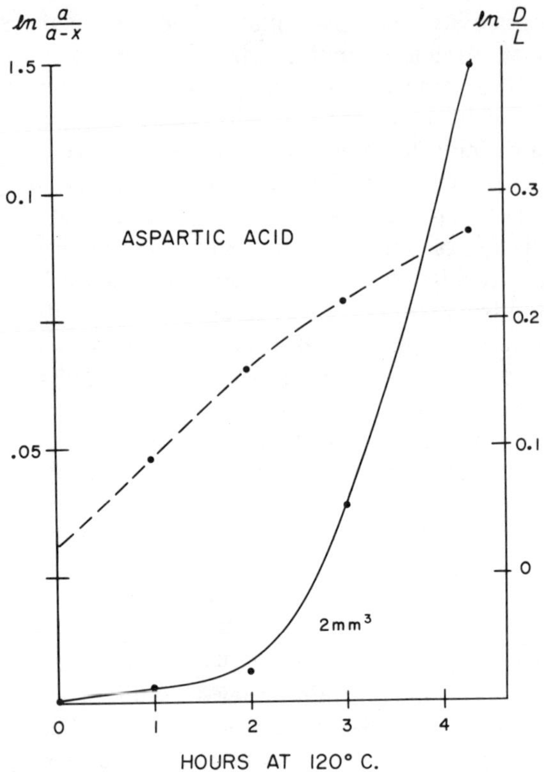


**Figure 2.** A plot of *ln a/a-x* (left ordinate) versus time for the release of aspartic acid from a 2 mm cube of bone heated at 120 C. The right ordinate of *ln* D/L represents the degree of racemization of aspartic acid that has appeared in the solution.

samples are taken, the data will also depend on where along the hydrolysis-racemization-diffusion continuem the sample was removed.

In summary, it has been shown that hydrolysis of bone protein proceeds at a rate inversely proportional to the size of bone. It has also been shown that racemization is not complete in peptides and amino acids that have already diffused from bone and reached the environment. A diffusion gradient within bone has been inferred. The experimental results have been accounted for by considering disruption of the native protein-mineral bond and protein hydrolysis to be the primary events in the alteration of bone. These events proceed at a nonlinear rate and are greatly influenced by the postmortem environment (e.g., water content, pH) of bone. For instance, mild acids easily dissolve bone mineral and catalyze hydrolytic reactions. These observations are directly translatable to field data in which the same results are to be expected.

## *ACKNOWLEDGMENTS*

Many thanks are due Dr. P. E. Hare, who not only discussed this project with me but also generously allowed me to use his laboratory facilities. Thanks are also due Mrs. A. I. Stix, who conducted the experiments and typed the manuscript. Mrs. M. D. Bakry provided the illustrations.

## *REFERENCES*

Abelson, P. H., Organic constituents of fossils, *Geol. Soc. Amer. Mem.*, **67**, 87–92, 1957.

Abelson, P. H., Geochemistry of organic substances, in *Researches in Geochemistry,* P. H. Abelson, ed., Wiley, New York, 1959.

Abelson, P. H., Geochemistry of amino acids, in *Organic Geochemistry,* I. A. Berger, ed., Macmillan, New york, 1963.

Bada, J. L., B. P. Luyendyk, and J. B. Maynard, Marine sediments: Dating by the racemization of amino acids, *Science,* **170,** 730–732, 1970.

Bada, J. L., Kinetics of racemization of amino acids as a function of pH, *J. Amer. Chem. Soc.,* **94,** 1371–1373, 1972a.

Bada, J. L., The dating of fossil bones using the racemization of isoleucine, *Earth Planet. Sci. Lett.,* **15,** 223–231, 1972b.

Bada, J. L. and P. M. Helfman, Amino acid racemization dating of fossil bones, *World Archaeol.,* **7,** 160–173, 1975.

Bada, J. L. and R. A. Schroeder, Racemization of isoleucine in calcareous marine sediments: Kinetics and mechanisms, *Earth Planet. Sci. Lett.,* **15,** 1–11.

Benson, J. R. and P. E. Hare, *o*-Phthalaldehyde: Fluorogenic detection of primary amines in the picomole range. Comparison with fluorescamine and ninhydrin, *Proc. Nat. Acad. Sci.,* **72,** 619–622.

Dungworth, G., J. A. Th. Vrenken, and A. W. Schwartz, Amino acid compositions of pleistocene collagens, *Comp. Biochem. Physiol.,* **51 B,** 331–335, 1975.

Hare, P. E., Geochemistry of proteins, peptides, and amino acids, in *Organic Geochemistry,* G. Eglinton and M. Morphy, Sr., eds., Springer-Verlag, Berlin, 1969.

Hare, P. E., Ultrasensitive amino acid analyzer, *Carnegie Inst. Wash. Year Book 70,* 268–269, 1973.

Hare, P. E., Amino acid dating — a history and an evaluation, *University of Pennsylvania Museum of Archeology News,* **10,** 4–8, 1974.

Hare, P. E. and R. M. Mitterer, Laboratory simulation of amino acid diagenesis in fossils, *Carnegie Inst. Wash. Year Book 67,* 205–208.

Konig, W. A. and W. Rahn, Gas chromatographic separation of diastereoisomeric amino acid derivatives on glass capillaries: The use of Pentafluoropropionyl-amino acid (+)-3-Methyl-2-Butyl esters, *J. Chromatog.,* **133,** 141–146.

Kriausakul, N. and R. M. Mitterer, Isoleucine epimerization in peptides and proteins: Kinetic factors and application to fossil proteins, *Science,* **201,** 1011–1014, 1978.

Kvenvolden, K. A., E. Peterson, and G. E. Pollock, Geochemistry of amino acid enantiomers: Gas chromatography of their diastereomeric derivatives, *Advances in Organic*

*Geochemistry 1971,* H. von Gaertner and H. Wehner, eds, Pergamon, Braunschweig, 1972.

Lee, C., J. L. Bada, and E. Peterson, Amino acids in modern and fossil woods, *Nature,* **259,** 183–186, 1976.

Moore, S., Amino acid analysis: Aqueous dimethyl sulfoxide as solvent for the ninhydrin reaction, *J. Biol. Chem.,* **243,** 6281–6283, 1968.

Ortner, D. J., D. W. Von Endt, and M. S. Robinson, The effect of temperature on protein decay in bone: Its significance in nitrogen dating of archaeological specimens, *Amer. Antiquity,* **37,** 514–520, 1972.

Von Endt, D. W. and D. J. Ortner, Experimental studies on chemical changes likely to occur in buried bone, *J. Archeol. Sci.,* in press, 1979.

Von Endt, D. W., P. E. Hare, D. J. Ortner, and A. I. Stix, Amino acid isomerization rates and their use in dating archeological bone, *Soc. Amer. Archeol. Abstr., 40th Annu. Mtg.,* 1975.

Von Endt, D. W., Techniques of amino acid dating, *Pre-Llano Cultures of the Americas,* R. Humphrey and D. Stanford, eds, Anthropological Society of Washington, 1979.

Wehmiller, J. and P. E. Hare, Racemization of amino acids in marine sediments, *Science,* **173,** 907–911, 1971.

Wyckoff, R. W. G., *The Biochemistry of Animal Fossils,* Scientechnica, Bristol, 1972.

# Inter- and Intrageneric Trends in Apparent Racemization Kinetics of Amino Acids in Quaternary Mollusks

KENNETH R. LAJOIE
*U.S. Geological Survey, Menlo Park, California*

JOHN F. WEHMILLER
*Department of Geology, University of Delaware*

GEORGE L. KENNEDY
*U.S. Geological Survey, Menlo Park, California*

## *ABSTRACT*

The following conclusions can be drawn from amino acid enantiomeric (D/L) ratios in coexisting genera of Quaternary mollusks: (1) Differences in *apparent racemization rates* are not detectable among species of the same genus, they may be detectable among genera in the same subfamily or family, and they are normally detected among genera in different families or at higher taxonomic levels. The apparent rate of racemization is slower in the bivalve family Veneridae (*Protothaca, Saxidomus, Tivela, Chione*) than in the bivalve families Lucinidae (*Epilucina*), Semelidae (*Cumingia*), and Tellinidae (*Macoma*), or in the gastropod family Trochidae (*Tegula*). (2) The order of apparent relative rates of racemization among five amino acids is the same in different genera (proline ≧ phenylalanine ≧ leucine > glutamic acid > valine). (3) Intersections of kinetic pathways during later stages of diagenesis occur, thereby producing reversals in apparent rates of racemization. (4) Statistical regressions of intergeneric and intrageneric relations in enantiomeric ratios permit interconversion of results among different genera for purposes of sample evaluation, comparison of relative enantiomeric ratios among different localities, or interpretation in terms of leucine kinetic models (Wehmiller and Belknap, 1978). (5) Kinetic curves for *Protothaca*, derived with independently dated samples, are consistent with modeled curves showing rapid initial racemization rates with a decrease in rates beyond about 70,000 years (for typical effective temperatures). (6) The common venerid bivalves *Protothaca* and *Saxidomus* yield the most consistent results and are most useful for chronological studies on the Pacific Coast.

## INTRODUCTION

Amino acid enantiomeric ratios in fossil mollusks provide a means of dating Quaternary marine terrace deposits. Early work along the Pacific coast of the U.S. showed that gastropods such as *Diodora* and *Polinices* yielded erratic results when compared with their known relative ages. Thick-shelled marine bivalves produced reproducible results, consistent with their known relative ages, and therefore have been used in all subsequent work (Wehmiller et al., 1977; Wehmiller et al., 1978a; Kennedy 1978).

Two approaches can be taken to the derivation of amino acid age estimates: (1) kinetic models (primarily for leucine in *Protothaca*) that quantify age/temperature relationships (Wehmiller and Belkap, 1978), or (2) relative age-correlation diagrams (Wehmiller et al., 1977) for individual amino acids in single genera, such as leucine in *Saxidomus* (Kennedy, 1978). Because no genus occurs in every fossil assemblage, interconversion of enantiomeric ratios in one genus to equivalent values in another genus becomes necessary for either relative age or kinetic model age estimation. This approach has been used for the conversion of *Macoma* and *Saxidomus* leucine results to "equivalent *Protothaca*" leucine results for kinetic model age estimation (Wehmiller et al., 1978a).

This paper presents a compilation of relative amino acid racemization kinetics observed in eight genera of marine mollusks from Quaternary localities along the Pacific coast of the United States. These relative kinetics are defined and discussed as follows: (1) intergeneric relative racemization kinetics, the apparent relative rates of racemization of a single amino acid in *coexisting* samples (at the same fossil locality) of different genera; and (2) intrageneric relative racemization kinetics, the apparent rates of racemization of different amino acids (all compared with leucine) in the total amino acid fraction of a single sample (samples of different ages and genera studied).

Relative intergeneric kinetics have been documented for foraminifera

(King and Hare, 1972; King and Neville, 1977) and for mollusks (Miller and Hare, 1975; Masters and Bada, 1977; Wehmiller et al., 1977). Relative intrageneric kinetics have been reported for foraminifera (Kvenvolden et al., 1973), mollusks (Hare and Abelson, 1968; Hare and Hoering, 1973), corals (Wehmiller et al., 1976), and bones (Bada et al., 1973; summary by Williams and Smith, 1977).

## SAMPLES

Most thick-shelled aragonitic marine mollusks used in this and earlier studies (Wehmiller et al., 1977; Kennedy, 1978) belong to the bivalve family Veneridae. The species analyzed and the subfamilies to which they belong are:

| | |
|---|---|
| **Subfamily Meretricinae:** | *Tivela stultorum* (Mawe) |
| **Subfamily Pitarinae:** | *Saxidomus giganteus* (Deshayes)<br>*Saxidomus nuttalli* Conrad |
| **Subfamily Chioninae:** | *Chione californiensis* (Broderip)<br>*Chione undatella* (Sowerby)<br>*Protothaca (s.s.) staminea* (Conrad), and the *situs* form *P. s.* var. *ruderata* (Deshayes)<br>*Protothaca (Callithaca) tenerrima* (Carpenter) |

Of these species, *Saxidomus giganteus, S. nuttalli,* and *Protothaca staminea* are the only ones that have been studied thoroughly enough to permit inter- and intrageneric comparisons. A few results for *Chione* and *Tivela* (Wehmiller et al., 1977) are cited where appropriate. *Protothaca tenerrima, Tivela stultorum,* and several *Chione* species are common in Pleistocene deposits in southern California and Baja California Norte, Mexico, and are potentially useful for dating. *Mercenaria* (subfamily Chioninae) does not occur naturally on the Pacific coast but has been used extensively for amino acid dating studies on the Atlantic coast (Mitterer, 1974, 1975; Belknap, 1979).

Three species of *Macoma* (family Tellinidae), *M. lipara* Dall, *M. inquinata* (Deshayes), and *M. nasuta* (Conrad) have also been analyzed, with most samples belonging to the last two species. Samples of the bivalves *Epilucina californica* (Conrad) (family Lucinidae), *Cumingia californica* (Conrad) (family Semelidae), and the archaeogastropods *Tegula funebralis* (Adams) and *T. gallina* (Forbes) (family Trochidae) have also been analyzed but in such small numbers that their intergeneric relative rates are only partially known.

## SAMPLE LOCALITIES

Sample localities range from San Diego, California, to Puget Sound, Washington (Figure 1). Locality information is given by Wehmiller et al.

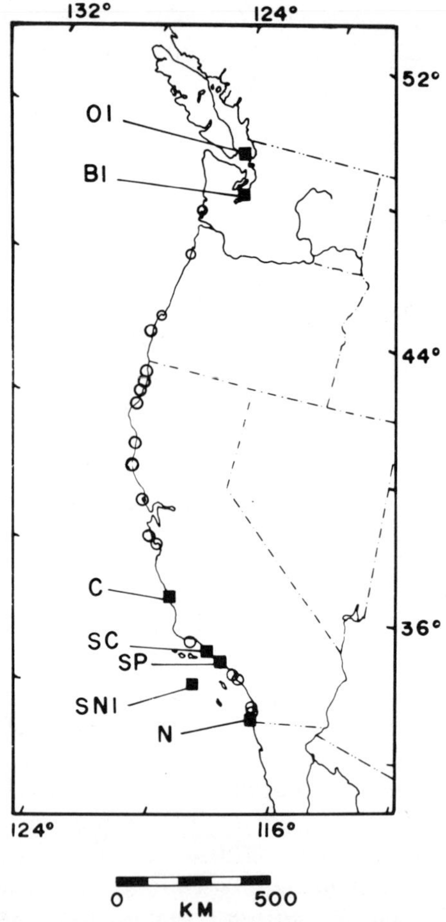


**Figure 1.** Map showing sample localities. Localities with independent age control (either radiometric or stratigraphic), summarized in Table 1, are labeled. Additional locality data can be found in Wehmiller et al. (1977) or Kennedy (1978). Abbreviations: OI, Orcas Island; BI, Bainbridge Island; C, Cayucos; SC, Sea Cliff; SP, San Pedro; SNI, San Nicolas Island; N, Nestor Terrace, San Diego (Bird Rock Terrace lies immediately below Nestor Terrace).

**Table 1. Sample Localities with Independent Age Control**[a]

| Amino Acid Sample Locality | Mean Ann. Temp.[b] (°C) | Age | Source for Locality and Independent Age Estimate |
|---|---|---|---|
| **Holocene** | | | |
| Sea Cliff, Ventura Co., California first terrace, USGS loc. M7288 | 15.3 | 1,820 ± 40 $^{14}C$ (USGS-400) | This study (USGS-400 and USGS-401 from USGS loc. M7228). |
| | | 3090 ± 40 $^{14}C$ (USGS-401) | |
| | | 4155 ± 90 $^{14}C$ (UM-1465) | This study (UM-1465 from USGS loc. M7292) |
| | | $<$5000 U-series (mollusks) | Kaufman et al., 1971: Table 7, sample 892 |
| Bainbridge Island, Washington, LACMNH loc. 5012, USGS loc. M6543 | 11.1 | 3260 ± 80 $^{14}C$ (USGS-7) | Robinson, 1977: USGS-7 from USGS loc. M6543. |
| **Latest Pleistocene** | | | |
| Orcas Island, Washington, Everson Glaciomarine drift, LACMNH loc. 5009 | 9.9 | 12,350 ± 400 $^{14}C$ (I-969) | Easterbrook 1969: Tables 2 and 5; I-969 from UW loc. 1215 |
| **Wisconsinan (Stage 3)** | | | |
| Sea Cliff, Ventura Co., California second terrace, USGS locs. M7229, M7230, M7245 | 15.3 | ca. 50,000 U-series (mollusks) | Kaufman et al., 1971; Table 7, samples 849C, 849F, 849G from locality equivalent to M7230; amino acid kinetic model age estimate of 45,000 ± 5,000 (Wehmiller et al., 1978a,b) |

**Table 1. Continued**

| Amino Acid Sample Locality | Mean Ann. Temp.[b] (°C) | Age | Source for Locality and Independent Age Estimate |
|---|---|---|---|
| **Sangamon Interglacial (Stage 5)** | | | |
| Point Loma, San Diego, California, Bird Rock Terrace, SDSU loc. 2523 | 16 | 85,000 ± 5,000 | Kern 1977. This terrace lies directly below Nestor Terrace, age estimated from geological and paleontological relationships, and amino acid data (Wehmiller et al., 1977: Table 13) |
| Cayucos, California, UCLA loc. 3393 and USGS loc. M5922 | 13.3 | 124,000 ± 27,000 U-series (corral) | Ku and Kern, 1977: Table 1; recalculation of U-series data of Veeh and Valentine 1967: Table 1, UCLA loc. 3447 |
| Point Loma, San Diego, California, Nestor Terrace, SDSU locs. 2520, 2521, 2527, USGS locs. M6702, M6703, M6706 | 16 | 109,000 ± 6,000<br>124,000 ± 7,000<br>131,000 ± 8,000<br>U-series (corals) | Ku and Kern 1974; Table 2, from SDSU loc. 2577 |
| San Nicolas Island, California, second terrace USGS locs. 21664, 21665 | 14.8 | >90,000 to ≥120,000 (SN-1)<br>87,000 ± 12,000 to 120,000 ± 20,000 (SN-13) | Valentine and Veeh 1969: Table 1; SN-1 from USGS loc. 21653, SN-13 from USGS loc. 21665 |
| | | 128,000 ± 25,000 (SNI-S-1)<br>125,00 ± 25,000 (SNI-S-2)<br>U-series (corals) | Szabo and Vedder 1971; SNI-S-1 from USGS loc. 21653, SNI-S-2 from USGS loc. 21664 |
| **Middle Pleistocene** | | | |
| San Nicolas Island, California,[c] tenth terrace, USGS locs. | 14.8 | >200,000 (SN-7) U-series (coral) | Valentine and Veeh 1969: Table 1; SN-7 from USGS loc. 21659 |

| | | | |
|---|---|---|---|
| | | 450,000 ± 100,000 | Amino acid kinetic model age estimate (Wehmiller et al., 1977: Table 13) |
| | | 770,000 ± 100,000 | This study: age estimate of tenth terrace based upon assumed long term uplift rate derived from U-series date and elevation of second terrace. |
| San Pedro, California,[c] San Pedro Sand, LACMNH loc. 332 | 16.9 | "Early Pleistocene" | Woodring, Bramlette, and Kew 1946; Kennedy 1975; classically considered as early Pleistocene on paleontological evidence. |
| | | <600,000 | Langenwalter, in Kennedy 1975: 37, 45, based on Rancholabrean fauna in San Pedro Sand |
| | | 350,000 to 450,000 | Amino acid kinetic model age estimate (Wehmiller et al., 1977: Table 13) |

[a] Age estimates derived from leucine kinetic models of racemization are included for comparison with independent chronological control.

[b] Mean annual air temperatures from summary presented in Wehmiller et al., 1977: Table 2.

[c] No radiometric or paleontologic control exists to assign relative ages to the San Pedro Sand versus the tenth terrace on San Nicolas Island; results presented here suggest that the San Pedro Sand (at LACMNH loc. 332) is younger than the tenth terrace on San Nicolas Island (see Wehmiller et al., 1977: 90, 98, Table 13).

Locality abbreviations:

LACMNH—Los Angeles County Museum of Natural History, Los Angeles, California.
SDSU—San Diego State University, San Diego, California.
UCLA—University of California, Los Angeles, California.
USGS—U. S. Geological Survey, Washington, D.C. and Meno Park, California (M series).
UW—University of Washington, Seattle, Washington.

(1977: Table 1 and Figs. 1, 2a–n) and Kennedy (1978: 753–776). Localities with independent age control are summarized in Table 1 and identified in Figure 1. Results for these control localities are shown in subsequent relative kinetic plots for illustrative purposes.

Most samples come from terrace deposits directly overlying a marine wave-cut platform in sea-cliff exposures. Burial depths are generally less than five meters. Many samples employed in this study were obtained from museum collections, and so were not specifically collected to avoid problems of surface temperature exposure. Examination of surface and buried samples from marine terrace localities indicates little effect (<3%) on observed enantiomeric ratios.

Geological, paleontological, radiometric and amino acid data indicate that the localities range in age from Holocene to middle and possibly early Pleistocene. Most of our localities are roughly 70,000 to 130,000 years in age (equivalent to marine isotopic Stage 5 of Shackleton and Opdyke, 1973: ≈ Sangamon Interglacial).

## ANALYTICAL PROCEDURES

Samples were processed according to techniques described in Kvenvolden et al. (1972, 1973). Analyses discussed here are those for the total amino acid fraction, which includes both free and hydrolyzable (in 6 *N* HCl, 22 hours, 110° C) amino acids recovered upon dissolution of shell material. To avoid possible variations within a single shell (Hare and Mitterer, 1969:206; Hare, 1977:1005) most samples were cut from the hinge or nymph area of the shell. In venerids this is usually the thickest part of the shell and most resistant to physical and chemical weathering. X-ray diffraction was used to test for possible aragonite-to-calcite alteration in most of the samples studied, and all samples reported here have less than 0.5% calcite.

Stereochemical measurements were obtained by capillary gas chromatographic analysis of N–TFA–(+)–2-butyl derivatives of the amino acid mixtures. Both stainless steel (150′ × 0.02″ and 0.03″) and glass (~300′ × 0.01″) columns of two different polarities have been used: OV225 and Carbowax K20M. No corrections for (−) -2-butyl impurities are made, since these corrections (≦0.01) are known to be less than the analytical precision of the method. Results are presented here for leucine, glutamic acid, valine, proline, and phenylalanine. Results for alanine, also determinable by the methods used here, are consistent with the presentations based upon the leu-glu-val-pro-phe results, but because alanine is usually extensively racemized (D/L ≧ 0.7) in the samples discussed here (except Holocene and latest Pleistocene samples), little kinetic information is available for this amino acid. Typical precision of multiple shell analyses is between 3% and 6%, depending on the amino acid. Examples of this precision are seen in intrageneric relative racemization figures shown below.

Because the gas chromatographic technique employed here has provided only moderate resolution of aspartic acid stereoisomeric derivatives, we report here only a few qualitative aspartic acid enantiomeric ratios for *Saxidomus*. A few high-resolution gas chromatographic determinations of aspartic acid enantiomeric ratios have been performed with the technique of Frank et al. (1977) using N-TFA-isopropyl derivatives on a 25 m glass capillary column coated with Chirasil-Val (Applied Science Laboratories, State College, PA). These results are discussed briefly in a subsequent section.

In all cases we report unfactored and uncorrected ratios of chromatogram peak heights. Analyses of equilibrium samples and racemic mixtures indicate that only in the case of phenylalanine does this procedure result in significant (>0.02) misrepresentation of actual enantiomeric ratios. For phenylalanine the measured peak height ratios are probably between 5% and 15% greater than the actual ratios. However, because the magnitude of this correction factor is not fully evaluated, we prefer to report (and use) only raw peak height data for phenylalanine. In spite of this uncertainty, intrageneric data presented here for phenylalanine implies a high degree of internal consistency in phenylalanine relative racemization rates.

Graphical formats are used for the presentation of both inter- and intrageneric systematic trends. Linear regression parameters have also been calculated for all these trends, for both partial and complete data sets. We prefer these graphical presentations because in many cases they (and the regression parameters) identify deviations from simple linear trends that might be inferred from numerical comparisons of relative rate constants. These deviations and their implications are fully discussed in following sections. Typical 95% confidence limits on slopes and intercepts calculated from these regressions are on the order of 0.04 to 0.06 for correlation coefficients of 0.96 to 0.98. Much of the data presented here is found in Wehmiller et al. (1977: Tables 3 and 4) and Kennedy (1978:791–796).

## DEFINITIONS

The following terminology is employed in discussions of inter- and intrageneric trends in enantiomeric ratios. These terms represent a distinction between "rate" (a specific slope of one parameter versus time or another parameter) and "kinetics" (a more general term to characterize the overall change of one parameter relative to another).

**Rate:** the slope of D/L values (or converted parameter) versus time.

**Relative rate:** the slope of one amino acid D/L value versus another amino acid D/L value in the same genus (intrageneric relative rate), *or* the slope of one genus versus another for the same amino acid (intergeneric relative rate).

**Apparent relative rates:** the relative order of observed enantiomeric ratios either within a single genus or among several genera (i.e., ala > leu > val, or *Macoma* > *Protothaca*: intrageneric and intergeneric apparent relative kinetics).

**Apparent kinetics:** the overall pathway an amino acid enantiomeric ratio will follow through time (apparent relative kinetics, the generalized pathway of one amino acid or one genus versus another).

**Diagenetic pathway:** the complex sequence of reactions that a calcified protein amino acid follows during diagenesis: hydrolysis, residence in short polypeptides, leaching, racemization, etc.

The necessity for these distinctions arises from the fact that only "apparent kinetics" of diagenetic racemization have actually been documented in a limited number of natural samples (i.e., those with independent chronologic control), and that specific *rates* of individual reactions involved in diagenetic racemization are poorly known. It seems likely that many of these individual reaction rates are time-dependent in the types of fossil systems being studied.

## AGE AND TEMPERATURE EFFECTS

Amino acid geochronology is based on the premise that enantiomeric ratios increase with age and/or diagenetic temperature. The age relationship has been demonstrated for mollusks (Hare and Mitterer, 1967; Hare and Abelson, 1968; Mitterer, 1974, 1975; Wehmiller et al., 1977; Miller and Hare, 1975a,b) and foraminifera (Wehmiller and Hare, 1971; King and Hare, 1972; Bada and Schroeder, 1972; King and Neville, 1977). The age relationship in our data is demonstrated in Figure 2, which is a plot of intrageneric glu-leu relationships in *Protothaca staminea* samples from the control localities summarized in Table 1.

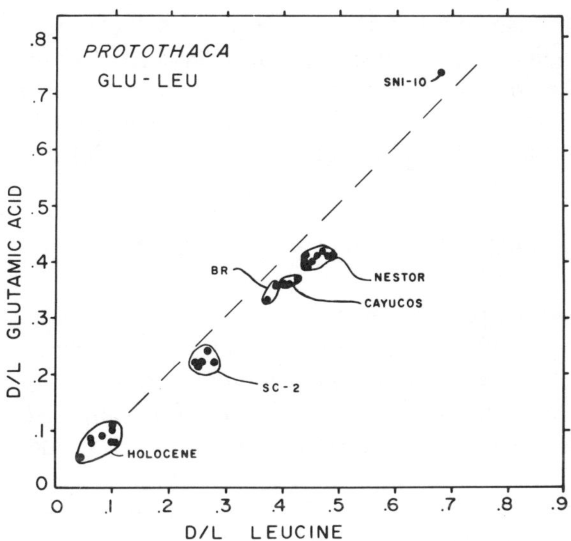


**Figure 2.** Plot of D/L glutamic acid versus D/L leucine in *Protothaca staminea* from independently dated localities to show the relationship of increasing enantiomeric ratios with age. Locality and age data are summarized in Table 1. The separation between the Nestor and Cayucos data points is a function of temperature.

The temperature effect is illustrated in Figure 2 by the separation of data points for Cayucos and the Nestor Terrace in San Diego. These localities are essentially the same age, but Cayucos is cooler by about 2.5° C (see Table 1). Additional demonstrations of this temperature effect between these two localities are seen in Wehmiller et al. (1977: Figs. 10–13) and Wehmiller and Belknap (1978: Fig. 1). Because the remaining localities plotted on Figure 2 lie between Cayucos and the Nestor Terrace, the latitude-temperature effect can be no greater than the difference between these two. Kennedy (1978) has presented an expanded version of this latitude-temperature effect for samples of equal age.

Although the age effect is clearly illustrated in relative D/L plots such as Figure 2, actual racemization kinetic curves cannot be derived from this type of diagram. A plot of D/L values for five amino acids in *Protothaca staminea* samples from our control localities (Table 1) with similar present-day mean annual temperatures shows that rapid initial rates of racemization are followed by gradually decreasing rates. The inversion of apparent relative rates in the oldest sample (SNI-10) may be an artifact of the scarcity of *Protothaca* data at this locality, as similar inversions are not observed in *Saxidomus* from this terrace (Wehmiller et al., 1977: Table 3). Sufficient data do not exist to permit a distinction between smooth or multisegmented kinetic curves, but the general trends shown in Figure 3 indicate that overall racemization kinetics of the several amino acids are quite similar. Comparable kinetic curves, constructed with the assumptions of kinetic models, are found elsewhere (Wehmiller et al., 1977; Figs. 14 and 17; Wehmiller and Belknap, 1978).

## INTERGENERIC RELATIVE RACEMIZATION KINETICS

In Figures 4–8 we present apparent relative racemization kinetics of leucine, glutamic acid, valine, proline, and phenylalanine in *Protothaca, Saxidomus, Macoma, Cumingia, Epilucina,* and *Tegula.* For convenience, *Protothaca* has been used as the basis for comparison, because this genus is most frequently available at the localities studied. *Tivela* and

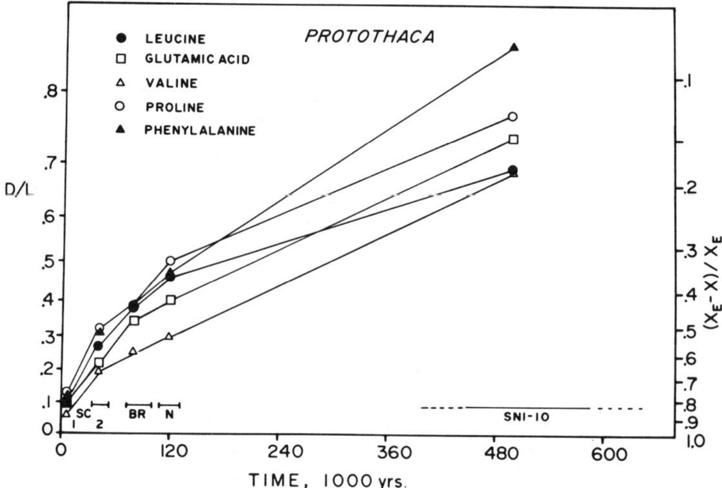


**Figure 3.** Plot of amino acid enantiomeric ratios in *Protothaca staminea* samples from southern California control localities listed in Table 1. Modern mean annual temperatures at these localities is between 14° and 16° C. Age control is summarized in Table 1, with probable age uncertainty as indicated along time axis of figure. Semilog format, with D/L values on left axis and corresponding $(X_E - X)/X_E$ values on right axis. ($X = D/(D + L)$ at time $t$, $X_E = 0.50$ for amino acids plotted here.) See Wehmiller et al., 1977: Fig. 14, for comparison. Multiple analyses involved in all data points except those for SNI-10, so apparent inversions might be an artifact of sampling deficiencies. Mean values plotted for all other data points. SC-1: Sea Cliff, first terrace. SC-2: Sea Cliff, second terrace. BR: Bird Rock terrace, San Diego. N: Nestor Terrace, San Diego. SNI-10: San Nicolas Island, tenth terrace (plotted with age estimate of ca. 500,000 years to demonstrate *minimum* curvature of kinetics; greater age possible if constant uplift rates invoked—see Table 1. Cooler temperature at SNI compared with other localities shown here complicates the interpretation of these curves).

*Chione* data are not presented in Figures 4–8 for purposes of clarity, but they have enantiomeric ratios within 5% of those for coexisting *Protothaca* (Wehmiller et al., 1977: Table 4).

Each data point plotted in Figures 4–8 represents the mean value for all determinations of a single amino acid enantiomeric ratio in pairs of

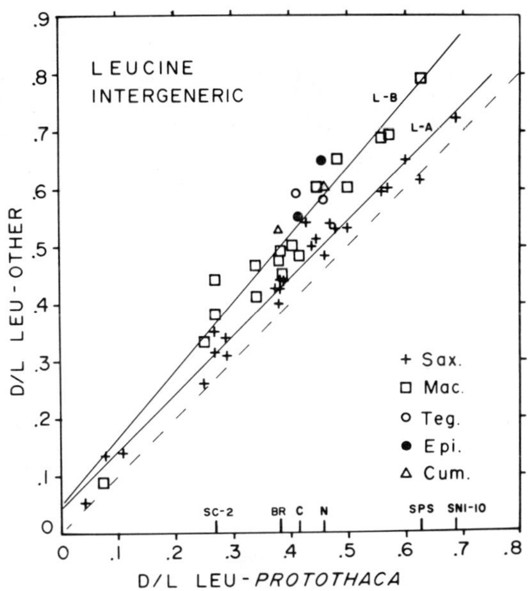


**Figure 4.** Leucine intergeneric trends (dashed line 1:1)
Regressions:

| | | | |
|---|---|---|---|
| | *Protothaca-Saxidomus* | | |
| (L-A) | All | P = 1.002S − 0.042 | $n$ = 26, $r$ = 0.988 |
| | Pleist. | P = 1.036S − 0.061 | $n$ = 23, $r$ = 0.978 |
| (O-force) | All | P = 0.918S | $n$ = 52, $r$ = 0.998 |
| | Holo. | P = 0.700S | $n$ = 6, $r$ = 0.990 |
| | *Protothaca-Macoma* | | |
| (L-B) | All | P = 0.847M − 0.038 | $n$ = 17, $r$ = 0.982 |
| (O-force) | All | P = 0.783M | $n$ = 34, $r$ = 0.998 |

**Figures 4–8.** Intergeneric trends in enantiomeric ratios: Dashed lines represent 1:1 slope; solid lines, identified as L-A, L-B, etc. correspond to whole data set linear regressions. Additional regression approaches also shown numerically. *Tegula, Epilucina,* and *Cumingia* results not included in any regression calculations. Locality abbreviations along the *Protothaca* axis correspond to Pleistocene control localities summarized in Table 1. Vertical bars indicate mean D/L values at these localities. Ranges of observed data for these localities can be inferred from Figures 9–12. Regressions are calculated with *Protothaca* as the dependent variable to permit conversion to an "equivalent *Protothaca*" for use in kinetic model age estimation (Wehmiller and Belknap, 1978). Regressions calculated using mean values for all analyses of a single amino acid in cases where multiple samples of a single genus have been available.

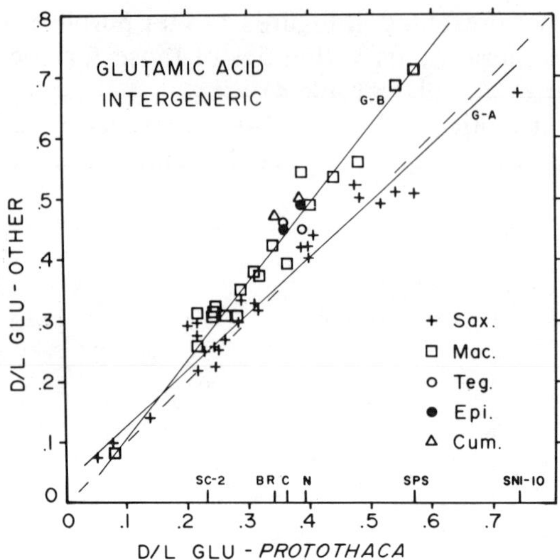


**Figure 5.** Glutamic acid intergeneric trends (dashed line 1 : 1)
Regressions:

| | | *Protothaca-Saxidomus* | |
|---|---|---|---|
| (G-A) | All | P = 1.082S − 0.035 | $n$ = 24, $r$ = .982 |
| | Pleist. | P = 1.125S − 0.054 | $n$ = 21, $r$ = .973 |
| (O-force) | All | P = 0.993S | $n$ = 48, $r$ = 0.996 |
| | Holo. | P = 0.909S | $n$ = 6, $r$ = 0.994 |
| | | *Protothaca-Macoma* | |
| (G-B) | All | P = 0.768M + 0.022 | $n$ = 16, $r$ = 0.975 |
| (O-force) | All | P = 0.812M | $n$ = 32, $r$ = 0.998 |

coexisting genera at a given locality. Because no *Protothaca* results are available for the SNI-2 locality, no intergeneric results for this locality are shown in Figures 4–8; however, the *Saxidomus-Epilucina* intergeneric relationships at SNI-2 are entirely consistent with the trends shown in Figures 4–8 (Wehmiller et al., 1977: Table 3).

Along the *Protothaca* axis in Figures 4–8 are shown the mean values for leucine, glutamic acid, valine, proline, and phenylalanine in the Pleistocene control localities listed in Table 1. Trends of increasing enantiomeric ratios with increasing age are seen for all of the amino acids, though proline and phenylalanine do not demonstrate the Nestor-Cayucos temperature effect that is seen for leucine, glutamic acid, and valine.

Regression lines shown in Figures 4–8 are those for the whole data set; because several of these regression trends indicate nonzero intercepts (and, by implication, nonlinear apparent relative kinetics), other approaches to the statistical analyses of these intergeneric results must also

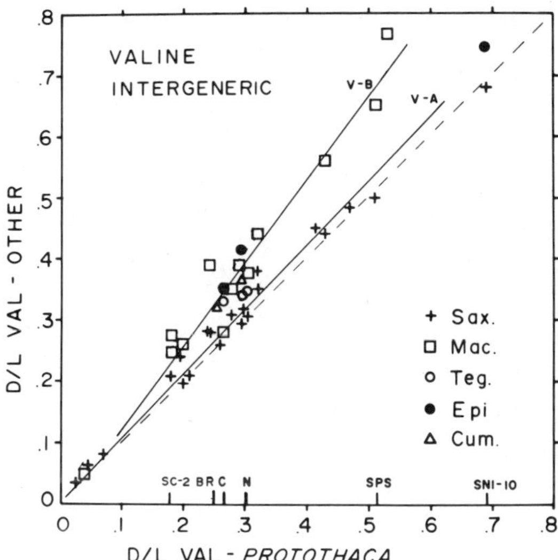


**Figure 6.** Valine intergeneric trends (dashed line 1:1)
Regressions:

| | | | |
|---|---|---|---|
| | *Protothaca-Saxidomus* | | |
| (V-A) | All | P = 1.014S − 0.023 | $n$ = 23, $r$ = 0.992 |
| | Pleist. | P = 1.04S − 0.035 | $n$ = 20, $r$ = 0.988 |
| (O-force) | All | P = 0.970S | $n$ = 46, $r$ = 0.995 |
| | Holo. | P = 0.815S | $n$ = 6, $r$ = 0.996 |
| | *Protothaca-Macoma* | | |
| (V-B) | All | P = 0.707M + 0.26 | $n$ = 13, $r$ = 0.981 |
| (O-force) | All | P = 0.761M | $n$ = 26, $r$ = 0.996 |

be considered. Two additional approaches are given in the captions to Figures 4–8: one approach represents the linear regression for only the Pleistocene samples (labeled "Pleist."), the second approach represents a "forcing" of the regression through a zero intercept (labeled "O-force"). The latter approach is shown for both the Holocene and the total data set regressions; forced regressions were calculated by inclusion of the negative counterparts of all the data points shown in Figures 4–8. Increases in the $n$ and $r$ values are simple artifacts of this manipulation.

Comparison of these three approaches (two in the case of *Macoma,* because only one Holocene data point is involved) to the regression analysis of the *Macoma, Saxidomus,* and *Protothaca* results indicates that among *Pleistocene* samples there is no significant difference (in comparison with typical data range for multiple shell analyses from a single locality) in the enantiomeric ratios predicted by the three regressions. Therefore, we conclude that, for purposes of conversion of enantiomeric

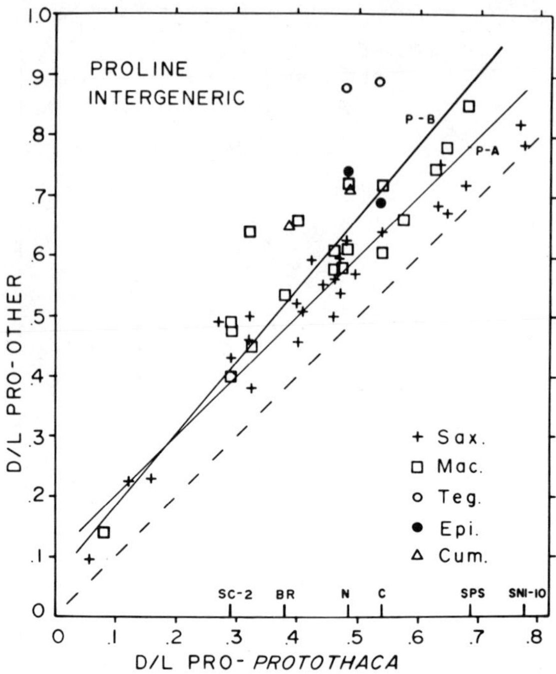


**Figure 7.** Proline intergeneric trends (dashed line 1 : 1)
Regressions:

| | | | |
|---|---|---|---|
| | *Protothaca-Saxidomus* | | |
| (P-A) | All | P = 1.020S − 0.106 | $n$ = 24, $r$ = 0.960 |
| | Pleist. | P = 1.138S − 0.173 | $n$ = 21, $r$ = 0.945 |
| (O-force) | All | P = 0.836S | $n$ = 48, $r$ = 0.992 |
| | Holo. | P = 0.63S | $n$ = 6, $r$ = 0.994 |
| | *Protothaca-Macoma* | | |
| (P-B) | All | P = 0.842M − 0.056 | $n$ = 16, $r$ = 0.911 |
| (O-force) | All | P = 0.757M | $n$ = 32, $r$ = 0.990 |

ratios in one genus to an "equivalent *Protothaca*" (i.e., in order to compare results among Pleistocene samples of *Macoma, Saxidomus,* and *Protothaca*), any of the three intergeneric regressions shown in Figures 4–8 would suffice. Wehmiller et al. (1978a) used whole data set intergeneric leucine regressions to interpret *Macoma* and *Saxidomus* in terms of the *Protothaca* leucine kinetic model.*

* The *Saxidomus-Protothaca* leucine intergeneric regression suggests an equivalent kinetic model temperature difference of approximately 1° C, rather than the 2° C originally calculated by Wehmiller and Belknap (1978).

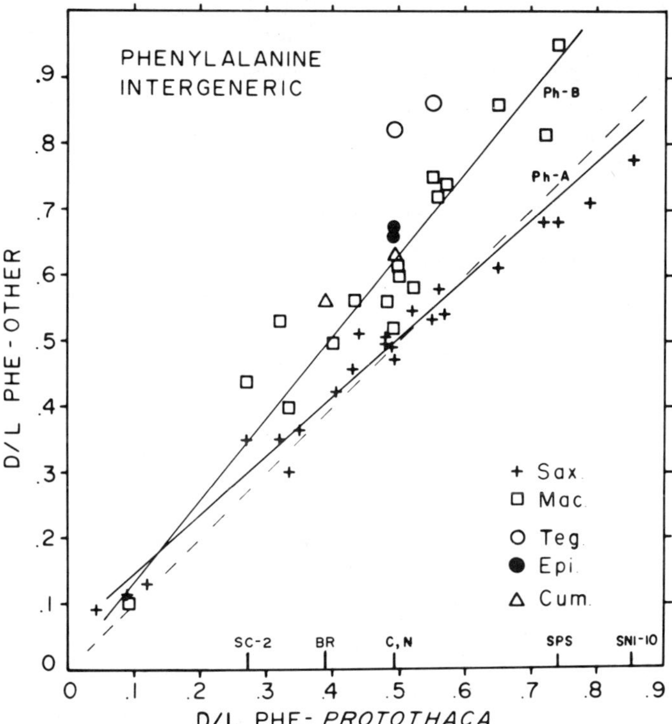


**Figure 8.** Phenylalanine intergeneric trends (dashed line 1:1)
Regressions:

| | *Protothaca-Saxidomus* | | |
|---|---|---|---|
| (Ph-A) | All | P = 1.13S − 0.064 | $n$ = 24, $r$ = 0.983 |
| | Pleist. | P = 1.20S − 0.106 | $n$ = 21, $r$ = 0.971 |
| (O-force) | All | P = 1.011S | $n$ = 48, $r$ = 0.996 |
| | Holo. | P = 0.772S | $n$ = 6, $r$ = 0.975 |
| | *Protothaca-Macoma* | | |
| (Ph-B) | All | P = 0.782M + 0.007 | $n$ = 17, $r$ = 0.955 |
| (O-force) | All | P = 0.790M | $n$ = 34, $r$ = 0.995 |

Though sufficient data for *Epilucina, Tegula,* and *Cumingia* are not available for regression calculations, it appears that the *Macoma* regression lines may adequately characterize these three genera, at least for leucine, glutamic acid, and valine. Only proline and phenylalanine in *Tegula* plot well outside the range of intergeneric results observed in the *Macoma* data.

## INTRAGENERIC RELATIVE RACEMIZATION KINETICS

In Figures 9–12 we summarize the intrageneric relative racemization kinetics of glutamic acid, valine, proline, and phenylalanine relative to leucine *within* the molluscan genera discussed here. For these figures, each

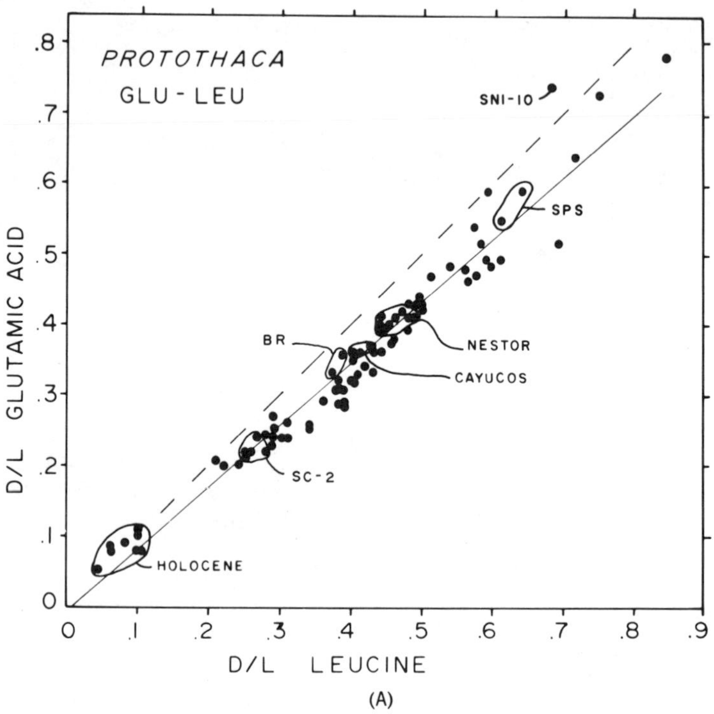


(A)

**Figure 9.** Glu-Leu intrageneric trends (dashed lines 1:1)

(A) *Protothaca*: Glu = 0.880Leu − 0.008
$n = 94, r = 0.974$

(B) *Saxidomus* (all): Glu = 0.892Leu − 0.051
$n = 108, r = 0.966$
*S. giganteus* Glu = 0.895Leu − 0.058
$n = 62, r = 0.973$
*S. nuttalli* Glu = 0.880Leu − 0.033
$n = 43, r = 0.963$

(C) *Macoma*: Glu = 0.888Leu − 0.032
$n = 59, r = 0.977$

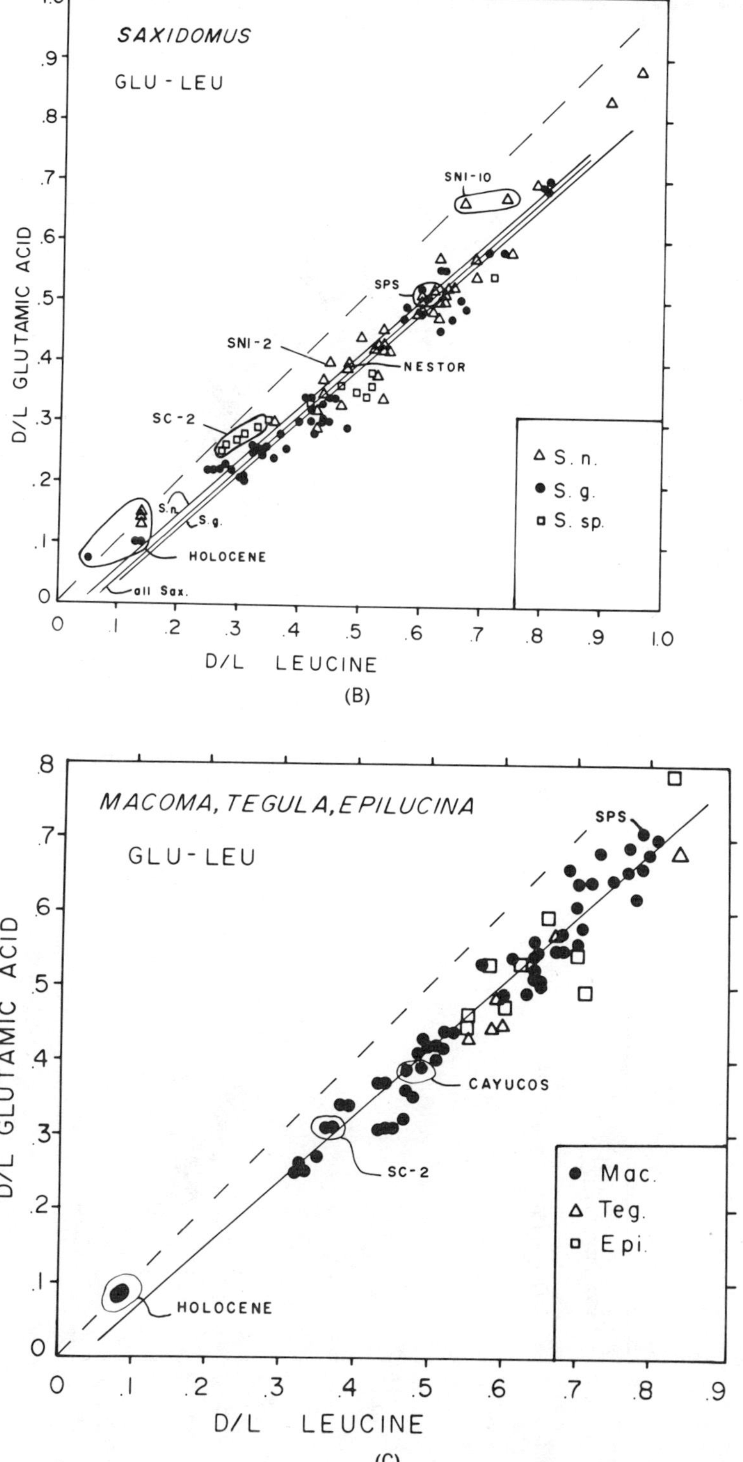

SAXIDOMUS
GLU - LEU
SNI-10
SPS
SNI-2
NESTOR
SC-2
S.n.
S.g.
HOLOCENE
all Sax.
S. n.
S. g.
S. sp.
D/L GLUTAMIC ACID
D/L LEUCINE
0
.1
.2
.3
.4
.5
.6
.7
.8
.9
1.0

(B)

MACOMA, TEGULA, EPILUCINA
GLU - LEU
SPS
CAYUCOS
SC-2
HOLOCENE
Mac.
Teg.
Epi.
D/L GLUTAMIC ACID
D/L LEUCINE
0
.1
.2
.3
.4
.5
.6
.7
.8
.9

(C)

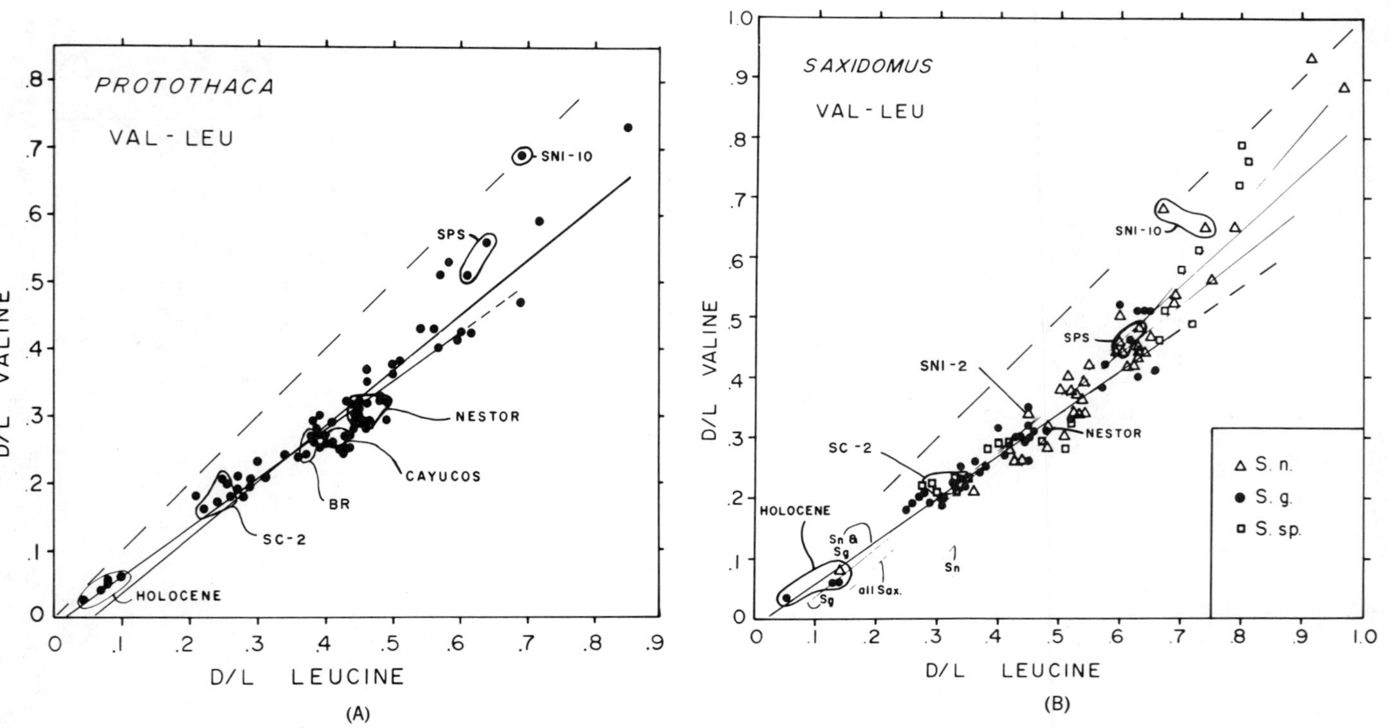
PROTOTHACA
VAL-LEU
SNI-10
SPS
NESTOR
CAYUCOS
BR
SC-2
HOLOCENE
D/L VALINE
D/L LEUCINE
(A)
SAXIDOMUS
VAL-LEU
SNI-10
SPS
SNI-2
NESTOR
SC-2
HOLOCENE
Sn & Sg
Sg
all Sax.
Sn
S. n.
S. g.
S. sp.
D/L VALINE
D/L LEUCINE
(B)

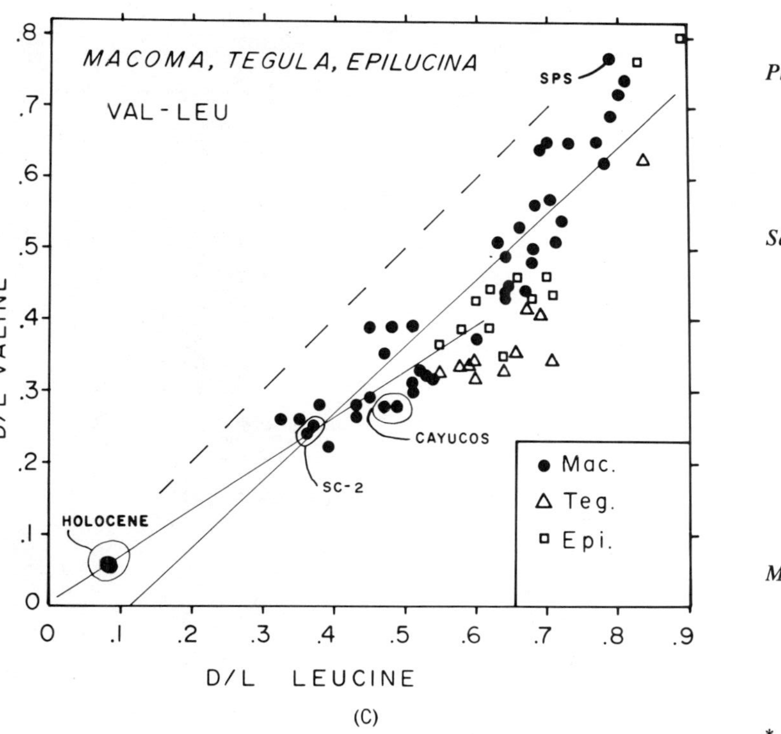


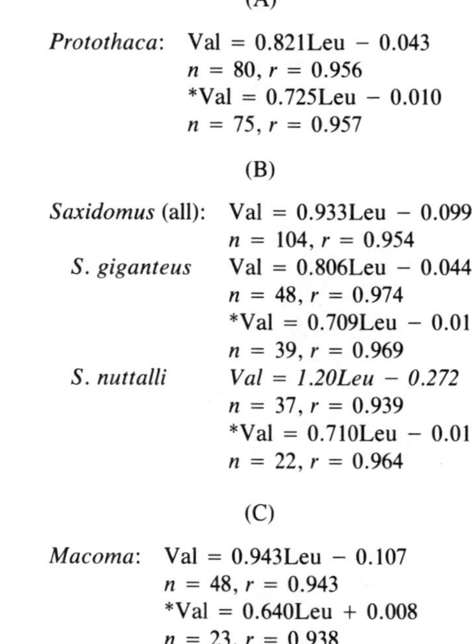

**Figure 10.** Val-Leu intrageneric trends (dashed lines 1:1)

(A)

*Protothaca*: Val = 0.821Leu − 0.043
$n$ = 80, $r$ = 0.956
*Val = 0.725Leu − 0.010
$n$ = 75, $r$ = 0.957

(B)

*Saxidomus* (all): Val = 0.933Leu − 0.099
$n$ = 104, $r$ = 0.954

*S. giganteus* Val = 0.806Leu − 0.044
$n$ = 48, $r$ = 0.974
*Val = 0.709Leu − 0.011
$n$ = 39, $r$ = 0.969

*S. nuttalli* *Val = 1.20Leu − 0.272*
$n$ = 37, $r$ = 0.939
*Val = 0.710Leu − 0.017
$n$ = 22, $r$ = 0.964

(C)

*Macoma*: Val = 0.943Leu − 0.107
$n$ = 48, $r$ = 0.943
*Val = 0.640Leu + 0.008
$n$ = 23, $r$ = 0.938

* Regression calculated with data for D/L leu ≦ 0.60

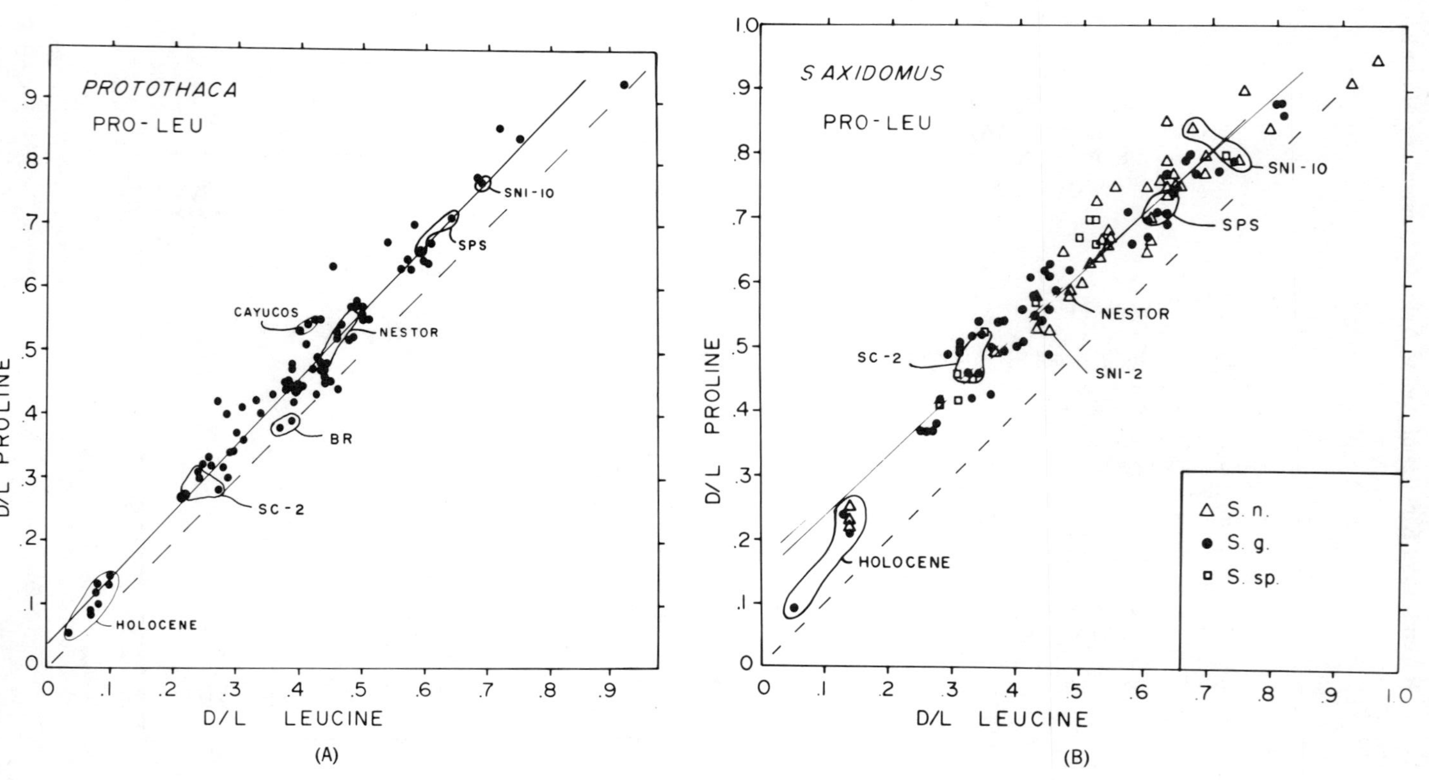

PROTOTHACA
PRO-LEU
D/L PROLINE
D/L LEUCINE
HOLOCENE
SC-2
BR
CAYUCOS
NESTOR
SPS
SNI-10
S AXIDOMUS
PRO-LEU
D/L PROLINE
D/L LEUCINE
HOLOCENE
SC-2
SNI-2
NESTOR
SPS
SNI-10
S. n.
S. g.
S. sp.


(A)

(B)

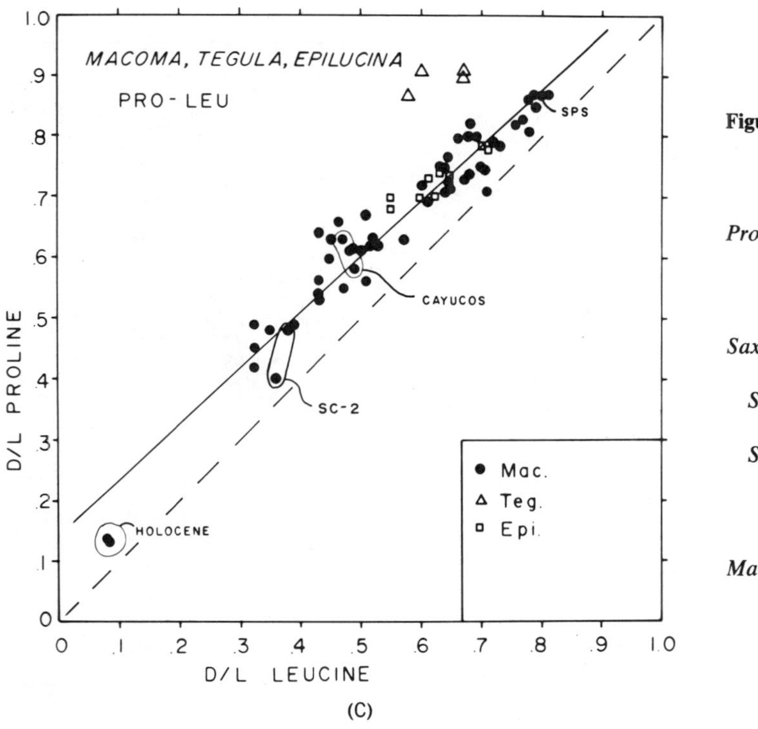


(C)

**Figure 11.** Pro-Leu intrageneric trends (dashed lines 1:1)

(A)

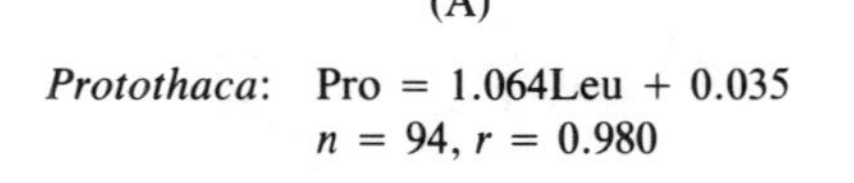

*Protothaca*: Pro = 1.064Leu + 0.035
$n = 94$, $r = 0.980$

(B)

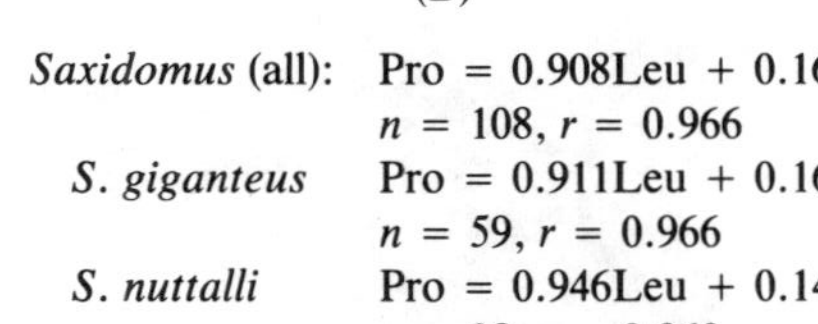

*Saxidomus* (all): Pro = 0.908Leu + 0.167
$n = 108$, $r = 0.966$

*S. giganteus* Pro = 0.911Leu + 0.165
$n = 59$, $r = 0.966$

*S. nuttalli* Pro = 0.946Leu + 0.143
$n = 38$, $r = 0.960$

(C)

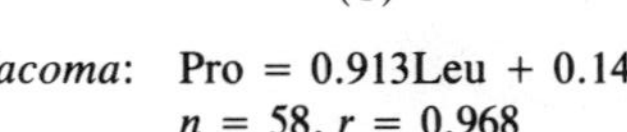

*Macoma*: Pro = 0.913Leu + 0.143
$n = 58$, $r = 0.968$

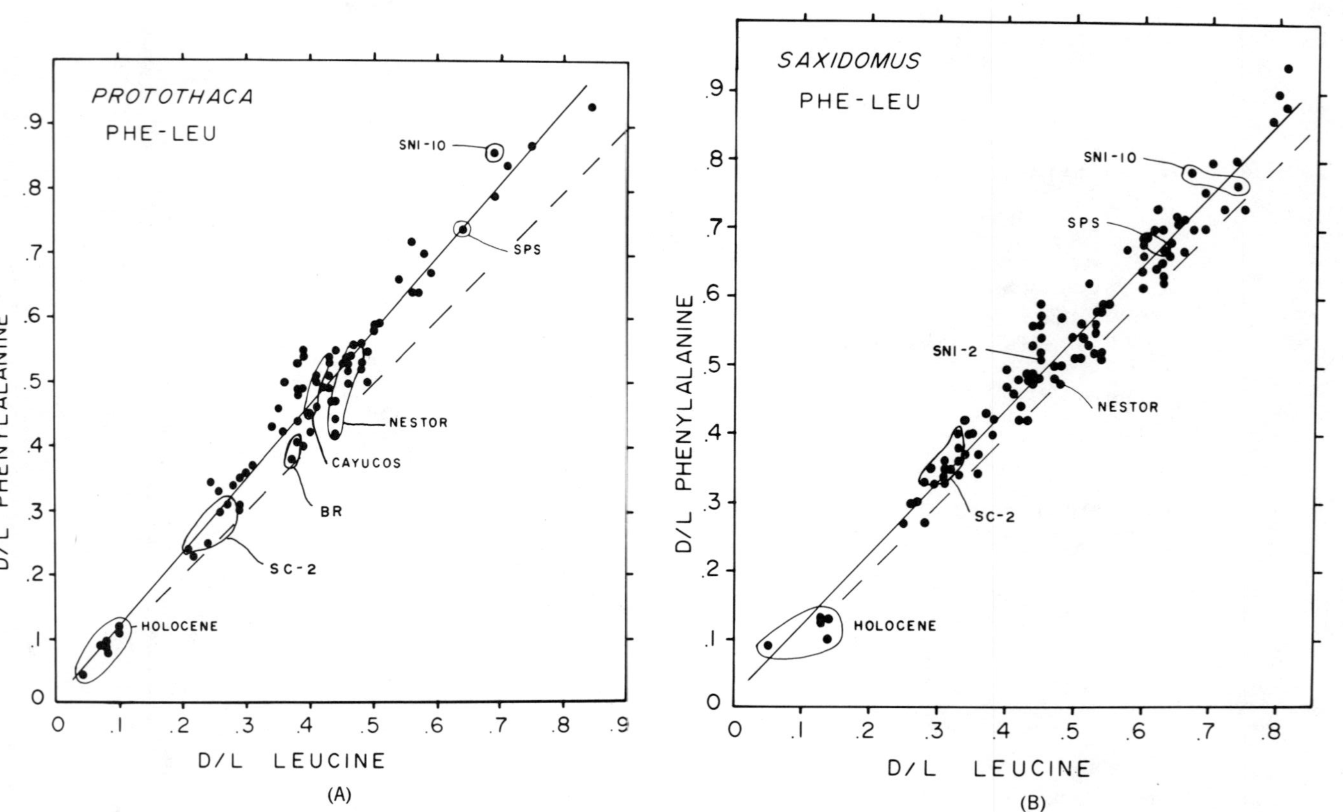
PROTOTHACA
PHE-LEU
SNI-10
SPS
NESTOR
CAYUCOS
BR
SC-2
HOLOCENE
D/L PHENYLALANINE
D/L LEUCINE
(A)
SAXIDOMUS
PHE-LEU
SNI-10
SPS
SNI-2
NESTOR
SC-2
HOLOCENE
D/L PHENYLALANINE
D/L LEUCINE
(B)

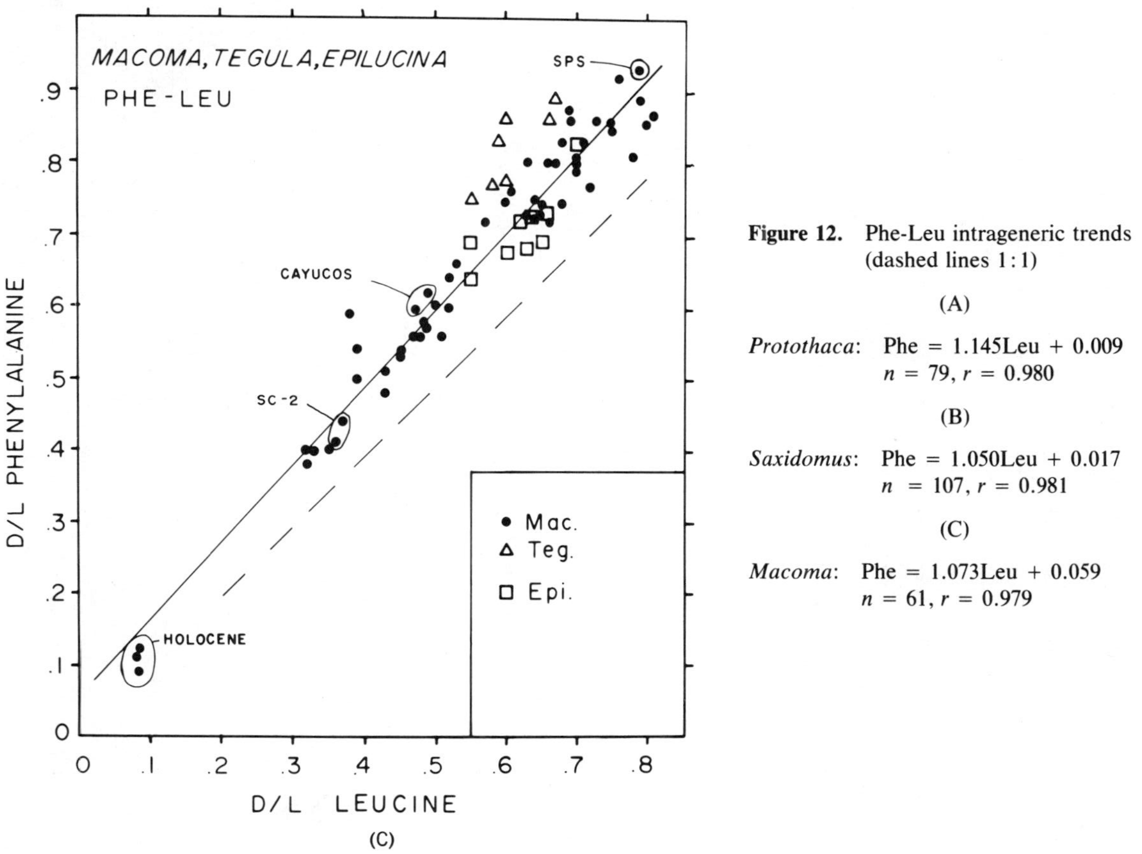


**Figure 12.** Phe-Leu intrageneric trends (dashed lines 1:1)

(A)

*Protothaca*: Phe = 1.145Leu + 0.009
$n = 79$, $r = 0.980$

(B)

*Saxidomus*: Phe = 1.050Leu + 0.017
$n = 107$, $r = 0.981$

(C)

*Macoma*: Phe = 1.073Leu + 0.059
$n = 61$, $r = 0.979$

data point represents a single sample. Linear regression trends for whole and partial data sets (in the case of valine) are listed with Figures 9–12; in the case of *Saxidomus*, we have been able to calculate these regressions for both species studied.

Shown in Figures 9–12 are the clusters of data points representing analyses from the control localities (Table 1). Age/temperature relationships can be inferred from these figures for all amino acid pairs, as was shown in Figure 2. Typical ranges of results observed in multiple sample analyses from a single locality are also represented by these data clusters.

The intrageneric regressions shown in Figures 9–12 indicate a high degree of similarity in apparent relative racemization kinetics of glutamic acid, valine, proline, and phenylalanine compared with leucine in the samples reported here. The intrageneric trends indicate the following apparent relative racemization rates:

$$\text{pro} \geqq \text{phe} \geqq \text{leu} > \text{glu} > \text{val}$$

Several of the whole data set linear regressions have significant nonzero intercepts, indicating that two-step (or convex- or concave-upward) regressions would be more accurate representations of the data. This is especially true for valine and proline (Figures 10 and 11). In a few cases, these nonzero intercepts for whole data set regressions are statistical artifacts of incomplete sample coverage (e.g., *S. gigantus* and *S. nuttalli* are not represented over equal ranges of enantiomeric ratios because of different latitude and age distribution). Therefore, the significance of nonzero intercepts can only be qualitatively evaluated unless there is a broad data base for comparison.

If only the early portions of the valine intrageneric regressions are compared (to leucine D/L $\leqq$ 0.60), there is little difference in the val-leu intrageneric relationships for *Macoma, Saxidomus* (both species), and *Protothaca* (Figure 10).

Significant nonlinearity of proline-leucine intrageneric relative rates is indicated for *Saxidomus* and *Macoma* (Figure 11). Holocene pro-leu slopes are approximately 1.4, whereas Pleistocene pro-leu slopes are approximately 0.91. This nonlinearity is not as pronounced for *Protothaca* pro-leu relations. This difference in initial relative rate of proline racemization in *Protothaca* is also reflected in the significant nonzero intercept for the proline intergeneric regressions (Figure 7). Because the apparent racemization kinetics of proline can be complicated by the diagenetic production of proline (racemic?) from arginine (Vallentyne, 1964; Williams and Smith, 1977), these differences in proline relative rates could be explained by intergeneric differences in arginine abundances. A similar explanation for the extensive racemization of proline in *Tegula* might be made.

Relative glutamic acid–leucine intrageneric trends might be characterized by slight concave-upward curves (Figure 9) with slight positive intercepts. Nevertheless, the high degree of similarity of the glu-leu intrageneric relationships suggests a similar diagenetic pathway for these two amino acids in the molluscan genera discussed here.

Phenylalanine-leucine intrageneric trends (Figure 12) are linear and indicate no difference among the genera studied.

With the exception of proline and phenylalanine in *Tegula,* it appears that intrageneric relationships similar to those discussed are appropriate for *Epilucina, Tegula,* and *Cumingia.* The smaller number of analyses and greater amount of scatter and the fewer sample analyses for these three genera do not permit valid regression comparisons to be made.

In Figure 13 aspartic acid–leucine intrageneric trends are shown for

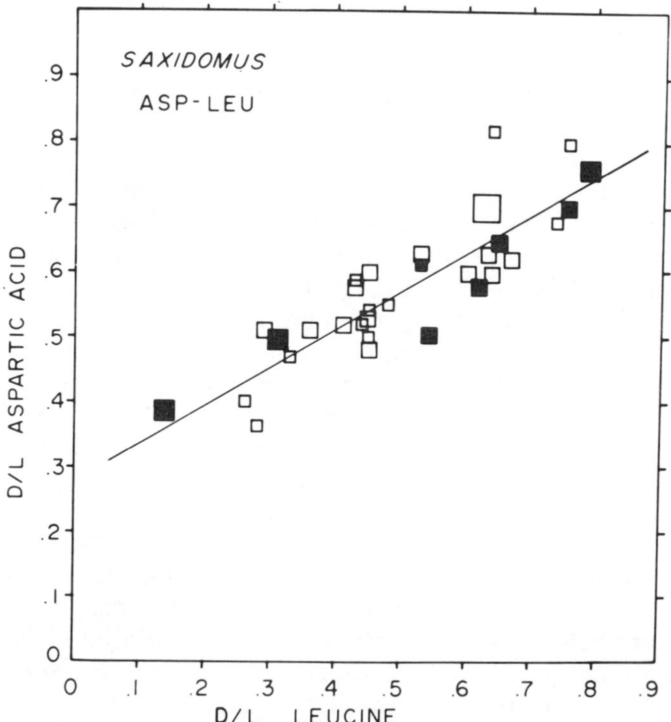


**Figure 13.** Aspartic acid–leucine intrageneric relation in *Saxidomus.* Size of squares represents estimated uncertainty of single sample analysis. Open squares represent analyses by moderate-resolution GC techniques (+2-butanol derivatives); solid squares represent analyses by high-resolution GC analyses (propanol derivatives, Chirasil-Val column). See text for further discussion. Regression line shown:

$$\text{Asp} = 0.591\text{Leu} + 0.276 \quad n = 33 \quad r = 0.873$$

*Saxidomus,* the only genus for which adequate aspartic acid data are available. Though analytical difficulties clearly contribute to the greater scatter in results ($r = 0.87$), the trend for aspartic acid–leucine in *Saxidomus* appears unique in that the large positive intercept is much greater than for other amino acids (see also *Mercenaria* data, Hare and Hoering, 1973). The slow relative rate (slope of 0.6) beyond the Holocene age range, combined with the range of aspartic acid results observed among samples of equal age, suggests that there is little chronological utility in molluscan aspartic acid determinations.

## TAXONOMIC LEVEL AT WHICH RACEMIZATION DIFFERENCES CAN BE DETECTED

The various regressions summarized in Figures 9–12 for individual species of *Saxidomus* (*S. giganteus* and *S. nuttalli*) indicate that when comparisons are made between comparable ranges of data, little or no species-level effect on apparent racemization kinetics can be observed for this genus. If kinetic differences in racemization cannot be seen for two species of the same genus, then at what taxonomic level do these differences become detectable? It is unlikely that the taxonomic differentiation is expressed in a systematic and hierarchical way in all groups of organisms. Even within the class Bivalvia, there may be cases where kinetic differences can be distinguished on the generic level, whereas in other cases differences may be distinguishable only at the superfamily level. Only multiple analyses of coexisting taxa will reveal the taxonomic level at which recognizable differences can be consistently detected.

Both *Protothaca* and *Saxidomus* belong to different subfamilies in the family Veneridae. Unfortunately we do not have enough data to adequately compare two genera in the same subfamily, although *Chione, Protothaca,* and *Mercenaria* do all belong to the subfamily Chioninae. Limited data (unpublished; Wehmiller et al., 1977: Table 4; Wehmiller, 1977) on enantiomeric ratios from coexisting *Protothaca, Saxidomus, Chione,* and *Tivela* indicate *Chione* and *Tivela* have enantiomeric ratios equal to or slightly greater than (within 5%) those seen in *Protothaca. Mercenaria* may also have kinetics similar to those of *Protothaca* (Wehmiller and Belknap, 1978). The available data suggest that within the Veneridae only very slight (5%–10%?) variations in relative racemization kinetics might be observed. Significantly greater differences (20%–35%) are observed in the comparison of Tellinidae and Veneridae.

The species-specific kinetics observed by King and Neville (1977) for two species of foraminifera may be only a generic or family effect; the two species belong to different families. Because we see some slight differences in racemization among genera, even within the same subfamily, we consider this to be a generic effect rather than a species effect. However, there may be more variation among genera in a subfamily than there is among certain other genera in different subfamilies. Additional work must be done to establish this relationship in any particular case.

## SUMMARY MODELS OF RACEMIZATION KINETIC PATHWAYS

From the results presented in Figures 3–12, it is possible to propose a series of general models of racemization kinetic pathways for either different genera or amino acids. These models are shown in Figure 14. All three models shown represent simple two-step interpretation of typical

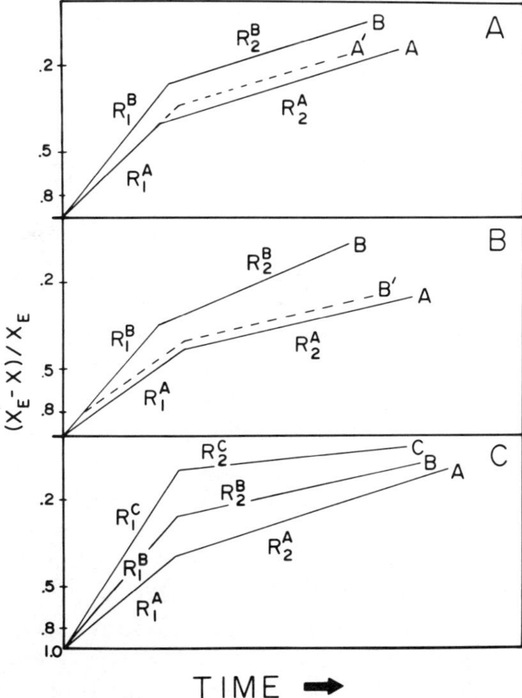


**Figure 14.** Conceptual models of racemization kinetic pathways, for either intergeneric or intrageneric relationships as seen in Figures 3–12. $R_1$, $R_2$, etc. = *rates*; semilog format like that of Figure 3 used so that first-order rate constants can be compared. Two-component models of racemization curves are used, though it is recognized that multisegmented or smooth—curve models might be better representations of actual data. Model A: unequal initial rates, equal final rates; no intersections of apparent kinetics; Model B, unequal initial and final rates, no intersections of apparent kinetics; Model C, unequal initial and final rates, intersections of apparent kinetics in late stages of racemization.

nonlinear kinetic pathways, with significantly different initial and final rates ($R_1$ and $R_2$: 5–10×). The transition zone between these two rates could be either a smooth curve or a multi-segmented zone (cf. Wehmiller and Hare, 1971; Schroeder and Bada, 1976; King and Neville, 1977; Wehmiller and Belknap, 1978; Kriausakul and Mitterer 1978). The multisegmented nature of the foraminifera kinetics (King and Neville, 1977) would introduce the type of scatter actually observed in molluscan intergeneric regression plots, so the foraminifera kinetics are probably a good analogy for any hypothetical multisegmented molluscan model that is proposed. With the information shown in Figure 3 and the paleotemperature constraints that can be placed on kinetic pathways (Wehmiller and Belknap, 1978) it is possible to establish the limits of the transition zone as being between approximately 30,000 and 70,000 years for the samples and temperature ranges discussed here. The transition zone would shift to greater ages at lower temperatures.

In Model A (Fig. 14), two curves are shown with different initial rates ($R_1^B > R_1^A$) and equal subsequent rates ($R_2^A = R_2^B$). The apparent relative rate of $B$ would always be greater than that of $A$. $A'$ represents the possible kinetic pathway where all rates are equal but the apparent relative rate diverges after the transition zone. An example of this case might be the foraminifera data of King and Hare (1972).

Model B (curves $A$ and $B$) represents the situation in which real rates and apparent relative rates always differ ($R_1^B > R_1^A$, $R_2^B > R_2^A$). These two curves could represent leucine in *Protothaca* and *Macoma,* respectively. Curve $B'$, Model B, represents equal rates (compared with A) beyond an initial period of rapid racemization. Curves $A$ and $B'$, Model B, could represent the typical *Saxidomus-Protothaca* (leucine) intergeneric trend, characterized by an early "offset" and subsequent parallelism (Figure 4). Because intersection of *Saxidomus-Protothaca* intergeneric relations are seen in extensively racemized samples (Figures 4–8), curve $B'$ could also be drawn to converge with curve $A$. Some precedent for convergence of intergeneric pathways may exist for extensively epimerized foraminifera (King and Hare, 1972: Fig. 142, "*Globorotalia* curve").

Model C represents the complex situation of inversion of rates through the transition zone: $R_1^C > R_1^B > R_1^A$; $R_2^C < R_2^B < R_2^A$. Though apparent relative kinetics might seem to be constant in this situation ($C > B > A$), intersection of these kinetics late in the time period represented is also predicted. The curves of Model C are approximations for the following relative kinetic pathways:

| | | | |
|---|---|---|---|
| Curve $A$ | *Macoma* - glu | ≅ | *Protothaca* - leu |
| Curve $B$ | *Macoma* - leu | ≅ | *Protothaca* - pro |
| Curve $C$ | *Macoma* - pro | | |

The model curves presented in Figure 14 are consistent with the intrageneric relationships shown in Figures 9–12 (curvilinear trends, nonzero intercepts, rapid initial rates). In spite of uncertainties regarding the nature and exact position of the transition zone, these model curves are also consistent with the available constraints (e.g., Figure 3) that require this transition zone to be roughly contemporaneous for all amino acids and all genera. In reality, it would seem likely that some combination of Models A, B, and C would be appropriate for molluscan amino–acid racemization kinetics.

The practical application of relative racemization kinetics in amino acid dating lies mainly in sample evaluation. Significant inversions in either intrageneric or intergeneric trends would indicate that a single analysis should be interpreted with caution. The apparent intersections in kinetic pathways indicate, however, that some inversions may be inherent to the diagenetic racemization process.

In addition to the type of inversion that might be expected because of intersecting kinetics, other inversions of results have occasionally been encountered. These inversions are divided into four categories:

1. Major intrageneric inversions: when observed intrageneric relations do not conform to the general trends described herein. This type of inversion is observed infrequently, and can usually be attributed to chemical contamination during diagenesis. Samples affected by contamination usually have high amounts of serine (thermally unstable) and enantiomeric ratios much lower than in uncontaminated samples of the same age.
2. Minor intrageneric inversions: when reproducible analyses for an amino acid contradict the trends inferred from other amino acids when comparing samples from two localities. These inversions are typically observed only in cases where temperature or age distinctions are subtle and can probably be explained by normal data scatter caused by perturbations in diagenetic histories. An example is the proline inversion between Cayucos and Nestor Terrace *Protothaca* samples (multiple samples at each locality). Enantiomeric ratios of leucine, glutamic acid, and valine all conform to the expected trend of increasing ratios between Cayucos and Nestor (latitude-temperature effect for samples of equal age), but the mean proline ratios are reversed (Figures 4–7). Phenylalanine (Figure 8) also does not conform to this expected trend, though the discrepancy is not so great as in proline.
3. Major intergeneric inversions: when significant inversions occur in enantiomeric ratios of *all* amino acids in a particular sample compared to other samples from the same locality but when intrageneric rela-

tions within this sample are maintained. The simplest explanation for this type of inversion is geologic reworking from one unit into another, younger unit. Cases such as this should be documented with multiple analyses of several genera, if possible, in order to reliably document the younger age. Geologic criteria for recognition include suitable nearby exposures which could act as a source area, mode of preservation (worn shell fragents *vs. in situ* articulated valves), radiometric dating. Several examples of reworking have been recognized in our work and substantiated by both geologic or radiometric evidence. We have also encountered other examples in our studies for which we propose reworking, without substantiation by either criteria 2 or 3, above.

4. Minor intergeneric inversions: minor inversions (ca. 5%) in intergeneric relations among all amino acids that could be caused by typical data scatter, slight age or temperature differences among samples, or intersecting kinetics, depending upon which amino acid or genus appears to be incongruent.

The existence of inversions in both intergeneric and intrageneric relationships demonstrates the importance of using data for several amino acids in several genera (if possible) for the evaluation of relative or absolute age estimates. Although no amino acid or genus can always be reliable or identified as being "correct," we have mostly relied upon leucine and/or glutamic acid in *Protothaca* and *Saxidomus* in our age estimates (Wehmiller et al., 1977; Kennedy, 1978: 69, Fig. 7). In situations where these genera are absent, it becomes necessary to employ other genera for age estimation. The intergeneric regressions presented here are one tool for the comparison of multiple genera. Because the relative frequency of apparently "anomalous" results appears to be greater among the "faster" racemizing genera studied here (occasionally *Macoma,* often *Epilucina*), some caution must be used in these conversions. In some cases, however, *Macoma* samples have provided analytically and stratigraphically consistent results for which converted *Protothaca* kinetic model ages are in good agreement with available chronological control (Wehmiller et al., 1978a).

The regression relationships calculated in Figures 4–12 do not include data that have been judged to be of the "major inversion" category. "Minor inversions," because they may represent normal data scatter, have been included in the regressions. Even in cases of very high correlation ($r$ = 0.98) examples of minor inversions can be recognized. Low correlation coefficients (e.g., regression PB, Fig. 7, $r$ = 0.91) indicate that frequent ambiguities will be encountered.

## CONCLUSIONS

1. The apparent rate of racemization is slower in the bivalve family Veneridae (*Protothaca, Saxidomus, Chione, Tivela*) than in the bivalve families Tellinidae (*Macoma*), Lucinidae, and Semelidae or in the gastropod family Trochidae.
2. The order of apparent rates of racemization (relative D/L values) among five amino acids is the same in different genera: proline $\geqq$ phenylalanine $\geqq$ leucine $>$ glutamic acid $>$ valine.
3. Kinetic pathways of racemization of different amino acids may intersect, especially in older samples, producing apparent inversions in apparent rates of racemization.
4. Slight differences in apparent racemization kinetics are detectable among genera of the same subfamily but are not detectable at the species level in *Saxidomus*. Differences detected at subfamily levels are small compared with those observed between some families.
5. The kinetic curve for any amino acid in *Protothaca staminea* (from independently dated localities) shows a rapid initial rate which gradually decreases with time. Intergeneric relationships show that kinetic curves for other genera are similar to those for *Protothaca*.
6. Significant deviations from either intergeneric or intrageneric relative kinetics can be used as indications of either geological (reworking) or chemical contamination.
7. Conversion of either intergeneric or intrageneric relative amino acid data, with regressions presented here, permits the reduction of a large number of data to a single, common reference amino acid (or genus) for comparison of relative ages or for use with the *Protothaca* leucine kinetic model of Wehmiller and Belknap (1978).

### *ACKNOWLEDGMENTS*

We are indebted to many museum and collection curators for allowing fossils under their care to be analyzed for amino acids. The bulk of this material was kindly loaned to us by Warren O. Addicott, U.S. Geological Survey, Menlo Park; J. Philip Kern, San Diego State University, San Diego; Roy F. Kohl, then at Humbolt State University, Arcata (Calif.); and Edward C. Wilson, Los Angeles County Museum of Natural History, Los Angeles. Amino acid analyses were performed at the University of Delaware with the support of U.S. Geological Survey Grant Nos. 14-08-0001-G-247 and 14-08-0001-G-248.

## REFERENCES


Bada, J. L., K. A. Kvenvolden, and E. Peterson, Racemization of amino acids in bones, *Nature,* **245,** 308–310, 1973.

Bada, J. L. and R. A. Schroeder, Racemization of isoleucine in calcareous marine sediments: Kinetics and mechanism, *Earth Planet. Sci. Lett.,* **15,** 1–11, 1972.

Belknap, D. F., *Application of amino acid geochronology to stratigraphy of late Cenozoic marine units of the Atlantic coastal plain,* Ph.D. dissertation, Dept. of Geology, Univ. of Delaware, 532 pp., 1979.

Easterbrook, D. J., Pleistocene chronology of the Puget Lowland and the San Juan Islands, Washington, *Geol. Soc. Amer. Bull.,* **80,** 2273–2286, 1969.

Frank, H., G. J. Nicholson, and E. Bayer, Rapid gas chromatographic separation of amino acid stereoisomers with a novel chiral stationary phase, *J. Chrom. Sci.,* **15,** 174–176, 1977.

Hare, P. E., Amino acid dating—limitations and potential, *Geol. Soc. Amer. Abstr. Progr.,* **9**(7), 1004–1005, 1977.

Hare, P. E. and P. H. Abelson, Racemization of amino acids in fossil shells, *Carnegie Inst. Wash. Year Book 66,* 526–528, 1968.

Hare, P. E. and T. C. Hoering, Separation of amino acid optical isomers by gas chromatography, *Carnegie Inst. Wash. Year Book 72,* 690–694, 1973.

Hare, P. E. and R. M. Mitterer, Non-protein amino acids in fossil shells, *Carnegie Inst. Wash. Year Book 65,* 362–364, 1967.

Hare, P. E. and R. M. Mitterer, Laboratory similation of amino acid diagenesis in fossils, *Carnegie Inst. Wash. Year Book 67,* 205–208, 1969.

Kaufman, A., W. S. Broecker, T. L. Ku, and D. L. Thurber, The status of U-series methods of mollusk dating, *Geochim. Cosmochim. Acta,* **35,** 1155–1183, 1971.

Kennedy, G. L., *Marine studies of San Pedro Bay, California, Part 9, Paleontology: Paleontologic record of areas adjacent to the Los Angeles and Long Beach Harbors, Los Angeles County, California,* Univ. of So. California Sea Grant Publ. USC-SG-4-75, D. F. Soule and M. Oguri, eds., pp. 1–119, 1975.

Kennedy, G. L., *Pleistocene paleoecology, zoogeography and geochronology of marine invertebrate faunas of the Pacific northwest coast (San Francisco Bay to Puget Sound),* Ph.D. dissertation, Univ. of California, Davis, 824 pp., 1978.

Kern, J. P., Origin and history of upper Pleistocene marine terraces, San Diego, California, *Geol. Soc. Amer. Bull.,* **88,** 1553–1566, 1977.

King, K., Jr. and P. E. Hare, Species effects on the epimerization of isoleucine in fossil planktonic foraminifera, *Carnegie Inst. Wash. Year Book 71,* 596–598, 1972.

King, K., Jr. and C. Neville, Isoleucine epimerization for dating marine sediments: Importance of analyzing monospecific foraminiferal samples, *Science,* **195,** 1333–1335, 1977.

Kriausakul, N. and R. M. Mitterer, Isoleucine epimerization in peptides and proteins: Kinetic factors and application to fossil proteins, *Science,* **201,** 1011–1014, 1978.

Ku, T. L. and J. P. Kern, Uranium-series age of the upper Pleistocene Nestor Terrace, San Diego, California, *Geol. Soc. Amer. Bull.,* **85,** 1713–1716, 1974.

Kvenvolden, K. A., E. Peterson, and G. E. Pollock, Geochemistry of amino acid enantiomers: Gas chromatography of their diastereomeric derivatives, in *Advances in Organic Geochemistry,* 1971, H. von Gaertner and H. Wehner, eds., Pergamon, New York, pp. 387–410, 1972.

Kvenvolden, K. A., E. Peterson, J. F. Wehmiller, and P. E. Hare, Racemization of amino acids in marine sediments determined by gas chromatography, *Geochim. Cosmochim. Acta,* **37,** 2215–2225, 1973.

Masters, P. M. and J. L. Bada, Racemization of isoleucine in fossil mollusks from Indian middens and interglacial marine terraces in southern California, *Earth Planet. Sci. Lett.,* **37,** 173–183, 1977.

Miller, G. H. and P. E. Hare, Use of amino acid reactions in some Arctic marine fossils as stratigraphic and geochronologic indicators, *Carnegie Inst. Wash. Year Book 74,* 612–617, 1975a.

Miller, G. H. and P. E. Hare, Use of amino acid reactions in some Quaternary Arctic marine fossils as stratigraphic and geochronologic indicators, *Geol. Soc. Amer. Abstr. Progr.,* **7,** (7), 1200, 1975b.

Mitterer, R. M., Pleistocene stratigraphy in southern Florida based on amino acid diagenesis in fossil *Mercenaria, Geology,* **2,** 425–428, 1974.

Mitterer, R. M., Ages and diagenetic temperatures of Pleistocene deposits of Florida based upon isoleucine epimerization in *Mercenaria, Earth Planet. Sci. Lett.,* **28,** 275–282, 1975.

Robinson, S. W., US Geological Survey, Menlo Park, California, radiocarbon measurements I, *Radiocarbon,* **19**(3), 460–464, 1977.

Schroeder, R. A. and J. L. Bada, A review of the geochemical applications of the amino acid racemization reaction, *Earth Sci. Rev.,* **12,** 347–391, 1976.

Shackleton, N. J. and N. D. Opdyke, Oxygen isotope and palaeomagnetic stratigraphy of equatorial Pacific core V28-238: Oxygen isotope temperatures and ice volumes on a $10^5$ year and $10^6$ year time scale, *Quat. Res.,* **3,** 39–55, 1973.

Szabo, B. J. and J. G. Vedder, Uranium series dating of some Pleistocene marine deposits in southern California, *Earth Planet. Sci. Lett.,* **11,** 283–290, 1971.

Valentine, J. W. and H. H. Veeh, Radiometric ages of Pleistocene terraces from San Nicolas Island, California, *Geol. Soc. Amer. Bull.,* **80,** 1415–1418, 1969.

Vallentyne, J. R., Biogeochemistry of organic matter—II: Thermal reaction kinetics and transformation products of amino compounds, *Geochim. Cosmochim. Acta,* **28,** 157–188, 1964.

Veeh, H. H. and J. W. Valentine, Radiometric ages of Pleistocene fossils from Cayucos, California, *Geol. Soc. Amer. Bull.,* **78,** 547–550, 1967.

Wehmiller, J. F., Amino acid studies of the Del Mar, California, midden site: Apparent rate constants, ground temperature models, and chronological implications, *Earth Planet. Sci. Lett.,* **37,** 184–196, 1977.

Wehmiller, J. F., Intergeneric differences in apparent racemization kinetics in mollusks and foraminifera: Implications for models of diagenetic racemization, this volume.

Wehmiller, J. F. and D. F. Belknap, Alternative kinetic models for the interpretation of amino acid enantiomeric ratios in Pleistocene mollusks: Examples from California, Washington, and Florida, *Quat. Res.,* **9,** 330–348, 1978.

Wehmiller, J. F. and P. E. Hare, Racemization of amino acids in marine sediments, *Science,* **173,** 907–911, 1971.

Wehmiller, J. F., P. E. Hare, and G. A. Kujala, Amino acids in fossil corals: Racemization (epimerization) reactions and their implications for diagenetic models and chronological studies, *Geochim. Cosmochim. Acta,* **40,** 763–776, 1976.

Wehmiller, J. F. et al., Correlation and chronology of Pacific coast marine terraces of continental United States by amino acid stereochemistry—technique evaluation, relative ages, kinetic model ages, and geologic implications, *U.S. Geological Survey Open–File Report 77-680,* 106 pp., 1977.

Wehmiller, J. F. et al., Amino acid racemization dating of Quaternary mollusks, Pacific coast United States, in *Short Papers of the Fourth International Conference, Geochronology, Cosmochronology, and Isotope Geology, 1978,* R. E. Zartman, ed., *U.S. Geological Survey Open–File Report 78-701,* pp. 445–448, (1978a).

Wehmiller, J. F., K. R. Lajoie, A. M. Sarna-Wojcicki, and R. F. Yerkes, Unusually high rates of crustal uplift in Ventura County, California inferred from Quaternary marine terrace chronology, *Geol. Soc. Amer. Abstr. Progr.,* **10**(7), 513, (1978b).

Williams, K. M. and G. G. Smith, A critical evaluation of the application of amino acid racemization to geochronology and geothermometry, *Origins of Life,* **8,** 91–144, 1977.

Woodring, W. P., M. N. Bramlette, and W. S. W. Kew, Geology and paleontology of Palos Verdes Hills, California, *U.S. Geological Survey Professional Paper 207,* 145 pp., 1946.

# Intergeneric Differences in Apparent Racemization Kinetics in Mollusks and Foraminifera: Implications for Models of Diagenetic Racemization

**JOHN F. WEHMILLER**
*Department of Geology, University of Delaware*

*ABSTRACT*

Systematic intergeneric differences in apparent racemization kinetics in both mollusks and foraminifera can be explained by a model that relates racemization rates to initial rates of hydrolysis during diagenesis. Information on the relative abundance of aspartic acid in calcareous organisms, combined with data indicating a particular stability of the aspartyl peptide bond in these mineral phases, indicates that relative aspartic acid abundances and relative racemization kinetics are inversely related. Genera (mollusks or foraminifera) with fast relative racemization kinetics have low aspartic acid abundances, and are assumed to have rapid initial hydrolysis rates. The similarities of these relations in both mollusks and foraminifera suggest that a common process is involved in the diagenetic racemization pathways in these two fossil types.

## INTRODUCTION

The role of hydrolysis of polypeptides in the overall pathway of diagenetic racemization of amino acids has been frequently discussed. It is now well documented that free amino acids (those extracted by simple dissolution in dilute acid) in fossil calcareous organisms (mollusks, corals, foraminifera) are more extensively racemized than their total (bound + free) or bound counterparts (Hare and Abelson 1968; Wehmiller and Hare 1971; Bada and Schroeder 1972; Hare and Hoering 1973; Kvenvolden et al., 1973; Miller and Hare 1975; Wehmiller et al., 1976; Kriausakul and Mitterer 1978). Since the apparent rates of racemization inferred from the free amino acid data are significantly greater than would be expected from kinetic data for free amino acids at neutral pH (e.g., Kriausakul and Mitterer 1978), several hypotheses have been proposed to explain this apparent acceleration of racemization rates. Williams and Smith (1977) have recently reviewed these hypotheses.

"Concerted hydrolysis/racemization" and metal-ion catalysis have been the primary focus of debate (Wehmiller and Hare 1971; Hare 1971; Bada and Schroeder 1972; Schroeder and Bada 1976). Evidence that the absence of water will limit apparent racemization in elevated temperature experiments (Hare and Mitterer 1969; Wehmiller 1971) indicates that water (hydrolysis) is needed for racemization to proceed. The nonlinearity of the apparent kinetics of racemization in foraminifera has also been related to the role that hydrolysis has in the accumulation of extensively racemized free amino acids in the total mixture (Wehmiller and Hare 1971; Kvenvolden et al., 1973). Assumptions about the nonlinearity of apparent racemization kinetics in mollusks derive from the hypothesis that hydrolytic reactions are of equal importance in both mollusk and foraminifera diagenetic racemization (Wehmiller and Belknap 1978).

Kriausakul and Mitterer (1978) have demonstrated that extensive racemization of peptide-bound amino acids occurs while the amino acids are in terminal positions—subsequent cleavage of these terminal amino acids could then result in an extensively racemized free amino acid population, even though the actual rate of free amino acid racemization under identical conditions might be significantly slower. Kriausakul and Mitterer's (1978) results demonstrate the importance of distinguishing between "rates" and "apparent rates" (see Lajoie et al., this volume, for discussion of these terms) in discussing enantiomeric ratio determinations on complex mixtures.

This paper demonstrates that systematic generic effects on apparent racemization kinetics in both foraminifera (King and Hare 1972a; King and Neville 1977) and mollusks (Hare and Mitterer 1967; Miller and Hare 1975; Masters and Bada 1977; Wehmiller et al., 1977; Lajoie et al., this volume) can be related to a simple model of diagenetic racemization controlled (at least in the initial portion of the reaction) by rates of hydrolysis of polypeptide bonds. Though many details of this model undoubtedly need to be clarified, the results summarized here suggest a high degree of similarity in diagenetic racemization pathways in various calcareous organisms. The specific issues of metal-ion catalysis and concerted hydrolysis/racemization are not considered, since the model presented here appears to be a simpler explanation of the general observations of racemization and hydrolysis kinetics in fossil calcareous organisms.

## GENERAL MODEL OF AMINO ACID DIAGENETIC PATHWAYS

In Figure 1 a generalized model of amino acid diagenetic pathways is proposed. This model is similar to those proposed by several previous

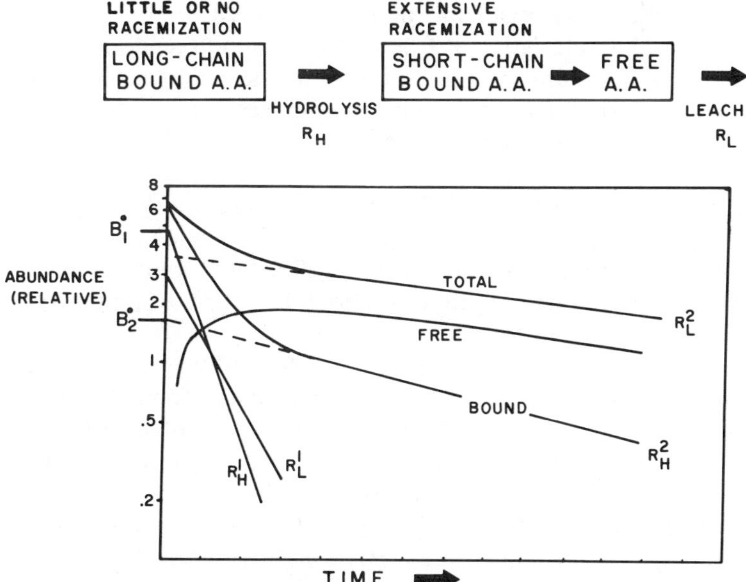


**Figure 1.** Conceptual model of diagenetic pathways of protein-bound amino acids in calcareous fossils, redrawn from Wehmiller (1971). Constraints on the model are the typical distributions of total and free amino acids observed through time. Logarithmic format used for convenience in separating complex curves into assumed first-order components. Analysis of the total and free curves in this manner suggests that hydrolysis of two bound amino acid components, combined with accumulation and leaching of free amino acids, dominates the overall reaction pathway. Each of these bound components, $B_1^0$ and $B_2^0$, has its own hydrolysis ($R_H$) and leaching ($R_L$) coefficients in this particular model. Early stages of diagenesis are characterized by rapid disappearance of more labile component $B_1$, with accumulation of free amino acids (and short peptides) from the hydrolysis of this component. Later stages of diagenesis would involve the slower hydrolysis of more resistant material ($B_2$) and slow leaching or diffusive loss of free amino acids from the fossil. The second, more resistant polypeptide ($B_2$) component could be original protein material (intimately involved in the mineral matrix) or short peptide components cleaved from the more abundant $B_1$ component during early diagenesis. See Figure 2 for discussion of relation of this model to racemization kinetic pathways.

authors (Bada and Schroeder 1972; Wehmiller and Hare 1971). Further details and optional approaches to this model are found elsewhere (Wehmiller 1971). Data on free and total amino acid abundances in foraminifera (King 1978; Wehmiller and Hare 1971; Wehmiller 1971), mollusks (Hare and Mitterer 1967; Hare and Abelson 1968) and corals (Wehmiller et al., 1976; Wehmiller 1971) are used as constraints on this model, though it is recognized that significant variability is usually encountered in quantitative analyses such as these (see King 1978, for example). It is assumed that bound amino acid abundances can be calculated by subtraction of free from total abundances, and it is further assumed that the model curves of Figure 1 would be appropriate for all the amino acids commonly abundant in the calcareous organisms being considered.

Two major characteristics of free and total amino acid abundances in fossil calcareous organisms are common to the published observations cited above: (1) In early diagenesis, a rapid decline in total amino acid abundance is observed; simultaneously, a rapid rise in free amino acid abundance (from essentially zero to about 25% of initial amino acid content) occurs. (2) In later stages of diagenesis, there is a "plateau" of very gradually declining total and free amino acid abundance levels, with free amino acids accounting for about 70 ± 20% of the amino acid mixture and with total amino acid content at about 40 ± 10% of initial total concentrations.

The characteristics of the overall diagenetic pathway can be explained by a combination of hydrolysis, leaching/diffusion, and decomposition reactions. For discussion purposes, the model presented in Figure 1 assumes that hydrolysis and diffusion (of *only* free amino acids) are the dominant reactions. Only with closed-system laboratory experiments can any distinction between diffusive loss (of either bound or free amino acids) and decomposition be made (e.g., Hare and Mitterer 1969). The population of extensively racemized amino acids is shown as including both free and short-chain peptide-bound amino acids; this particular approach would then be consistent with both "concerted hydrolysis" and "terminal-position" explanations for accelerated racemization kinetics.

The model shown in Figure 1 is presented in a logarithmic format so that "apparent first order rate constants" for the various reactions can be visualized by separation of complex curves into linear components. Two bound amino acid components ($B_1^0$ and $B_2^0$), each with their own hydrolysis ($R_H$) and leaching ($R_L$) characteristics, are considered in the model. Other approaches to the interpretation of the available constraints on the model (such as fewer components and a greater number of reactions and rate constants) are possible, but this two-component approach is preferred because of its consistency with observed intergeneric effects on racemization kinetics, to be discussed below. The basic equations used in the calculation of these model curves are:

$$B = \text{bound amino acids } (B_1 \text{ and } B_2)$$

$$F = \text{free amino acids } (F_1 \text{ and } F_2)$$

$$B = B^0 e^{-R_H t}$$

$$F = \frac{R_H}{R_L - R_H} B^0 (e^{-R_H t} - e^{-R_L t})$$

Component 1 of this model can be considered as a more labile component, hydrolyzing and diffusing from the calcareous matrix relatively quickly. The rapid hydrolysis of this component is responsible for the rapid accumulation of free amino acids in the mineral matrix. The second component, less abundant initially, would be more resistant to hydrolysis and leaching either because of its relation with the mineral phase or because of its particularly stable sequence of peptide bonds. (This second component could also be a series of short, stable peptides produced during early diagenetic cleavage of the first component).

Several important questions for further investigation are indicated in the model shown in Figure 1: (1) the significance of the plateau in free amino acids—its relation to shell structure and/or different initial polypeptide components. (2) The nature of the bound material in the later stages of diagenesis—residual peptides, or diagenetic (or native) humic complex of amino acids? (3) The nature and duration of the transition between the early and late stages of diagenesis; this is the period during which short-chain polypeptides are being both produced and destroyed by diagenetic hydrolysis, so the molecular weight distribution of amino acid material in this segment of time would be particularly interesting.

## RELATION BETWEEN INTERGENERIC EFFECTS ON APPARENT RACEMIZATION KINETICS AND APPARENT STABILITY OF FOSSIL POLYPEPTIDES

Systematic generic effects on apparent racemization kinetics in both foraminifera and mollusks can be related to the model shown in Figure 1 with qualitative information on the relative stability of fossilized polypeptide material in these samples. Relative peptide bond stabilities can apparently be correlated with relative aspartic acid abundances in the calcified polypeptides.

Intergeneric trends in relative racemization kinetics in mollusks, summarized from the relationships given in the preceeding paper (Lajoie et al., this volume), are shown in Figure 2. It is concluded that two kinetic categories of mollusks can be recognized with the available data: (1) "fast-racemizing" genera, consisting of *Macoma, Tegula, Epilucina,* and *Cumingia*; and (2) "slow-racemizing" genera, consisting of *Protothaca, Saxidomus, Tivela,* and *Chione*.

These categories might be subdivided if results for single amino acids are considered; but if the apparent relative rate results are considered as a whole (i.e., as shown in Fig. 2), then the following general conclusions can be drawn: (1) The fast category has average racemization rates 20–30% greater than the slow group. (2) In the slow category, most amino acid racemization rates are within 5% of a general mean for the group. Proline in *Saxidomus* is the significant exception to this generalization, though even for proline there is no overlap between the two categories.

Available data (Hare and Abelson 1968; Degens et al., 1967) indicate

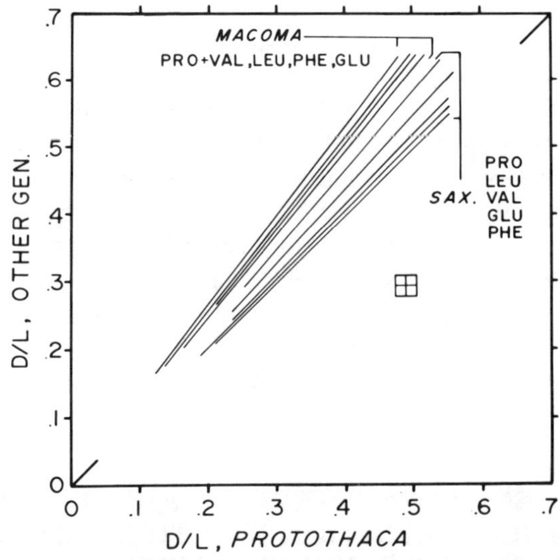


**Figure 2.** Summary of intergeneric regression trends, molluscan data (from previous paper, Lajoie et al., this volume). Comparison of 0-forced total data set linear regressions for *Macoma-Protothaca* and *Saxidomus-Protothaca* intergeneric relations, summarized from previous paper. With the exception of proline, other amino acids in *Saxidomus* show enantiomeric ratios within 5% of their counterparts in coexisting *Protothaca*; All amino acids in *Macoma* show enantiomeric ratios about 25% greater than their counterparts in coexisting *Protothaca*. These two clusters of relative rate trends define the "fast" (*Macoma*) and "slow" (*Saxidomus* and *Protothaca*) kinetic categories. Box represents ±3% analytical uncertainty.

that mollusks of the fast racemizing group have aspartic acid abundances on the order of 80 to 120 residues per 1000, while those of the slow racemizing group have aspartic acid abundances on the order of 200 to 250 residues per 1000. In the absence of quantitative ion-exchange chromatographic data for the fossil mollusk samples discussed here, data on relative aspartic acid abundance are presented in the form of the ASP/LEU value. This ratio is determinable, at least semi-quantitatively, by the capillary column gas chromatographic techniques employed for this study (see previous paper for technique description). ASP/LEU values determined by this method and ASP/LEU values from published sources are presented in Table 1 for the molluscan genera studied here.

From the ASP/LEU ratios summarized in Table 1, it is concluded that the fast racemizing group has ASP/LEU ratios always less than 3.0, with a mean value of approximately 2.0. Among the slow racemizing genera, there is somewhat more scatter in the GC data, but the more reliable ion-exchange data indicate that the slow racemizing group is characterized by an ASP/LEU value of between 5.0 and 7.0, with a mean of about 6.0. Among the foraminifera genera that have been studied for racemization (epimerization) kinetics (King and Hare 1972a; King and Neville 1977) the slowest racemizing genus (*Globigerinoides*) has an ASP/LEU value of 8.1, and the fastest racemizing genus (*Globoritalia*) has an ASP/LEU value of 3.9 (data for modern samples only, King and Hare 1972b). Foraminifera samples, either single or mixed genera, with intermediate racemization kinetics, have ASP/LEU values between 4.9 and 6.4 (King and Hare 1972a; 1972b; Wehmiller and Hare 1971; Wehmiller 1971).

The link between aspartic acid abundance and relative racemization rates involves the stability of aspartic acid peptide bonds in calcified matrices. Contrary to the lability of the aspartyl peptide bond in dilute acid solution (Hill 1965), these peptide bonds appear to be quite stable during artificial diagenesis of mollusk shell protein (Hare et al., 1975). This "anomalous" behavior of aspartic acid is probably a consequence of the close interaction of aspartic acid with the mineral phase during both the living and fossil stages of the sample (see Weiner and Hood 1975, and references therein). It follows from these observations that genera (either mollusks or foraminifera) with a greater relative abundance of aspartic acid would contain a greater proportion of hydrolysis-resistant peptide bonds than would genera that were deficient in aspartic acid. The more slowly that free amino acids and short peptides were produced during early diagenesis, the more slowly extensively racemized amino acids would appear in the total amino acid mixture. In later stages of diagenesis, when most peptide bonds had been cleaved, individual free amino acid racemization rates would dominate the rates observed for total analyses

**Table 1. Relative Aspartic Acid Abundance in Molluscan Genera Studied for Intergeneric Relative Racemization Kinetics: Abundance Expressed as ASP/LEU Ratio**

| Genus | ASP/LEU by GC[a] Mean | ASP/LEU by GC[a] Range | Number of Samples | ASP/LEU by Ion-Exchange[b] | Reference |
|---|---|---|---|---|---|
| "Fast racemizers" | | | | | |
| *Tegula* | 2.6 | 1.7–3.2 | 6 | | Wehmiller et al., 1977 |
| *Macoma* | 1.7 | 0.7–2.3 | 15 | 2.2 | Wehmiller et al., 1977; unpublished: Degens et al., 1967 |
| *Epilucina* | 1.4 | 1.2–1.6 | 6 | | Wehmiller et al., 1977 |
| *Cumingia* | 1.7 | 1.5–1.9 | 2 | | Wehmiller et al., 1977 |
| "Slow racemizers" | | | | | |
| *Protothaca* | 6.2 | 3.2–8.0 | 17 | 6.0 | Wehmiller et al., 1977; unpublished; Hare and Abelson 1965 |
| *Saxidomus* | 9.1 | 6.2–12.5 | 17 | 5.2 | Wehmiller et al., 1977; unpublished; Degens et al., 1967 |
| *Tivela* | 10.0 | 8.5–12.0 | 4 | 7.3 | Wehmiller et al. 1977; Hare and Abelson 1965 |
| *Chione* | 4.8 | 4.3–5.2 | 4 | | Wehmiller et al., 1977 |

[a] Ratio determined by peak area integration of unresolved D–L ASP peak and resolved D and L leucine peaks, with no corrections for variable molar responses; ratio determined on fossil samples only.
[b] Ratio determined on modern samples only.

(Wehmiller and Hare 1971; Bada and Schroeder 1972; Kriausakul and Mitterer 1978), though apparent relative rates of racemization might remain discordant with free racemization rates because of enantiomeric ratios inherited from the early hydrolysis stages. The transition zone between the fast and slow portions of the racemization curve would then be expected to coincide with the transition zone in hydrolysis/leaching kinetics, roughly the point at which free amino acids reach maximum abundance. Foraminifera results (King 1978; Wehmiller 1971) are consistent with this hypothesis. The position of the transition zone in the racemization kinetics (expressed as a percentage of the total reaction to racemic equilibrium) would then be inversely related to the relative abundance of aspartyl peptide bonds.

The model shown in Figure 1 can be easily related to the model curves for diagenetic racemization, shown in Figure 14 of Lajoie et al., in this volume, and reproduced here as Figure 3. As an example, the leucine racemization kinetic pathways in *Protothaca* and *Macoma* (curves A and B, respectively, of either Model A or B, Figure 3) are considered. During early diagenesis, both apparent racemization and hydrolysis rates would be greater for *Macoma* than for *Protothaca*. Beyond the transition zone, if racemization kinetics are truly dominated by free amino acid racemization rates, then it would be expected that observed racemization rates for *Protothaca* and *Macoma* would be equal ($R_2^B = R_2^A$, Model A, Figure 3). The foraminifera data of King and Hare (1972a) are consistent with this option. If late stage hydrolysis rates for *Macoma* remain greater than those for *Protothaca*, then faster racemization rates for *Macoma* would still be expected beyond the transition zone ($R_2^B > R_2^A$, Model B, Figure 3). This latter option is more consistent with the *Protothaca-Macoma* intergeneric trends seen in Lajoie et al., this volume, Figs. 4–8). Slight variations in the shape, timing, and duration of the transition zones (Figures 1 and 3) would be expected for different genera if they contained variable amounts of the two bound components ($B_1^0$ and $B_2^0$) depicted in Figure 1. Different activation energies for the hydrolysis and racemization reactions would also affect the dependence of the transition zone on temperature. The foraminifera data (King 1978) indicate that both the racemization and hydrolysis transition zones overlap in the period between approximately 100,000 and 300,000 years at low temperature (about 1°C). As suggested elsewhere (Lajoie et al., this volume; Wehmiller and Belknap 1978), the racemization transition zone probably occurs between 30,000 and 70,000 years in the mollusk samples discussed here (typical effective temperatures between 10° and 13°C.).

Intrageneric relative racemization kinetics in mollusks and foraminifera (data summarized in Figures 4 and 5) could also be explained by the

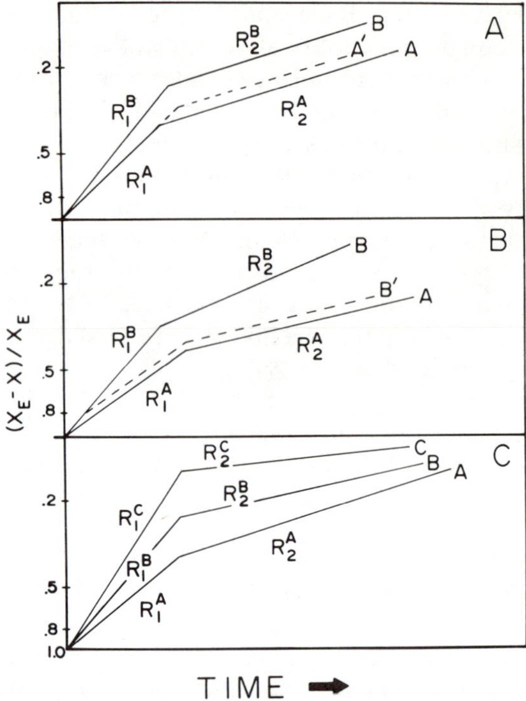


**Figure 3.** Models of racemization kinetic pathways in mollusks or foraminifera, for comparison with diagenetic pathway model shown in Figure 1. Models A, B, and C fully discussed in previous paper. Transition zone between two components in racemization curves coincides with period when rate of production of extensively racemized amino acids (either as short peptides or free amino acids) decreases, probably at about the time when free amino acid abundance goes through its maximum. Different intergeneric rates of racemization can be related to different rates of hydrolysis and/or accumulation of free amino acids. Further discussion of these models in Lajoie et al., this volume.

hydrolysis/racemization pathways proposed herein. Of primary importance is the observation that relative rates of racemization and relative rates of diagenetic hydrolysis follow the same general trends: those that racemize quickly also appear quickly in the free amino acid population. This observation is inferred from fragmentary data from several studies of free/total and D/L ratios in both foraminifera and mollusks (Hare and Abelson 1968; Wehmiller 1971; Wehmiller and Hare 1971; Kvenvolden et al., 1973; Hare et al., 1975). This relationship does not prove that hydrolysis is the major control on early diagenetic racemization, since the observed relative rates of racemization of most of these amino acids (as-

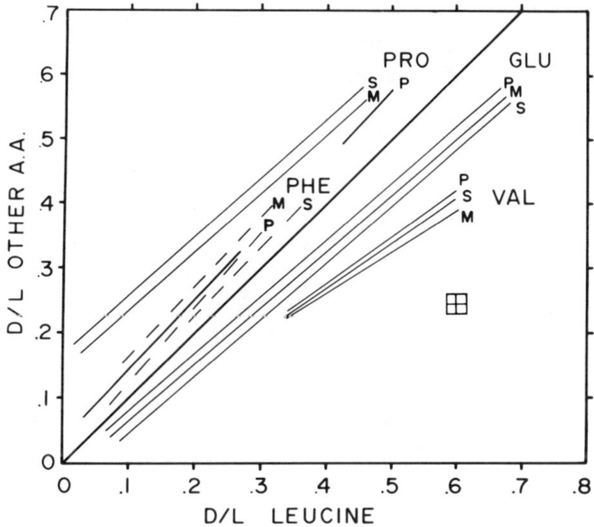


**Figure 4.** Summary of intrageneric linear regressions in mollusks, drawn from data shown in Lajoie et al., this volume. P, S, M = *Protothaca, Saxidomus* (all), *Macoma*, respectively. Box represents 3% uncertainty. Valine regressions are those for leucine ≤ 0.60. With the exception of proline, it appears that generic effects on intrageneric relative racemization kinetics are not recognizable within analytical (or statistical) uncertainties. 95% confidence intervals on the values of slopes and intercepts in these regressions are between 0.04 and 0.06 for correlation coefficients between 0.96 and 0.98.

partic acid and phenylalanine being the major exceptions) are also consistent with their relative racemization rates in aqueous solution (Schroeder and Bada 1976: Table III). The similarity of these intrageneric trends among the different kinetic categories of mollusks (Figure 4) and between mollusks and foraminifera (Figure 5) also suggests that a common process (hydrolysis?) is affecting the diagenetic racemization pathways of these different sample types. Minor variations in intrageneric apparent racemization kinetics could be explained by "sequence effects" on racemization, as demonstrated by Kriausakul and Mitterer (1978) for di- and tripeptides. Since the variability observed in intrageneric relative racemization kinetics in both mollusks and foraminifera is so slight compared with the sequence effects observed in short peptides (Kriausakul and Mitterer 1978), it must be concluded that these effects are being averaged by the long polypeptide chains involved in the calcified matrices of mollusks and foraminifera.

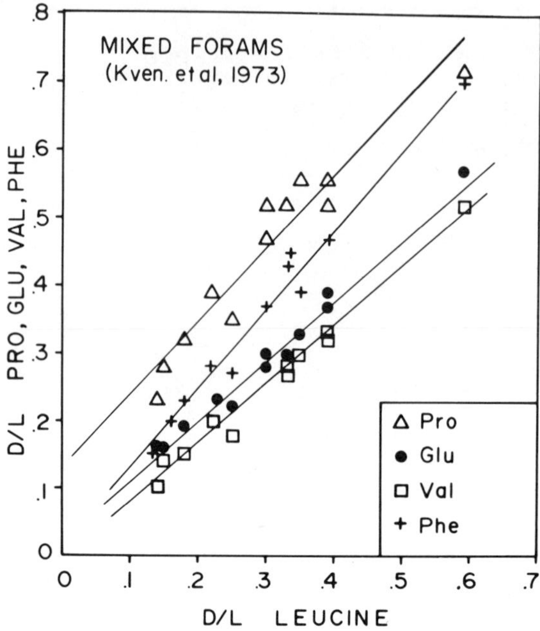


**Figure 5.** Intrageneric relative racemization kinetics of proline, glutamic acid, valine, and phenylalanine, compared with leucine, in mixed foraminifera samples with ages from ca. 0.01 to 2.0 million years. Data from Kvenvolden et al. (1973). Regression lines as follows:

Pro = 1.080Leu + .132; n = 13 r = .958
Glu = .896Leu + .021; n = 13 r = .991
Val = .884Leu − .009; n = 13 r = .988
Phe = 1.178Leu + .013; n = 12, r = .986

Alanine and Aspartic acid intrageneric regressions, not shown here, are:

Ala = 1.185Leu + .063; n = 13, r = .945
Asp = .890Leu + .203; n = 13, r = .956

## CONCLUSIONS

1. The general model presented here, combined with information on intergeneric and intrageneric relative racemization kinetics and relative aspartic acid abundance, suggests that diagenetic racemization in both mollusks and foraminifera is controlled by identical reactions involving hydrolysis of polypeptides and the accumulation of extensively racemized amino acids in either free or short polypeptide form. Intergeneric differences in apparent racemization kinetics can be related to initial rates of hydrolysis of long-chain polypeptide material. The level to

which free amino acids accumulate and the nature of bound amino acid material during later diagenesis will also affect apparent racemization kinetics. Other factors that have been considered in discussion of diagenetic racemization mechanisms, such as metal-ion catalysis, concerted hydrolysis/racemization, and sequence effects on apparent rates, remain as significant details for further investigation. Nevertheless, the major conclusion of this work is that the similarity of racemization/hydrolysis trends in several types of calcareous organisms identifies a basic chemical process that is a common feature of the diagenetic racemization reaction.

2. A basic component of the nonlinear kinetic model for leucine racemization in mollusks (Wehmiller and Belknap 1978) is the assumption that nonlinear kinetics observed in foraminifera (Wehmiller and Hare 1971; Bada and Schroeder 1972; King and Hare 1972a; King and Neville 1977) are a useful model for the general kinetic pathway for mollusk racemization. The assumption has been made (Wehmiller and Belknap 1978) that a general curve for isoleucine epimerization in mixed foraminifera samples can be extrapolated to higher ambient temperatures for use in estimation of mollusk ages. Particular kinetic curves developed in this manner must be calibrated and evaluated with data for individual well-dated genera of both Holocene and late Pleistocene age. Critical to this model development is the position of the transition zone, or slope change in racemization kinetics, in relation to the overall reaction to racemic equilibrium. Both paleoclimatic and chronologic evidence (Wehmiller and Belknap 1978; Wehmiller et al., 1977; Lajoie et al., this volume) support the conclusion that the transition zone for *Protothaca* (and the entire group of slow-racemizing mollusks) occurs at about the same relative position as in mixed foraminifera samples (the kinetic curve of Wehmiller and Hare 1971, as discussed by Wehmiller and Balknap 1978). This conclusion is further supported by the similarity in relative aspartic acid abundances in the slow racemizing category of mollusks and in the mixed foraminifera from which the mollusk racemization model was developed (Wehmiller and Hare 1971; Wehmiller and Belknap 1978).

3. Many of the hypotheses and models presented are based upon fragmentary and/or qualitative data. Important questions for investigation appear to be: (1) molecular weight distribution of amino acid material in fossils of different ages; (2) the nature of aspartyl bond stability in calcareous matrices; (3) temperature dependence of hydrolysis reactions in calcareous organisms; and (4) parallelism, divergence, or intersection of intergeneric and intrageneric kinetic pathways of racemization.

## *ACKNOWLEDGMENT*

This research was supported by U. S. Geological Survey Grants, Nos. 14-08-0001-G-247 and 14-08-0001-G-248.

## *REFERENCES*

Bada, J. l. and R. A. Schroeder, Racemization of isoleucine in calcareous marine sediments: Kinetics and mechanism, *Earth Planet. Sci. Lett.*, **15,** 1–11, 1972.

Degens, E. T., D. W. Spencer, and R. H. Parker, Paleobiochemistry of molluscan shell proteins, *Comp. Biochem. Physiol.*, **20,** 553–579, 1967.

Hare, P. E., Effects of hydrolysis on the racemization rate of amino acids, *Carnegie Inst. Wash. Year Book 70,* 256–258, 1971.

Hare, P. E. and P. H. Abelson, Amino acid composition of some calcified proteins, *Carnegie Inst. Wash. Year Book 64,* 223–232, 1965.

Hare, P. E. and P. H. Abelson, Racemization of amino acids in fossil shells, *Carnegie Inst. Wash. Year Book 66,* 526–528, 1968.

Hare, P. E. and T. C. Hoering, Separation of amino acid optical isomers by gas chromatography, *Carnegie Inst. Wash. Year Book 72,* 690–694, 1973.

Hare, P. E. and R. M. Mitterer, Non-protein amino acids in fossil shells, *Carnegie Inst. Wash. Year Book 65,* 362–364, 1967.

Hare, P. E. and R. M. Mitterer, Laboratory simulation of amion acid diagenesis in fossils, *Carnegie Inst. of Wash. Year Book 67,* 205–208, 1969.

Hare, P. E., G. H. Miller, and N. C. Tuross, Simulation of natural hydrolysis of proteins in fossils, *Carnegie Inst. Wash. Year Book 74,* 609–612, 1975.

Hill, R. L., Hydrolysis of proteins, in *Advances in Protein Chemistry,* Vol. 20, C. B. Anfinsin, Jr., M. L. Anson, J. T. Edsall, and F. M. Richards, eds., Academic Press, New York, pp. 37–107, 1965.

King, K., Jr., Amino acid diagenesis in a single species of planktonic foraminifera: Implications for dating of marine sediments, *Geol. Soc. Amer. Abstr. Prog.*, **10**(7), 435, 1978.

King, K., Jr. and P. E. Hare, Species effect on the epimerization of isoleucine in fossil planktonic foraminifera, *Carnegie Inst. Wash. Year Book 71,* 596–598, 1972a.

King, K., Jr. and P. E. Hare, Amino acid composition of the test as a taxonomic character for living and fossil planktonic foraminifera, *Micropaleontology,* **18,** 285–293, 1972b.

King, K., Jr. and C. Neville, Isoleucine epimerization for dating marine sediments: Importance of analyzing monospecific foraminiferal samples, *Science,* **195,** 1333–1335, 1977.

Kriausakul, N. and R. M. Mitterer, Isoleucine epimerization in peptides and proteins: Kinetic factors and application to fossil proteins, *Science,* **201,** 1011–1014, 1978.

Kvenvolden, K. A., E. Peterson, J. F. Wehmiller, and P. E. Hare, Racemization of amino acids in marine sediments determined by gas chromatography, *Geochim. Cosmochim. Acta,* **37,** 2215–2225, 1973.

Lajoie, K. R., G. L. Kennedy, and J. F. Wehmiller, Inter- and intrageneric trends in apparent amino acid relative racemization kinetics in Quaternary mollusks, this volume.

Masters, P. M. and J. L. Bada, Racemization of isoleucine in fossil mollusks from Indian middens and interglacial terraces in southern California, *Earth Planet. Sci. Lett.*, **37**, 173–183, 1977.

Miller, G. H. and P. E. Hare, Use of amino acid reactions in some Arctic marine fossils as stratigraphic and geochronologic indicators, *Carnegie Inst. Wash. Year Book 74,* 612–617, 1975.

Schroeder, R. A. and J. L. Bada, A review of the geochemical applications of the amino acid racemization reaction, *Earth Sci. Rev.*, **12**, 347–391, 1976.

Wehmiller, J. F., Amino acid diagenesis in fossil calcareous organisms, Ph.D. dissertation, Columbia University, New York, 328 pp., 1971.

Wehmiller, J. F. and P. E. Hare, Racemization of amino acids in marine sediments, *Science,* **173,** 907–911, 1971.

Wehmiller, J. F., P. E. Hare, and G. J. Kujala, Amino acids in fossil corals: Racemization (epimerization) reactions and their implications for diagenetic models and chronological studies, *Geochim. Cosmochim. Acta,* **40,** 763–776, 1976.

Wehmiller, J. F. et al., Correlation and chronology of Pacific coast marine terraces of continental United States by amino acid stereochemistry—technique evaluation, relative ages, kinetic model ages, and geologic implications, *U. S. Geological Survey Open-File Report 77-680,* 106 pp., 1977.

Wehmiller, J. F. and D. F. Belknap, Alternative kinetic models for the interpretation of amino acid enantiomeric ratios in Pleistocene mollusks: Example from California, Washington, and Florida, *Quaternary Research,* **9,** 330–348, 1978.

Weiner, S. and L. Hood, Soluble protein of the organic matrix of mollusk shells: A potential template for shell formation, *Science,* **190,** 987–989, 1975.

Williams, K. M. and G. G. Smith, A critical review of the application of amino acids racemization to geochronology and geothermometry. *Origins of Life,* **8,** 91–144, 1977.

# The Radioracemization of Amino Acids by Ionizing Radiation: Geochemical and Cosmochemical Implications

**WILLIAM A. BONNER**
**NEAL E. BLAIR**
***Department of Chemistry, Stanford University***

**RICHARD M. LEMMON**
***Lawrence Berkeley Laboratory, University of California***

---

A portion of this research was described at the 144th National Meeting of the American Association for the Advancement of Science, Washington, D.C., Feb. 12–17, 1978; at the Fourth College Park Colloquium on Chemical Evolution, University of Maryland, College Park, Maryland, Oct. 18–20, 1978; and at the Carnegie Institution of Washington Conference: Advances in the Biogeochemistry of Amino Acids, Airlie House, Warrenton, Virginia, Oct. 29–Nov. 1, 1978. The present chapter first appeared in *Origins of Life*, 1979.

## INTRODUCTION

In recent years we have been engaged in a variety of experimental investigations (Bonner 1974; Bonner and Flores 1975; Bonner et al., 1975, 1976/77, 1978; Noyes et al., 1977) probing the possible validity of the Vester-Ulbricht mechanism (Vester et al., 1959; Ulbricht 1959; Ulbricht and Vester 1962) for the origin of optical activity in nature *via* parity violation during the β-decay of radionuclides. In one series of such experiments (Bonner et al., 1978) a number of 17–25 year old samples of crystalline $^{14}C$-labeled amino acids, both racemic and optically active, of high specific radioactivity were examined by gas chromatography (g.c.) for both gross radiolysis and possible stereoselective (asymmetric) degradation. The racemic samples, which had undergone self β-radiolysis during the several decades since their original preparation and purification (Bernstein et al., 1972), showed as much as 67% gross degradation but no asymmetric radiolysis. Several of the optically active $^{14}C$-labeled amino acids, however, gave g.c. analyses which suggested that radiation-induced racemization might also have accompanied their gross radiolysis. Since the possibility of such *radioracemization* had not been considered in any of the earlier experiments involving attempted stereoselective β-radiolysis, and since such a possibility had been only scarcely suggested in earlier literature (Feng and Tobey 1959; Evans 1966; Evans et al., 1968), we decided to undertake a systematic preliminary investigation of the phenomenon. Our general plan was to subject a number of crystalline or dissolved optically pure D- and L-amino acids to heavy doses of ionizing radiation, then to examine the partially destroyed residue by g.c. for both gross radiolysis and possible radioracemization.

## EXPERIMENTAL

The samples irradiated consisted of 10–17 mg portions, either solid or dissolved, of the optically pure (as determined by g.c.) amino acids listed in the tables. The samples were of the purest quality available from Aldrich Chemical Co., Calbiochem, Mann Reserach Laboratory, Nutritional Biochemicals, or Sigma Chemical Co. For hydrochloride salt irradiations, the weighed amino acid was first dissolved in an equivalent amount of 0.1 *N* aqueous HCl, and for sodium salt irradiations each sample was dissolved in one equivalent volume of 0.1 *N* NaOH. The solid or dissolved samples were placed in 1 × 4 cm glass vials stoppered with Bakelite caps fitted with Teflon gaskets. The vials were placed in a 3000-Ci $^{60}$Co γ-ray source at the Lawrence Berkeley Laboratory, a source designed to deliver high dose rates (4–10 × $10^6$ rads/hr) to small samples. Irradiations proceeded for time periods (5–90 hr) sufficient to afford the total radiation doses shown in the following tables and to provide about 50–70% gross radiolysis.

After irradiation, each solid sample was dissolved and each liquid sample diluted to a volume of 5.00 ml with water or dilute HCl; they were then quantitatively divided in half volumetrically, and all solutions were finally rotary-evaporated to dryness under vacuum. One of the 50% aliquots of each sample was then treated with a weighed quantity of the corresponding enantiomeric (or racemic) amino acid, so as to permit determination of the percent degradation of the sample by the "enantiomeric marker" technique (Bonner 1973). The amino acid samples were then converted into their N-trifluoroacetyl (N-TFA) isopropyl ester derivatives (I) for g.c. analysis (Bonner et al., 1974), by first esterifying with 2-PrOH/HCl, then acylating with trifluoroacetic anhydride. In the case of the g.c.

analyses of irradiated isovaline

$$\begin{array}{c} R \\ | \\ F_3CCONH\text{—}CH\text{—}COOCH(CH_3)_2 \end{array} \qquad \begin{array}{c} CH_3 \qquad\qquad CH_2CH(CH_3)_2 \\ | \qquad\qquad\qquad | \\ F_3CCONH\text{—}C\text{—}CONH\text{—}CH\text{—}COOCH(CH_3)_2 \\ | \\ CH_2CH_3 \end{array}$$

I 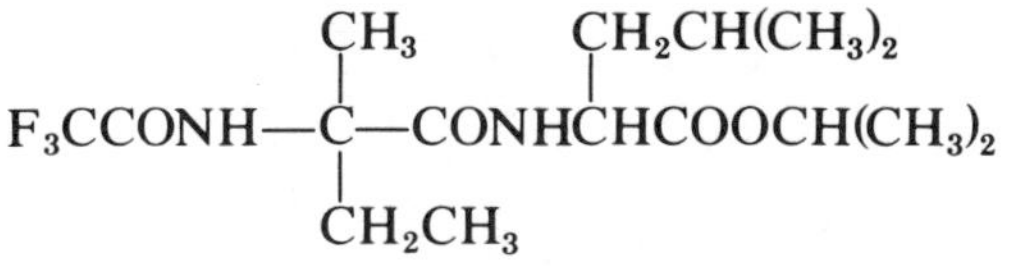 II

samples, however, enantiomeric derivatives of the type I proved unsatisfortory in their resolvability characteristics. Suitable quantitative analyses (i.e., baseline resolution) could be achieved, however, by converting the isovaline first into its diastereomeric N-TFA-isovalyl-D(or -L)-leucine isopropyl ester derivatives (II) (Flores et al., 1977; Bonner et al., 1979b).

Each derivatized sample was then quantitatively analyzed by g.c. for its enantiomeric composition, using 46 m × 0.5mm stainless steel capillary columns coated with one of the two optically active enantiomeric g.c. phases, N-docosanoyl-D(or L)-valine *tert*-butylamide (Bonner and Blair 1979). The columns were installed in a Hewlett-Packard 5700 A gas chromatograph, and peak area integration was accomplished with the aid of a Hewlett-Packard 3380A digital electronic integrator-recorder, affording the analytical reproducibility and precision previously described (Bonner et al., 1974).

To assess the effect of increasing radiation dosage on the extent of both gross radiolysis and racemization, a number of D- or L-leucine samples (15–17 mg) were dissolved in the equivalent volumes of 0.1 *N* HCl or NaOH, and the solutions were irradiated in the above γ-ray source for increasing time periods, so as to achieve the increasing dosages shown in Table 4. The data in the tables have been presented separately in several publications (Bonner and Lemmon 1978a,b; Bonner et al., 1979a,b).

## RESULTS

Tables 1 and 2 indicate that γ-radiation not only causes the previously documented radiolysis of amino acids (Garrison 1968, 1972) but also engenders significant radioracemization—both in the solid state and as sodium salts in aqueous solution. Extensive radiolysis also occurs with the dissolved hydrochloride salts of amino acids (Table 3) but in contrast to the sodium salts (Table 2), little or no concomittant racemization takes place. Table 4 indicates that increasing radiation dosage causes increasing radiolysis of leucine salts in aqueous solution, as well as increasing radioracemization of the sodium salt. Again the hydrochloride salt samples

**Table 1.** **$\gamma$-Radiolysis and Radioracemization of Solid Amino Acids**

| No. | Amino Acid | Radiation Dose (rads × $10^{-8}$) | Decomposition % | Residual Amino Acid | | |
|---|---|---|---|---|---|---|
| | | | | %D[a] | %L[a] | Racemization, %[b] |
| 1. | L-Alanine | 8.1 | 38.6 | 1.9 | 98.1 | 3.8 |
| 2. | D-2-Aminobutyric acid | 8.1 | 55.8 | 99.2 | 0.8 | 1.6 |
| 3. | L-Norvaline | 8.1 | 66.1 | 1.6 | 98.4 | 3.2 |
| 4. | L-Norleucine | 8.1 | 63.1 | 1.3 | 98.7 | 2.6 |
| 5. | D-Leucine | 8.1 | 67.9 | 97.2 | 2.8 | 5.6 |
| 6. | L-Leucine | 8.1 | 68.0 | 2.5 | 97.5 | 5.0 |
| 7. | D-Leucine | 10.2 | 96.1 | 96.2 | 3.8 | 7.6 |
| 8. | L-Leucine | 10.2 | 93.2 | 6.8 | 93.2 | 13.6 |

[a] Standard deviations for averages of 2–4 g.c. analyses were generally ±0.1–0.3%. Same in other tables.
[b] Two times the percent of inverted enantiomer.

**Table 2.** **$\gamma$-Radiolysis and Radioracemization of Amino Acid Sodium Salts (0.10 *M* aqueous solution)**

| | | | | Residual Amino Acid | | |
|---|---|---|---|---|---|---|
| No. | Amino Acid | Radiation Dose (rads $\times$ $10^{-7}$) | Decomposition % | %D | %L | Racemization, % |
| 1. | L-Alanine | 1.7 | 65.8 | 5.8 | 94.2 | 11.6 |
| 2. | D-2-Aminobutryic acid | 1.8 | 58.0 | 96.0 | 4.0 | 8.0 |
| 3. | D-2-Aminobutyric acid | 1.8 | 59.4 | 95.8 | 4.2 | 8.4 |
| 4. | L-Norvaline | 1.7 | 59.6 | 4.2 | 95.8 | 8.4 |
| 5. | L-Norleucine | 1.7 | 52.8 | 5.3 | 94.7 | 10.6 |
| 6. | L-Valine | 1.7 | 55.3 | 5.5 | 94.5 | 11.0 |
| 7. | L-Leucine | 1.7 | 54.4 | 2.6 | 97.4 | 5.2 |

**Table 3.** **$\gamma$-Radiolysis and Radioracemization of Amino Acid Hydrochloride Salts (0.10 *M* aqueous solution)**

| | | | | Residual Amino Acid | | |
|---|---|---|---|---|---|---|
| No. | Amino Acid | Radiation Dose (rads × $10^{-7}$) | Decomposition % | %D | %L | Racemization, % |
| 1. | L-Alanine | 2.2 | 53.4 | 0.2 | 99.8 | 0.4 |
| 2. | D-2-Aminobutyric acid | 2.2 | 52.1 | 100.0 | 0.0 | 0.0 |
| 3. | L-Norvaline | 2.2 | 57.9 | 0.0 | 100.0 | 0.0 |
| 4. | L-Norleucine | 2.2 | 63.5 | 0.0 | 100.0 | 0.0 |
| 5. | L-Valine | 2.2 | 54.8 | 0.0 | 100.0 | 0.0 |
| 6. | L-Leucine | 2.2 | 55.3 | 0.0 | 100.0 | 0.0 |

**Table 4.** **γ-Radiolysis and Radioracemization of Leucine Salts (0.10 *M* aqueous solution) as a Function of γ-Ray Dosage**

| Enantiomer | Salt | Radiation Dose (rads × $10^{-6}$) | Decomposition % | Residual Leucine | | |
|---|---|---|---|---|---|---|
| | | | | %D | %L | Racemization, % |
| D | Na | 1.0 | 1.8 | 99.2 | 0.8 | 1.6 |
| D | Na | 3.0 | 9.4 | 99.2 | 0.8 | 1.6 |
| D | Na | 14.0 | 41.7 | 97.9 | 2.1 | 4.2 |
| L | Na | 17.3 | 54.4 | 2.6 | 97.4 | 5.2 |
| D | Na | 20.5 | 63.1 | 96.2 | 3.8 | 7.6 |
| D | Na | 27.0 | 73.6 | 95.4 | 4.6 | 9.2 |
| D | HCl | 1.0 | 3.3 | 99.8 | 0.2 | 0.4 |
| D | HCl | 3.0 | 7.6 | 99.7 | 0.3 | 0.6 |
| D | HCl | 14.0 | 32.5 | 99.2 | 0.8 | 1.6 |
| L | HCl | 22.0 | 55.3 | 0.0 | 100.0 | 0.0 |
| D | HCl | 22.1 | 57.6 | —[a] | n.d.[b] | n.d |
| D | HCl | 36.1 | 75.7 | —[a] | n.d. | n.d |

[a] G.C. traces obscured by peaks for degradation products.
[b] n.d. = not determined.

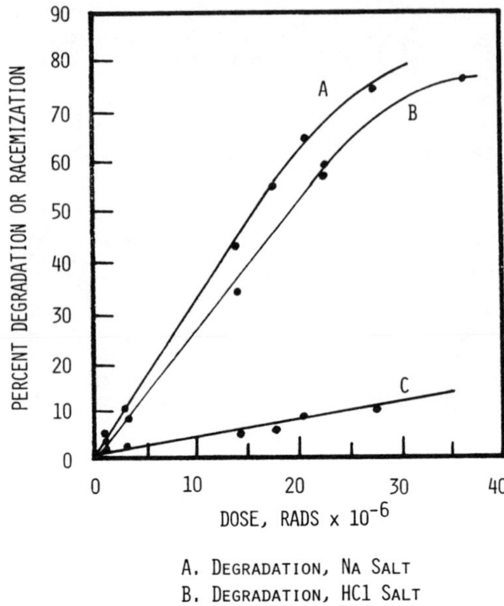


**Figure 1.** Radiolysis and radioracemization of leucine salts (0.1 *M* aqueous solution) by γ-radiation.

underwent little or no racemization. A plot (Fig. 1) of the data in Table 4 shows that radiolysis of the sodium salt of leucine occurs slightly faster than radiolysis of its hydrochloride salt, and that for both salts the extent of radiolysis is roughly proportional to total dosage up to about 60% degradation. Tolbert et al. (1962) have found a similar initial linear relation between radiolysis and dose for crystalline amino acids, and we have previously noted a linear decomposition-dose relationship during the first 80% of the γ-radiolysis of solid leucine (Bonner 1973). Figure 1 also indicates an approximately linear relationship between the extent of radioracemization and the dosage for the sodium salt of leucine within the dose range studied. A rough indication of the overall reproducibility of the radiolysis and racemization data in the above tables is seen in numbers 2 and 3 of Table 2, duplicate experiments in which identically sized samples of the dissolved sodium salt of 2-aminobutyric acid were irradiated simultaneously.

For reasons discussed below, it became desirable to investigate the γ-radiolysis and possible radioracemization of isovaline (α-amino-α-methyl-butyric acid, III), and to compare these with the results for the ordinary

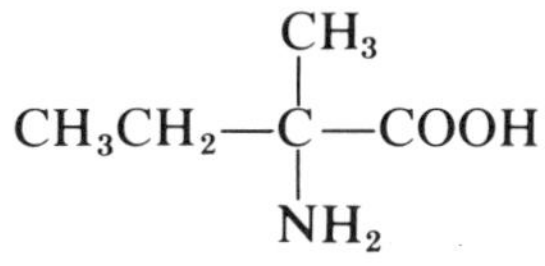

$$\begin{array}{c} CH_3 \\ | \\ CH_3CH_2\text{—}C\text{—}COOH \\ | \\ NH_2 \end{array}$$

III

amino acids in Tables 1–3. The principal results of these experiments are summarized in Table 5, where we observe that isovaline samples in the solid state, either hydrated or anhydrous, or as the solid sodium (No. 7) or hydrochloride salt (No. 8), undergo significant radioracemization accompanying their γ-radiolysis. Again (Nos. 3,4 versus 1,2) as with solid (Table 1) or dissolved (Table 4) leucine, increasing radiation doses cause an increase in both radiolysis and racemization. Isovaline in the hydrated crystal form (Nos. 5,6) appears slightly less susceptible to radioracemization than does the anyhdrous form (Nos. 1,2) for the same radiation dose, though the two forms are about equally prone to radiolysis. While the radiolyses and radioracemizations of isovaline in the solid state are roughly comparable to those of the other amino acids in Table 1, the complete absence of radioracemization for the aqueous sodium salt of isovaline (Nos. 9,10) is noteworthy in its contrast to the facile racemizations summarized in Table 2. This difference is discussed below. Again, the approximate reproducibility of our radiolysis and radioracemization data is indicated by the effects noted at equivalent dosages for the various enantiomeric pairs in Table 5 (Nos. 1,2; 3,4; 5,6; 9,10).

## MECHANISTIC SPECULATIONS

We have not experimentally established any mechanisms whereby the radioracemization of solid amino acids or their aqueous sodium salts might occur, nor have we characterized the radiolysis products responsible for the several extraneous peaks in the various g.c. traces of our enantiomeric composition analyses. One can only speculate on mechanisms for radioracemization as they might be superimposed upon currently accepted mechanisms for the radiolysis of solid amino acids or their aqueous solutions. Garrison (1968, 1972) has recently discussed such mechanisms in detail.

The initial step in the γ-radiolysis of a solid amino acid zwitterion (IV) is thought to be carbon-hydrogen scission at the α-carbon atom, producing an α-radical (V), a proton, and a secondary electron, $e_s^-$ (Eq. 1). The sec-

**Table 5. γ-Radiolysis and Radioracemization of Isovaline**

| No. | Substrate | Radiation Dose (rads × $10^{-8}$) | Degradation, % | Racemization, % |
|---|---|---|---|---|
| 1. | D-Ival[a] | 9.0 | 79.6 | 4.9 |
| 2. | L-Ival[a] | 9.0 | 78.0 | 4.7 |
| 3. | D-Ival[a] | 12.0 | 86.5 | 6.1 |
| 4. | L-Ival[a] | 12.0 | 86.8 | 6.4 |
| 5. | D-Ival·$H_2O$[b] | 9.0 | 80.0 | 2.5 |
| 6. | L-Ival·$H_2O$[b] | 9.0 | 79.0 | 3.2 |
| 7. | D-Ival Na salt[a] | 9.0 | 87.7 | 3.8 |
| 8. | L-Ival HCl salt[a] | 9.0 | 91.8 | 3.3 |
| 9. | D-Ival Na salt (aq)[c] | 0.25 | 68.1 | 0.0 |
| 10. | L-Ival Na salt (aq)[c] | 0.25 | 68.8 | 0.0 |

[a] Solid, anhydrous.
[b] Monohydrate crystals.
[c] 0.1 *M* aqueous solution.

ondary electron as well as the α-radical V then attack both IV and other

$$\underset{\text{IV}}{\text{R—}\overset{\overset{+}{\text{N}}\text{H}_3}{\text{CH}}\text{—COO}^-} \xrightarrow{\gamma\text{-photon}} \underset{\text{V}}{\text{R—}\overset{\overset{+}{\text{N}}\text{H}_3}{\dot{\text{C}}}\text{—COO}^- + \text{H}^+ + \text{e}^-_{\text{s}}} \quad (1)$$

intermediate species trapped in the crystal lattice, ultimately affording the observed radiolysis products. In such a reaction sequence, to the extent that the α-radical V abstracts hydrogen atoms from other constituents of the crystal lattice to re-form IV, the regenerated IV should be totally or partially racemic, since V is presumably unable fully to maintain its original stereochemical configuration. Clearly, some similar sequence of degradative reactions, one or more of which regenerates the original but racemized amino acid, must apply also to the radioracemization of solid isovaline. Since isovaline has no α-hydrogen, however, the initial radiation-induced α-carbon scission must involve a C-C rather than a C-H bond.

The radiolysis of amino acids in aqueous solution is believed (Garrison 1968, 1972) to be initiated by hydrated electrons, $e^-_{aq}$, or hydroxyl radicals, HO·, produced by prior radiolysis of the water solvent (Eq. 2). These species

$$H_2O \xrightarrow{\gamma\text{-photon}} e^-_{aq} + HO\cdot + H^+ + H_2 + H_2O_2 \quad (2)$$

in turn attack the amino acid and initiate a series of degradation reactions similar to those postulated for solid state radiolysis, ultimately affording the observed fatty acid and α-keto acid degradation products. In particular, α-hydrogen abstraction by HO· can produce a key radiolysis intermediate, α-radical V( Eq. 3). Reversal of reaction 3 would clearly provide a

$$\text{HO}\cdot + \underset{\text{IV}}{\text{R—}\overset{\overset{+}{\text{N}}\text{H}_3}{\text{CH}}\text{—COO}^-} \longrightarrow \text{H}_2\text{O} + \underset{\text{V}}{\text{R—}\overset{\overset{+}{\text{N}}\text{H}_3}{\dot{\text{C}}}\text{—COO}^-} \quad (3)$$

facile mechanism for the radioracemization of zwitterion IV in aqueous solution.

Our solution radiolyses, however, were conducted using sodium (VI) or hydrochloride salts (VII) of the amino acids in question. Here the

mechanism

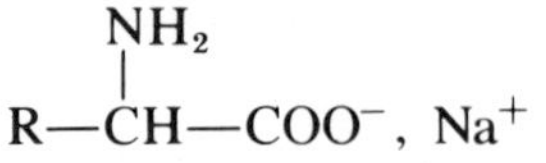

$$\underset{\text{VI}}{\text{R—}\overset{\overset{\text{NH}_2}{|}}{\text{CH}}\text{—COO}^-\text{, Na}^+} \qquad \underset{\text{VII}}{\text{R—}\overset{\overset{^+\text{NH}_3\text{, Cl}^-}{|}}{\text{CH}}\text{—COOH}}$$

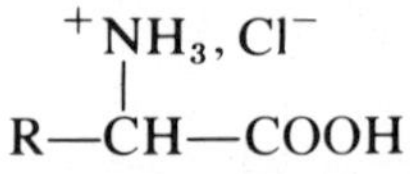

for radioracemization would again presumably be HO· attack to form a stereochemically labile α-radical analogous to V, followed by solvent attack to regenerate racemized VI or VII, akin to the reversal of Eq. 3. However, our findings were that the sodium salts VI were particularly susceptible to radioracemization, while the hydrochloride salts VII under similar conditions were essentially unracemized. The facile radioracemization of the aqueous sodium salts of amino acids has been rationalized (Bonner and Lemmon 1978b) as resulting from the ready formation of the α-radical from the anion of salt VI because of its stabilization as the highly symmetrical resonance hybride VIII. Such a resonance stabilized intermediate cannot arise from the

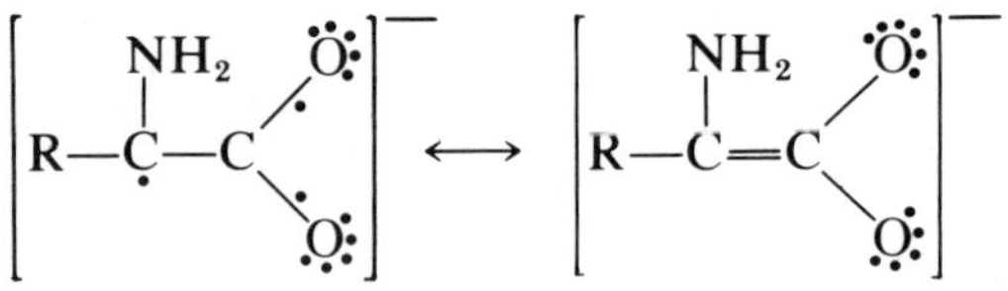

$$\underset{\text{VIII}}{\left[\text{R—}\overset{\overset{\text{NH}_2}{|}}{\dot{\text{C}}}\text{—C}\begin{matrix}\diagup \ddot{\text{O}}\!:\\ \diagdown \ddot{\text{O}}\!:\end{matrix}\right]^- \longleftrightarrow \left[\text{R—}\overset{\overset{\text{NH}_2}{|}}{\text{C}}\text{=C}\begin{matrix}\diagup \ddot{\text{O}}\!:\\ \diagdown \ddot{\text{O}}\!:\end{matrix}\right]^-} \qquad \underset{\text{IX}}{\text{R—}\overset{\overset{^+\text{NH}_3}{|}}{\dot{\text{C}}}\text{—C}\begin{matrix}\diagup\!\!\diagup \text{O}\\ \diagdown \text{OH}\end{matrix}}$$

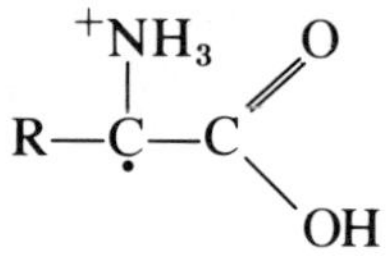

corresponding α-radical (IX) from the cation of VII, thus explaining its apparent slow formation and the lack of accompanying radioracemization. Other speculations regarding the radioracemization of VI and the optical stability of VII have been advanced by Bonner and Lemmon (1978b)

The complete absence of radioracemization for the sodium salt of aqueous isovaline (Table 5) is noteworthy in its contrast to the extensive racemization of the sodium salts of common amino acids (Table 2). This observation is understandable in terms of Eq. 3. If the radioracemization of the sodium salts of amino acids is initiated by analogous α-hydrogen abstraction by HO· to form the resonance stabilized α-radical VIII, such a mechanism is not available to isovaline, as it lacks the requisite α-hydrogen atom. The sodium salt of isovaline is thus immune to radioracemization in aqueous solution (but not in the solid state), although gross radiolysis (by other reaction paths) is extensive (Table 5).

## GEOCHEMICAL AND COSMOCHEMICAL IMPLICATIONS

The racemization of natural amino acids under diagenetic environmental conditions is a phenomenon of considerable interest to geochemists, paleontologists, and archeologists, since it has been widely assumed that the D/L ratios for residual amino acids isolated from ancient specimens may provide a measure of the age of the specimen. Assuming that the amino acids isolated from a prehistoric sample are of protein origin and were therefore originally of the L-configuration and that the simple first-order kinetics applicable to the racemization of amino acids in solution are similarly valid for the natural diagenetic racemization of these amino acids, it is possible—after establishing the D/L ratios of several amino acids in a sample of known age—to calculate the specific rate constants for the diagenetic racemization of those amino acids. Assuming also a uniform external environment at the given sample site within the geological epoch in question, one can then use the calculated rate constants to estimate the ages of the prehistoric samples from the same general area, again by determining the D/L ratios of the same amino acids isolated from the latter samples. In this general way D/L enantiomer ratios have been utilized during the past decade or so for the age-dating of ancient specimens of geological sediments (Bada et al., 1970; Wehmiller and Hare 1971), shells (Hare and Mitterer 1968), bones (Bada 1972; Bada and Protsch 1973; Dungworth et al., 1974), teeth (Helfman and Bada 1975, 1976), and corals (Wehmiller et al., 1976). In addition, having once independently established the age of a particular prehistoric bone sample (e.g., by $^{14}C$ dating), and having then also determined in the laboratory the temperature effects on the rates at which certain amino acids in modern bones racemize or epimerize, several investigators have thereby used subsequently determined enantiomer or epimer ratios of these amino acids in the sample to estimate the temperatures prevailing over the past geological lifetime of the sample, that is, as paleotemperature indicators (Bada et al., 1973; Schroeder and Bada 1973; Mitterer 1975).

The validity of applying amino acid racemization criteria to geochronology and geothermometry has recently been critically questioned (Dungworth 1976; Williams and Smith 1977), and a host of previously ignored environmental and other factors have been enumerated which could cause serious errors in estimates of the rate constants for the diagenetic racemization of amino acids. Such errors could in turn seriously invalidate geochronological or geothermometric conclusions based on the simple measurement of D/L enantiomer ratios for the amino acids in question. To these formidable pitfalls challenging the validity of D/L ratios alone as criteria for estimating geological ages or temperatures must now be added the possibility of radioracemization during the epoch in question. While uniform radioactivity in the earth's crust might conceivably contribute relatively uniformly to the diagenetic racemization of amino acids in ancient samples, the proximity of additional radioactive material, either in an indigenous mineral matrix or dissolved in ground waters, would clearly induce additional indeterminate amounts of radioracemization in the samples in question, and thus suggest their spurious antiquity. Furthermore—a question we are now investigating experimentally—it seems possible that clay minerals (in conjunction with radioactive sources) might enhance the effectiveness of radioracemization, as they apparently do for thermal racemization (Kroepelin 1968; Flores and Bonner 1974), thus shortening further the time required for the diagenetic racemization of ancient amino acid samples. In summary, we argue that unless specific knowledge of the radiation exposure history of a given ancient amino acid sample is available, radioracemization might cause considerable error in geochronological or geothermometric conclusions based solely on the measurement of the simple D/L enantiomer (or epimer) ratio of the amino acid in question.

Our subsequent interest (Bonner et al., 1979a,b) in the possible radioracemization of isovaline (III) was occasioned by the cosmological importance of this nonprotein amino acid. In September 1969 a type II carbonaceous chondrite fell to earth near Murchison, Victoria, Australia, and shortly thereafter extensive investigations were undertaken into the organic constituents of this fragment, using g.c. and other analytical techniques (Kvenvolden et al., 1970, 1971; Oró et al., 1971a,b; Cronin and Moore 1971; Lawless 1973; Lawless and Peterson 1975). As early as 1971 Kvenvolden et al., reported 18 amino acids to be present in the Murchison chondrite, 12 of which, including isovaline, were nonprotein and 6 of which were common to terrestrial proteins; and by 1973 Lawless had extended the number of amino acids present to 45. During their g.c. analyses, the optically active amino acids from the Murchison meteorite were found to consist of approximately equal amounts of D- and L-enantiomers (Kvenvolden et al., 1971, 1972; Oró et al., 1971b), a fact interpreted as indicating their probable abiotic origin. The presence of nonprotein amino acids in the Murchison and Murray meteorites has led to the same interpretation (Kvenvolden et al., 1971; Lawless et al., 1971).

Of the nonprotein amino acids from the Murchison chondrite, isovaline has been of particular interest cosmologically because, in contrast to other amino acids thus far isolated, it lacks a hydrogen atom on its $\alpha$-carbon atom. Accordingly, isovaline cannot undergo racemization by the known mechanisms responsible for the racemization of common amino acids, namely, reactions involving reversible scission of the α-C-H bond (Pollock et al., 1975). For this reason, it has been argued (Lawless 1973) that the enantiomeric composition of the Murchison isovaline should be that which prevailed at the time of its original synthesis in the meteorite, thus giving a clue to the primordial enantiomeric composition of other amino acids in the meteorite. In 1975 Pollock et al., achieved a partial g.c. resolution of authentic D,L-isovaline as well as of the isovaline from the Murchison meteorite, obtaining comparable analytical results from each (D-III, ca. 52%; L-III, ca. 48%). This led to the conclusion that the Murchison isovaline was in fact racemic and that therefore it and the other amino acids in this meteorite had originated as racemic mixtures by abiotic, extraterrestrial syntheses. Since these conclusions were based on the demonstrated (Pollock et al., 1975) nonracemization of isovaline under ordinary laboratory conditions and did not take into account the then unknown possibility of radioracemization, the possible radioracemization susceptibility of isovaline clearly became a cosmologically pertinent question.

Our finding that solid isovaline (Table 5) is approximately as susceptible to radioracemization as are the solid common amino acids having α-hydrogen atoms (Table 1) indicates that this structural difference does not pertain to the racemization of solid amino acids by ionizing radiation. The probable mechanistic reason, in terms of its lack of an α-hydrogen atom, for the immunity to radioracemization of isovaline as its dissolved sodium salt, however, has been discussed above. While the enantiomeric composition of the isovaline in the Murchison meteorite was found (Pollock et al., 1975) to be *approximately* D:L/50:50, our observed radioracemization of solid isovaline suggests the need to reevaluate the earlier conclusion (Pollock et al., 1975; Lawless 1973) that the primordial enantiomeric composition of the isovaline and other amino acids in the Murchison meteorite must therefore have been racemic. Although the time since the Murchison chondrite fragmented from its parent body is only about $1.2 \times 10^6$ years (Cressy and Bogard 1976), resulting in a cosmic ray radiation dose of about $3 \times 10^7$ rads during this period (based on $10^8$ rads during $3.5 \times 10^6$ years for the Orgueil meteorite (Studier et al., 1965)), the natural radioactivity of a meteorite parent body would have provided an integrated dose of about $5 \times 10^8$ rads during the $4.5 \times 10^9$ years of its existence (Anders 1961; Studier et al., 1965). Thus, the isovaline in the Murchison chondrite has received a total radiation dose of about $5.3 \times 10^8$ rads, some 60% of the dose that caused 4.8% racemization of anhydrous isovaline (Table 5). Although it is not known how the mineral matrix of the meteorite might alter the efficacy of radioracemization, it seems probable that significant radioracemization of any nonracemic amino acids indigenous to meteorite parent bodies could have occurred during the 4.5 × $10^9$ years since their origin. These speculations, of course, carry no implication that the racemic amino acids found in present-day meteorites were originally optically active or of biological origin. Nevertheless we believe that the phenomenon of radioracemization, in principle, makes the question of the primordial enantiomeric composition of amino acids in meteorites a fundamentally indeterminate one.

## *ACKNOWLEDGMENTS*

We are indebted to the National Aeronautics and Space Administration (W.A.B., N.E.B.) and to the U.S. Department of Energy (R.M.L.) for their support of portions of the above studies.

## *REFERENCES*

Anders, E., Meteoritic hydrocarbons and extraterrestrial life, *Ann. New York. Acad. Sci.*, **93,** 651–657, 1961.

Bada, J. L., Dating of fossil bones using the racemization of isoleucine, *Earth Planet. Sci. Lett.*, **15,** 223–231, 1972.

Bada, J. L., B. P. Luyendyk, and J. B. Maynard, Marine sediments: Dating by the racemization of amino acids, *Science,* **170,** 730–731, 1970.

Bada, J. L. and R. Protsch, Racemization reaction of aspartic acid and its use in dating fossil bones, *Proc. Nat. Acad. Sci. U.S.A.*, **70,** 1331–1334, 1973.

Bada, J. L., R. Protsch, and R. A. Schroeder, The racemization reaction of isoleucine used as a paleotemperature indicator, *Nature,* **241,** 394–395, 1973.

Bernstein, W. J., R. M. Lemmon, and M. Calvin, An investigation of the possible differential radiolysis of amino acid optical isomers by $^{14}C$ betas, in *Molecular Evolution, Prebiological and Biological,* D. L. Rolfing and A. I Oparin, eds., Plenum, New York, pp. 151–155, 1972.

Bonner, W. A., Enantiomeric markers in the quantitative gas chromatographic analysis of optical isomers. Application to the estimation of amino acid degradation, *J. Chromatogr. Sc.*, **11,** 101–104, 1973.

Bonner, W. A., Experiments on the origin of molecular chirality by parity non-conservation during β-decay, *J. Molec. Evol.*, **4,** 23–39, 1974.

Bonner, W. A. and N. E. Blair, Enantiomeric phases in analytical gas chromatography, *J. Chromatogr.* **169,** 153–159 (1979).

Bonner, W. A., N. E. Blair, and R. M. Lemmon, The racemization of isovaline by γ-radiation, *J. Amer. Chem. Soc.* **101,** 1049, 1979a

Bonner, W. A., N. E. Blair, R. M. Lemmon, J. J. Flores, and G. E. Pollock, The radioracemization of isovaline. Cosmochemical implications. *Geochim. Cosmochim. Acta* **43,** 1841–1846 (1979b).

Bonner, W. A. and J. J. Flores, Experiments on the origins of optical activity, *Origins of Life*, **6,** 187–194, 1975.

Bonner, W. A. and R. M. Lemmon, Radiolysis, racemization, and the origin of molecular asymmetry in the biosphere, *J. Molec. Evol.*, **11,** 95–99, 1978a.

Bonner, W. A. and R. M. Lemmon, Radiolysis, racemization, and the origin of optical activity, *Bioorg. Chem.*, **7,** 175–187, 1978b.

Bonner, W. A., R. M. Lemmon, and H. P. Noyes, β-Radiolysis of crystalline $^{14}C$-labeled amino acids, *J. Org. Chem.*, **43,** 522–524, 1978.

Bonner, W. A., M. A. Van Dort, and J. J. Flores, Quantitative gas chromatographic analysis of leucine enantiomers. A comparative study, *Anal. Chem.*, **46,** 2104–2107, 1974.

Bonner, W. A., M. A. Van Dort, and M. R. Yearian, Asymmetric degradation of DL-leucine with longitudinally polarized electrons, *Nature*, **258,** 419–421, 1975.

Bonner, W. A., M. A. Van Dort, M. R. Yearian, H. D. Zeman, and G. C. Li, Polarized electrons and the origin of optical activity, *Israel J. Chem.*, **15,** 89–95, 1976/77.

Cressy, P. J., Jr. and D. D. Bogard, On the calculation of cosmic-ray exposure ages of stone meteorites, *Geochim. Cosmochim. Acta*, **40,** 749–762, 1976.

Cronin, J. R. and C. B. Moore, Amino acid analyses of the Murchison, Murray, and Allende carbonaceous chondrites, *Science*, **172,** 1327–1329, 1971.

Dungworth, G., Optical configuration and the racemization of amino acids in sediments and in fossils—A review, *Chem. Geol.*, **17,** 135–153, 1976.

Dungworth, G., N. J. Vincken, and A. W. Schwartz, Composition of fossil collagens. Analysis by gas-liquid chromatography, *Comp. Biochem. Physiol.*, **47B,** 391–399, 1974.

Evans, E. A., Control of self-irradiation decomposition of tritium-labeled compounds at high specific activity, *Nature*, **209,** 169–171, 1966.

Evans, E. A., R. H. Green, W. R. Waterfield, Stability and preparation, mainly by catalyzed exchange reactions, of generally and specifically tritiated amino acids and their behavior with biological oxidases, in *Proceedings of the International Conference on Methods of Preparation and Storage of Labelled Compounds*, 2nd, pp. 1019–1036, 1968.

Feng, P. Y. and S. W. Tobey, Radiation induced racemization of l-mandelic acid in aqueous solution, *J. Phys. Chem.*, **63,** 759–760, 1959.

Flores, J. J. and W. A. Bonner, On the asymmetric polymerization of aspartic acid enantiomers on kaolin, *J. Molec. Evol.*, **3,** 49–56, 1974.

Flores, J. J., W. A. Bonner, and M. A. Van Dort, The gas chromatographic resolution of DL-isovaline, *J. Chromatogr.*, **132,** 152–154, 1977.

Garrison, W. M., Radiation chemistry of organo-nitrogen compounds, in *Current Topics in Radiation Research*, Vol. IV, M. Ebert and A. Howard, eds., North-Holland, Amsterdam, pp. 45–94, 1968.

Garrison, W. M., Radiation-induced reactions of amino acids and peptides, *Radiation Res. Rev.*, **3,** 305–326, 1972.

Hare, P. E. and R. M. Mitterer, Laboratory simulation of amino acid diagenesis in fossils, *Carnegie Inst. Wash. Year Book 67*, 205–208, 1968.

Helfman, P. M. and J. L. Bada, Aspartic acid racemization in tooth enamel from living humans, *Proc. Nat. Acad. Sci. U.S.A.*, **72,** 2891–2894, 1975.

Helfman, P. M. and J. L. Bada, Aspartic acid racemization in dentine as a measure of aging, *Nature*, **262,** 279–281, 1976.

Kroepelin, H., Racemization of amino acids on silicates, in *Advances in Organic Geochemistry, Proceedings of the Fourth International Meeting*, P. A. Schlenck, ed., Pergamon, Oxford, pp. 535–542, 1968.

Kvenvolden, K. et al., Evidence for extraterrestrial amino acids and hydrocarbons in the Murchison meteorite, *Nature*, **228**, 923–926, 1970.

Kvenvolden, K. A., J. G. Lawless, and C. Ponnamperuma, Non-protein amino acids in the Murchison meteorite, *Proc. Nat. Acad. Sci. U.S.A.*, **68**, 486–490, 1971.

Kvenvolden, K., E. Peterson, and G. E. Pollock, Geochemistry of amino acid enantiomers: Gas chromatography of their diastereomeric derivatives, *Adv. Org. Geochem., Proc. Int. Meeting, 5th*, Hanover, Germany, Pergamon, Oxford-Braunschweig, pp. 387–401, 1972.

Lawless, J. G., Amino acids in the Murchison meteorite, *Geochim. Cosmochim. Acta*, **37**, 2207–2212, 1973.

Lawless, J. G., K. A. Kvenvolden, E. Peterson, C. Ponnamperuma, and C. Moore, Amino acids indigenous to the Murray meteorite, *Science*, **173**, 626–627, 1971.

Lawless, J. G. and E. Peterson, Amino acids in carbonaceous chondrites, *Origins of Life*, **6**, 38–43, 1975.

Mitterer, R. M., Ages and diagenetic temperatures of Pleistocene deposits of Florida based on isoleucine epimerization in Mercenaria, *Earth Planet. Sci. Lett.*, **28**, 275–282, 1975.

Noyes, H. P., W. A. Bonner, and J. A. Tomlin, On the origin of biological chirality via natural beta-decay, *Origins of Life*, **8**, 21–23, 1977.

Oró, J., J. Gilbert, H. Lichtenstein, S. Wikstrom, and D. A. Flory, Amino acids, aliphatic and aromatic hydrocarbons in the Murchison meteorite, *Nature*, **230**, 105–106, 1971a.

Oró, J., S. Nakaparksin, H. Lichtenstein, and E. Gil-Av, Configuration of amino acids in carbonaceous chondrites and a pre-Cambrian chert, *Nature*, **230**, 107–108, 1971b.

Pollock, G. E., C. N. Cheng, S. E. Cronin, and K. A. Kvenvolden, Stereoisomers of isovaline in the Murchison meteorite, *Geochim. Cosmochim. Acta*, **39**, 1571–1573, 1975.

Schroeder, R. A. and J. L. Bada, Glacial-postglacial temperature difference deduced from aspartic acid racemization in fossil bones, *Science*, **182**, 479–481, 1973.

Studier, M. H., R. Hayatsu, and E. Anders, Organic compounds in carbonaceous chondrites, *Science*, **149**, 1455–1459, 1965.

Tolbert, B. M., M. H. Krinks, J. A. Castrillan, L. E. Henderson, and M. B. Finch, Radiation chemistry of crystalline amino acids, unpublished manuscript, 1962.

Ulbricht, T. L. V., Asymmetry: The non-conservation of parity and optical activity, *Quart. Rev.*, **13**, 48–60, 1959.

Ulbricht, T. L. V. and F. Vester, Attempts to induce optical activity with polarized β-radiation, *Tetrahedron*, **18**, 629–637, 1962.

Vester, F., T. L. V. Ulbricht, and H. Krauch, Optical activity and parity violation in β-decay, *Naturwiss.*, **46**, 68, 1959.

Wehmiller, J. F. and P. E. Hare, Racemization of amino acids in marine sediments, *Science*, **173**, 907–911, 1971.

Wehmiller, J. F., P. E. Hare, and G. A. Kajula, Amino acids in fossil corals: Racemization (epimerization) reactions and their implications for diagenetic models and geochronological studies, *Geochim. Cosmochim. Acta*, **40**, 763–776, 1976.

Williams, K. M. and G. G. Smith, A critical evaluation of the application of amino acid racemization to geochronology and geothermometry, *Origins of Life*, **8**, 91–144, 1977.

# IV

# APPLICATIONS: GEOLOGICAL, ARCHEOLOGICAL, AND BIOLOGICAL

A long-standing goal of amino acid biogeochemistry is to use reaction rates in the geological environment as a tool for geochronology. Amino acids have a wide range of stabilities, and some of them have intrinsic reaction times that lie in the range of 50,000 to a few million years. It is just this critical range that is not well covered by other common geochronometric techniques. It is too old for radiocarbon methods and too young for potassium-argon dating. Other amino acid reactions appear suitable in a time frame of a few thousand years, a period that overlaps the radio carbon technique. A major advantage of amino acid dating lies in the extremely small sample required for an analysis; a property not shared by other methods.

The papers presented in this section illustrate two approaches to amino acid geochronology. The first one attempts to arrive at a quantitative estimate of time since the death of a living organism. The second uses amino acid reactions for relative dating, time correlation, and stratigraphy. The papers in this volume describe the difficult problems with absolute dating. However, good progress has been shown, especially in the use of amino acid racemization reactions. Relative dating and amino acid stratigraphy are on a firmer basis. A practical application is based on the fact that amino acid reaction rates depend on both time and temperature. In cases where precise independent measurements of time can be made, paleotemperatures can be estimated.

The practical applications of amino acid geochemistry presented in papers of this section show how the field has grown. Correlation of marine terrace deposits, estimates of rates of uplift, and screening of samples for radiocarbon dating are but a few examples of how the method has become a working tool in the earth sciences.

# Applications of Amino Acid Biogeochemistry for Marine Sediments

**KENNETH KING, JR.**
*Lamont-Doherty Geological Observatory of Columbia University*

*ABSTRACT*

Amino acid analysis of monospecific samples of fossil planktonic foraminifera is the most reliable approach currently available for applying amino acid biogeochemistry to biological and geological problems of marine sediments. The four key properties of preserved foraminiferal organic matrices and their corresponding applications are: (1) species-specific amino acid composition (taxonomy), (2) essentially constant amino acid composition over geologic time (phylogenetic relationships), (3) inverse relationship of amino acid concentration with carbonate dissolution intensity (carbonate dissolution index), and (4) species-specific racemization rates (geochronology). Increased research effort to characterize the various molecular species of amino acids generated by progressive diagenesis is needed to provide the mechanistic data for constructing a valid model of the racemization process.

## INTRODUCTION

Within two years of the discovery of preserved amino acids in fossils by Abelson (1954), Erdman, Marlett, and Hanson (1956) published a paper entitled "Survival of Amino Acids in Marine Sediments." They compared the amino acids content of a Recent shallow-water marine deposit with that of a similar sediment of Oligocene age. Although measurable amounts of amino acids were found in the Oligocene sample, their total concentration was less than that of the Recent sample by a factor of 6. The surviving acids were the same ones found in older fossils by Abelson.

In the intervening 22 years between that early paper and this conference, there has been a sustained interest in the amino acid content of marine sediments. The primary emphasis of the work was shifted noticeably, however, from distribution-type studies in the 1960s to applications-oriented investigations in the 1970s. Work on geochronological applications has expanded at a particularly rapid rate under the impetus of (1) the finding by Hare and Mitterer (1967) and Hare and Abelson (1968) that amino acid racemization reactions proceed over geologic time and (2) the suggestion by Hare and Mitterer (1969) that the resulting D/L ratios should be useful tools for age determinations, paleotemperature estimates and stratigraphic correlations.

The primary purpose of this paper, however, is not to trace the historical development of the application of amino acid techniques to marine sediments. Its main objective is to focus attention on one technique used at the Lamont-Doherty Geological Observatory for attacking a number of geological and biological problems associated with marine sediments. This method is based on the study of the proteinaceous skeletal matrices preserved in single species of deep-sea microfossils. Although suitable organic matrices have been found in all calcareous (King and Hare 1972a; King 1977) and siliceous (King 1974, 1977) microfossil groups analyzed, most of the applications work has involved the calcareous foraminifera (King and Hare 1972b).

Planktonic foraminifera are amoeboid, calcitic protozoans that occur in great abundance and diversity in deep-sea sediments. Their stratigraphic record extends back to early Cretaceous time (~130 m.y. B.P.) and contains worldwide species distribution with morphological diversity expressed in 40 or more genera and nearly 1000 species, most of which are extinct.

Three amino acid parameters were measured at the picomole level of sensitivity in the tests (skeletons) of fossil planktonic foraminifera: amino acid composition, amino acid concentration, and the degree of isoleucine epimerization (King and Neville 1977). These molecular data were then analyzed and interpreted in relation to selected problems in evolution, geochronology, and diagenesis.

## FORAMINIFERAL PROPERTIES AND APPLICATIONS

Four striking characteristics or properties of foraminiferal amino acid biogeochemistry have emerged from the applications work in the problem areas of evolution, geochronology, and diagenesis.

**1. Species-specific amino acid compositions are observed in foraminiferal tests.** Each species has a unique amino acid pattern that differs to a greater or lesser degree from compositions of all other species. In addition, this composition differs from others in a manner which parallels morphology and thereby reflects the genotype.

**2. Planktonic foraminifera exhibit remarkably constant amino acid composition over geologic time.** This property permits the use of amino acid composition as a chemotaxonomic character for tracing evolutionary lineages back through geologic time in the deep-sea fossil record.

**3. There is an inverse relationship of amino acid concentration with carbonate dissolution in fossil planktonic foraminifera.** This observation is the result of the routine use in downcore studies of the parameter amino acid concentration (expressed in terms of nanomoles amino acid / gram of skeletal material). This correlation is potentially useful for quantifying the intensity of deep-sea carbonate dissolution.

**4. There is a marked species dependency in racemization rates.** Racemization rates in different planktonic foraminifera may vary by more than a factor of 2. To measure rates with acceptable accuracy and precision, it is therefore essential to sample individual species for analysis rather than bulk sediment, or even mixed foraminiferal assemblages.

Since these four properties of the foraminiferal organic matrix are basic to the approach under discussion, the key data from which each was derived will be reviewed.

The first set of data (Figure 1) relates to the species-specific nature of foraminiferal amino acid composition. It is in the form of a plot generated by a *Q*-mode factor analysis of the amino acid compositions of 16 Recent planktonic foraminiferal species isolated from the tops of deep-sea sediment cores (King and Hare 1972a). The symbols represent groups of morphologically similar species according to a current taxonomic scheme. In addition to highlighting the unique nature of each species' composition, this three-factor solution clusters the species in a manner closely paralleling a widely accepted taxonomic scheme based on morphology. This is most obvious at the family level where foraminiferal tests have been classically separated into the spinose and nonspinose groups. The nonspinose species are tightly clustered by amino acid composition in the lower left vertex. At the species level, a detailed comparison of species pairs shows a remarkable agreement between the degree of morphological similarity and the degree of compositional relatedness. Thus the amino acid composition of the test appears to reflect the genotype.

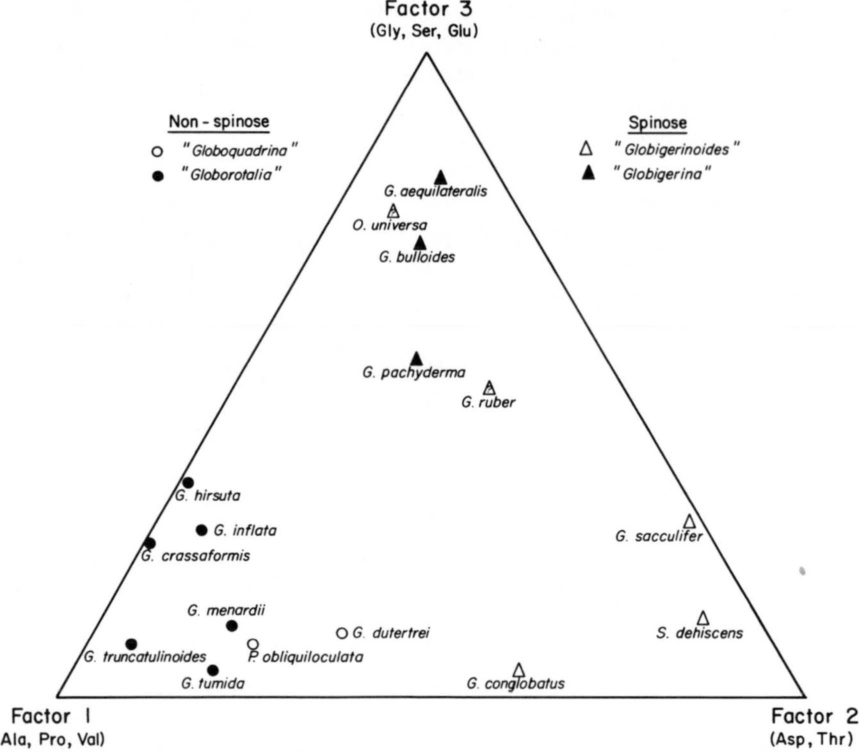


**Figure 1.** Classification of 16 Recent planktonic foraminiferal species by amino acid composition by means of *Q*-mode factor analysis. The symbols represent groups of morphologically similar species in a current taxonomic scheme.

The second important feature of the amino acid content in planktonic foraminifera is its constancy in composition through geologic time.

Figure 2 shows the amino acid compositions of two extinct Early Miocene foraminiferal species isolated from the same deep-sea core sample (King and Hare 1972a). Although of the same genus, *Globoquadrina,* these two species are characterized by significant morphological differences. Based on the parallelism found in Recent specimens, one would predict from the morphologies that the amino acid patterns should be similar but differ significantly in the more sensitive, more variable amino acids such as aspartic acid. The plot in Figure 2 confirms this prediction. Thus, diagenesis over geologic time does not obliterate significant differences in original amino acid composition. Subsequent down-core studies, which involved close-interval sampling of the same species over more recent time intervals (i.e., 0–700,000 yr and 1.7–2.0 m.y.), also indicate a con-

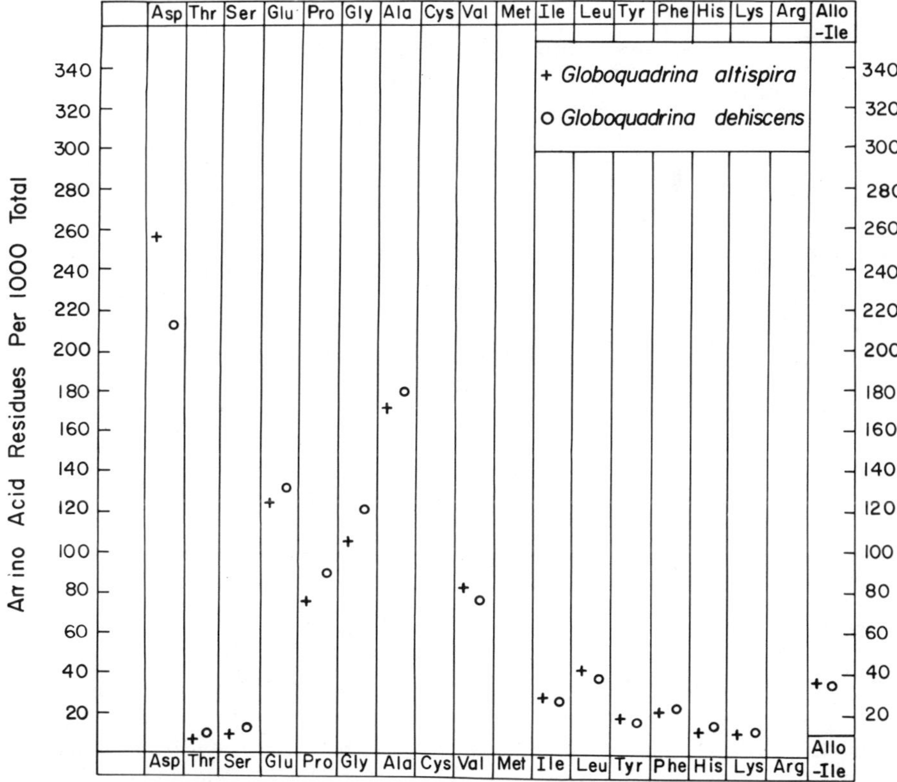


**Figure 2.** Comparison of the amino acid compositions of two morphologically distinct Early Miocene species of the same genus.

stancy of amino acid composition for a given species (King 1976, 1978). It is, of course, necessary to exclude the thermally unstable amino acids (e.g., threonine, serine, methionine) and their decomposition products from such compositional comparisons over significant periods of geologic time.

The next three sets of data (Figures 3,4,5) are related to the third property of the foraminiferal organic matrix — the inverse correlation of amino acid concentration with carbonate dissolution.

A plot of amino acid concentration in the extinct *Globigerinoides fistulosus* versus depth in deep-sea core V28-239 is shown in Figure 3 (King 1976). Percent coarse fraction in the same samples is also plotted. A low value of percent coarse fraction is indicative of intense carbonate dissolution. Four different measures of dissolution were used to identify the peaks and troughs of carbonate dissolution for sampling purposes: average

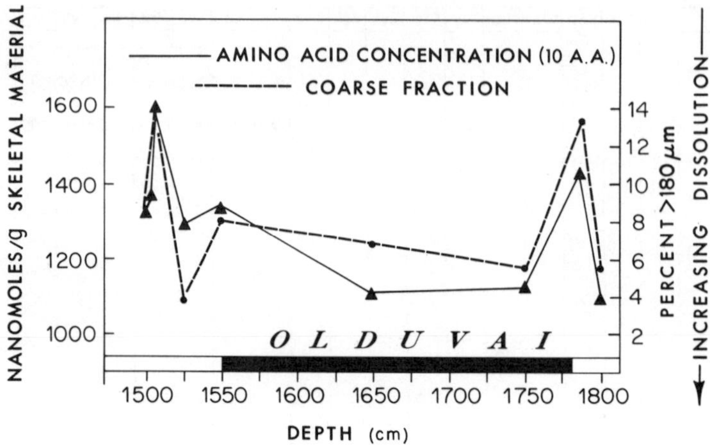


**Figure 3.** Plot of amino acid concentration in *Globigerinoides fistulosus* vs. depth in deep-sea core V28-239. Percent coarse fraction is also plotted. Percent coarse fraction decreases with increasing intensity of carbonate dissolution.

weight per foraminiferal test, percent coarse fraction, fragments per gram, and faunal abundance.

As shown in Figure 3, amino acid concentration does not decrease monotonically through diagenesis over the 300,000 years represented by this segment of the core. Instead, this parameter covaries with percent coarse fraction, which indicates an inverse correlation with intensity of dissolution. The maximum percent decrease in amino acid concentration in *G. fistulosus* over a single dissolution cycle was 40%.

A similar plot (Figure 4) of *G. fistulosus* amino acid concentration versus another dissolution measure, average weight per individual foraminiferal skeleton, shows the same inverse relationship between this amino acid parameter and intensity of carbonate dissolution.

The data shown in Figure 5 was generated by a systematic study of down-core diagenesis in the free and peptide-bound amino acid fractions of the foraminifer *Globorotalia tumida* isolated from deep-sea core V28-238. The concentration curve is for the total hydrolyzate (the sum of 10 most abundant amino acids) and is therefore the same parameter plotted in Figures 3 and 4. Percent coarse fraction is again plotted as a co-parameter. The time interval involved in this study, however, is from 0 to 700,000 years (or 0–1200 cm depth in core V28-238).

If the previously observed inverse relationship between amino acid concentration and carbonate dissolution holds for this core, one would expect the amino acid concentration to co-vary with percent coarse fraction. As shown in Figure 5, concentration parallels the coarse fraction

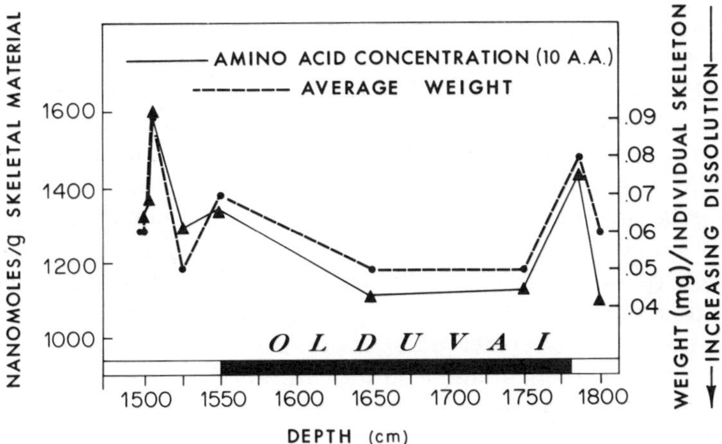


**Figure 4.** Plot of amino acid concentration in *Globigerinoides fistulosus* vs. average weight per individual foraminiferal skeleton.

curve from the top down to 500 cm and then again from 800 to 1200 cm. The two curves appear to be out of phase between the 500 and 800 cm levels. The reason for this is not clear. It is interesting to note, however, that significant anomalies have been observed during roughly the same time period in the paleoclimatic studies of other cores (P. R. Thompson, personal communication). These include unexpectedly high carbonate contents and a break-down in faunal transfer equations. Tentatively, therefore, carbonate dissolution is considered to be the controlling factor

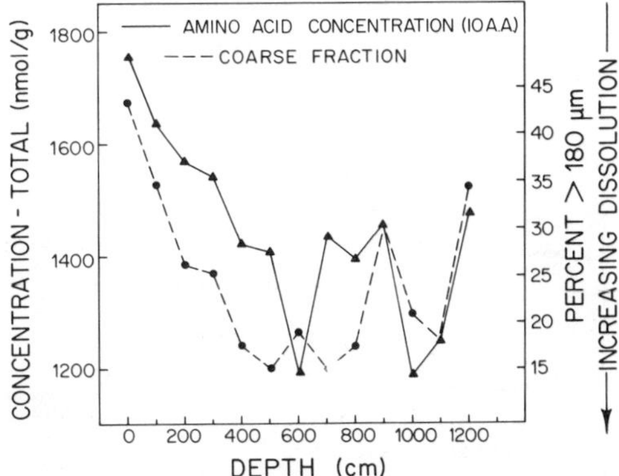


**Figure 5.** Plot of amino acid concentration in *Globorotalia tumida* vs. depth in deep-sea core V28-238. Percent coarse fraction is also plotted. Percent coarse fraction decreases with increasing intensity of carbonate dissolution.

in determining the observed fluctuations in amino acid concentration in this core also.

This apparent dissolution effect in down-core samples was corroborated by analyses (King 1976) of core-top samples of *Globigerinoides sacculifer* and *Globorotalia tumida* (Figure 6). These species were isolated from two pairs of cores (one from the Atlantic/Caribbean and the other from the Pacific) with each pair represented by one core of a minimum degree of dissolution (excellent preservation) and another of a maximum degree of dissolution (poor preservation). Water depth determined the selected preservation states, with the best preserved cores being taken at about 1700 m and the worst preserved at about 4500 m. *G. sacculifer* showed a maximum variation in amino acid concentration of 45% and *G. tumida* 50%, when samples representing extremes in dissolution intensity were compared.

What is the causal mechanism for this observed inverse relationship between amino acid concentration and dissolution intensity? A possible explanation is suggested by combining the results of two different techniques — amino acid analysis and transmission electron microscopy. The relevant amino acid data plotted in Figure 7 indicates that the total hydrolysate curve (previously correlated with dissolution intensity in Figure 5) is closely paralleled by the diagenetic curve of the bound fraction.

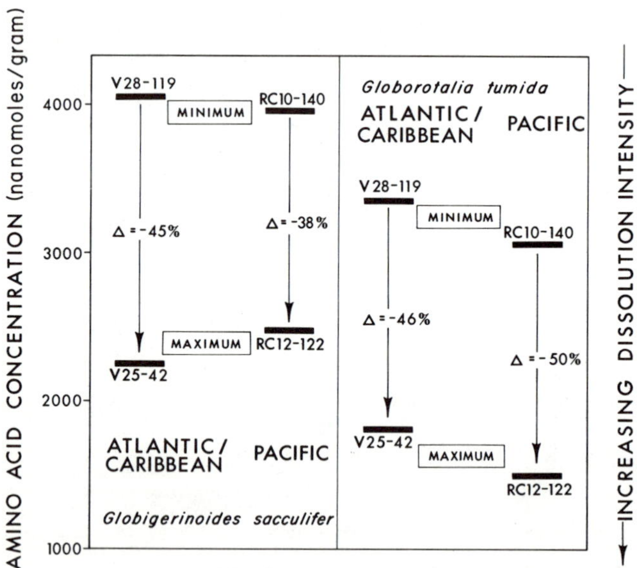


**Figure 6.** Core top study: amino acid concentration vs. dissolution intensity — maxima and minima (2 species, Atlantic/ Caribbean vs. Pacific).

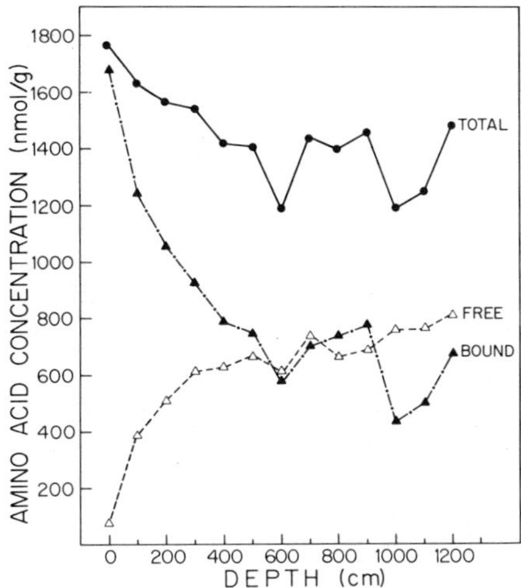


**Figure 7.** Plot of amino acid concentration in three fractions (total, free, bound) of *Globorotalia tumida* vs. depth in deep-sea core V28-238. Total curve is also shown in Fig. 5 plotted against coarse fraction (dissolution intensity).

Because the free fraction concentration is relatively stable below 300 cm, fluctuations in the total hydrolysate concentration appear to be primarily controlled by variations in bound fraction concentration.

Transmission electron micrographs (Figure 8) provide visual evidence of peptide-bound amino acids in foraminiferal tests as organic lamellae. Under environmental conditions of minimal dissolution intensity, these discrete organic layers would be well protected in fossil samples by the enclosing mineral phase and would account for a substantial fraction of the total amino acid concentration. However, more corrosive conditions associated with significant calcite dissolution would lower the absolute concentration of the bound fraction in the test (and thereby reduce the total hydrolysate concentration) by concomitant removal of the organic-rich lamellae. Although details of this process are not known, it is likely that dissolution would proceed preferentially along the lines of mechanical weakness caused by the organic lamellae intersecting the inorganic phase (K. M. Towe, personal communication).

These data suggest that the amino acid concentration in fossil planktonic foraminifera is a sensitive quantitative parameter for measuring the degree of carbonate dissolution in deep-sea sediments. This technique should be particularly useful in studying carbonate dissolution cycles in higher latitude cores having relatively solution-resistant, low diversity faunas which do not readily reflect changes in dissolution intensity.

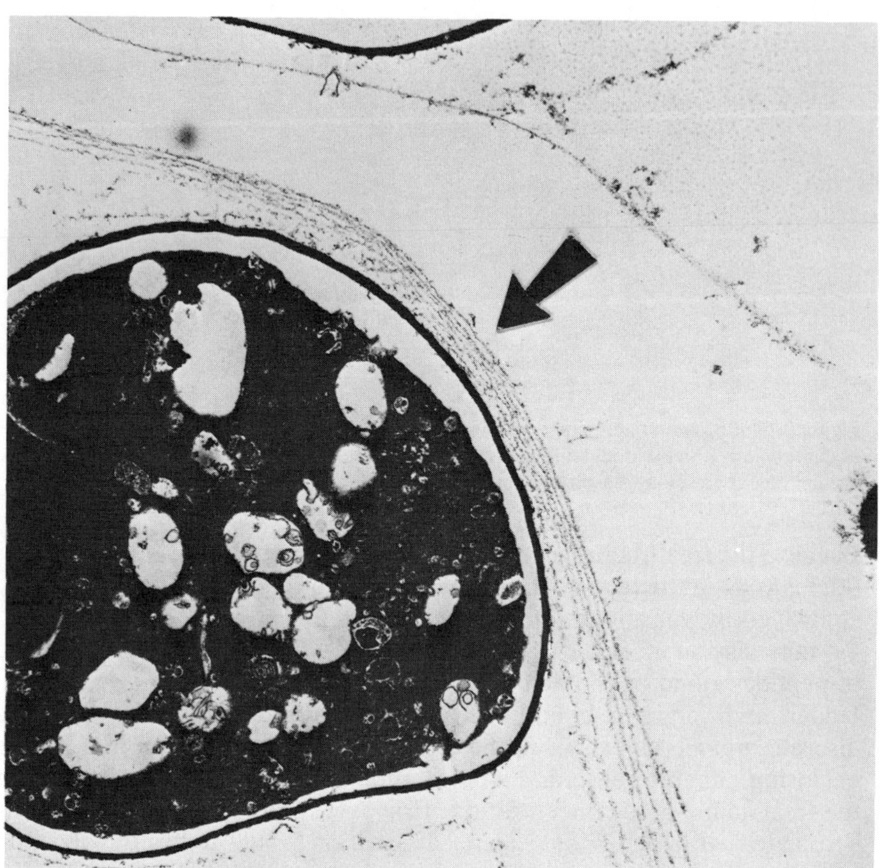

**Figure 8.** Transmission electron micrograph of live *Globorotalia menardii*. Arrow indicates protein-rich lamellae which may be removed by progressive dissolution. Photomicrograph by A. W. H. Bé times 12,500.

The next data shown in Figures 9 and 10 are related to the fourth major property of foraminiferal organic matrices, that is, epimerization/racemization rates are species-dependent.

Figure 9 summarizes the results of a study of the isoleucine epimerization reaction in individual foraminiferal species isolated from several deep-sea cores covering the time period of about 700,000 to 10 m.y. (King and Hare 1972c).

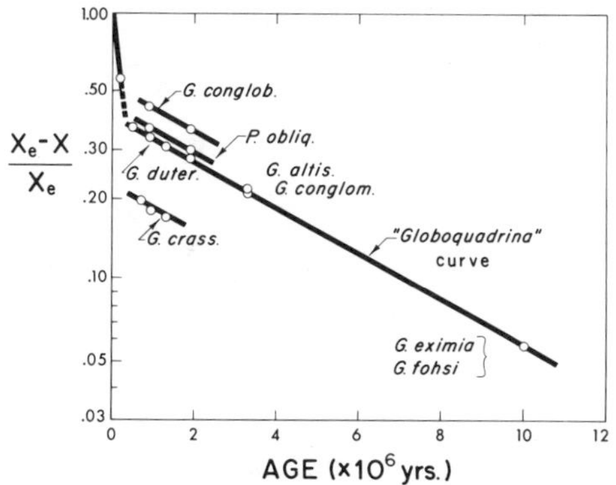


**Figure 9.** Isoleucine epimerization curves for selected fossil planktonic foraminifera isolated from deep-sea cores (700,000 yr to 10 m.y.) Species: G. conglob., *Globigerinoides conglobatus;* P. obliq., *Pulleniatina obliquiloculata;* G. duter., *Globoquadrina dutertrei;* G. crass., *Globorotalia crassaformis;* G. altis., *Globoquadrina altispira;* G. conglom., *Globoquadrina conglomerata;* G. eximia, *Globoquadrina eximia;* G. fohsi, *Globorotalia fohsi lobata.*

A marked species effect is noted in this semi-log plot of ($X_e$ minus $X$)/$X_e$ versus sample age. Each species exhibits a different degree of epimerization at a given time horizon, and four straight lines having essentially the same slope are generated over the time period of 0.7–1.86 m.y.

On the basis of analyses of four additional species of genus *Globoquadrina* isolated from the 3.3 m.y. and 10 m.y. levels of older cores, the *Globoquadrina* curve was extended back to the Middle Miocene with its linearity maintained.

Although only two measurements for one species were made in this study on samples younger than 700,000 yr, the positions of the species curves for the older samples dictate upper segments of accelerated rates for samples less than 700,000 yr in age. Because of the resulting high precision pontentially attainable for dating purposes, we initiated a detailed study of two common species over the time period 0–700,000 yr B.P.

In Figure 10 two isoleucine epimerization parameters [($X_e$ minus $X$)/$X_e$ and $A/I$] are plotted against depth in the equatorial Pacific deep-sea core V28-238 (King and Neville 1977). Samples of two species (*G. sacculifer* and *G. tumida* ) were taken at 100-cm intervals from the top of the core to the 1200 cm level, which was dated paleomagnetically at 700,000 yr B.P. As in Figure 9, the striking down-core species differences in epimerization rates clearly show why epimerization/racemization studies of marine sediments must be based on monospecific samples. The total assemblage of mixed foraminiferal species typically changes in faunal composition over time and is therefore an unsatisfactory sampling basis for

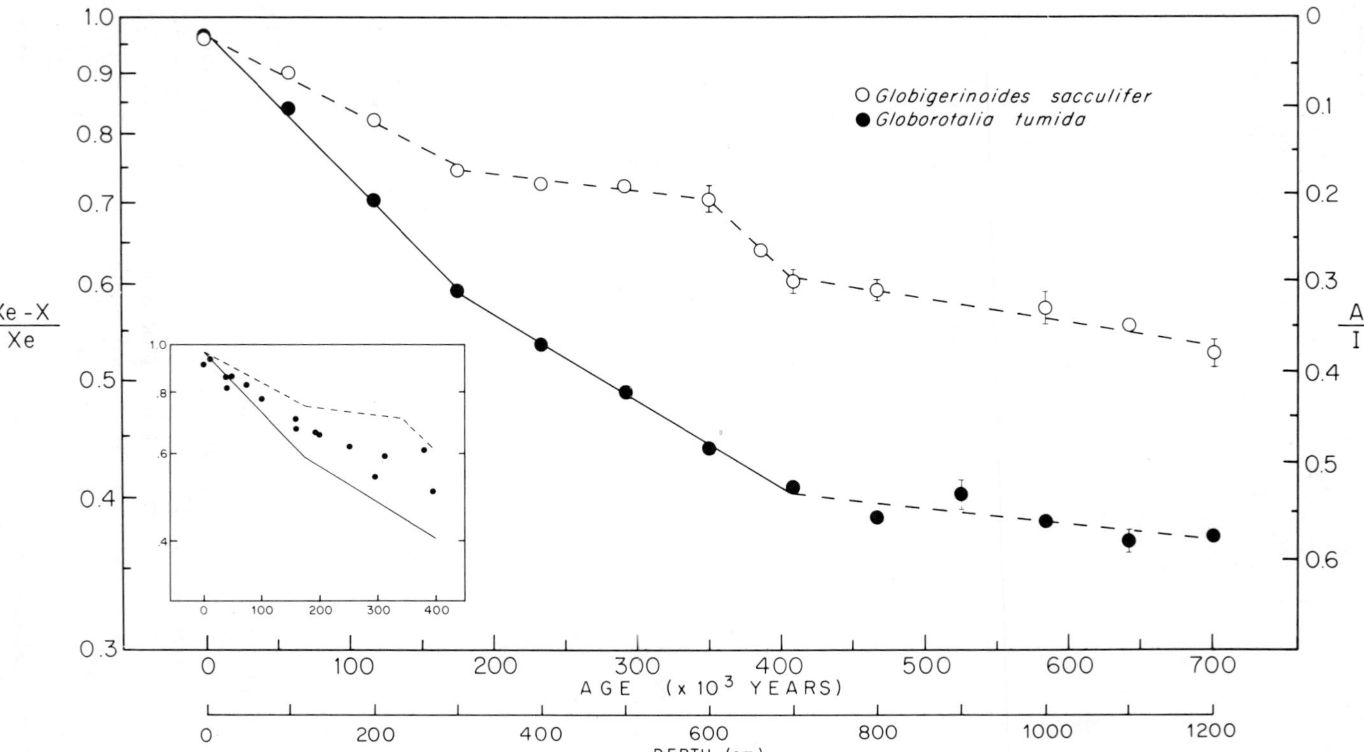


**Figure 10.** Isoleucine epimerization curves for two species of planktonic foraminifera isolated from V28-238 (0 – 700,000 yr). The solid lines indicate segments of potential value in dating marine sediments ~50,000 to 400,000 years old. Symbols (circles) enclose the standard deviations of duplicate analyses, except where noted by uncertainty bars. Each line segment was generated by a least-squares fit. The inset shows the mixed species data (0 to 400,000 years) of Wehmiller and Hare (1971) plotted with the species curves.

accurately measuring racemization rates. If intercore correlations now being studied are successful, the *G. tumida* curve for V28-238 could be used as an empirical calibration curve for dating marine sediments at other locations.

## SUMMARY AND CONCLUSIONS

Four important biogeochemical properties of the skeletal organic matrices within single species of planktonic foraminifera have been determined in a series of empirical studies: (1) Species-specific amino acid composition; (2) Essentially constant amino acid composition over geologic time (excluding thermally unstable acids); (3) Inverse relationship of species-specific amino acid concentration with carbonate dissolution intensity; and (4) Species-specific epimerization/racemization rates.

Considered individually, each property represents a quantitative molecular parameter that can be applied to biological and geological problems associated with marine sediments. For example, the properties of species-specific amino acid composition and time-constancy of amino acid composition can be utilized to answer biological questions of taxonomy and phylogenetic relationships, respectively, among both Recent and fossil foraminifera. Similarly, the parameters of amino acid concentration and racemization rates can be used to solve problems of deep-sea carbonate dissolution and geochronology, respectively.

Considered collectively, these four properties clearly establish the use of monospecific samples as the preferred approach for applying amino acid biogeochemical techniques to foraminifera-containing marine sediments. Further, the available amino acid data on other microfossil organic matrices (King 1977) strongly suggest that a monospecific approach is required for all biomineralized microfossils found in deep-sea sediments.

## FUTURE RESEARCH

Although the empirical studies described above have revealed major properties of the foraminiferal organic matrix and thereby provided a sound experimental approach to the amino acid biogeochemistry of marine sediments, the resulting data have raised a number of fundamental questions: (1) What are the sources of the observed variability in amino acid composition among foraminiferal species? (2) Why does each species exhibit a unique epimerization/racemization rate? (3) Why does the epimerization/racemization rate of a given species vary over geologic time?

It now appears highly probable that these questions will only be resolved by coupling a detailed characterization of the amino acid polymers (proteins and peptides) in the Recent foraminiferal skeletal matrix with a similar study of their diagenetic products generated over geologic time within fossil foraminifera. By isolating the various molecular species of amino acids from monospecific samples in progressive stages of downcore diagenesis and measuring their key parameters (including D/L ratios), major diagenetic pathways could be traced and the underlying mechanism of the associated racemization process could also be elucidated.

## *ACKNOWLEDGMENTS*

The author thanks P. E. Hare for helpful discussions. This research was supported by National Science Foundation Grants GA-34139, GA-35463 and OCE75-18136 and fellowships from the Geophysical Laboratory, Carnegie Institution of Washington. Contribution no. 2944 of Lamont-Doherty Geological Observatory of Columbia University.

## *REFERENCES*

Abelson, P. H., Organic constituents of fossils, *Carnegie Inst. Wash. Year Book 53*, 97–101, 1954.

Erdman, J. G., E. M. Marlett, and W. E. Hanson, Survival of amino acids in marine sediments, *Science,* **124,** 1026, 1956.

Hare, P. E. and P. H. Abelson, Racemization of amino acids in fossil shells, *Carnegie Inst. Wash. Year Book 66,* 526, 1968.

Hare, P. E. and R. M. Mitterer, Nonprotein amino acids in fossil shells, *Carnegie Inst. Wash. Year Book 65,* 362–364, 1967.

Hare, P. E. and R. M. Mitterer, Laboratory simulation of amino-acid diagenesis in fossils, *Carnegie Inst. Wash. Year Book 67,* 205–208, 1969.

King, K., Jr., Preserved amino acids from silicified protein in fossil Radiolaria, *Nature,* **252,** 690–692, 1974.

King, K., Jr., Amino acid concentration in planktonic foraminifera: A new approach for studying the dissolution of deep-sea carbonates, *Annual Meeting of the Geological Society of America,* **8,** 955, 1976.

King, K., Jr., Amino acid survey of recent calcareous and siliceous deep-sea microfossils, *Micropaleontology,* **23,** 180–193, 1977.

King, K., Jr., Amino acid diagenesis in a single species of planktonic foraminifera: Implications for dating marine sediments, *Annual Meeting of the Geological Society of America,* **10,** 435, 1978.

King, K., Jr. and P. E. Hare, Amino acid composition of planktonic foraminifera: A paleobiochemical approach to evolution, *Science,* **175,** 1461–1463, 1972a.

King, K., Jr. and P.E. Hare, Amino acid composition of the test as a taxonomic character for living and fossil planktonic foraminifera, *Micropaleontology,* **18,** 285–293, 1972b.

King, K., Jr. and P. E. Hare, Species effects in the epimerization of L-isoleucine in fossil planktonic foraminifera, *Carnegie Inst. Wash. Year Book 71,* 596–598, 1972c.

King, K., Jr. and C. Neville, Isoleucine epimerization for dating marine sediments: Importance of analyzing monospecific foraminiferal samples, *Science,* **195,** 1333–1335, 1977.

Wehmiller, J. and P. E. Hare, Racemization of amino acids in marine sediments, *Science,* **173,** 907, 1971.

# Amino Acid Dating of *Saxidomus giganteus* at Willapa Bay, Washington, by Racemization of Glutamic Acid

**KEITH A. KVENVOLDEN**
**DAVID J. BLUNT**
***U.S. Geological Survey, Menlo Park, California***

***ABSTRACT***

The extent of racemization of glutamic acid in shells of the mollusk *Saxidomus giganteus* (Deshayes) is used to estimate the ages of estuarine shell deposits at Willapa Bay, Washington. By applying first-order kinetics and making assumptions about diagenetic temperatures, two distinct ages of shell deposits, 110,000 ± 20,000 years and 190,000 ± 40,000 years old, can be identified. These ages correspond to interglacial periods having high stands of sea level.

Amino acid racemization (or epimerization) reactions have been successfully applied to correlating and dating Pleistocene marine shell deposits from the arctic (Miller and Hare 1975; Miller et al., 1977), eastern (Mitterer 1974, 1975; Wehmiller and Belknap 1978), and western (Wehmiller et al., 1977) coasts of North America. In that work attention was given mainly to either the epimerization of isoleucine or the racemization of leucine. Recently, Kvenvolden et al. (1979) measured the racemization of leucine, aspartic acid, and glutamic acid in an effort to correlate and date estuarine shell deposits at Willapa Bay, Washington. We now focus on glutamic acid to demonstrate how this amino acid can be applied to dating *Saxidomus giganteus* (Deshayes), one of several pelecypod species found in shell deposits exposed in cliffs along the east side of Willapa Bay (Addicott 1966; Mallory et al., 1977).

## SAMPLING AND ANALYSIS

Specimens of *Saxidomus* were collected as part of a larger study (Kvenvolden et al., 1979) at 15 localities between Goose Point and Lynn Point (Fig. 1). Articulated valves were selected wherever possible to avoid contamination from older, possibly reworked shells.

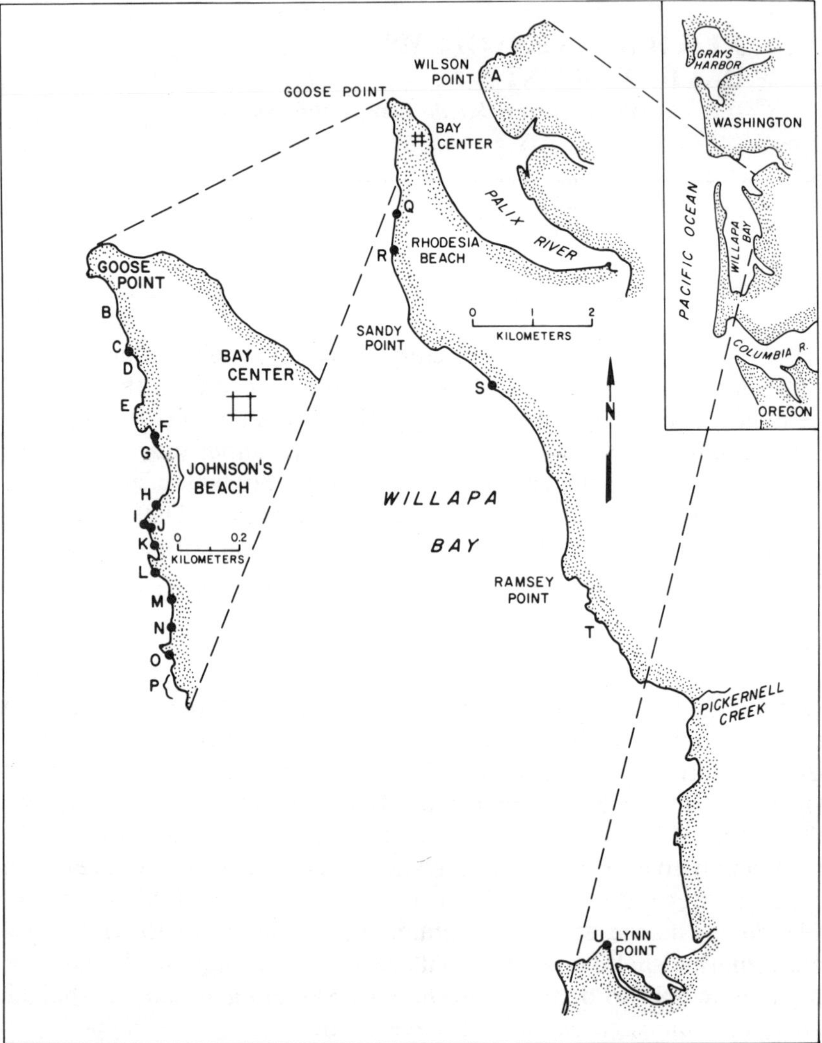


**Figure 1.** Location map showing 21 sample sites, lettered A through U (Kvenvolden et al., 1979), on the east side of Willapa Bay, Washington. *Saxidomus giganteus* is found at the 15 sites marked (●).

In the laboratory the shells were cleaned, and amino acids were recovered by acid hydrolysis and analyzed by gas chromatography. Details about sampling and analysis can be found in Kvenvolden et al. (1979). Adhering sediment was mechanically removed from the shells, which were then acid-etched. A sample weighing from 0.4 to 1 g was removed from the shell and placed in a tube where the sample was hydrolyzed in 6 *N* HCl under nitrogen for 20 hr at 110°C. The resulting mixture was desalted on a cation exchange resin. Derivatives of amino acids were prepared as N-pentafluoropropionyl-(+)-2-butyl esters, which were analyzed by gas chromatography on two different capillary columns, one coated with UCON 75 H 90,000 and the other with Carbowax 20 M. Quantitation of the extent of racemization (D/L ratios of the enantiomers of glutamic acid) was obtained by measuring peak heights on gas chromatograms. No corrections were applied to the data.

## RESULTS AND DISCUSSION

The extent of racemization (D/L ratios) of glutamic acid in the total hydrolysate from 33 samples (29 individual shells and 4 duplicates) of *Saxidomus* collected at 15 sites (Fig. 1) was measured. Table 1 records the

**Table 1. D/L Ratios of Glutamic Acid in Total Hydrolysates of *Saxidomus giganteus* at 15 Sites (Fig. 1)**

| Site Designation | Number of Samples | Number of Analyses | D/L Glutamic Acid[a] |
|---|---|---|---|
| C | 3 | 5 | 0.40 ± 0.01 |
| F | 2 | 4 | .27 ± .01 |
| H | 4 | 7 | .22 ± .01 |
| I | 1 | 2 | .22 ± .01 |
| J | 1 | 2 | .25 ± .01 |
| K | 1 | 2 | .25 ± .01 |
| L | 2 | 4 | .23 ± .03 |
| M | 1 | 2 | .26 ± .01 |
| N | 1 | 2 | .24 ± .01 |
| O | 2 | 4 | .25 ± 0.3 |
| P | 5 | 9 | .24 ± .01 |
| Q | 2 | 4 | .21 ± .01 |
| R | 3 | 6 | .26 ± .02 |
| S | 1 | 2 | .24 ± .01 |
| U | 4 | 8 | .37 ± .02 |

[a] Mean ± standard deviation. Where the mean is calculated from two D/L ratios a range of values is given rather than a standard deviation.

mean D/L ratios of glutamic acid from shells at each of the 15 sites. The shell bed exposed at the nine sites between H and P is a single, continuous stratigraphic entity, and it is assumed that at least the articulated shells in this unit must be virtually the same geologic age. The mean and standard deviations of the D/L ratios of glutamic acid from these nine sites are 0.24 ± 0.01, and this information is incorporated into Figure 2, where the mean D/L ratios from all sites are plotted. This figure provides a way of comparing the average D/L glutamic acid ratios from *Saxidomus* from sites H through P with ratios from isolated deposits at sites C, F, Q, R, S, and U. Mean D/L ratios for *Saxidomus* at four sites (F, Q, R, and S) fall within three standard deviations of the mean ratio from sites H through P. Application of the t test shows that as a group the D/L ratios from the same four sites are statistically within the 95% confidence interval of the ratios from sites H through P. The D/L ratios from sites C and U lie twelve and nine standard deviations, respectively, away from the mean D/L ratio from sites H through P. It is likely that the shell deposits at sites F, Q, R, and S correlate with and are approximately the same age as those at sites H through P and that at sites C and U the shell deposits are older.

Several assumptions are required in order to determine the ages of the younger and older shell deposits by the extent of racemization of glutamic acid:

1. Linear first-order kinetics describe the racemization of glutamic acid, and the following expression, derived by Bada and Schroeder (1972),

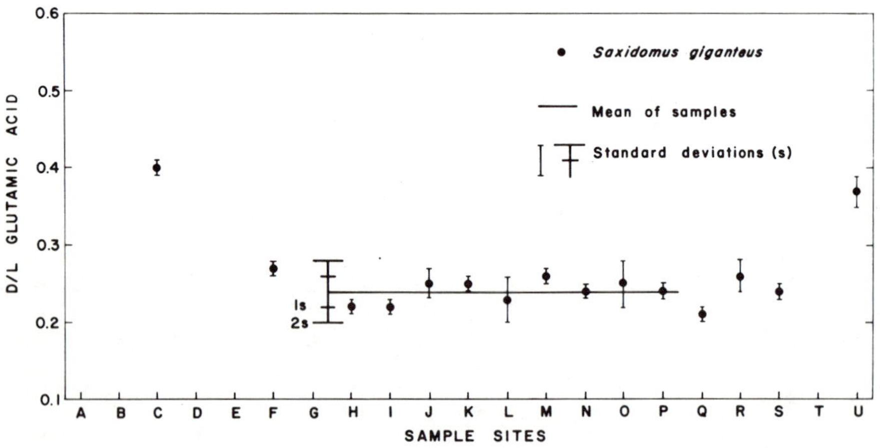


**Figure 2.** D/L ratios of glutamic acid in *Saxidomus giganteus* at 15 sampling sites noted on Figure 1. *Saxidomus* was not found at the sites labeled A, B, D, E, F, and T.

adequately relates D/L ratios, rate constants and time:

$$\ln\left[\frac{1+\frac{D}{L}}{1-\frac{D}{L}}\right]_t - \ln\left[\frac{1+\frac{D}{L}}{1-\frac{D}{L}}\right]_{t=0} = 2\,k\,t$$

where k = the rate constant for the racemization of glutamic acid, t = time, and the D/L ratio of glutamic acid at t = 0 is 0.03.

2. The diagenetic temperature in the area of Willapa Bay is about 7°C, the difference between the present, long-term mean annual air temperature of 10.7°C and an average temperature reduction of about 3.6°C based on considerations of the palynological record of the past 47,000 years by Heusser (1977).
3. The independently determined ages of 68,000 to 100,000 years for the lowest terrace at Santa Cruz, California (Bradley and Addicott 1968; Szabo and Vedder 1971), the D/L ratio for glutamic acid of 0.34 from *Saxidomus* at this terrace (Wehmiller et al., 1977), and the assumed diagenetic temperature of 10.7°C for this area provide adequate information to permit this locality to serve as a calibration site for determining age by amino acid racemization as proposed by Bada and Protsch (1973) and Bada et al. (1974).
4. The Arrhenius equation

$$\ln k = \ln A - \frac{E_a}{RT},$$

where A is the pre-exponential factor, $E_a$ is the activation energy, R is the gas constant, and T is the absolute temperature, adequately relates the rate constant k and the temperature. A further assumption is that Ea for the calculation is 29.4 kcal/mole. This activation energy was determined by Mitterer (1975) for the epimerization of isoleucine in the mollusk *Mercenaria*. Bada et al. (1973) showed that activation energies for the epimerization of isoleucine and the racemization of other amino acids are essentially the same. Therefore, an activation energy of 29.4 kcal/mole for the racemization of glutamic acid in *Saxidomus* seems reasonable.

From these assumptions, and using D/L glutamic acid ratios of 0.24 and 0.38, the ages of *Saxidomus* in the younger and older shell deposits at Willapa Bay can be estimated. Shells in the younger deposits exposed at sites F, H through P, Q, R, and S are calculated to be about 110,000

± 20,000 years old, and those at sites C and U are about 190,000 ± 40,000 years old. (Ranges given include all ages calculated using both calibration ages from assumption 3.) In the more detailed study involving the racemization of leucine, aspartic acid, and glutamic acid, Kvenvolden et al. (1979) estimate the age of the younger deposit to be about 120,000 ± 40,000 years old and the other deposit to be 190,000 ± 40,000 years old.

The glutamic acid age estimates are consistent with observations made by Mallory et al. (1977), who suggested that these shell deposits are older than 57,000 years based on radiocarbon-dated wood and older than 100,000 years based on their evaluation of local marine terraces. Our estimated ages fit to some extent with the oxygen isotope stages 5 (75,000–128,000 years) and 7 (195,000–251,000 years) of the marine isotope record of Shackleton and Opdyke (1973). Both of these stages represent periods of time when sea level was high. Therefore, it is possible that the estuarine shell beds studied here were deposited during two distinct interglacial periods. Finally, our calculated ages are also consistent with the general geologic framework of the area where the older age measurements are derived from shells in stratigraphic units which underlie the unit containing the younger shell deposits. The geologic relationships of the estuarine and nonmarine deposits suggest that the shells were deposited at two different times when sea level was high; nonmarine channel and channel-fill deposits suggest that a lowering of sea level occurred between the times of deposition of the units containing the shell deposits.

## REFERENCES

Addicott, W. O., Late Pleistocene marine paleontology and zoogeography in Central California, *U.S. Geol. Survey Prof. Paper 523-C,* C1–C21, 1966.

Bada, J. L. and R. Protsch, Racemization reaction of aspartic acid and its use in dating fossil bones, *Proc. Nat. Acad. Sci. U.S.A.,* **70,** 1331–1334, 1973.

Bada, J. L. and R. A. Schroeder, Racemization of isoleucine in calcareous marine sediments: Kinetics and mechanisms, *Earth Planet. Sci. Lett.,* **15,** 1–11, 1972.

Bada, J. L., K. A. Kvenvolden, and E. Peterson, Racemization of amino acids in bones, *Nature,* **245,** 308–310, 1973.

Bada, J. L., R. A. Schroeder, and G. F. Carter, New evidence for the antiquity of man in North America deduced from aspartic acid racemization, *Science,* **184,** 791–793, 1974.

Bradley, W. C. and W. O. Addicott, Age of the first marine terrace near Santa Cruz, California, *Geol. Soc. Amer. Bull.,* **79,** 1203–1210, 1968.

Heusser, C. J., Quaternary palynology of the Pacific slope of Washington, *Quaternary Research,* **8,** 282–306, 1977.

Kvenvolden, K. A., D. J. Blunt, and H. E. Clifton, Amino-acid racemization in Quaternary shell deposits at Willapa Bay, Washington, *Geochim. Cosmochim. Acta,* **43,** 1505–1520, 1979.

Mallory, V. S. et al., Pre-Wisconsin biostratigraphy, paleoecology, and sedimentation at Bay Center, Southwestern Washington, *Geol. Soc. America, Abst. with Programs,* **9,** 459–460, 1977.

Miller, G. H. and P. E. Hare, Use of amino acid reactions in some arctic marine fossils as stratigraphic and geochronological indicators, *Carnegie Inst. Wash. Year Book 74,* 612–617, 1975.

Miller, G. H., J. T. Andrews, and S. K. Short, The last interglacial-glacial cycle, Clyde Foreland, Baffin Island, N.W.T.: Stratigraphy, biostratigraphy, and chronology, *Can. J. Earth Sci.,* **14,** 2824–2857, 1977.

Mitterer, R. M., Pleistocene stratigraphy in southern Florida based on amino acid diagenesis in fossil *Mercenaria, Geology,* **2,** 425–428, 1974.

Mitterer, R. M., Ages and diagenetic temperatures of Pleistocene deposits of Florida based on isoleucine epimerization in *Mercenaria, Earth Planet. Sci. Lett.,* **28,** 275–282, 1975.

Shackleton, N. J. and N. O. Opdyke, Oxygen isotope and paleomagnetic stratigraphy of equatorial Pacific core V 28-238: Oxygen isotope temperature and ice volumes on a $10^5$ year and $10^6$ year scale, *Quaternary Research,* **3,** 39–55, 1973.

Szabo, B. J. and J. G. Vedder, Uranium series dating of some Pleistocene marine deposits in southern California, *Earth Planet. Sci. Lett.,* **11,** 283–290, 1971.

Wehmiller, J. F. and D. F. Belknap, Alternate kinetic models for the interpretation of amino acid enantiomeric ratios in Pleistocene mollusks: Examples from California, Washington, and Florida, *Quaternary Research,* **9,** 330–348, 1978.

Wehmiller, J. F. et al., Correlation and chronology of Pacific coast marine terrace deposits of continental United States by fossil amino acid stereochemistry-technique evaluation, relative ages, kinetic model ages, and geologic implications, *U.S. Geol. Survey Open-file Report 77-680,* 197 pp., 1977.

# Amino Acid Racemization in Quaternary Mollusks: Examples from Delaware, Maryland, and Virginia

DANIEL F. BELKNAP
JOHN F. WEHMILLER
*Department of Geology, University of Delaware*

## INTRODUCTION

In this paper we present examples of application of amino acid racemization to dating of Quaternary units in the Delmarva Peninsula and the shores of Chesapeake Bay. The discussion is limited to the quahog *Mercenaria,* although 16 other genera have been analyzed. A discussion of mollusks other than *Mercenaria* and generic effects on racemization is to be found in Belknap (1979) and will be covered in future papers. In addition, only those locations and samples described and collected by the authors in the course of Belknap's dissertation (Belknap 1979) are discussed.

Amino acid racemization methods are particularly well suited to relative chronological problems such as these, where isolated fossiliferous outcrops (or cores) are found within a localized temperature region. The amino acid age estimates (either ralative or kinetic model ages) presented here suggest an unexpected complexity in the chronology of Quaternary marine deposition in the Delmarva-Chesapeake Bay region. This chronology can be interpreted in terms of the marine isotopic record of ice-volume (sea-level) fluctuations (Shackleton and Opdyke 1973) and Holocene examples of coastal environmental lithosomes.

## METHODS

Samples of *Mercenaria* and other fossils were collected during 1975–1978 from Quaternary and Tertiary sediments in the study area. Articulated life-position valves were selected wherever possible. Collection was from excavated outcrop to avoid possible solar heating effects in exposure, or in a few cases from cores. Figure 1 is a map of sample localities. Circumscribed dots are those discussed herein. Information including detailed location, lithologic desciptions, measured sections and samples taken is included in Belknap (1979).

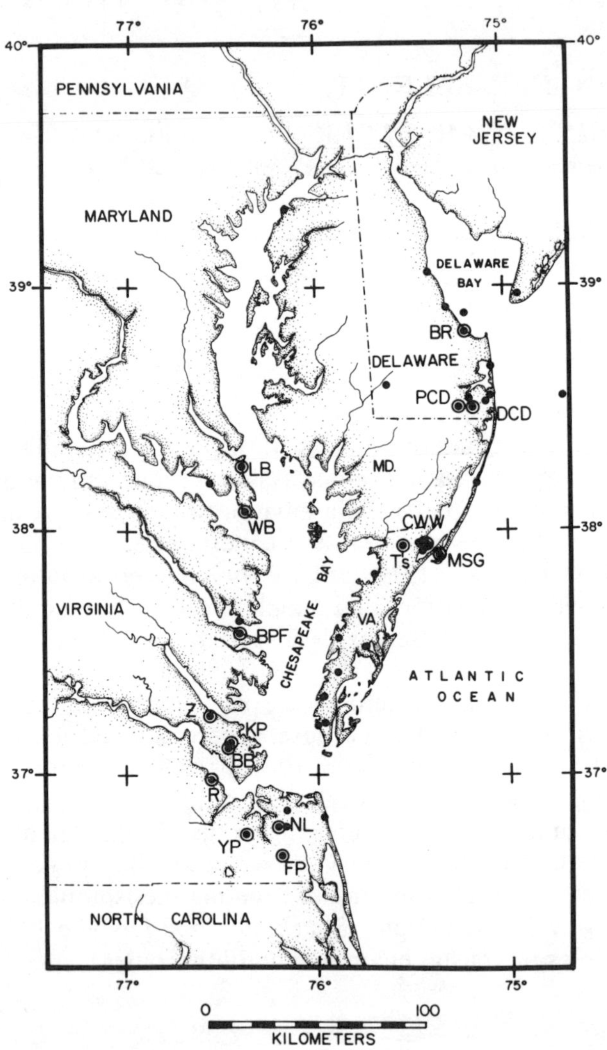


**Figure 1.** Sample location map, Delmarva Peninsula–Chesapeake Bay region. Circumscribed dots are localities discussed in the text. Other localities are described in Belknap (1979). All are outcrops in bluffs or excavations, with the exceptions of MSG and CWW, which are drill holes.

Sample preparation in the laboratory was carried out under very clean conditions. Gloves were worn at all times. In the first step, samples were cleaned of sediments and then stripped of outer shell layers with a dental saw. Only the inner layer of the three-layer *Mercenaria* shell from near the hinge was used to avoid variations in amino acid ratios between layers (Hare 1969; Hare and Mitterer 1969). One-gram subsamples were cut for amino acid extraction, and identical, adjacent subsamples were taken for X-ray analysis of aragonite/calcite ratio. No X-ray analysis showed more than 1% calcite. The fragments were cleaned with dilute, ultraclean HCl and ultraclean (4 times) distilled water. All reagents used in the extraction and desalting steps were prepared from pure gases (HCl and $NH_3$) bubbled into 4 times distilled water. All glassware was acid-cleaned and stored at 300°C for at least 24 hours before use. After sample cleaning, the fragments were dissolved in the stoichiometric amount of concentrated HCl to make the final solution 6 *N*, then hydrolyzed for 22 hours at 110°C, in sealed tubes with an $N_2$ atmosphere. Samples were desalted on cation exchange resin. They were then esterified in acidified (+) 2 butanol and acylated with a mixture of trifluoroaceticanhydride and dichloromethane, resulting in N-trifluoroacetyl (+) 2 butyl ester derivatives of the contained amino acids. (Kvenvolden et al., 1972).

After preparation samples were analyzed by capillary column gas chromatography, using stainless steel columns 45 m × 0.5 and 0.8 mm coated with the phases OV225 and K20M. Typical results are displayed in Figure 2. Amino acid ratios were determined by measuring peak heights. The three chromatograms demonstrate differences in samples ranging from living (DFB-3) to late Pleistocene (DFB-132, locality WB Wailes Bluff, Maryland) to Pliocene (?) (DFB-131, locality YP, Yadkin, Virginia, Yorktown Formation). Increase of D/L ratio with age is evident, as are changes in relative proportions of amino acids.

Data were plotted by D/Lratio for eight amino acids. Wehmiller et al. (1977) and Belknap (1979) have used D/L leucine ratios as primary referents. In addition, the "average extent of racemization" (the mean of D/L alanine, valine, leucine, proline, phenylalanine, and glutamic acid) can be used to represent relative data in a simplified format. Belknap (1979, Figure D-12) has shown through regression analysis that the mean values of these six amino acids closely follow trends for LEU in *Mercenaria*, by the equation:

$$\text{D/L mean} = 0.045 + 0.99\ \text{D/L LEU}, \quad r = 0.99, \quad n = 75$$

The mean value is used for relative relationships in this discussion; plots of individual amino acids are shown in Belknap (1979, Figures 14 and D1-D6).

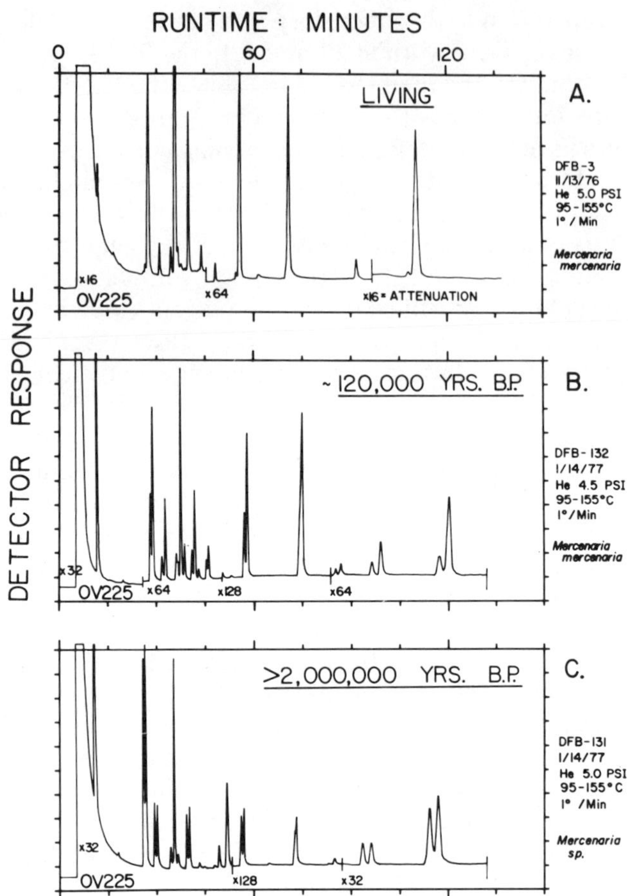


**Figure 2.** Sample chromatograms of amino acid extracts from *Mercenaria* specimens from the study area. Elution order starting at ~20 min: D,L-ALA, D,L-VAL, (THR,ALEU,GLY,ILEU), D,L-LEU, D,L-SER, D,L-PRO, D+L-ASP, D,L-PHE, D,L-GLU. From Belknap (1979, Figure 11).

Since amino acid racemization is temperature dependent, the thermal history of a deposit must be known in order to allow kinetic calculations. Models of temperature change were based on the method of Shackleton and Opdyke (1973), CLIMAP (1976) models of full glacial temperature reductions, and NOAA data of present mean annual temperatures. These calculations are discussed in detail by Wehmiller and Belknap (1978) and Belknap (1979).

Using D/L ratios and assumed temperature history, age estimates were made based on kinetic models of leucine racemization presented by Wehmiller and Belknap (1978) and modified by Belknap (1979). In this discussion emphasis is placed on correlation by extent of racemization (mean D/L) rather than age estimates, since these estimates may be refined by better models of kinetics, but ratios of amino acid enantiomers are immutable data points.

## DISCUSSION

### Analysis of Data Precision and Reliability

To use amino acid racemization values as a correlation tool and in estimating ages, limits of reliability must be understood. Figure 3 demonstrates that amino acid D/L ratios generally increase with increasing age

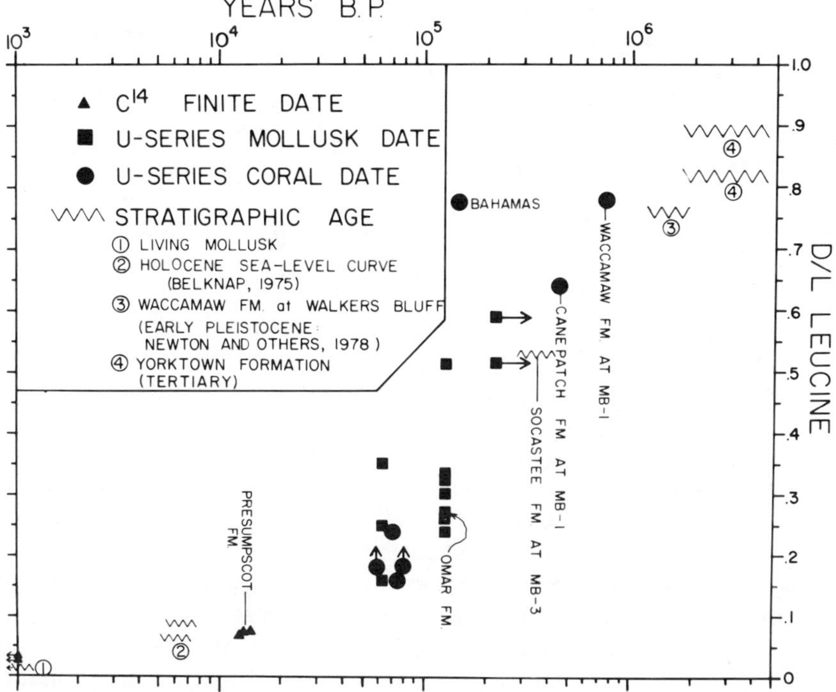


**Figure 3.** Plot of D/L leucine values versus independently determined age, showing increase of D/L ratio with increasing age. The same trend is evident for D/L mean data. Not corrected for temperature or generic effects. All are *Mercenaria* with the exception of those labeled (2), (3), Presumpscot Fm., and the Bahamas sample. From Belknap (1979, Figure 2).

until reaching an equilibrium mix. The temperature effect is evident in the large difference in D/L LEU ratio of a sample from the Bahamas and those from the Omar Formation in Delaware.

Multiple analyses of the same shell and multiple samples from the same outcrop yield estimates of uncertainty. Repeats on the same shell and repeats of chromatograms from the same preparation indicate an overall analytical uncertainty of ± 3% (Wehmiller et al. 1977; Belknap 1979). Multiple samples from within a single bed occasionally show greater variation. Probable causes include finite duration of depositional events in a bed, possible reworking from older beds, and variations in diagenesis (primarily leaching).

Precision of kinetic model ages is stated to within approximately ± 20%, which corresponds to ± 1°C in the model.

## Results

It has been shown in Figure 3 that in general D/L ratios increase with increasing age. A living *Mercenaria* (Figure 2a) yielded a D/L mean ratio of 0.01, a late Pleistocene *Mercenaria* (Figure 2b) yielded a ratio of 0.39, and that of a Tertiary *Mercenaria* (Figure 2c) was 0.90, nearly at equilibrium. Figure 4 presents data for sampled mid-Atlantic coast localities: extent of racemization versus present mean annual temperature. Multiple analyses are represented by mean and standard deviation as well as the number of analyses. There are three major groups of ratios: Holocene, Pleistocene, and Tertiary. The histogram demonstrates that the Pleistocene enantiomeric ratios can be further subdivided into perhaps as many as five groups. Groupings of relative ages are emphasized by the shading. The slope of the regions is due to temperature effects on racemization. Similar graphical analysis of individual amino acid data yields the same relative age groupings as seen for the mean D/L value (Belknap 1979).

Based on a U-series dated coral at NL (70,000 yr B.P., B. J. Szabo, personal communication) and the leucine kinetic model age estimates, the histogram is correlated to the oxygen isotopic record of ice-volume (sea-level) changes of Shackleton and Opdyke (1973). The gap between Holocene and stage 5 samples is due to non-deposition during Wisconsin time over most of this region. Possible early stage 3 materials are represented elsewhere in the study area (Belknap 1979). The relative width of this gap is due to initial rapid reaction rate of racemization kinetics ("nonlinear kinetics" — Wehmiller and Belknap 1978). Later groupings tend to "crowd together" due to slower reaction rates. Stage 5 *may* be divisible into three high sea-level substages (a,c,e), as is shown in Figure 4, although a division into early and late groups may be more defensible.

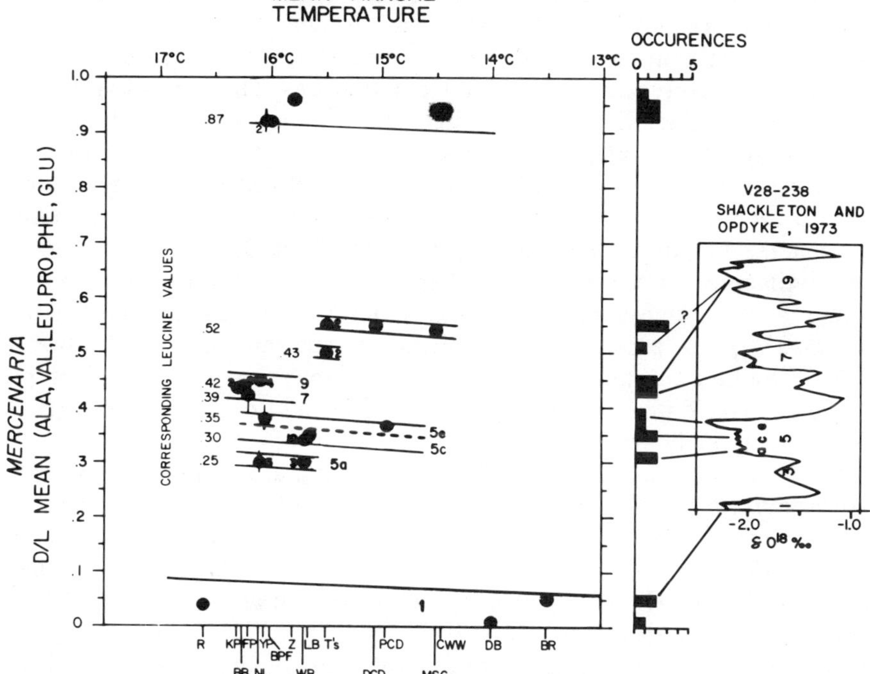


**Figure 4.** Extent of racemization, represented as the mean D/L ratio of six amino acids, versus present mean annual temperature. Histogram is number of occurrences of a ratio, multiple analyses at a single locality counting as one, interval 0.02. Correlation is with part of Shackleton and Opdyke (1973) oxygen-isotopic record for V28-238. Dots represent mean standard deviation shown by vertical lines, number of analyses shown.

Stages 7 and 9 are not definitely separable by this method. Older stages are also represented in the Pleistocene record. The large gap from D/L mean of 0.6 to 0.8 appears to be a consequence of a lack of preservation of material or lack of sampling of shells of this age. Only a few samples of this apparent age have been sampled in an extensive study of Atlantic coastal plain Quaternary marine sediments (Belknap 1979), and the deficiency of data in this region is interpreted as a real gap in the geologic record.

## Kinetic Model Age Assignments

Based on leucine kinetic models developed by Wehmiller and Belknap (1978) and modified by Belknap (1979), kinetic model ages can be assigned to the groups. For purposes of comparison, equivalent leucine ratios are included in Figure 4. Effective temperatures, which integrate full-glacial temperature reductions with short periods near or above present, as shown by the isotopic record, have a somewhat steeper gradient across the region than present mean annual temperatures. They are also different for each interglacial. On the average, $T_{eff}$ for the right side of Figure 4 range near 8.5°C, while the left side has $T_{eff}$ of 11.5°C (Belknap 1979)

Group one is composed of Holocene shells, with effective temperatures near present mean annual temperature, all less than 5000 years old based on leucine kinetics. The second group, at mean D/L ratios of 0.30, is approximately 70,000–90,000 yr in age, and may correlate with isotopic stage 5a. This age is supported by a U-series date on coral at NL of 70,000 yr (B. J. Szabo, personal communication). The third group, at ratios of 0.35, may be approximately 110,000 yr and may correlate with isotopic stage 5c. This date is supported by a U-series mollusk date at WB (B. J. Szabo, personal communication). The fourth group, at mean D/L of 0.37, represented by *Mercenaria* at PCD and YP and also supported by *Crassostrea virginica* data (Belknap 1979, not shown here) at PCD and DCD, may represent ages 120,000–140,000 B.P., stage 5e. The next grouping *may* be divisible into two, at 0.42 and 0.45, leucine kinetic model ages of 250,000 and 300,000–350,000, respectively. An age of 500,000 B.P. is proposed for the group at 0.50 D/L mean, with a $T_{eff}$ of 9.0°C. Based on leucine ratios alone it would appear difficult to separate the T's and NL samples, but with $T_{eff}$ of 9.0 and 11.5°C, respectively, the kinetic model ages are significantly different. The grouping of D/L mean ratios near 0.56 is assigned a leucine model age of 0.7–1.0 million years. The highest ratios, 0.8 and above, are calculated to be greater than 2 million years in age.

## Stratigraphic Applications

Racemization data can be applied to specific stratigraphic problems in the Delmarva Peninsula–Chesapeake Bay region. Geologic conditions here have been greatly affected by Quaternary sea-level fluctuations. Large amounts of coarse-grained sediments of uncertain age — the Columbia Formation of Delaware (Pleistocene: Jordan 1964, 1974) and the Upland Gravels of Maryland (Schlee 1957) were deposited on Tertiary and older marine sediments in the northern part of the region. These deposits are found in the central and southern parts of the region. During lowered sea-levels major fluvial systems such as the Susquehanna and Delaware, as well as lesser rivers, cut major valleys, eroding the area. Unraveling the stratigraphy of depositional and erosional events in the area has proceeded by lithostratigraphy, biostratigraphy, and recently more accurate chronostratigraphy based on "absolute" dating. Lithostratigraphy is not the optimal single approach for unraveling the sequence of events in a regional sense, since lithosomes are essentially identical in character from each succeeding high sea-level event. Relative stability of the region has resulted in complex juxtaposition of deposits of similar genesis of various ages (Belknap 1979). Biostratigraphy is hampered by rates of evolution which are slow relative to the rapid succession of glacial-interglacial cycles.

Oaks et al. (1974) have summarized extensive work in the Norfolk, Virginia, area. They recognize several formations, including the widespread Norfolk and Kempsville Formations, and several facies of each, especially the Norfolk Formation. They believe these deposits were laid down in a single late Pleistocene transgression. Figure 4 shows amino acid ratios at FP (Fentress, Va.), NL(New Light, Va.), and YP (Yadkin, Va.) borrow pits in the Norfolk area which contain these deposits. More time than a single high sea-level event is indicated. At NL there is a distinct unconformity with articulated, life-position *Mercenaria* below which yield mean D/L ratios of 0.45, and articulated, life-position *Mercenaria* above with mean D/L of 0.30. These represent model ages of 300,000–350,000 yr (stage 9) and 70,000–90,000 yr (late stage 5), respectively. This disconformity *may* correspond to a break between the lower sandy facies of the Norfolk Formation and the overlying Kempsville Formation of Oaks and Coch (1973). It represents a hiatus of over 200,000 years. At FP there are shells of mean D/L ratio 0.42, which may also belong to the Norfolk Formation, with a model age of 250,000–300,000 yr (stage 7 or 9), slightly younger than at NL. At YP there is an intermediate group with mean value of 0.39, which is found in sediments also attributed to the Norfolk Formation by Valentine (1971). These shells are in two beds which may represent reworking of a shell from 300,000to 350,000 yr units into a basal shell lag and deposition of an in-place shell at about 100,000 yr B.P. in a bed 1 m higher in the section. Underlying deposits of the Yorktown Formation yield two shells of ratio near 0.91, greater than 2 million years old. It appears that the several facies of the Norfolk Formation of Oaks and Coch (1973) are in reality deposits of several high sea-level events, rather than a single transgression.

On the York-James Peninsula, Johnson (1969, 1976) has described Pleistocene sediments of the Norfolk and Tabb Formations as well as the Yorktown Formation which he places in the Pliocene. Tabb Formation shells at BB (Big Bethel) and KP (Krause Pit) in the Williamsburg, Virginia, area yield ratios of D/L mean of 0.43, essentially identical to the lower beds at NL and at FP, and are assigned a kinetic model age of 300-

350 ka (stage 9). A Yorktown Formation *Mercenaria* at Z (Zooks Pit, Yorktown, Va.) gave a ratio of 0.96 D/L mean, a kinetic model age of greater than 2 million years. The R (Rescue, Va.) shell, with a ratio of 0.03, is Holocene spoil.

A single *Mercenaria* from BPF (Bushy Park Farm, Wilton, Va.), on the south bank of the Rappahannock River, yielded a mean value of 0.92, near equilibrium, and is estimated to be 1.8 million (± 250 thousand) years old. This shell comes from a basal gravelly shell zone over an unconformity on the Yorktown Formation.

On the Potomac-Patuxent Peninsula there are deposits of a Pleistocene estuarine unit that are described by Thompson (1972). He designates this unit the Pamlico Formation. The deposits underlie a distinct terrace cut into the Upland Gravels of Schlee (1957), which is at an elevation of 3–12 m above present sea level. Elevation of the shell beds is a maximum of 1–2 meters in the exposure, overlain by alluvium. Localities WB (Wailes Bluff, Md.) and LB (Langleys Bluff, Md.) are within the Pamlico Formation. They correlate lithologically, biostratigraphically, and chronologically. WB is the most extensively analyzed locality in the study, with 31 analyses on 18 specimens of 3 molluscan genera. It has also been analyzed by Hare and Mitterer (1967). At WB there is a disconformity between a lower bed containing articulated growth-position *Mercenaria* and an upper bed containing primarily oysters (*Crassostrea virginica*). Two subpopulations of *Mercenaria* data are discernible. Ten analyses cluster with a mean value for extent of racemization (D/L mean) at 0.34, and three cluster at 0.30. Although within the limits of precision, these clusters suggest a real age separation. This may indicate two depositional events, at approximately 125,000 yr B.P. and about 110,000–115,000 yr B.P. (leucine calculated kinetic model ages). *Crassostrea* data from WB suggest similar clusters (Belknap 1979). Correlation with substages of the Shackleton and Opdyke (1973) curve is not definite, but two or more of the substages 5a, 5c, and 5e may occur in this outcrop, Assignment varies slightly due to clustering differences between leucine and extent of racemization results. The *Mercenaria* D/L mean value of 0.35 at LB has a model age of about 115,000 yr B.P., correlative with the articulated *Mercenaria* bed at WB.

On the southern Delmarva Peninsula *Mercenaria* were collected from drill holes (MSG-1 and MSG-2, Assateague Island, Va.; CWW, Chincoteague town water well) and T's, an excavation at T's Corner, Hallwood, Va. The MSG drill holes yielded not-in-place shells with D/L mean ratios of 0.54 and 0.94. The higher ratio is interpreted as a pre-Pleistocene shell older than 2 million years old. The 0.54 mean D/L ratio gives an age estimate of about 1 million years B.P. CWW is a large water well. A

*Mercenaria* from −52 m, presumably within the Tertiary, yielded a mean D/L ratio of 0.94, greater than 2 million years B.P. Two shells collected from in place by the authors at T's Corner excavation yielded a mean ratio of 0.50, estimated leucine model age of about 520,000 yr B.P. Two other shells from within 100 m of this locality were supplied by R. B. Mixon, of the U.S. Geological Survey, from dragline spoil. They gave a mean ratio of 0.55 and a leucine kinetic model age of about 900,000 yr B.P. U-series analysis of one of these shells gave an age about 120,000 yr (B. J. Szabo, personal communication), but similar materials at Parksley, Virginia, nearby, yielded U-series ages greater than 220,000 yr (B. J. Szabo, personal communication). The observed ranges of analytical results remains unexplained*. It appears that the 0.55 ratio shells at T's and the 0.54 ratio shells at MSG came from similar age source beds.

Localities PCD (Pepper Creek Ditch, Delaware) and DCD (Dirickson Creek Agricultural Ditch, Delaware) are outcrops in drainage ditches in southern Delaware. They are within the Omar Formation, which has been described by Jordan (1964, 1974). The Omar Formation has been interpreted as having been deposited in lagoonal, beach, and estuarine environments at a sea level at least 6 m above present. A *Mercenaria* at PCD yielded a ratio of 0.36 D/L mean, a model age of about 130,000 yr B.P. *Crassostrea* values are in agreement with this age (Belknap 1979). In addition B. J. Szabo (personal communication) found an age of about 120,000 yr B.P. on an oyster at PCD. *Crassostrea* from DCD also cluster near this value. A *Mercenaria* from the base of the exposed oysters gave a ratio of 0.55 D/L mean, an assigned age of about 1 million years B.P. Recent study by J. M. Demarest and E. M. Keenan (personal communication) has shown that there is a lower shell bed at DCD, 1 meter below the outcrop, with similar high enantiomeric ratios. Also, similar ratios are found at a site 4.5 km to the northwest. It appears that the two different ages are present at DCD. The older materials correlate with MSG and T's, the younger with PCD, WB, LB, and YP (Figure 4). Radiocarbon dates ranging from 28,000 to more than 40,000 yr in these units (Jordan 1974; Belknap 1975; Demarest et al., 1979) are thought to indicate problems of contamination by modern $^{14}C$ in groundwater.

The remaining two localities are from Delaware Bay. DB is an indefinite location, a living *Mercenaria* bought at a fish market. The mean ratio of 0.01 is consistent with a lack of significant racemization in this living quahog. BR (Broadkill River, Del.) is a *Mercenaria* from a shell midden. J. C. Kraft (personal communication) reports a $^{14}C$ date of 345 ± 85 yr

* Analysis of ten additional samples from T's Corner indicate that the results presented here represent the extremes of a broad range of data, from which a range of kinetic model ages between 500,000 and 900,000 yrs. would be inferred.

B.P. for this deposit. The ratio of 0.05 D/L mean is consistent with the relative youth of this shell.

A general picture of the evolution of the region emerges from these interpretations. Some time before 1 million years B.P. a large mass of coarse-grained fluvial sediments was deposited in Delaware and Maryland: the Columbia Formation and the Upland Gravels. About 1 million years B.P. barrier, back-barrier, and estuarine sediments were deposited, forming parts of the Accomack and Omar Formations, which are now the core of the lower Delmarva Peninsula. Approximately 0.5 million years B.P. there was another phase of deposition in the T's Corner area. During stages 9 and 7 (350,000–300,000 and about 250,000 yr B.P.) there were periods of deposition in southeastern Virginia, resulting in the Tabb Formation and at least parts of the Norfolk Formation. The best represented phase of deposition in this region is stage 5, which may be separable into its substages 5e, 5c, and 5a at about 125,000, 110,000 and 70,000–90,000 yr, respectively. Sea levels during stage 5e reached at least 6 meters above present (PCD), in agreement with eustatic data (Bloom et al., 1974 for example). Possible stage 3 deposition in this area is discussed by Belknap (1979). After a period of emergence and erosion, the ongoing Holocene sea-level rise has reached levels occupied by previous high sea stands. It seems likely that an ancestral Chesapeake Bay has been present for at least 125,000 years from deposits at WB and LB, and possibly up to 1 million years B.P. from deposits at T's and DCD. In addition, deposits in Norfolk represent more than a single transgression. In all, the record is one of persistence of similar environments of deposition through a long series of sea-level fluctuations for more than 1 million years. Details of the geologic evolution of the mid-Atlantic coast are provided by Belknap (1979).

## CONCLUSIONS

We have shown that amino acid racemization ratios occur in the correct stratigraphic order in single outcrops, that ratios increase from near zero to near equilibrium with increasing age, and that when temperature is included in the calculations, the ratios can be used as a correlation tool. Assignment of ages to the ratios depends on models of kinetics, which are not well understood at present. Further work to check kinetics under realistic geologic conditions, as well as independent age checks on fossil samples dated radiometrically, will refine age estimates. In all cases but one, U-series dates and amino acid age estimates are in good agreement. At present we have confidence within ± 20% for the stated ages. The use of the extent of racemization value (D/L mean) rather than a single amino acid for relative age comparisons simplifies presentation of relative enantiomeric ratios. This approach is only valid, however, when the average trend is consistent with individual amino acid trends, as it is in this case.

The trend of groupings of amino acids compares favorably with the Shackleton and Opdyke (1973) curve of ice-volume changes. Interpretations of geologic history which envision a simple four ice-age Pleistocene record are likely to be oversimplified.

Analysis of localities based on the ratios, knowledge of Holocene coastal environments, and models of sea-level fluctuations have resulted in new, useful models of evolution in this area (Belknap 1979). In some cases the geologic history is more complex than previously realized, such as in the Norfolk area and in southern Delaware, but in other cases lithologically defined time assumptions have been confirmed (WB-LB, BB-KP, DCD-PCD).

## ACKNOWLEDGMENTS

We wish to acknowledge valuable contributions by J. M. Demarest, M. S. Goettle, G. H. Johnson, E. M. Keenan, J. C. Kraft, R. M. Mixon, W. L. Newell, J. P. Owens, and B. J. Szabo. This work was supported by U. S. Geological Survey grants 14-08-0001-G-247 and 14-08-0001-G-248.

## REFERENCES

Abelson, P. H., Amino acids in fossils, *Science,* **119,** 1954.

Bada, J. L., B. P. Luyendyk, and J. B. Maynard, Marine sediments: Dating by the racemization of amino acids, *Science,* **170,** 1970.

Belknap, D. F., Dating of late Pleistocene and Holocene relative sea levels in coastal Delaware, unpublished M.S. thesis, Dept. of Geology, Univ. of Delaware, Newark, 95 pp., 1975.

Belknap, D. F., Application of amino acid geochronology to stratigraphy of late Cenozoic marine units of the Atlantic Coastal Plain, unpublished Ph.D. dissertation, Dept. of Geology, Univ. of Delaware, Newark, 550 pp., 1979.

Bloom, A. L., W. S. Broecker, J. M. A. Chappell, R. K. Matthews, and K. J. Mesolella, Quaternary sea level fluctuations on a tectonic coast: New $^{230}Th/^{234}U$ dates from the Huon Peninsula, New Guinea, *Quaternary Research,* **4,** 1974.

CLIMAP Project members: A. McIntyre et al., The surface of the Ice-Age Earth, *Science,* **191,** 1976.

Demarest, J. M., R. B. Biggs, and J. C. Kraft, The interpretation of Pleistocene Epoch coastal barrier complexes with Holocene models, southeastern Delaware: American Assoc. Petroleum Geologists–Soc. Economic Paleontologists Mineralogists Ann. Conv., Houston, Texas, April 1–4, Abstract, 1979.

Hare, P. E., Geochemistry of proteins, peptides, and amino acids, in G. Eglinton and M. T. J. Murphy, eds., *Organic Geochemistry,* New York, Springer Verlag, 1969.

Hare, P. E. and R. M. Mitterer, Nonprotein amino acids in fossil shells, *Carnegie Inst. Wash. Year Book 65,* 1967.

Hare, P. E., Laboratory simulation of amino acid diagenesis in fossils, *Carnegie Inst. Wash. Year Book 67,* 1969.

Johnson, G. H., Geology of the lower York-James Peninsula and south bank of the James River, College of William and Mary, Dept. of Geology, *Guidebook 1,* Williamsburg, Va., 1969.

Johnson, G. H., Geology of the Mulberry Island, Newport News North, and Hampton Quadrangles, Virginia: Virginia Div. Mineral Resources *Rept. Invest. No. 41,* Charlottesville, Va., 1976.

Jordan, R. R., Columbia (Pleistocene) sediments of Delaware, *Delaware Geological Survey Bull. No. 12,* Newark, 1964.

Jordan, R. R., Local straigraphic studies Pleistocene deposits of Delaware, in R. Q. Oaks Jr. and J. R. DuBar, eds., *Post-Miocene Stratigraphy Central and Southern Atlantic Coastal Plain,* Logan, Utah State University Press, 1974.

Kvenvolden, K. A., E. Peterson, and G. E. Pollock, Geochemistry of amino acid enantiomers: Gas chromatography of their diasteriomeric derivatives, *Advances in Organic Geochemistry 1971,* 1972.

Miller, G. H. and P. E. Hare, Use of amino acid reactions in some arctic marine fossils as stratigraphic and geochronological indicators, *Carnegie Inst. Wash. Year Book 74,* 1975.

Mitterer, R. M., Pleistocene stratigraphy in southern Florida based on amino acid diagenesis in fossil *Mercenaria, Geology,* **2,** 1974.

Mitterer, R. M., Age and diagenetic temperatures of Pleistocene deposits of Florida based on isoleucine epimerization in *Mercenaria, Earth Planet. Sci. Lett.* **28,** 1975.

Newton, C. R., D. F. Belknap, and G. W. Lynts, Early Pleistocene (Calabrian) age of the Waccamaw Formation at Walkers Bluff, Elizabethtown, North Carolina *Geological Soc. American Abstr. w. Progr.,* **10,** 1978.

Oaks, R. Q., Jr. and N. K. Coch, Post-Miocene stratigraphy and morphology, southeastern Virginia, *Virginia Division Mineral Resources Bull.,* **82,** 1973.

Oaks, R. Q., Jr., J. E. Sanders, and R. F. Flint, Post-Miocene shorelines and sea-levels, southeastern Virginia, in R. Q. Oaks, Jr., and J. R. DuBar, eds., *Post-Miocene Stratigraphy Central and Southern Atlantic Coastal Plain,* Logan, Utah State University Press, 1974.

Schlee, John, Upland gravels of southern Maryland, *Geol. Soc. Amer. Bull.,* **68,** 1957.

Shackleton, N. J. and N. D. Opdyke, Oxygen isotope and paleomagnetic stratigraphy of equatorial Pacific core V28-238: Oxygen isotope temperatures and ice volumes on a $10^5$ and $10^6$ year scale, *Quaternary Research,* **3,** 1973.

Thompson, D. E., Paleoecology of the Pamlico Formation, Saint Mary's County, Maryland: unpublished Ph.D. dissertation, Dept. of Botany, Rutgers University, Rutgers, NJ, 1972.

Wehmiller, J. F. and D. F.Belknap, Alternative kinetic models for the interpretation of amino acid enantiomeric ratios in Pleistocene mollusks: Examples from California, Washington, and Florida, *Quaternary Research,* **9,** 1978.

Wehmiller, J. F. and P. E. Hare, Racemization of amino acids in marine sediments, *Science,* **173,** 1971.

Wehmiller, J. F., et al., Correlation and chronology of Pacific coast marine terrace deposits by fossil amino acid stereochemistry — technique evaluation, relative ages, and geologic implications, *U. S. Geological Survey Open File Rept. 77-680,* 1977.

# Amino Acid Geochronology: Integrity of the Carbonate Matrix and Potential of Molluscan Fossils

**GIFFORD H. MILLER**
*Institute of Arctic and Alpine Research and Department of Geological Sciences, University of Colorado*

**P. E. HARE**
*Geophysical Laboratory, Carnegie Institution of Washington, Washington, D.C.*

## *ABSTRACT*

Analyses of amino acid assemblages in Pleistocene mollusks indicates that the carbonate matrix of many species approximates a closed system capable of retaining free amino acids and peptide chains derived from indigenous proteinaceous material over geologic time spans of at least $10^5$ to $10^6$ years. Changes in the relative abundance of each amino acid over time is primarily related to the interplay between the strength of the peptide bond and the stability of the amino acid in the free state. In general, with increasing age (or temperature), aspartic acid, threonine, serine, and arginine are relatively depleted; glutamic acid, valine, and the leucines remain relatively unchanged; and alanine is relatively enriched from various decomposition reactions. An exceptional natural experiment on the Pribilof Islands, Alaska, confirms that the incorporation of secondary amino acids from extra-shell sources is negligible over a time range of more than $10^5$ years.

The most reliable time-dependent reactions are racemization (or epimerization) in the total (free plus peptide bound) amino acid fraction and to a lesser degree in the free fraction. During the early stages of diagenesis, the proportion of free to bound amino acids and relative destruction of threonine and serine are also broadly time-dependent reactions.

In a region of uniform paleotemperature variations, amino acid ratios may be used directly as stratigraphic correlation tools. We term this usage aminostratigraphy. Aminostratigraphy can be used not only to correlate between disjunct fossiliferous strata but also to provide relative age differences between strata and as a base for a regional chronostratigraphy. Marine, freshwater, and terrestrial mollusks appear to be equally suited to such applications. Quantification of the parameters required for absolute age evaluation remains elusive; consequently, racemization "dates" should probably be regarded only as preliminary estimates unless corroborated by other independent dating criteria.

## INTRODUCTION

The existence of amino acids in fossil shells was first demonstrated by Abelson (1955) only two decades ago and the potential time-dependency of the racemization reaction in a carbonate matrix was documented just a decade ago (Hare and Mitterer 1966). Amino acid geochronology has expanded rapidly since that time. In some areas of Quaternary research, it is accepted as a standard chronostratigraphic tool, although a more precise understanding of the reaction kinetics is needed (Bender 1974; Williams and Smith 1977).

Of the various materials potentially useful for amino acid geochronological studies carbonate fossils are well suited because of the relatively impermeable nature of the shell structure, the natural buffering effect of the carbonate matrix and the relatively large size for ease in sample preparation. Although we have concentrated on marine bivalves, recent studies have shown that amino acid ratios in freshwater bivalves and nonmarine gastropods are equally reliable chronological indices.

The purpose of this paper is to document the capability of the carbonate matrix to preserve indigenous amino acids over geologic time ($10^6$–$10^8$ years), to define the natural range of various amino acid ratios within and between shells of the same species, to identify the most reliable reactions for geochronological purposes, and to provide examples of applications in amino acid geochronology.

All analyses were performed on an automated ion-exchange liquid chromatographic amino acid analyzer utilizing o-Phthalaldehyde (OPA, Benson and Hare 1975) fluorescence detection. The instrument is capable of detecting amino acid residues as low as $10^{-12}$ moles. Details of the analyzer are given by Benson and Hare 1975; sample preparation procedures are outlined in Appendix 1.

## Sample Selection

Perhaps the most important step in any geochronological study is the selection of appropriate samples. First, it is well to bear in mind that the dating of any material can never have more significance than allowed by the original stratigraphic description of the collection site. Second, the datable material must be unambiguously associated with a sedimentary stratum. And third, the samples should be preferably in situ, or at least penecontemporaneous with the deposition of the associated sediments. In the case of bivalve mollusks, paired valves are likely to be in situ and are the preferred material to date. Whole unabraded shells may be suitable for dating, but the results on abraded, worn, or highly leached shells must be viewed with caution. For shells of aragonitic structure, it may be useful to determine if recrystallization to the more stable calcite form has occurred, although it has not been shown that such recrystallization involves a release or uptake of amino acids.

Amino acid composition, relative abundances, and the rates of various amino acid reactions in invertebrate shells are known to be species dependent. The importance of selecting at least monogeneric samples has been emphasized in several studies. The discrepancies in D/L ratios in different genera from the same locality are generally less than a factor of 2, but even greater differences have been reported. King and Hare (1972) found allo/iso ratios for three foraminifera species of the same age differed by less than 25%, whereas a fourth species had allo/iso ratios more than 50% lower than in the fastest racemizing group. A more detailed study (King and Neville 1977) revealed similar species-dependent reaction rates. Wehmiller et al. (1977) found that enantiomeric ratios in four molluscan genera from marine terraces in California yielded similar ratios, whereas a fifth genus had ratios 15% to 20% higher, and the same ratios in two additional genera were consistently 40% to 50% greater than in the first group. Miller and Hare (1975) found allo/iso ratios that varied by more than a factor of 2 between different marine mollusk genera in Arctic Canada. Table 1 lists similar results for a different suite of samples from the same region. Isoleucine epimerization ratios in seven genera of nonmarine mollusks from lacustrine sediments in the Searles Valley, California, showed slightly less genus-dependent variation (Table 2). N. W. Rutter (personal communication, 1978) found genus-dependent racemi-

**Table 1. Genus-Dependency of Racemization Ratios in Pelecypod Mollusks of Last Interglacial Age, Clyde Forelands, Baffin Island, NWT, Canada**

| Species | Allo : Iso Total | Allo : Iso Free |
|---|---|---|
| *Hiatella arctica* | 0.16 | 0.66 |
| *Mya truncata* | 0.14 | 0.62 |
| *Astarte* cf *striata* | 0.09 | 0.79 |
| *Serripes groenlandicus* | 0.09 | 0.62 |
| *Clinocardium ciliatum* | 0.06 | 0.68 |

zation rates in nonmarine gastropods from the Old Crow Plains, Yukon Territory, Canada.

Clearly, it is imperative that identification at least to the generic level be undertaken on all carbonate fossils and that for comparative purposes the genus selected for analyses should be present in reasonable numbers in shell collections from the area under study. For some regions it may be possible to use two or three common genera with overlapping distributions so that the reaction rates may be intergenerically calibrated where two genera are present in the same collection. In this way, collections devoid of the primary genus may be compared by utilizing amino acid ratios in a calibrated genus.

**Table 2. Isoluecine Epimerization Ratios in Nonmarine Mollusks from Searles Lake Deposits, California**[a]

| Sample ID | Genus | Class | Allo : Iso[b] |
|---|---|---|---|
| AAL-852 | *Sphaerium* | Pelecypoda | 0.55 |
| AAL-851 | *Pisidium* | Pelecypoda | 0.54 |
| AAL-836 | *Parapholyx* | Gastropoda | 0.53 |
| AAL-888 | *Valvata* | Gastropoda | 0.49 |
| AAL-889 | *Lymnaea* | Gastropoda | 0.38 |
| AAL-887 | *Physa* | Gastropoda | 0.38 |
| AAL-890 | *Gyraulus* | Gastropoda | 0.38 |

[a] Samples collected from a single horizon in unit AB3, Poison Canyon section (G. I. Smith, Menlo Park, California, unpublished mimeo for Friends of the Pleistocene excursion, November 1978).

[b] Mean value of D-alloisoleucine to L-isoleucine determined in the total acid hydrolysate of three individuals of each genus. Collection and analyses by G. H. Miller and W. D. McCoy, University of Colorado.

Because amino acid reaction rates are temperature dependent, it is preferred that samples are collected from excavations some depth in the enclosing sediment to minimize the effects of diurnal and seasonal temperature extremes. Samples that have been exposed at the surface for a considerable time may yield spurious results. In general, samples from 2 m depth are adequately insulated from short-term temperature anomalies, but even burial depths as shallow as 0.5 to 1.0 m are acceptable.

### Preservation of Indigenous Amino Acids in Molluscan Fossils

In their review article, Schroeder and Bada (1976) suggest that molluscan shells lose substantial quantities of amino acids with time, in contrast with foraminifera, which more closely approximate a closed system. Some possible mechanisms suggested for such losses include microfracturing of the shells in high energy depositional environments and alteration during recrystallization of aragonite to calcite. However, neither mechanism has been confirmed by controlled experiments. Analyses of amino acid assemblages and isoleucine epimerization ratios in partially silicified mid-Pleistocene shells from Arctic Canada yielded results consistent with their estimated age and were similar to unsilicified shells of the same age in the region. Wehmiller et al. (1977) found enantiomeric ratios to be similar in recrystallized and unaltered samples but did not attempt an exhaustive study of this phenomenon.

However, our data indicate that there is a rapid decrease in total amino acid content of molluscan fossils during the earliest stages of diagenesis. Within a few thousand to a few tens of thousands of years, amino acid concentrations stabilize, after which there is a very gradual decrease in amino acid concentrations in progressively older shells of a given species during the next 1 to 2 $\times$ $10^6$ years. Some molluscan genera yield remarkably unchanged compositional patterns over this time range. In sedimentary environments of exceptional preservation, the stable amino acids are preserved in relatively unchanged concentrations for $10^7$ to $10^8$ years.

### Retention of Indigenous Amino Acids

We have obtained collections of the fresh water pelecypod *Corbicula fluminalis* from estuarine deposits in Great Britain that span most of the Pleistocene. Based on biological and geological considerations, these deposits have been assigned to the Ipswichian (*ca.* 125,000 BP) and Hoxnian (*ca.* 300,000 $\pm$ 100,000 BP) interglacials and to the Icenian Crag (middle–early Pleistocene). Amino acid concentrations were determined for three separate valves from each locality. Table 3 lists the average composition in two localities each of the Ipswichian and Hoxnian interglacial deposits and a single locality from the Icenian Crag. A more complete presentation of the data is in Miller et al. (1979).

In general, the average concentrations for Ipswichian and Hoxnian deposits are remarkably similar. There is a slight depletion of threonine and serine, two of the least stable amino acids, and a significant increase in the alanine content in the Hoxnian shells. Isoleucine epimerization values (allo/iso ratios) clearly indicate that the Hoxnian deposits are the older of the two, yet there is no apparent lowering in the amino acid concentrations over this time range. There is, however, a decrease in amino acid concentrations between shells of Ipswichian and Hoxnian age and those from the Icenian Crag. In general, the less stable amino acids (asp, thr, ser) are greatly depleted, whereas the more stable amino acids (glu, gly, val, leu) have been reduced by only 25%. Alanine, however, is present in concentrations similar to those in Hoxnian shells because of its chemical stability and relative enrichment from the decomposition of serine and methionine. Isoleucine is a special case, as its depletion is also caused by epimerization to alloisoleucine.

A series of the pulmonate gastropod *Ashmunella rhyssa* collected from the Sacramento Mountains, southern New Mexico, was obtained from Dr. A. L. Metcalf, University of Texas, El Paso. These samples span the last $10^6$ years. Amino acid concentrations and allo/iso ratios are listed in Table 4. The reduction in amino acid concentrations in this genus is most rapid between the modern and 11,000 BP samples, after which the depletion rate is substantially less (Fig. 1).

Some caution must be exercised when interpreting these data, as the amino acid concentrations per gram shell weight and their relative abundances are known to show significant departures under different environmental stresses (Degens et al., 1967). Insufficient data are available to quantify how paleoenvironmental differences in the sample suites analyzed might have affected the initial amino acid concentrations.

The retention of indigenous amino acids in mollusk shells over substantially longer periods of geologic time is not well documented. At midlatitude temperatures, even the slowest racemizing amino acids should reach equilibrium within 5 to 20 $\times$ $10^6$ years, and hydrolysis of the shell protein should have reduced the constituent amino acids to the free state. Hare and Abelson (1967) reported essentially racemic assemblages in Miocene *Mercenaria,* and Miller et al. (1979) report isoleucine at equilibrium in Pliocene *Mya* from England, suggesting uptake of secondary amino acids in these samples is negligible. However, amino acid residues in early Cenozoic oyster and pecten shells reported by Matter et al. (1969)

**Table 3. Amino Acid Residues in *Corbicula fluminalis* from the Thames Estuary and Adjacent Regions, England**

| Relative Age | Absolute Age | Amino Acid Residues (pmole/mg shell) | | | | | | | | | Allo:Iso (total) |
|---|---|---|---|---|---|---|---|---|---|---|---|
| | | Asp | Thr | Ser | Glu | Gly | Ala | Val | Iso | Leu | |
| Ipswichian (2 sites) | 125,000 | 498 | 20 | 33 | 134 | 218 | 128 | 58 | 24 | 43 | 0.20 |
| Hoxnian (2 sites) | ca. 300,000 ± 100,000 | 408 | 25 | 26 | 137 | 220 | 181 | 64 | 24 | 47 | 0.29 |
| Icenian Crag | 0.7 to $1.0 \times 10^6$ | 192 | 10 | 12 | 100 | 161 | 188 | 37 | 12 | 32 | 0.75 |

**Table 4. Amino Acid Residues in *Ashmunella rhyssa* from the Sacramento Mountains, New Mexico**

| ID | Estimated Age | Amino Acid Residues (pmole/mg shell) | | | | | | | | | Allo:Iso (total) |
|---|---|---|---|---|---|---|---|---|---|---|---|
| | | Asp | Thr | Ser | Glu | Gly | Ala | Val | Iso | Leu | |
| AAL-187 | Modern | 471 | 206 | 202 | 399 | 496 | 245 | 124 | 65 | 283 | 0.013 |
| AAL-185 | 11,000 ($^{14}C$) | 324 | 72 | 87 | 275 | 232 | 149 | 75 | 46 | 137 | 0.045 |
| AAL-184 | late Wisconsin | 275 | 50 | 53 | 232 | 259 | 171 | 65 | 42 | 125 | 0.12 |
| AAL-181 | >40,000 ($^{14}C$) | 309 | 55 | 63 | 263 | 359 | 171 | 89 | 44 | 157 | 0.16 |
| AAL-183 | pre-late Wisconsin | 236 | 53 | 23 | 246 | 211 | 186 | 90 | 35 | 134 | 0.38 |
| AAL-186 | pre-Wisconsin | 119 | 3 | 1 | 117 | 102 | 91 | 40 | 13 | 60 | 0.41 ± |
| AAL-182 | Kansan | 150 | 14 | 8 | 131 | 145 | 144 | 55 | 13 | 74 | 0.83 |

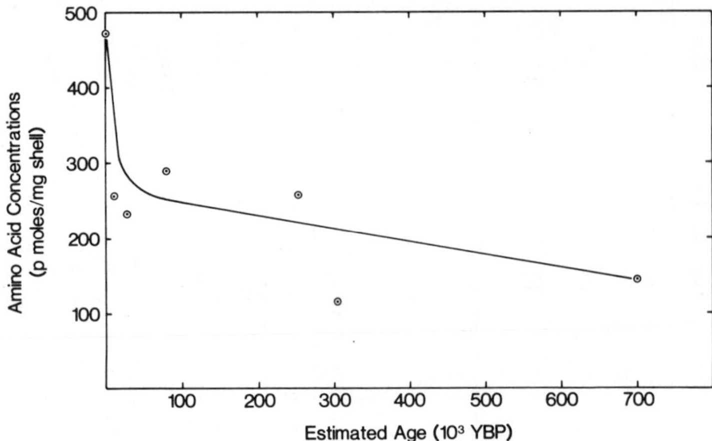


**Figure 1.** Depletion of stable amino acids (Val + Iso + Leu) in the pulmonate gastropod *Ashmunella rhyssa*, Sacramento Mountains, New Mexico. The initial rapid depletion rate is succeeded by a substantially slower rate after about 10,000 years.

and Akiyama and Wyckoff (1970) do not show equilibrium epimerization ratios and contain considerable quantities of serine. Either a special mechanism of preservation influenced these samples, or secondary contaminants had been incorporated into the shell matrices. In contrast, individual shells from a remarkably well preserved suite of molluscan fossils recovered from the Coon Creek Formation, Tennessee, of late Cretaceous age (70 $\times$ $10^6$ YBP) supplied by Dr. E. Kaufmann, Smithsonian Institution, yielded compositional patterns consistent with the trends established in the late Tertiary shells. In certain genera, amino acid concentrations in the free and total (free plus peptide-bound) fractions are nearly indistinguishable with isoleucine epimerization values at equilibrium (1.3). As would be predicted from the studies of Vallentyne (1964, 1969), aspartic acid, threonine, serine and arginine are present in negligible concentrations, whereas glutamic acid, glycine, alanine valine, isoleucine and leucine are present in concentrations of 50 to 500 pmoles/mg shell, levels similar to those in late Pleistocene molluscan shells (see Tables 3 and 4). Genera with high relative levels of serine and low allo/iso ratios were found to contain low overall amino acid concentrations (10 to 30 pmoles/mg shell); secondary contaminants are suspected in these samples.

## Incorporation of Secondary Amino Acids

Having shown that the loss of amino acids in Pleistocene molluscan fossils is slight, the inverse question is the possibility of postdepositional incorporation of secondary amino acids by the carbonate matrix. Hare and Mitterer (1968) heated *Mercenaria* in an aqueous solution containing 10 $\mu M$/ml of the nonprotein amino acid norleucine. After 100 hours at 165°C norleucine amounted to 1.3% of the total amino acid assemblage in the outer portion of the shell and 0.5% in the inner portion. The identification of racemic mixtures in shells more than $5 \times 10^6$ years old also suggests that the carbonate matrix effectively excludes secondary amino acids. In addition, Miller and Hopkins (this volume) report on a unique natural experiment on the Pribilof Islands, Alaska, that provides additional documentation of the integrity of carbonate fossils. On St. Paul Island in the Pribilof group, marine sediments of last interglacial age containing pelecypod shells of the genus *Mya* are overlain by lava flows deposited shortly after shell deposition. Transient heat flow penetrating the sediments as the lavas cooled, accelerated the amino acid reactions. In two samples, the hydrolysis and racemization reactions were driven to completion, and an even more intense heat flow in a third sample resulted in the decomposition of all amino acids (Table 5). The retention of exclusively free amino acids in the former case, and the lack of incorporated secondary amino acids in the latter case are evidence for the integrity of the carbonate matrix.

These data sets substantiate the integrity of the carbonate matrix as an agent for the preservation of indigenous amino acids. At mid-latitudes, amino acid concentrations in carbonate fossils undergo an initial reduction during the first $10^3$ to $10^4$ years of diagenesis, after which concentrations are reduced at a substantially lower rate that amounts to a 20% to 50% depletion in early Pleistocene samples. The depletion rates are derived from shells that were of only average preservation; hence these rates may be taken as representative for most deposits. The uptake of secondary amino acids by the carbonate matrix has been demonstrated to be negligible over a time range of more than $10^5$ years, and even the maximum values are insufficient to alter amino acid ratios of Pleistocene fossils beyond normal analytical uncertainty. It is concluded that mid-latitude and high-latitude carbonate fossils can in general be expected to faithfully preserve indigenous amino acids and exclude secondary amino acids that may be present in the sedimentary environment for more than $10^6$ years.

## Sample Evaluation

The suitability of a mollusk sample must be evaluated in physical and analytical terms. Physical prerequisites have been discussed previously. It is possible to use the amino acid data to provide an additional qualitative index of contamination. Our analyses of suites of monogeneric samples

**Table 5. Amino Acid Ratios in *Mya* from the Pribilof Islands, Alaska**

| Sample ID | Location | Age (YBP) | Association with lavas | Allo : Iso (total) |
|---|---|---|---|---|
| AAL-27 (2) | Bootlegger Cove Clay | 14,000 (14C) | None, thermal history = +2°C | 0.034 |
| AAL-31 (1) | Tolstoi Point | 125,000 | Subaerial flow 5 m above shells | 0.18 |
| AAL-30 (3) | Einahnuhto Bluffs (lower bed) | 125,000 | Subaqueous lava 1–2 m above shells | 1.26 |
| AAL-28 (2) | Einahnuhto Bluffs (Einahnuhto transgression) | >125,000 | Subaqueous flow within 1.5 m of shells | 1.22 |
| AAL-29 (2) | Einahnuhto Bluffs (upper bed) | 125,000 | Subaerial flow 1–2 m above shells | <1 pmole/mg isoleucine present |

covering a broad spectrum of geologic time, coupled with pyrolysis experiments on modern valves of several species reveal the following trends:

1. Each genus has a characteristic amino acid assemblage, or "signature," that distinguishes it from most other genera. The signature is altered in a regular way as diagenetic reactions alter the compositional pattern over time. If the amino acid signature of a shell deviates noticeably from the established genus-specific pattern, a problem is indicated (such as incorrect taxonomic identification, incomplete removal of periostracum, contaminants in the sample or introduced during sample preparation), and the amino acid ratios are probably not reliable.
2. Isoleucine epimerization (and other D/L) ratios and the relative proportions of free to bound amino acids increase with increasing sample age and/or thermal histories.
3. The concentrations of the less stable amino acids, notably threonine and serine, show a steady decrease with increasing time/temperature. This relationship holds whether determined in an absolute sense (moles/g shell weight) or relative to a stable amino acid such as glutamic acid or leucine. Anomalously high serine residues are generally considered to indicate contamination.
4. Alanine is relatively enriched with time/temperature from various decomposition reactions and because of its chemical stability.
5. α–Aminobutyric acid (α–ABA) appears in generally increasing proportions in progressively older samples. The appearance of this nonprotein amino acid is attributed to the decomposition of threonine.

If suites of samples known to be of increasing geologic age do not show these trends, contamination should be suspected.

## AGE-DEPENDENT AMINO ACID RATIOS

Protein protected from rapid decomposition by the carbonate matrix of invertebrate shells undergoes a variety of chemical reactions including hydrolysis of the peptide bonds to release smaller polypeptides and free amino acids, decarboxylation, deamination, oxidation, and racemization/epimerization. Of these, the racemization reaction has received the widest attention in geochronological studies. In addition to the extent of racemization in the total acid hydrolysate (total fraction), Miller and Hare (1975) and Miller et al. (1977) have shown that the racemization in the free amino acid fraction is less variable during the early stages of protein diagenesis (up to total allo/iso ratio of 0.1 and free allo/iso ratio of 0.6). In older samples, free racemization levels are more variable than those in the total fraction. In addition, Miller and Hare (1975) found that the proportion of free to peptide-bound amino acids increased with sample age in a broadly time dependent pattern. A summary of their data is given in Table 6. The hydrolysis data show more scatter between individual shells of the same species than do the racemization ratios, in part because of inhomogeneities in the ratio of protein to carbonate in the shell structure. In general, the racemization data yield the most reliable ratios, whereas the other ratios are useful as corroborative evidence.

In addition to these reactions, we have studied the time dependency of the threonine decomposition reaction. Bada et al. (1978) suggested possible diagenetic pathways for the decomposition reactions of the hydroxy amino acids. Our data show a relative enrichment in alanine with increasing age (either in concentration or as the glycine/alanine ratio) and support their contention that dehydration is the dominant decomposition reaction for threonine and serine. The depletion of threonine or serine in mollusks can be monitered both in relative (threonine/glutamic acid) and absolute (pmoles/mg shell) concentrations. Plots of ln (thr : glu) versus time from pyrolysis experiments on *H. arctica* at 152°, 144° and 110°C were all nonlinear, whereas the absolute concentration of threonine versus time closely approximates a linear fit on a semilog plot (Fig. 2). The graphs suggest that the threonine decomposition reaction follows irreversible first order kinetics. Although preliminary, the results indicate that the activation energy for this reaction is considerably greater than for the isoleucine epimerization reaction (*ca.* 37 kcal/mole vs. 29 kcal/mole), sug-

**Table 6. Amino Acid Ratios[a] in *Hiatella arctica*, Qivitu, Baffin Island, Arctic Canada**

| Age | Allo : Iso | | Percent Free Amino Acids | | | Thr : Glu |
|---|---|---|---|---|---|---|
| | Total | Free | Val | Iso | Leu | (total) |
| Modern | 0.018 | N.D. | 0.0 | 0.0 | 0.0 | 1.2 |
| 9,000 ($^{14}$C) | 0.02 | N.D. | 1 | 1 | 2 | 1.0 |
| *ca.* 70,000 | 0.045 | 0.29 | 7 | 4 | 8 | 0.78 |
| *ca.* 130,000 | 0.15 | 0.71 | 18 | 11 | 20 | 0.60 |
| >130,000 | 0.23 | 0.86 | 22 | 17 | 38 | 0.39 |
| >200,000 | 0.34 | 1.06 | 34 | 22 | 41 | 0.31 |
| >200,000 | 0.48 | 1.11 | 65 | 54 | — | 0.26 |

[a] Data from Miller (unpublished) and Nelson (1978).

gesting the intriguing possibility of writing simultaneous equations for the threonine decomposition and epimerization reactions that would provide a unique solution for both time and temperature. The most serious constraint at present is that the initial threonine concentration of the shell, which is required to evaluate the extent of decomposition, is known to show considerable variation in modern shells (the absolute abundance varies for all amino acids in shells, while the relative abundances are constant).

## REPRODUCIBILITY

A basic test for the scientific reliability of any experimental data is the degree to which the results can be reproduced. To demonstrate the constancy of certain age-dependent amino acid ratios, we have performed multirun, multi-valve experiments and compared ratios from various pelecypod genera collected from a single locality. Individual ratios derived from chromatographic analyses are not considered reliable past two significant figures.

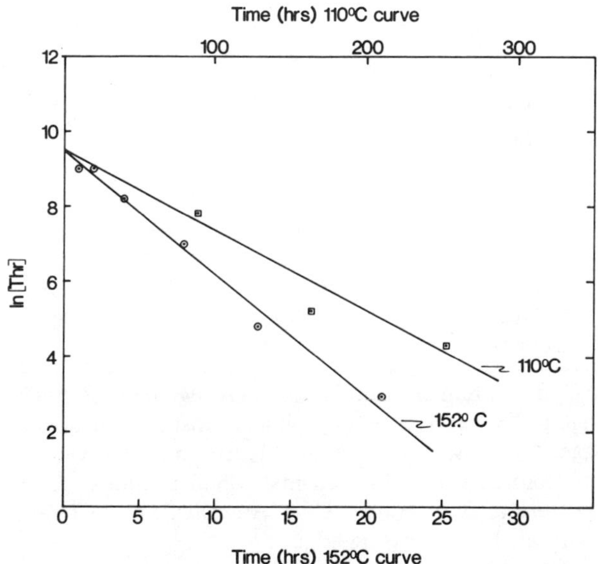


**Figure 2.** Decomposition of threonine during pyrolysis of modern valves of the pelecypod *Hiatella arctica* at 152° and 110°C.

## Instrumental Error

Back-to-back analyses generally agree within ±1 digit of the second significant figure. Analyses performed more than a few days apart may differ up to 5%.

## Intrashell Variation

Multiple preparations of single valves of the pelecypod *H. arctica* revealed that the variation within a valve is similar to the variation between valves (Table 7) and shows no consistent trend with the portion of the

**Table 7. Intravalve and Intervalve Variations in the Isoleucine Epimerization Ratio in *H. arctica* from Arctic Canada[a]**

| Sample ID | Allo : Iso (Total) | Mean |
|---|---|---|
| AAL-806 B | 0.054 | 0.053 |
| C | 0.055 | |
| D | 0.053 | |
| E | 0.049 | |
| AAL-807 A | 0.058 | |
| B | 0.049 | |
| C | 0.052 | 0.051 |
| D | 0.054 | |
| E | 0.042 | |
| AAL-808 A | 0.057 | 0.057 |
| AAL-809 A | 0.051 | |
| B | 0.057 | 0.054 |
| C | 0.057 | |
| D | 0.060 | |
| Mean | 0.0545 | 0.0538 |
| Standard Deviation | ±0.0066 | ±0.0025 |

[a] Radiocarbon age of this sample is >46,000 BP; $^{230}$Th age of 86,000 ± 5000 is probably closer to the actual age. Amino acid analyses by J. K. Brigham, University of Colorado (unpublished ms). Shell portions: A = Posterior, B = Hinge, C = Anterior, D = Central portion, E = Growing edge.

Two-way analysis of variance indicates no statistically significant differences in allo-iso ratios between individual valves or between different portions of the shells.

shell sampled. Similar analyses on other genera have suggested that the hinge area may yield slightly higher ratios.

### Intershell Variations

The range of isoleucine epimerization ratios between monospecific samples from a single spectrum is perhaps most critical in the evaluation of amino acid geochronology. Table 8 lists several multivalve analyses on a variety of mulluscan genera. The range of allo/iso ratios in columns A–C, Table 8, are typical of molluscan samples from marine and nonmarine sites. On rare occasions we have found extreme variability in a single shell as noted in column D. Because of the possible range in allo/iso ratios (and presumably other D/L values) within a single collection, we believe that for amino acid ratios to be effectively used as geochronological indices a minimum of three individuals must be analyzed from each locality. If the racemization data agree within a reasonable value (±10%), the mean figure may be considered to reliably characterize the deposit as a whole. The relationship between the number of valves analyzed and the precision with which a unit may be characterized is given by Dixon and Massy (1957) as:

$$N = \left[\frac{(Z_1 - \frac{1}{2}\alpha)\sigma}{d}\right]^2$$

where: $N$ = the number of individual shell analyses required $(Z_1 - \frac{1}{2}\alpha)$ is related to the confidence level. For a 95% confidence level, this term is approximately equal to 2. The standard deviation of the population is $\sigma$, and $d$ is the desired precision of the mean.

For example, if allo/iso ratios in shells from a stratigraphic unit yield a mean ratio of 0.20 and a standard deviation of ±0.015, and it is necessary to characterize the unit to the third significant figure (±0.001) at the 95% confidence level, it would require

$$\left(\frac{(2)(.015)}{0.001}\right)^2 = \text{analysis of 900 individuals,}$$

whereas characterization to two significant figures would require only nine analyses.

## TEMPERATURE SENSITIVITY

As for all chemical reactions, the reactions involved in the diagenesis of shell protein are highly temperature dependent. In an attempt to determine

**Table 8. Racemization Variations in Monogeneric Samples**

| | A | | B | C | D |
|---|---|---|---|---|---|
| Sample ID | AAL-238 | | AAL-115 | AAL-636 | AAL-699 |
| Sample Location | Arctic Canada | | Arctic Canada | Searles Lake, California | Thames Estuary, UK |
| Genus | *Hiatella* | | *Hiatella* | *Parapholyx* | *Corbicula* |
| Absolute Age | ≥50,000 | | >130,000 | *ca*. 29,000 | *ca*. 125,000 |
| Amino/acid Fraction | Free | Total | Total | Total | Total |
| Allo/Iso Ratios | 0.28 | 0.037 | 0.19 | 0.48 | 0.34 |
| | 0.29 | 0.043 | 0.17 | 0.57 | 0.30 |
| | 0.29 | 0.047 | 0.17 | 0.52 | 0.27 |
| | 0.33 | 0.044 | 0.18 | 0.53 | 0.24 |
| | 0.27 | 0.031 | 0.20 | | |
| Mean | 0.292 | 0.042 | 0.182 | 0.525 | 0.29 |
| Standard Deviation | ±0.023 | ±0.0066 | ±0.013 | ±0.0037 | ±0.043 |

whether temperature is the only significant environmental parameter, we analyzed a suite of $^{14}$C-dated early postglacial *Hiatella arctica* samples from a wide variety of current climatic regimes. Data on the extent of natural hydrolysis and isoleucine epimerization are listed on Table 9. Most samples were collected *in situ* from marine deposit that emerged from the sea shortly after their formation. The enclosing sediments range from silts to coarse sands, with all samples collected below 100 m asl. Temperature data are the mean annual temperature of the last one to five decades from the nearest representative meteorological station. Despite the potential effects of variables other than temperature (such as pH, sedimentary matrix, ground water), there is a fairly regular increase in allo/iso ratios for samples of similar age with increasing thermal histories. This relationship identifies temperature as the primary, if not only, rate-controlling variable in the diagenesis of shell protein.

## ABSOLUTE AGE DETERMINATIONS

The kinetics of amino acid reactions in a carbonate matrix are complex and as yet incompletely understood. Detailed pyrolysis experiments (Mitterer 1978; this volume) indicate that racemization in carbonate fossils does not follow simple first-order reversible kinetics, primarily because of the retention of free amino acids by the carbonate structure and the different reaction rates in the free and bound (both terminal and internal positions) fractions. In addition, the proportion of carbonate to protein may influence the racemization rate with higher carbonate: protein shells yielding the lowest racemization rates. Nearly all pyrolysis experiments on molluscan samples indicate an initial period of relatively rapid racemization succeeded by a slower racemization rate. In field studies for which independent ages are available for both young and old sites (e.g., early postglacial and last interglacial), the rate constants derived from the young site when extrapolated to the older site result in anomalously young ages or require unrealistically low temperatures.

Wehmiller and Belknap (1978) argued that ages of West Coast mollusks derived from linear kinetics models conflicted with available geologic and paleoclimatic data. They presented a nonlinear model that agreed more closely with the known ages and paleotemperature reconstructions.

We had previously derived the Arrhenius parameters for the isoleucine epimerization reaction in *H. arctica* (Fig. 3) based on pyrolysis experiments and the data in Table 10. The calculated activation energy (29.0 kcal mole$^{-1}$) is similar to other reported values. However, allo/iso ratios in last interglacial *H. arctica* from Arctic Canada (0.15 in the total fraction,

**Table 9. Temperature Sensitivity of Amino Acid Reactions in Dated Early Postglacial *Hiatella arctica* Samples**

| Location | $^{14}C$ age | Temp.[a] °C | Species[b] | Percent Free Amino Acids[c] | | | | Allo:Iso[d] | | Thr:Glu[e] |
|---|---|---|---|---|---|---|---|---|---|---|
| | | | | Ala | Val | Iso | Leu | Total | Free | |
| Washington | 13,010 | +10 | H.a. | — | 15 | 15 | 23 | 0.078 | 0.27 | 0.56 |
| Denmark | 13,000 | +7.7 | H.a. | 19 | 8 | 7 | 11 | 0.053 | 0.21 | 0.5 |
| Maine | 12,230 | +7 | H.a. | 16 | 5 | 5 | 8 | 0.050 | 0.21 | 0.57 |
| New Brunswick | 12,500 | +5 | H.a. | 13 | 3 | 3 | 7 | 0.043 | 0.18 | 0.71 |
| SE Alaska | 10,640 | | H.a. | 20 | — | 4 | 8 | 0.040 | 0.15 | 0.76 |
| Anchorage | 14,160 | +2.1 | M.t. | | | | | 0.034 | 0.16 | |
| S. Greenland | 13,380 | −1 | H.a. | | 1.5 | 1.2 | 2.1 | 0.027 | ≤0.09 | 0.85 |
| S. Baffin Is. | 10,740 | −7 | M.t. | | | | | 0.024 | ≤0.1 | |
| Spitzbergen | 11,000 | −8 | M.t. | | | | | 0.022 | ≤0.1 | |
| N. Baffin Is. | 10,095 | −12 | H.a. | 4 | 0.5 | 0.2 | 1 | 0.020 | ND | 1.2 |
| Somerset Is. | 9,000 | −16 | H.a. | 2.4 | 0.1 | 0.3 | 0.7 | 0.018 | ND | 1.2 |
| Modern | 0 | — | H.a. | 0.0 | 0.0 | 0.0 | 0.0 | 0.018 | ND | 1.2 |

[a] Mean annual temperature of the past 1 to 5 decades based on records of the nearest representative weather station.

[b] H.a. = *Hiatella arctica*; M.t. = *Mya truncata*. Hydrolysis rates in *Mya* are not directly comparable with those in *Hiatella*. For most localities, three or more separate valves were analyzed; ratios given here are mean values.

[c] 1 − (p − x/p) where p = residues in the total fraction and x = residues in the free fraction; Ala = alanine, Val = valine, Iso = isoleucine, Leu = leucine.

[d] Ratio of D-alloisoleucine to L-isoleucine; ND = no detectable concentration of alloisoleucine.

[e] Ratio of threonine to glutamic acid in the total amino acid fraction.

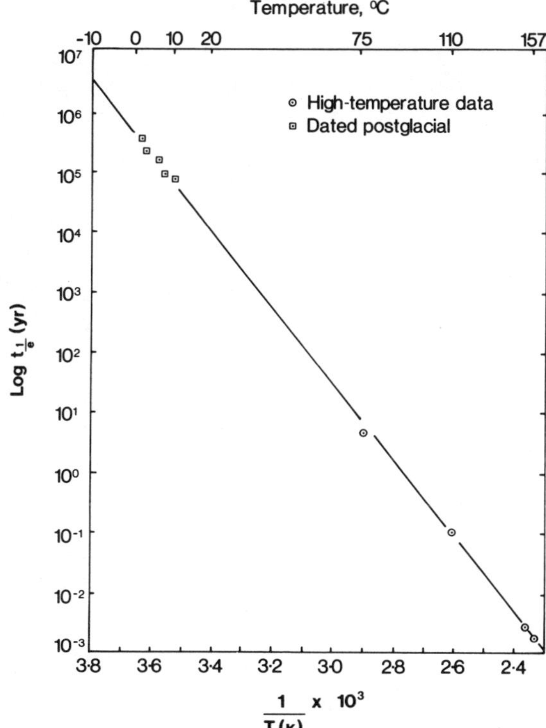


**Figure 3.** Arrhenius plot of isoleucine epimerization in *Hiatella arctica* derived from pyrolysis experiments at 75°, 110°, 152°, and 157°C and dated early postglacial samples (data in Table 9). The activation energy is calculated to be 29.0 kcal $mole^{-1}$

Miller et al., 1977) would require an average diagenetic temperature of −1°C, 10°C higher than the current interglacial temperature. Clearly the linear kinetic model cannot be reliably applied to molluscan carbonate fossils.

Because of the large uncertainties in the derivation of the Arrhenius parameters and a general inability to precisely define the integrated thermal history of a sample, it is unlikely that absolute ages derived from mollusk racemization data are reliable.

## APPLICATIONS

As a result of the development of radiocarbon dating, the chronology of events during the most recent 30,000 or so years of earth history is reasonably well known. As the focus of Quaternary research turns to events

**Table 10. Threonine Decomposition during Pyrolysis of Modern *Hiatella arctica***

| Sample ID | Temperature | Duration | [Thr] (pm/mg shell) | ln[Thr] | Allo : Iso (total) | Thr : Glu |
|---|---|---|---|---|---|---|
| AAL-545 | 100°C | 0 hr. | 1130 | 7.0 | 0.02 | |
| | | 720 | 525 | 6.3 | 0.12 | |
| | | 1440 | 232 | 5.1 | 0.13 | |
| | | 2160 | 68 | 4.2 | 0.23 | |
| AAL-76 | 110°C | 0 hr. | — | — | 0.02 | 1.2 |
| | | 164 | 184 | 5.2 | 0.60 | .08 |
| | | 253 | 76 | 4.3 | 0.69 | .07 |
| AAL-541 | 144°C | 0 hr. | 1130 | 7.0 | 0.02 | |
| | | 10 | 412 | 6.0 | 0.14 | |
| | | 25 | 133 | 4.9 | 0.34 | |
| | | 50 | 90 | 4.5 | 0.42 | |
| | | 70 | 60 | 4.1 | 0.49 | |
| AAL-76 | 152°C | 0 hr. | — | — | 0.02 | 1.2 |
| | | 1 | 7977 | 9.0 | 0.04 | 0.64 |
| | | 2 | 7944 | 9.0 | 0.06 | 0.51 |
| | | 4 | 3156 | 8.1 | 0.10 | 0.31 |
| | | 8 | 1087 | 7.0 | 0.17 | 0.13 |
| | | 12.8 | 120 | 4.8 | 0.30 | 0.062 |
| | | 21 | 18 | 2.9 | 0.58 | 0.024 |
| | | 27.8 | — | — | 0.64 | 0.018 |
| | | 34 | — | — | — | 0.009 |
| | | 58 | — | — | 0.81 | <0.001 |

earlier during the last glaciation and to previous glaciations, periods beyond the limits of $^{14}C$ capabilities, the need for a geochronological method capable of reliable relative age determinations is of critical importance. Recent reevaluations of the type sequences in midwestern and western U.S. (Dreimanis 1977; Lineback and Wickman 1978; Burke and Birkeland in press) and revisions of the pluvial lake sequences (Morrison 1975) underscore the tenuous nature of existing chronocorrelations without adequate dating control.

Carbonate fossils are preserved in a wide variety of shallow marine and nonmarine deposits. The potential applications of amino acid geochronology to these fossils are listed below.

### Resolution of Mixed Populations

Quaternary continental and shallow marine deposits are characterized by their discontinuous nature, rapid facies changes, and incorporation of previously deposited sediment. The reworked sediment frequently contains molluscan fossils that are not readily distinguishable from indigenous shells. The advantage of the amino acid method is that individual shells can be analyzed, and if two or more populations of significantly different ages are present, this will be apparent in the clustering of the amino acid ratios. An example, given in Table 11, is of a fossiliferous glaciomarine

**Table 11. Resolution of Mixed Shell Populations Using Racemization Data**

| | | Allo : Iso | | |
|---|---|---|---|---|
| ID | Species | Total | Free | Comment |
| AAL-109 | H.a. | 0.055 | 0.41 | Single Valve |
| AAL-110 | H.a. | 0.068 | 0.35 | Single Valve |
| AAL-111A | H.a. | 0.056 | 0.43 | Mixture of thin-and-thick-walled shells suggested mixed population; 111C is clearly older than the other 4 shells. $^{14}C$ age is >40,000 (date on younger shells). |
| B | H.a. | 0.052 | 0.30 | |
| C | H.a. | 0.11 | 0.68 | |
| AAL-131 | H.a. | 0.14 | 0.66 | Recollection at same locality revealed pocket of "old" shells believed to be glacially transported. This shell from the "old" population. |
| AAL-883 | H.a. | 0.14 | 0.61 | Later recollection from same locality as above collections. |

drift suspected of containing glacially transported and indigenous shells. The amino acid ratios clearly distinguish two populations. Nelson (1978) also found a large number of mixed populations in a nearby region. It is probable that mixed populations are not infrequent in Quaternary deposits.

## Relative Chronostratigraphy

Amino acid ratios from monogeneric mollusk samples in deposits within a similar climatic regime may be applied directly as stratigraphic correlation indices and to develop a regional relative chronostratigraphic framework. Miller et al. (1977) and Andrews et al. (1976) used this method to subdivide and correlate deposits of the last glaciation in the eastern Canadian Arctic (Fig. 4). Nelson (1978) working in the same general region, applied computerized statistical procedures to subdivide amino acid data on 118 samples into eight aminozones.

A study of isoleucine epimerization ratios in *Corbicula fluminalis* from the Thames estuary, U.K. (Table 3), yielded four age clusters tentatively associated with Ipswichian, Hoxnian, pre-Hoxnian and Icenian Crag units (Miller et al., 1979). Assuming that the deposits represent high stands of the sea, and that the Ipswichian deposits are of last interglacial age, then probable correlations with the deep-sea oxygen isotope stratigraphy can be suggested for the older samples.

## Absolute Chronology

Absolute age calculations require an understanding of the racemization kinetics and an evaluation of the integrated thermal history or an independent dating control. Mitterer (1975, 1977) attempted to date a series of raised marine terraces on the SE Atlantic Seaboard based on isoleucine epimerization in *Mercenaria* using last interglacial deposits as a calibration point. Wehmiller et al. (1977) attempted a similar chronostratographic subdivision from enantiomeric ratios in marine bivalves in the California terrace sequences but invoked an empirically derived nonlinear kinetic model. A reevaluation of Mitterer's ages using nonlinear kinetics increased his proposed time scale by a factor of 2 to 3. Kvenvolden et al. (1977) estimated an age of about 125,000 for *Saxidomus* shells in Washington state but indicated a large uncertainty because of inaccuracies in the thermal history term.

| LITHOSTRATIGRAPHY[1] Fmtn | Member | Informal Subdivision | BIOSTRATIGRAPHY[2] | CHRONOSTRATIGRAPHY | | | AMINO ACID[3] F | AMINO ACID[3] C |
|---|---|---|---|---|---|---|---|---|
| Clyde Foreland Formation | Eglinton Member | Scott Inlet eolian sands | | Holocene | Neoglaciation | Modern | — | .017±.001 |
| | | | | | | 2000–5000 BP | | |
| | | peat/soil | | | | | | |
| | | Ravenscraig marine sediments | | | late | 8000–10000 BP | < .08 | .020±.001 |
| | | peat/soil | | | | | | |
| | Kogalu Member | upper marine sediments | | Foxe Glaciation | mid | 40000–48000 BP ($^{14}C$) | .29±.04 | .045±.007 |
| | | Ayr Lake till | drift 1? | | | | | |
| | | lower marine sediments | Islandiella islandica subzone | | | | .29±.04 | .045±.007 |
| | Kuvinilk Member | upper marine sediments | | | early | | .56±.05 | .10±.015 |
| | | Clyde till | drift 2 ? | | | | | |
| | | lower marine sediments | | | | | .56±.05 | .10±.015 |
| | Cape Christian Member | soil / Cape Christian soil | Cassidulina teretis subzone | interglaciation | | ca. 130 000 BP ($^{230}Th$) | .66±.05 | .15±.025 |
| | | Cape Christian marine sediments | | | | | | |
| | | Sledgepointer till | drift 3 | pre-last glaciation | | | | |
| | older unnamed members | not subdivided | Rotundatus orbiculare zone | | | | 0.8 | 0.18 |
| | | | Nonion tallahattensis zone | | | | .9±.1 | 0.23 |
| | | | | | | | up to 1.10 | up to 0.42 |

**Figure 4.** Regional chronostratigraphic framework derived from amino acid ratios in *Hiatella arctica* samples from eastern Baffin Island, Arctic Canada (from Miller et al., 1977; reprinted with permission). Amino acid data (3) are the ratio of D-alloisoleucine to L-isoleucine in the free (F) and total (C: free plus peptide bound) amino acid assemblages. Note that in the younger samples the standard deviations are smaller for the free ratios than for the total ratios, but in the older samples the free ratios are less precise.

### Aid to Radiocarbon Dating

Radiocarbon dates on carbonate materials are not considered reliable for deposits older than about 30,000 years. Amino acid analyses can be performed more rapidly and are less costly than a $^{14}C$ date; hence it is sometimes practical to screen molluscan samples prior to submission by analyzing for the associated amino acid ratios (see Blake, this volume). In addition, because the incorporation of secondary amino acids by carbonate fossils is negligible, it may be feasible to isolate indigenous organic carbon from molluscan fossils for dating by the highly sensitive accelerator method. By that method it may prove possible to extend $^{14}C$ chronology of carbonate fossils into the 50,000 year time range (see also Miller and Hopkins, this volume).

## SUMMARY AND EVALUATION

The following conclusions have been reached regarding the diagenesis of proteinaceous material in mollusks preserved in the sedimentary environment:

1. The primary factor affecting the rate constants of the various amino acid reactions in molluscan shells is temperature. The buffering effect and the integrity of the carbonate matrix effectively exclude effects of other environmental parameters.
2. Although there appears to be an initial reduction in amino acid concentrations during the early stages of protein diagenesis in molluscan fossils, the relative abundances of amino acids are not altered, and this brief period is followed by a greatly reduced depletion rate over the next $10^6$ or more years, during which the shell approximates a closed system. Available evidence indicates that there is no significant incorporation of secondary amino acids by the carbonate matrix over $10^5$ to $10^6$ years, and in cases of exceptional preservation for $10^7$ to $10^8$ years.
3. Intershell isoleucine epimerization ratios in monogeneric samples commonly differ by up to 20%. Hence it is necessary to analyze at least three individuals from a given stratum to obtain a representative ratio. If the allo/iso (total fraction) ratios differ by more than $\pm 10\%$ of the mean, it may be necessary to analyze additional shells to determine if the variation is due to an inherent variability in the sample or to a sample population of more than one age.

Our evaluation of the status of amino acid geochronology in Pleistocene mollusks is that racemization data, supplemented by various other amino acid ratios resulting from hydrolysis and decomposition reactions, can be used as follows:

1. If the ratios determined from three or more in situ mollusks do not deviate from their respective means by more than about 10%, then the mean value can be considered to characterize the deposit. Racemization in the total acid hydrolysate has proven to be the most reliable and consistent ratio, but for young and/or polar samples, racemization in the free amino acid fraction can be a more sensitive parameter. The proportion of free alanine is the most sensitive parameter at the onset of diagenesis (up to an allo/iso ratio of 0.05); but in older samples the proportion of free valine, isoluecine, or leucine is more reliable. Threonine/glutamic acid ratios in the total fraction show a steady decrease with increasing sample age. The hydrolysis and decomposition reactions are particularly informative in samples for which the total allo/iso ratios differ markedly between monogeneric shells. In these instances, the other ratios can be used to independently confirm or deny the apparent age differences suggested by the epimerization data.
2. Over any limited geographic region characterized by similar mean annual and seasonal temperature regimes and uniformly affected by paleoclimatic changes, amino acid ratios in monogeneric mollusk samples can be used directly to correlate between disjunct fossiliferous strata and to develop a regional relative chronostratigraphy. We term this usage *aminostratigraphy*; subdivisions have been termed aminozones (Nelson 1978). This usage is the least ambiguous application of amino acid geochronology, and it has the advantages of remaining independent of the mechanisms and kinetics of racemization and eliminating the thermal history term.
3. In areas for which independent absolute age data are available, it is possible to calibrate the reaction rates and provide an *approximate* absolute time scale. Extrapolation based on $^{14}C$ dated samples to older deposits must be considered tenuous because of the nonlinearity of the racemization kinetics and uncertainties in the temperature term. A more reliable time base would be the Brunhes/Matayama paleomagnetic reversal, one of the Pearlette ashes or similar time-stratigraphic marker bed of mid-Pleistocene age. It is probably reasonable to assume that the average temperature of the last $10^6$ years at a given site is applicable to any sample over $10^5$ years old. Thus the extent of racemization in a mollusk genus of mid-Pleistocene age may be

used as a base to interpolate/extrapolate approximate ages for adjacent deposits containing the same genus in the range of $10^5$ to $2 \times 10^6$ years.

## *ACKNOWLEDGMENTS*

Support for this project was provided by a postdoctoral fellowship to GHM (1974–1975) by the Geophysical Laboratory of Carnegie Institution of Washington, and a grant from the National Science Foundation (EAR-74-01857) to evaluate the potential of amino acid geochronology utilizing Arctic mollusks. The original incentive for the study of Arctic materials came from Dr. J. T. Andrews, University of Colorado, and we would like to thank him for his continued encouragement of the program. Analyses were made at the Geophysical Laboratory, Washington, D.C., and INSTARR, University of Colorado. Support for the analyses of nonmarine mollusks was provided by NSF Grant EAR-78-08222. We would like to thank J. K. Brigham, W. D. McCoy, and W. N. Mode for assistance in the laboratory. Dated mollusks used to construct Table 9 were supplied by D. M. Hopkins (Anchorage), R. D. Miller (SE Alaska), M. Stuiver (Maine), A. Weidick (Greenland), J. T. Andrews (Washington), D. R. Grant (New Brunswick), G. S. Boulton (Spitzbergen), A. S. Dyke (Somerset Is.), H. Krog (Denmark), W. Blake, Jr. (Modern).

## *APPENDIX*

Depending on the size of a sample, a shell is cleaned by one of two procedures. Shells weighing more than 100 mg are processed by Procedure 1, and shells weighing less than 100 mg require Procedure 2.

**Procedure 1.** All periostracum and adhering sediment are removed by brush and dental tool if required. A fragment weighing 100 to 400 mg (generally taken from the growing edge of a shell) is separated and sufficient 2 *N* HCl added to dissolve one third of the shell by weight. Completion of the reaction is ensured by placing the sample in a vacuum centrifuge for 5 minutes. The supernatant is discarded and the sample vigorously rinsed three times.

**Procedure 2.** Shells are fragmented to expose interior chambers (if needed), then placed in sturdy glass jars with excess sodium metaphosphate solution (50 mg/l) and sonicated for five minutes. Supernatant is discarded, and the procedure is repeated four cycles, followed by five rinse cycles with purified water.

**Preparation of free and total fractions.** A portion of a shell cleaned as in 1 or 2 above weighing about 50 mg (recorded to the nearest 0.1 mg) is transferred to a sterile culture tube. 1.0 ml cold, purified 7 *N* HCl is added and the sample is placed in the vacuum centrifuge for 5 minutes. The resultant HCl solution is about 6 *N*. Exactly one half of this solution is transferred by pipet to a 1-dram vial and immediately dried in a heated block (85°C) under a stream of $N_2$ (30 minutes) or in the vacuum centrifuge at room temperature (8 hours). This is the free amino acid fraction. The culture tube containing the remaining solution is flushed with $N_2$, sealed, and heated at 110°C for 22 hours. When cooled, the solution is transferred to a 1-dram vial and dried as for the free fraction. This is the total (free plus peptide bound fraction). The dried samples are rehydrated in a pH 1.9 solution (2 ml 6 *N* HCl/500 ml purified $H_2O$) containing 3.125 M/l norleucine. Samples are rehydrated at a ratio of 25 mg shell weight/ml rehydration solution. For most samples 40 to 80 μl free solution and 10 to 30 μl total solution contains sufficient amino acids for detection by the analyzer. To optimize resolution of alloisoleucine, the column temperature is maintained at 62°C and the second buffer solution is at a pH of 3.86. The ratio of OPA to column effluent is approximately 1.2.

## *REFERENCES*

Abelson, P. H., Organic constituents of fossils, *Carnegie Inst. Wash. Year Book 54,* 107–109, 1955.

Akiyama, M. and R. W. G. Wyckoff, The total amino acid content of fossil pecten shells, *Proc. Nat. Acad. Sci. U.S.A.,* **67,** 1097–1100, 1970.

Andrews, J. T., R. W. Feyling-Hanssen, G. H. Miller, C. Schlüchter, M. Stuiver, and B. J. Szabo, Alternative models of early and middle Wisconsin events, Broughton Island, Northwest Territories of Canada: Toward a Quaternary chronology, *IGCP Project 73-1-24, Report No. 3,* D. J. Easterbrook and V. Sibrava, eds., 28–61, Bellingham, Washington, 1976.

Bada, J. L., M-Y. Shou, E. H. Man, and R. A. Schroeder, Decomposition of hydroxy amino acids in foraminiferal tests; kinetics, mechanism and geochronological implications, *Earth Planet. Sci. Lett.,* **41,** 67–76, 1978.

Bender, M. L., Reliability of amino acid racemization dating and paleotemperature analysis on bones, *Nature,* **252,** 378–380, 1974.

Benson, J. R. and P. E. Hare, O-Phthalaldehyde: Fluorogenic detection of primary amines in the picomole range: Comparison with fluorescamine and ninhydrin, *Proc. Nat. Acad. Sci. U.S.A.,* **72,** 612–622, 1975.

Birkeland, P. W. and R. E. Burke, Re-evaluation of multiple relative dating techniques and their application to the glacial sequence along the eastern escarpment of the Sierra Nevada, California, *Quat. Res.*, **10,** in press.

Degens, E. T., D. W. Spencer, and R. H. Parker, Paleobiochemistry of molluscan shell proteins, *Comp. Biochem. Physiol.*, **20,** 553–579, 1967.

Dixon, W. J. and F. J. Massey, Jr., *Introduction to Statistical Analysis,* McGraw-Hill, NY, 488 pp., 1957.

Dreimanis, A., Correlation of Wisconsin glacial events between the eastern Great Lakes and the St. Lawrence Lowlands, *Géogr. Phys. Quat.*, **31,** 37–51, 1977.

Hare, P. E. and P. H. Abelson, Racemization of amino acids in fossil shells, *Carnegie Inst. Wash. Year Book 66,* 526–528, 1967.

Hare, P. E. and R. M. Mitterer, Nonprotein amino acids in fossil shells, *Carnegie Inst. Wash. Year Book 65,* 362–364, 1966.

Hare, P. E. and R. M. Mitterer, Laboratory simulation of amino acid diagenesis in fossils, *Carnegie Inst. Wash. Year Book 67,* 205–208, 1968.

King, K., Jr. and P. E. Hare, Species effects in the epimerization of L-isoleucine in fossil planktonic foraminifera, *Carnegie Inst. Wash. Year Book 71,* 596–598, 1972.

King, K., Jr. and C. Neville, Isoleucine epimerization for dating marine sediments: Importance of analyzing monospecific foraminifera samples, *Science,* **195,** 1333–1335, 1977.

Kriausakul, N. and R. M. Mitterer, Isoleucine epimerization in peptides and proteins: Kinetic factors and applications to fossil proteins, *Science,* **201,** 1011–1014, 1978.

Kvenvolden, K. A., D. J. Blunt, and H. E. Clifton, Application of amino acid stereochemistry to the correlation of late Pleistocene deposits at Willipa Bay, Washington, *Geol. Soc. Amer. Abst. Progr.*, **9,** 1062–1063, 1977.

Lineback, J. A. and J. T. Wickham, Is the Illinoian a superstage? *Geol. Soc. Amer. Abst. Progr.,* **10,** 444–445, 1978.

Matter, P. III, F. D. Davidsen, and R. W. G. Wyckoff, The composition of fossil oyster shell proteins, *Proc. Nat. Acad. Sci. U.S.A.*, **64,** 970–972, 1969.

Miller, G. H. and P. E. Hare, Use of amino acid reactions in some Arctic marine fossils as stratigraphic and geochronological indicators, *Carnegie Inst. Wash. Year Book 74,* 612–617, 1975.

Miller, G. H., J. T. Andrews, and S. K. Short, The last interglacial-glacial cycle, Clyde Foreland, Baffin Island, N.W.T.: Stratigraphy, biostratigraphy and chronology, *Can. J. Earth Sci.*, **14,** 2824–2857, 1977.

Miller, G. H., A. R. Nelson, W. D. McCoy, and A. L. Metcalf, Amino acid geochronology utilizing non-marine molluscan fossils, *Geological Soc. Amer. Abst. Progr.*, **11,** 1979.

Miller, G. H., J. T. Hollin, and J. T. Andrews, Aminostraitigraphy of UK Pleistocene deposits, *Nature,* **281,** 539–543, 1979.

Mitterer, R. M., Pleistocene stratigraphy in southern Florida based on amino acid diagenesis in fossil *Mercenaria, Geology,* **2,** 425–428, 1974.

Mitterer, R. M., Ages and diagenetic temperatures of Pleistocene deposits of Florida based on isoleucine epimerization, *Earth Planet. Sci. Lett.*, **28,** 275–282, 1976.

Morrison, R. B., Predecessors of Great Salt Lake, *Geol. Soc. Amer. Abst. Progr.*, **7,** 1206, 1975.

Nelson, A. R., Quaternary glacial and marine stratigraphy of the Qivitu Peninsula, northern Cumberland Peninsula, Baffin Island, Canada, Ph.D. dissertation, Univ. of Colorado, Boulder, CO, 298 pp., 1978.

Schroeder, R. A. and J. L. Bada, A review of the geochemical applications of the amino acid racemization reaction, *Earth Sci. Rev.*, **12**, 347–391, 1976.

Vallentyne, J. R., Biogeochemistry of organic matter—II, Thermal reaction kinetics and transformation products of amino compounds, *Geochim. Cosmochim. Acta,* **28,** 157–188, 1964.

Vallentyne, J. R., Pyrolysis of amino acids in Pleistocene *Mercenaria* shells, *Geochim. Cosmochim. Acta,* **33,** 1453–1458.

Wehmiller, J. F. et al., Correlation and chronology of pacific coast marine terrace deposits of continental United States by fossil amino acid stereochemistry-technique evaluation, relative ages, kinetic model ages, and geologic implications, *U.S.G.S. Open File Rept. 77-680,* 106 pp., 1977.

Wehmiller, J. R. and D. F. Belknap, Alternative kinetic models for the interpretation of amino acid enantiomeric ratios in Pleistocene mollusks: Examples from California, Washington and Florida, *Quat. Res.*, **9**, 330–348, 1978.

Williams, K. M. and G. G. Smith, A critical evaluation of the application of amino acid racemization to geochronology and geothermometry, *Origins of Life,* **8,** 91–144, 1977.

# Degradation of Molluscan Shell Protein by Lava-Induced Transient Heat Flow, Pribilof Islands, Alaska: Implications for Amino Acid Geochronology and Radiocarbon Dating

**GIFFORD H. MILLER**
*Institute of Alpine and Arctic Research and Department of Geological Sciences, University of Colorado*

**DAVID M. HOPKINS**
*Branch of Alaskan Geology, United States Geological Survey, Menlo Park, California*

*ABSTRACT*

Mollusk shells of last interglacial (and older) age are found in association with lava flows and pillow lavas on St. Paul Island, Pribilof Islands, Alaska. The amino acid reactions in these shells are far more advanced than would be expected under normal thermal conditions. It is concluded that lava-induced transient heat flow accelerated the amino acid reactions during a short time span (<50 yr) shortly after shell deposition. In two shell collections, the hydrolysis and racemization reactions reached equilbrium during heating and in a third collection the heat was sufficiently strong to decompose all amino acids. The retention of exclusively low-molecular-weight free amino acids and the absence of secondary amino acids in these shells confirms the integrity of the carbonate matrix and has implications for direct-count radiocarbon dating of indigenous organic carbon in carbonate fossils.

## INTRODUCTION

The Pribilof Islands form a small archipelago on the outer continental shelf of the Bering Sea off the west coast of Alaska. St. Paul Island, the northernmost island of the Pribilof group and the focus of this report, has undergone recurrent faulting and volcanic activity throughout much of the middle and upper Quaternary. The island consists primarily of basaltic rocks with lesser amounts of interbedded and underlying sediments of upper Tertiary and Quaternary age (Barth 1956). The Quaternary geology was mapped by Hopkins and Einarsson (unpublished) and age assignments were obtained by K-Ar dates (Cox et al., 1966). Because the only source rocks on the island are of basaltic composition, layers of quartz and mica-bearing eolian sand and silt must have been deflated from shelf sediment exposed during episodes of glacioeustatically lowered sea levels. Conversely, marine sands and gravels and pillow lavas record high stands of the sea that are inferred to represent interglacial sea level highs. Molluscan fauna are associated with many of the raised marine units, and valves of the pelecypods *Mya truncata* and *M. elegans* were analyzed for their amino acid content and the extent of various diagenetic amino acid reactions as part of a larger program on the geochronology of sea level fluctuations throughout coastal Alaska.

## SAMPLE LOCALITIES

Molluscan fossils were obtained from exposures in marine deposits at Tolstoi Point and Einahnuhto Bluffs. The samples from Tolstoi Point were separated from an overlying lava flow by 2.5 m of wind-blown sand and volcanic ash.

The section at Einahnuhto Bluffs consists of marine sediments interbedded with subaerial lava flows, pillow lavas, bedded tuffs and eolian sand and silt, and has been described by Cox et al. (1966) and Hopkins (1967, Figure 5; 1973, Figure 8). At least two distinct marine transgressions separated by low sea stands and periglacial climatic conditions are represented in the sequence. A thick basal marine unit, designated by Hopkins (1967) as the Einahnuhto transgression, lies unconformably on a buried wave-abrasion platform containing erosional remnants of thinly bedded lava flows that represent buried sea stacks. The marine sands and gravels contain interbedded pillow lavas. A mollusk collection containing valves of *Mya truncata* (AAL-28) was obtained from the marine sediments below the pillow lavas.

A younger sequence of interglacial marine sand, gravel and pillow lavas unconformably overlie the Einahnuhto beds and are overlain by the Rush Hill lava flow. This younger marine sequence was originally correlated with the middle Pleistocene Kotzebuan marine transgression (Hopkins, 1967, 1973), but a reevaluation of the stratigraphy indicates rather that it was deposited during the last interglacial (Pelukian) high stand of world sea level. Within the last interglacial unit, two marine beds have been recognized. The upper bed consists of fossiliferous marine sand and gravel overlain by 0.5 m eolian silts that are overlain by 2 m of ash on which rests the Rush Hill flow. Valves of *Mya elegans* (AAL-30) were collected from the marine sands. The lower marine beds are associated with pillow lavas and they unconformably overlie the lava capping the Einahnuhto beds. Sample AAL-29, containing valves of *M. truncata,* was collected from the marine sands in direct association with the pillow lavas.

K-Ar age determinations have been reported from several of the lava members (Cox et al., 1966). The oldest basalt flow, exposed as remnant sea stacks at the base of the Einhanuhto Bluffs has yielded K-Ar dates of about 160,000 and 320,000 yr BP. The Rush Hill flow, capping the Einahnuhto Bluff sequence, has been dated by the K-Ar method several times with a wide scatter of resultant ages. (Cox and others, 1966; M. L. Silberman, Albert Yang, and Minze Stuiver, unpubl. data, 1976–79). Despite the scatter, there is little doubt that the flow is near 100,000 years old; the underlying marine beds are evidently of last interglacial age and about 120,000 years old (Shackleton and Opdyke, 1973), and the oldest marine unit in the Einahnuhto Bluffs represent an earlier interglaciation.

## AMINO ACID ANALYSES

The amino acid assemblages in valves of *Mya* from the above collections were determined on an automated ion-exchange liquid chromatography amino acid analyzer (sample preparation procedure and analyzer description are given in Miller and Hare (this volume, appendix). The concentrations of amino acid residues and isoleucine epimerization ratios in these samples are given in Table 1. For comparative purposes we have also included the same data for *Mya truncata* (AAL-27) from the Bootlegger Clay, Anchorage, that has been radiocarbon dated at 13,000 ± BP.

The differences in amino acid ratios and concentrations in the three samples of last interglacial age (AAL-28, -29 and -31) are extreme (Table 1) and suggest that time is not the primary variable controlling the degree of protein diagenesis in these samples. The differences between the allo/iso ratio in the Bootlegger Clay sample (AAL-27) and the one from the Tolstoi Point sample (AAL-31) might be explained by the greater time since death of the Tolstoi Point shells, although the allo/iso ratio is some-

**Table 1. Amino Acid Residues in *Mya* from the Pribilof Islands, Alaska**

| Sample ID[a] | Location | Age | Thermal History | T/F[b] | Amino Acid Residues (Pmole/mg shell) | | | | | | | | | Allo: Iso |
|---|---|---|---|---|---|---|---|---|---|---|---|---|---|---|
| | | | | | Asp | Thr | Ser | Glu | Gly | Ala | Val | Iso | Leu | |
| AAL-27 (2) | Bootlegger Cove Clay | 14,000 BP (14C) | +2.1°C | T | 3490 | 936 | 1055 | 1442 | 1866 | 754 | 525 | 382 | 390 | 0.034 |
| | | | | F | 410 | 195 | 344 | 181 | 546 | 250 | 79 | 29 | 64 | 0.16 |
| AAL-31 (2) | Tolstoi Point | *ca.* 125,000 YBP | Current temp. = +0.5°C shells 4.3–5.0 m below subaerial lava flow | T | 2332 | 356 | 182 | 884 | 1324 | 1088 | 650 | 306 | 412 | 0.18 |
| | | | | F | 853 | 169 | 104 | 139 | 485 | 619 | 302 | 116 | 183 | 0.38 |
| AAL-29 (2) | Einahnuhto Bluffs (upper marine bed of last interglacial sediments) | *ca.* 125,000 YBP | 1.0 to 2.5 m below subaerial lava flow | T | 9 | 0 | 1 | 26 | 16 | 8 | 4 | 1 | 2 | — |
| | | | | F | 5 | 1 | 3 | 13 | 15 | 7 | 3 | 0.5 | 0.5 | — |
| AAL-30 (3) | Einahnuhto Bluffs (lower marine bed of last interglacial sediments) | *ca.* 125,000 YBP | <1.2 m below pillow lava | T | 90 | 4 | 10 | 930 | 660 | 1161 | 336 | 106 | 150 | 1.26 |
| | | | | F | 68 | 1 | 3 | 179 | 611 | 1032 | 263 | 74 | 90 | 1.31 |
| AAL-28 (2) | Einahnuhto Bluffs (basal marine beds; next-to-last interglacial) | >125,000 YBP | 0.3 to 1.5 m below pillow lava | T | 210 | 1 | 2 | 668 | 652 | 880 | 300 | 79 | 116 | 1.22 |
| | | | | F | 141 | 0 | 0 | 205 | 691 | 938 | 272 | 69 | 111 | 1.30 |

[a] Numbers in parentheses are the numbers of individual shells analyzed.
[b] T = total amino acid assemblage.
F = free amino acid fraction.

what higher than in Pelukian *Mya* from the mainland near Nome (0.18 vs. 0.04) (but note that the mean annual temperature at Nome is $-4.7°C$ as opposed to $+0.5°C$ at St. Paul).

The amino acid assemblages in all three samples from the Einahnuhto Bluffs sequence (AAL-28, -29, -30) show considerably advanced states of protein degradation that cannot be attributed to age alone and are not compatible with their stratigraphic positions. Shells from the upper bed of the last interglacial sediments (AAL-28) and from the Einahnuhto transgression deposit (AAL-30) have similar amino acid compositions, concentrations and isoleucine epimerization ratios. The amino acid assemblages in the free and total (free plus peptide bound) amino acid fractions in these samples show only minor differences, and the allo/iso ratios are between 1.22 and 1.31, suggesting that the hydrolysis reaction is at equilibrium and the mixture is racemic. Threonine and serine, the two least stable amino acids are present only at very low levels, representing a reduction of 2 to 3 orders of magnitude from their initial concentrations. The detected residues of these amino acids may have resulted in part from slight contamination in the laboratory reagents. Aspartic acid has also been reduced to less than 10% of its original concentration. On the other hand, Alanine is present in higher levels than in the early postglacial samples because of its stability in the free state and its relative enrichment from various decomposition reactions. The other stable amino acids (Glu, Gly, Val, Leu, Iso + Allo) have been reduced to about half their initial concentrations in the postglacial sample. The proportion of free glutamic acid that forms pyroglutamic acid is undetectable in the free amino acid fractions but reverts to glutamic acid during the acid hydrolysis step causing the large discrepancies between the free and total glutamic acid residues in Table 1.

In contrast to AAL-28 and -30, the shells from the lower last interglacial beds (AAL-29) in the bluffs are essentially devoid of amino acids. It is uncertain whether the detected residues were derived from the shells or from laboratory contaminants, but all amino acid residues have been reduced to 1% or less of their original concentrations, and the leucines have been reduced to less than 0.1%.

We propose that these three samples have experienced a short-term intense heat flux during deposition of lava flows associated with the marine units. The lavas at Tolstoi Point are separated from the shell collections by intervening sediment; hence the influence of transient heat flow in the shells was minimal. The heat released during the cooling of the lavas accelerated the various amino acid reactions, resulting in nearly complete hydrolysis and racemization and decomposition of the less stable amino acids. Sample AAL-29 is most closely associated with lavas and the transient heat flow experienced by these shells was sufficient to drive all reactions to completion, including decomposition of even the most stable amino acids.

Based on the Arrhenius parameters in Miller and Hare (this volume, Figure 3) we have computed possible time/temperature histories required to yield the isoleucine epimerization ratios of Table 1. Assuming the lavas were insignificant heat sources, a mean annual temperature of at least 25°C for the last 100,000 years would be required to yield the assemblages for AAL-28 and -30. Although the geothermal gradient beneath the Pribilof Islands is somewhat elevated, the required temperature is an order of magnitude greater than could be explained by the geothermal gradient.

A greater average temperature over a shorter time span could equally well account for the observed assemblages. Based on the same Arrhenius parameters used above, an average temperature of 100°C over a decade or 160°C for a few days would produce similar results. Dr. A. Lachenbruch (USGS, Menlo Park) has made calculations on the transient heat effects of lava flows overlying sedimentary deposits, and he concludes that elevated temperatures of the range and durations suggested above are reasonable within a few meters of a thick basalt flow. We conclude, therefore, that the shells in samples AAL-28, -29, and -30 experienced elevated temperatures induced by overlying lavas that greatly accelerated the various amino acid reactions within a decade or two of lava deposition. Since that time, negligible diagenesis has occurred.

## IMPLICATIONS FOR AMINO ACID GEOCHRONOLOGY AND $^{14}C$ DATING

The exceptional thermal histories of the samples from St. Paul Island provide documentation of the integrity of the carbonate structure. The shells in AAL-28 and -30 received enough lava-induced transient heat to drive the hydrolysis reaction to completion shortly after deposition. Since that time, the amino acids have existed in the free state; yet despite their low molecular weight (relative to protein and polypeptides), there has been little appreciable loss of amino acids. The low residues of the less stable amino acids are attributed to normal decomposition reactions (note, for example, that Alanine is present at greater levels and valine at 60% of its level in the early postglacial *Mya*.)

Secondly, shells in sample AAL-29 were so severely heated that more than 99% of all the indigenous proteinaceous material was destroyed during the transient heating that must have occurred shortly after shell deposition, as shells in the overlying unit that were also deposited during the last interglacial did not experience such intense heating. Consequently, the shells in AAL-29 have existed in an emergent, sandy marine unit for more than $10^5$ yr, without indigenous amino acids. If carbonate fossils incorporate significant levels of secondary amino acids from extrashell sources, they would be apparent in the analyses. The very low levels of amino acids detected in the two shells from these beds may represent the extent of incorporation of secondary amino acids in that period, but it is likely that at least some portion are the product of minor impurities in the laboratory reagents. It is noteworthy that the concentration of isoleucine in these shells (the amino acid most frequently utilized in geochronological analyses) was only 0.1% of its level in the postglacial samples.

Although carbonate fossils can be expected to faithfully preserve indigenous amino acids over long periods of geological time, they are notorious for the extent of carbonate exchange. Hence radiocarbon dates on carbonate shells are generally not considered reliable beyond about 30,000 years. However, recent developments in radiometric dating by accelerator methods open a new dimension into the dating of carbonate fossils. The advantage of the accelerator methods lies primarily in the capability to detect $^{14}C$ activity in samples of only a few milligrams. The data presented above documents that the incorporation of secondary (younger) amino acids in $10^5$ yr is well under 1% and for certain amino acids, less than 0.1%. These levels (assuming all contaminants are modern) correspond to radiocarbon ages of 38,000 and 55,000 years, respectively. By isolating specific amino acids and eliminating laboratory contaminants, it would be possible to determine the $^{14}C$ activity in the indigenous organic compounds of carbonate fossils and to obtain reliable ages in excess of 50,000 years. Because of the abundance of carbonate fossils (relative to wood) in terrestrial and shallow marine deposits, we suggest that this method may provide a significant improvement on the chronology of events during the middle period of the last glaciation.

Another intriguing possibility is to obtain a series of shells at various depths beneath a lava flow and attempt to reconstruct the transient heat flux through the sediment based on differences in the extent of the amino acid reactions.

## *REFERENCES*

Cox, Allan, Hopkins, D. M., and Dalrymple, G. R., 1966. Geomagnetic polarity epochs: Pribilof Island, Alaska: Geol. Soc. America Bull, v. 77, p. 1089–1108.

Hopkins, D. M., 1967. Quaternary marine transgression in Alaska: *in* D. M. Hopkins, ed., The Bering Land Bridge, Stanford Univ. Press, p. 47–86.

Hopkins, D. M., 1973. Sea level history in Beringia during the last 250,000 years: Quaternary Res., v. 3, p. 520–540.

Shackleton, N. J. and Opdyke, N. D., 1973. Oxygen isotope and paleomagnetic stratigraphy of equatorial Pacific core V28-238: oxygen isotope temperatures and ice volumes on a $10^5$ and $10^6$ year scale: Quaternary Research, v. 3, p. 39–55.

# Application of Amino Acid Ratios to Studies of Quaternary Geology in the High Arctic

**WESTON BLAKE, JR.**
*Terrain Sciences Division, Geological Survey of Canada, Ottawa, Ontario*

## INTRODUCTION

The advent of amino acid "dating" has provided students of Quaternary geology with yet another useful tool for deciphering chronological problems beyond the limit of radiocarbon dating (roughly 50,000 years or less for most laboratories). Determination of amino acid ratios is proving to be of particular value in areas such as the Canadian Arctic Archipelago, where abundant deposits of mollusk-bearing marine sediments are exposed at various elevations above sea level. The intent of this report is to mention just two of the ways in which amino acid ratios in mollusks are being utilized within the Geological Survey of Canada (GSC). The reader is referred to an earlier report by Miller and Hare (1975a) for a discussion of amino acid diagenesis in Arctic shells and the applicability of this dating method to correlating marine sediments.

## USE IN CONJUNCTION WITH RADIOCARBON DATING

Although determinations of amino acid ratios are not being carried out at the GSC, a Radiocarbon Dating Laboratory has been in operation since 1961 (Dyck 1967; Lowdon et al., 1977). One of the problems that confronts the GSC laboratory, or any laboratory involved in dating mollusks, is the validity of $^{14}C$ ages on carbonate material beyond, say, 20,000 years. This topic has been discussed at length by a number of authors, and the reader is referred to the useful summaries provided by Olsson (1968, 1972) and Stuckenrath (1977). For the purposes of the present report it is sufficient to note that only 1% of contamination with modern carbon is sufficient to give an apparent age of 35,000 years for a sample that is in reality more than 44,000 years old (Olsson and Blake 1962).

On many of the islands in the Canadian Arctic Archipelago mollusk shells (mainly pelecypods) have been collected above the limit of Holocene marine submergence. Most of these shells occur as fragments or occasional intact valves in till or till-like materials, whereas others occur in apparently undisturbed marine deposits which, because of their high elevation above the present level of the sea, are interpreted as being of interglacial age. Similar undisturbed marine deposits also occur at elevations closer to present sea level, in many cases beneath a veneer of Holocene-age deposits. Radiocarbon age determinations on these "old" mollusk shells, particularly samples collected from the ground surface, are apt to result in finite values in the 25,000 to 40,000 year range. Yet at the present state of knowledge it seems doubtful if many, if indeed any, of these determinations can be accepted as giving the true ages of the mollusks. There is no problem when a "greater than" result is obtained; it is the finite determinations that are difficult to interpret.

In recent years we have tried to attack this problem in various ways. One approach, when we are lucky enough to be dealing with large, robust pelecypods such as *Mya truncata, Hiatella arctica,* or *Astarte borealis* (preferably 10 g or more in weight so that the outer 10% of shell can be removed by HCl leaching, although smaller samples have been dated), is to use a single valve or a single pair of pelecypods for the radiocarbon age determination. In this way the possibility is eliminated that the sample contains shells of various ages.

Another approach has been to analyze a single pelecypod valve or several valves within a collection to determine amino acid ratios. For Arctic samples the ratios of D-alloisoleucine to L-isoleucine in both the free and total (free plus peptide-bound)* fractions have proven to be most useful (cf. Miller et al., 1977); the ratios increase with increasing age (Miller and Hare 1975b). Three representative valves or portions of valves, as only a fraction of a gram is needed, normally are tested in each collection. If there appear to be two distinct populations of shells in terms of size, robustness, state of preservation (e.g., pitting or incrustations), etc., then valves from both populations will be analyzed. Most of this work has been carried out under contract by Dr. G. H. Miller of INSTAAR, University of Colorado, and for each area it has been our practice to provide him with a control sample, either pelecypods of known Holocene age which

* This fraction has on occasion been referred to as "combined", but "total" is preferred (personal communication from G. H. Miller, November 1978).

have been dated in the GSC Radiocarbon Dating Laboratory or well-preserved valves from the modern storm beach which obviously have been living within the last few years.

If the ratios of D-alloisoleucine to L-isoleucine suggest that the age of the sample is beyond the limit of radiocarbon dating, it is unlikely that additional analyses will be undertaken. If, on the other hand, the ratios indicate that the sample is of Holocene age, then a sample is prepared for radiocarbon dating. Should the sample appear to contain mollusk shells of more than one age, including Holocene material, then it may be necessary to test additional valves in order to select a sample suitable for processing in the $^{14}C$ laboratory.

Utilizing amino acid ratios in this way has already saved the GSC Radiocarbon Dating Laboratory a considerable amount of valuable time. Furthermore, the proliferation of finite $^{14}C$ ages in the range of 25,000 to 40,000 years has been slowed, which seems prudent until more insight is gained into the significance of these "old" ages and how they may best be interpreted (cf. discussion in Barr 1971; Blake 1975a).

## CORRELATION OF TILLS BETWEEN ISLANDS

One of the characteristics of the Canadian Arctic Archipelago is that on a number of islands, as noted earlier, fragments of mollusks and less occasionally intact shells occur in till or till-like deposits. They are thus a type of erratic, as they have been excavated from the sea floor "up-glacier" (Blake 1975a) and redeposited together with the fines and the rock fragments of assorted sizes that comprise the till at a given site. Shelly till occurs in virtually all situations in the archipelago where glaciers had a chance to impinge on an island or on a peninsula after crossing a bay or flowing through a strait. Following a location map, Figure 1, two such sites are illustrated in Figures 2 and 3. Both sites are in Makinson Inlet, a major fiord indenting the east coast of Ellesmere Island.

Despite the fact that the underlying bedrock is quite different, red granite/granulite on Bowman Island (Frisch 1977) and Paleozoic gray dolomite on Swinnerton Peninsula (Christie 1962), both localities are characterized by a thin veneer (in many places $<0.5$ m thick) of shell-bearing calcareous till which overlies the striated surface of the bedrock. Similar alloisoleucine/isoleucine ratios, in both free and total fractions, have been obtained on fragments of *Hiatella arctica* (L.) from each site, as summarized in Table 1. The ratios (total allo/iso) differ markedly from the values obtained on Holocene shells of the same species collected nearby.

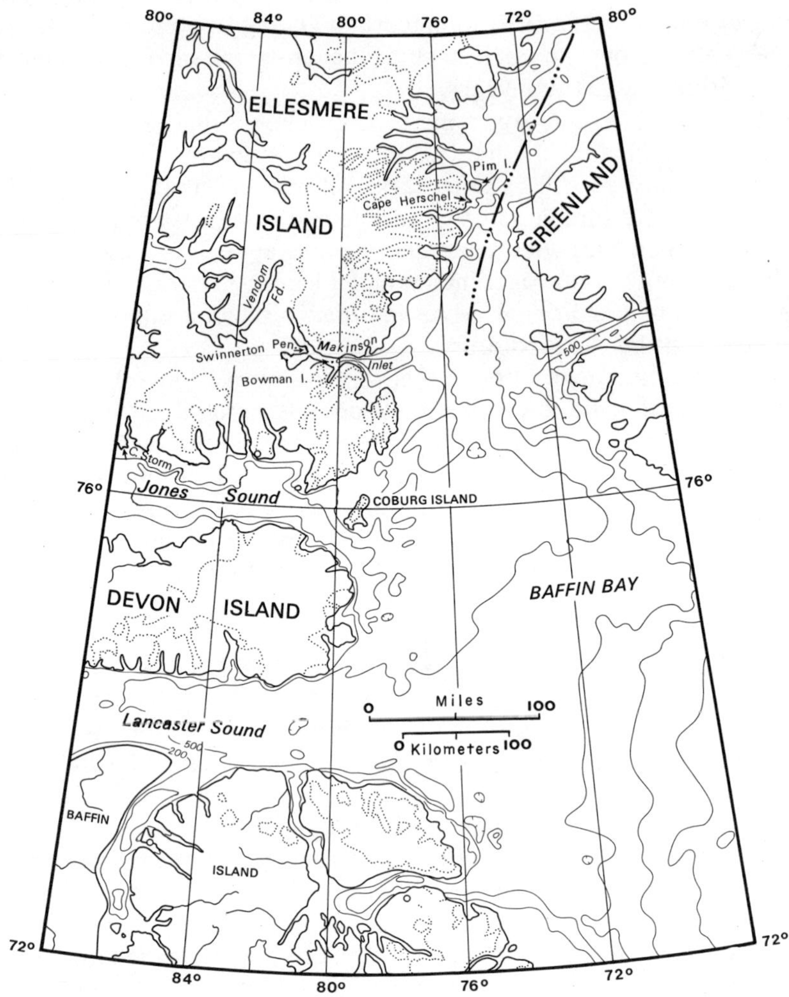


**Figure 1.** Location map, southeastern Queen Elizabeth Islands. Ice-covered areas are outlined by dots; bathymetric contours are in meters.

In one respect, such good correspondence between the ages of individual valves is to be expected, for at the time that Makinson Inlet was filled by an outlet glacier draining eastward, the valleys leading to it were occupied by glaciers tributary to the main ice stream. The dimensions of these glaciers were impressive. The shallowest water depth recorded on the north side of Bowman Island is approximately 400 m (Sadler 1973), but the shell fragments analyzed were collected at 250 to 270 m a.s.l.;

**Figure 2.** Telephoto view north at the eastern tip of Swinnerton Peninsula and the north arm of Makinson Inlet. The arrow indicates the site at an elevation of 345 m where pelecypod shell fragments were abundant in the till overlying striated dolomite. July 7, 1977. GSC photograph no. 173607.

**Figure 3.** Telephoto view southeast at Bowman Island, Makinson Inlet. The top of the island is approximately 570 m a.s.l. The shoulder at an elevation of 250–270 m, where shells abound in calcareous till overlying striated granite, is indicated by the arrow. July 9, 1977. GSC photograph no. 173639.

**Table 1. Analyses on *Hiatella arctica* (L.) shells, Makinson Inlet, Ellesmere Island**

| Field Sample No. | Site Location | Approximate Elevation (meters a.s.l.) | Radiocarbon Laboratory No. | Age ($^{14}C$ years)[a] | Amino Acid Laboratory No. | Alloisoleucine : isoleucine[b] | |
|---|---|---|---|---|---|---|---|
| | | | | | | Free | Total |
| BS-77-256 | Swinnerton Peninsula, south coast | 40 | GSC-2519 | 8930 ± 100 | AAL-522 | N.D. | 0.016 |
| | | | | | | N.D. | 0.017 |
| | | | | | | N.D. | 0.017 |
| BS-77-52 | Swinnerton Peninsula, northeast corner | 345 | | | AAL-520 | 0.26 | 0.053 |
| | | | | | | — | 0.053 |
| | | | | | | — | 0.053 |
| BS-77-73 | Bowman Island | 250–270 | | | AAL-521 | 0.19 | 0.058 |
| | | | | | | 0.26 | 0.045 |
| | | | | | | 0.18 | 0.040 |

[a] All age determinations from the Radiocarbon Dating Laboratory, Geological Survey of Canada, are based on a $^{14}C$ half-life of 5568 ± 30 years and 0.95 of the activity of the NBS oxalic acid standard. Ages are quoted in conventional radiocarbon years before present (B.P.), where "present" is taken to be 1950. $\delta^{13}C = +1.4‰$ for GSC-2519.

[b] Ratio of D-alloisoleucine to L-isoleucine in the free amino acid assemblage (Free) and after conventional acid hydrolysis (Total). ND = not detectable. Standard procedure at the Amino Acid Laboratory (AAL), INSTAAR, University of Colorado, is to analyze three valves from each sample (G. H. Miller, personal communication).

All shells were determined to be aragonitic by powder x-ray diffraction (courtesy of A. C. Roberts, Mineralogy Section, Geological Survey of Canada).

hence the glacier here must have exceeded 700 m in thickness. Likewise, a depth of roughly 200 m has been recorded off the northeast corner of Swinnerton Peninsula (Sadler 1973), and the shells were found at 340 to 345 m a.s.l.. A minimum ice thickness of 550 m is thus required, and of course the ice surface here had to be at a higher elevation than at Bowman Island, 'down-glacier' some 25 km to the east.

But since it is reasonable to assume that glaciers have flowed out Makinson Inlet repreatedly during Quaternary time, it would not have been surprising to find that the till at both sites contained shell fragments of various ages. This may indeed be the case, and additional analyses for amino acids may reveal the presence of shells that are older or younger than the few valves tested so far. Although Godwin (1969: 730–731) cautioned against radiocarbon dating of organic materials or shells in till because of the likelihood that such a deposit will "contain admixtures of material of different ages," using amino acid ratios is a much more promising technique. In a situation where a shell assemblage of mixed age is shown to be present in till, it is the age of the youngest shells that is critical. This age puts a limit on the timing of the latest glacial event that could have transported the shells to their present position. Of course the oldest shells may have undergone several cycles of deposition and erosion, and even the youngest shells may have been transported more than once. Some guidance as to relative age can be obtained from the physical condition of the shells themselves; intact valves, large and/or thin, are more likely candidates for amino acid work than are thick individuals with rounded corners or fragments whose diagnostic features have been abraded away. Pitted or encrusted shells should be avoided wherever possible, and the mineralogical composition should be checked by means of powder x-ray diffraction.

The absolute ages of the shell fragments in the till at the two Makinson Inlet sites are less easy to determine, and the solution to that problem is beyond the scope of this paper. However, since the thermal history of the shells is important in terms of the rate of amino acid hydrolysis and racemization, it is worth noting that the mean daily July isotherm for 4.4°C(40°F) passes near Clyde, on Baffin Island, as well as along the outer east coast of Ellesmere Island (Canadian Hydrographic Service 1970). Thus the two areas are not as dissimilar as might be expected from the difference in latitude of nearly 7°. The ratios for the shell-bearing till on Swinnerton Peninsula and Bowman Island correspond, for instance, to the ratios reported from the Kogalu Member of the Clyde Foreland Formation, Baffin Island (Miller et al., 1977), whereas a fragment of *Hiatella arctica* in till at 390 m a.s.l. above Søndre Strømfjord, West Greenland had an alloisoleucine/isoleucine ratio (total) of 0.22 ± 0.02, suggesting that the shells are "at least of early Wisconsin age and possibly date from the last interglacial or earlier" (Sugden and Miller 1976: 400; cf. also Miller et al., 1977, Table 4, for comparable ratios from the Clyde area of Baffin Island).

What is needed now, by way of control on deposits of the last glaciation in the high Arctic Island, is to discover deposits of peat or wood (*in situ* Arctic willow, for instance) on which it is possible to obtain finite radiocarbon ages in the range of 25,000 to 40,000 years. A second criterion is that sites must be located—and they may well be offshore—where such terrestrial deposits are intercalated with shell-bearing marine beds. Alternatively, further attempts should be made to find marine deposits of pre-Holocene age which contain driftwood or other land-derived organic debris suitable for $^{14}C$ dating.* So far all the "old" peat deposits in the Queen Elizabeth Islands have proven to be beyond the limit of radiocarbon dating (Blake 1974, 1975b). The most recent determination, from a site near the head of Makinson Inlet some 37 km north of Swinnerton Peninsula, resulted in a value of more than 52,000 years (GSC-2677; Blake and Matthews 1979). None of the 10 sites investigated are associated with either undisturbed deposits of shelly till or marine strata. After more control is established (preferably at a number of sites) by relating alloisoleucine/isoleucine ratios on shells to dated terrestrial organic deposits, then amino acid dating will become an even more powerful tool for resolving problems of glacial history.

## ACKNOWLEDGMENTS

It is a pleasure to record appreciation for logistical support in the field from Dr. T. Frisch, Regional and Economic Geology Division, GSC, and from the Polar Continental Shelf Project (G. D. Hobson, Director). R. J. Richardson assisted in making the collections on which this article is based. The radiocarbon age determinations were done under the direction of J. A. Lowdon and I. M. Robertson, and the amino acid analyses were carried out by Dr. G. H. Miller, INSTAAR, University of Colorado, Boulder, under DSS Contract OSQ77-00231.

## REFERENCES

Barr, W., Postglacial isostatic movement in northeastern Devon Island: A reappraisal, *Arctic*, **24**, 249–268, 1971.

* Marine deposits containing ash layers, which could be dated by the fission track method, or caves in which speleothems (datable by uranium series methods) are intercalated with marine shell-bearing strata, would enable calibration of older Quaternary deposits with regard to amino acid ratios.

Blake, W., Jr., Studies of glacial history in Arctic Canada II. Interglacial peat deposits on Bathurst Island, *Can. J. Earth Sci.,* **11,** 1025–1042, 1974.

Blake, W., Jr., Radiocarbon age determinations and postglacial emergence at Cape Storm, southern Ellesmere Island, Arctic Canada, *Geogr. Annaler,* **57A,** 1–71, 1975a.

Blake, W., Jr., Studies of glacial history in the Queen Elizabeth Islands, Canadian Arctic Archipelago, *Naturgeografiska Institutionen, Stockholms Universitet, Forskningsrapport,* **21,** 1–14, 1975b.

Blake, W., Jr. and Matthews, J. V., Jr., New data on an interglacial peat deposit near Makinson Inlet, Ellesmere Island, District of Franklin, *Current Research, Part A. Geol. Surv. Can., Paper 79-1A,* 157–164, 1979.

Canadian Hydrographic Service, Climate of the Canadian Arctic, *Can. Hydrographic Ser.,* Marine Sci. Branch, Ottawa, 1-71, 1970.

Christie, R. L., Geology, southeast Ellesmere Island, District of Franklin, *Geol. Surv. Can. Map 12-1962,* 1962.

Dyck, W., The geological survey of Canada radiocarbon dating laboratory, *Geol. Surv. Can., Paper 66-45,* 1–45, 1967.

Frisch, T., Northernmost part of Precambrian Shield undergoes reconnaissance geology mapping, *Northern Miner (Annual Review Number),* **63,** C20, 1977.

Godwin, H., The value of plant materials for radiocarbon dating, *Am. J. Bot., 56,* 723–731, 1969.

Lowdon, J. A., Robertson, I. M., and Blake, W., Jr., Geological survey of Canada radiocarbon dates XVII, *Geol. Surv. Can., Paper 77-7,* 1–25, 1977.

Miller, G. H. and Hare, P. E., Use of amino acid reactions in some Arctic marine fossils as stratigraphic and geochronological indicators, *Carnegie Inst. Wash. Year Book 74,* 612–617, 1975a.

Miller, G. H. and Hare, P. E., Use of amino acid reactions in some Arctic Quaternary marine fossils as stratigraphic and geochronological indicators, *Geol. Soc. Amer., Abstracts witm Programs,* **7,** p. 1200, 1975b.

Miller, G. H., Andrews, J. T., and Short, S. K., The last interglacial-glacial cycle, Clyde Foreland, Baffin Island, N.W.T.: Stratigraphy, biostratigraphy, and chronology, *Can. J. Earth Sci.,* **14,** 2824–2857, 1977.

Olsson, I. U., Modern aspects of radiocarbon datings, *Earth-Sci. Rev.,* **4,** 203–218, 1968.

Olsson, I. U., The pretreatment of samples and the interpretation of the results of $^{14}C$ determinations, in climatic changes in Arctic areas during the last ten-thousand years, Y. Vasari, H. Hyvärinen, and S. Hicks, eds., *Acta Universitatis Ouluensis, Ser. A, Sci. Rer. Nat. 3, Geol. 1,* 9–37, 1972.

Olsson, I. U. and Blake, W., Jr., Problems of radiocarbon dating of raised beaches, based on experience in Spitsbergen, *Norsk Geogr. Tidsskrift,* **18,** 47–62, 1962.

Sadler, H. E., On the oceanography of Makinson Inlet, *Arctic,* **26,** 76–77, 1973.

Stuckenrath, R., Radiocarbon: some notes from Merlin's diary, in Amerinds and their paleoenvironments in northeastern North America, W. H. Newman and B. Salwen, eds., *Annals New York Acad. Sci.,* **288,** 181–188, 1977.

Sugden, D. E. and Miller, G. H., Interglacial or early Wisconsin shell fragments in till on the flanks of Søndre Strømfjord, West Greenland, *Arctic and Alpine Res.,* **8,** 399–401, 1976.

# Correlation and Relative Age Dating of Quaternary Strata in the Continuous Premafrost Zone of Northern Yukon with D/L Ratios of Aspartic Acid of Wood, Freshwater Molluscs, and Bone

**N. W. RUTTER**
*Department of Geology, University of Alberta, Edmonton, Canada*
**R. J. CRAWFORD**
*Department of Chemistry, University of Alberta, Edmonton, Canada*
**R. HAMILTON**
*Department of Chemistry, Fort Lewis College, Durango, Colorado*

*ABSTRACT*

D/L ratios of amino acids were determined for wood, bone, and molluscs from Quaternary sedimentary units from sections located in northern Yukon that have had a long history of permafrost conditions. The objective was to evaluate the usefulness of D/L ratios as an aid in correlating and relative age-dating equivalent units between sections, and to gain insight into the magnitude of racemization to be expected from specimens of various ages under these conditions. D/L ratios of total amino acids were determined by gas chromatography, with aspartic acid proving to be the most useful. D/L ratios of aspartic acid from spruce taken from a unit more than 53,000 years old varied between 0.17 and 0.31 and averaged 0.24. Poplar and willow from a unit less than 10,000 years old yielded ratios of between 0.01 and 0.08 and averaged 0.06. The results indicate that wood offers potential as a stratigraphic tool and apparently can be used, at least in a general way, by comparing results of different genera. Eight genera of freshwater molluscs were identified from units of different ages. Racemization ratios appear to vary among genera, so that it is necessary to compare ratios from the same genus in detailed stratigraphic studies, although it was demonstrated that two genera analyzed together may give useful results. Bone analyses did not support the geological interpretation of the age and sequence of deposits. The inconsistences can be explained geologically by the redeposition of older bones in younger sediments or by the chemistry of digenesis. The bone is fragmented and could not always be identified. The discrepancies are larger than would be expected for different taxa, bone type, and/or variations within the same bone type of the same genus. Therefore, bone has been ruled out as a stratigraphic aid in the present work.

## INTRODUCTION

The Old Crow, Bluefish, and Bell Basins are located in the unglaciated region of northern Yukon within the Continuous Permafrost Zone (Fig. 1, Brown 1969). Rivers within the basins have exposed thick sequences of Quaternary sediments containing abundant fossils and organic material. These factors have generated detailed investigations by a number of Quaternary scientists anticipating a nearly complete record of Quaternary environments and events including elucidation of migration of man into North America (Hughes 1972; Irving 1978; Morlan 1978; Morlan and Matthews 1978). Major problems in stratigraphic investigations have been in correlating and dating equivalent units between sections. This is hampered by the small number of widespread stratigraphic markers such as volcanic ash, the similarity of units of various ages, the lateral facies changes of equivalent units, and the fact that most units are beyond the range of $^{14}C$ dating. These circumstances offer a unique opportunity to evaluate the usefulness of D/L ratios of amino acids of a variety of material in solving these problems in a region subjected to a long period of permafrost conditions. The D/L ratios of aspartic acid proved the most useful. The objectives here are to evaluate D/L ratios of aspartic acid of wood, freshwater molluscs, and bone as stratigraphic aids and to give an indication of the magnitude of racemization to be expected from fossils of various ages under these conditions.

## ANALYTICAL METHODS

Specimens were prepared and analyzed according to the general scheme of Hare (1969) with some modifications. Samples were cleaned to remove any surface contamination with 2 *N* HCl and then washed in double-distilled water with the aid of an ultrasonic cleaner. Total amino acids were extracted by dissolution in 6 *N* HCl and then heated at 150°C for 15 minutes. $Ca^{2+}$ and any other inorganic salts were removed by HF precipitation or with ion-exchange columns (resin - AG 50W - X8 ($H^+$)). Samples were then esterified using 0.1 ml of 4.0 *N* HCl in (+) 2 butanol. The excess reagent was removed with the aid of a Biodrier. Acylation was accomplished by adding 0.1 ml pentafluoroproprionic anhydride (PFPA), 0.3 methylene chloride, and then heated at 100°C for 5 minutes.

The n-pentafluoropropionyl (+) 2-butyl ester derivatives were chromatographed on a Hewlett-Packard Model 5840 gas chromatograph. Shell and bone samples were analyzed using a 50-meter capillary column coated with 20 M carbowax; wood was analyzed by utilizing a 56-meter EGA capillary. D/L ratios were determined from peak area integrations and corrected for the use of 96% (+) 2 butanol.

D/L ratios of alanine, valine, leucine, phenylalanine, proline, and aspartic acid were routinely determined. Aspartic acid proved the most useful because of the relatively fast rate of racemization. Therefore, only D/L ratios of aspartic acid are reported here.

## WOOD

### Introduction

There are abundant wood fragments in a wide range of stratigraphic positions in the Bell, Bluefish, and Old Crow Basins. Because of the abundance, the potential for correlating and determining relative ages of deposits from D/L ratios from within and between basins is better than other fossils found. The wood fragments consist of pieces of tree branches and twigs varying in size from 0.5 cm to 0.3 cm in diameter and from 5 cm to 20 cm in length. Wood specimens were identified as *Picea, Salix,* and *Populus* (L.D. Farley-Gill, R. Mott, Geologicaal Survey of Canada, personal communication, 1979). The paucity of information on analytical methods and racemization rates of various wood taxa dictated the approach taken here (Lee et al., 1976; Engel et al., 1977). We decided to determine and compare D/L ratios of aspartic acid of wood from various horizons of a stratigraphic unit known to be more than 53,000 years old (GSC - 2373-3) in the Bluefish Basin to that of wood from a Holocene (postglacial) unit in the Bell Basin. We felt that these results would aid in determining the feasibility of using wood for correlation purposes and give us an approximation of the magnitude of ratios to be expected from fossils of various ages, subjected to a long period of permafrost conditions.

## Bluefish River Section

The section on the Bluefish River in the Bluefish Basin is located about 40 km southwest of the settlement of Old Crow (Fig. 1) The unit consists of about 11 m of alluvial silt and sand with gravel near the base (Fig. 2) Wood pieces are plentiful throughout. This unit overlies Tertiary coal and underlies, unconformably, about 6 m of fluvial gravel and sand that underlines about 6 m of lacustrine silt. All that can be said of the age of the silt, sand unit is that it is older than 53,000 years. This is based upon a $^{14}C$ date on wood from the overlying gravel and sand unit. Paleoclimatic

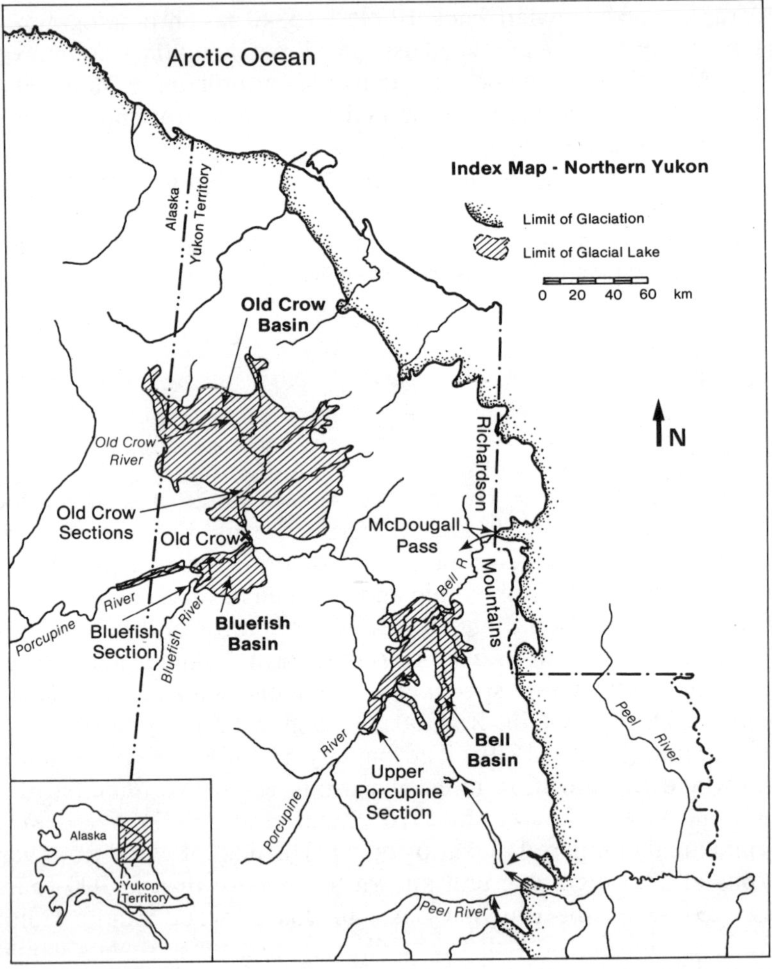


**Figure 1.** Index map of northern Yukon showing the locations of the Bluefish, Upper Porcupine, and Old Crow sections.

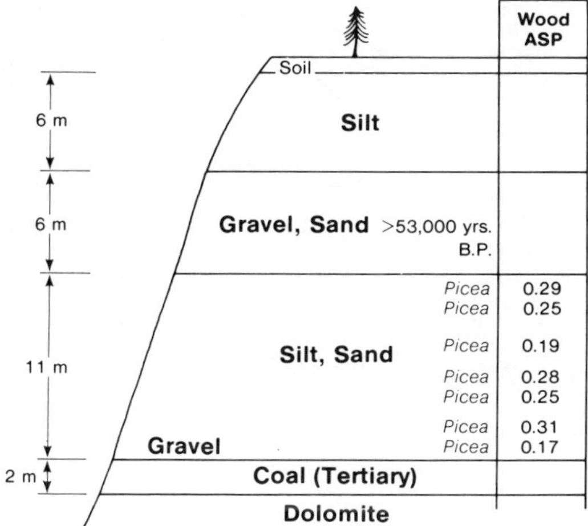


**Figure 2.** Bluefish River section showing the approximate postion of wood (*Picea*) samples and corresponding D/L ratios of aspartic acid.

and geological interpretation of other sections in the area, coupled with the abundant wood found in the silt, sand unit, suggests that the alluvium was deposited at a time warmer than today, most likely during the Sangamon Interglaciation, or more than 100,000 years ago. Although it is difficult to speculate on the temperature history of the unit since deposition, climatic evidence suggests that the deposits have been subjected to a long period of permafrost conditions with occasional thawing of unknown duration. Thawing, at least, took place during the time of deposition of the two overlying units.

## Upper Porcupine Section

A Holocene section was sampled down a terrace face located beside the Upper Porcupine River in the Bell Basin about 130 km southeast of the settlement of Old Crow, or 135 km from the Bluefish section (Fig. 1). The unit consists of a 9-m section of fluvial sand and silt with abundant wood overlain by about 3 m of peat (Fig. 3) The only $^{14}C$ date available for dating the deposit is derived from wood taken from some of the oldest sediments within the unit. Therefore, the date of 9190 years B.P. (GSC 2461) is a maximum date for the initiation of deposition of the material that forms the terrace. The paucity of trees in the region today compared with the abundance of wood found throughout this section, suggests that these sediments were deposited during a period warmer than today. After

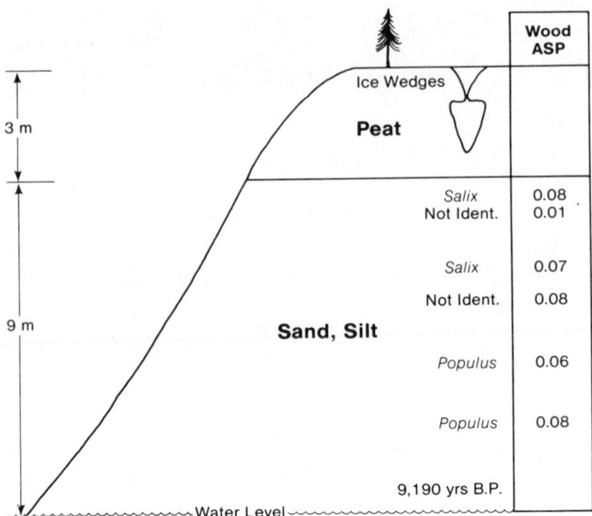


**Figure 3.** Upper Porcupine River section showing the approximate position of wood (*Salix* and *Populus*) samples and corresponding D/L ratios of aspartic acid.

the construction of the terrace, most of the unit has probably been continually frozen. The peat has thawed at least once as evidenced by the superimposed set of ice wedges.

## Results

Wood was sampled systematically from the lower to the upper part of the silt, sand unit of the Bluefish section (Fig. 2). Results show a variation of D/L ratios of aspartic acid from 0.17 to 0.31 (Fig. 2). The distribution of the ratios within the section does not show an increase of racemization with increasing sediment age but rather a random distribution. This can be explained by redeposition of older wood or different racemization rates for various components of wood. From a stratigraphic point of view, D/L ratios of aspartic acid averaging about 0.24 are probably typical for wood from sediments that have been under permafrost conditions for most of their history that are on the order of 100,000 years old; at any rate, older than 53,000 years.

Wood from the Holocene section was systematically sampled in a manner similar to that of the Bluefish section (Fig. 3). D/L ratios of aspartic acid varied from 0.01 to 0.08. The random distribution with no apparent increase of D/L ratios with increasing age, as implied by stratigraphic data, can probably be explained by the reasons given above. One cannot, however, expect much variation in samples that are less than 9000 years old. The average D/L ratio of 0.06 for aspartic acid is probably typical for wood deposited during the early Holocene and subjected to a long period of permafrost conditions.

Comparing the aspartic acid D/L ratios for the two units, probably the most important observation is the relatively low D/L ratios associated with sediments on the order of 100,000 years old and ratios of about one third the magnitude, associated with sediments less than 10,000 years old, deposited under similar conditions located in the same climatic region. This supports the geological inference that the old deposits have been under permafrost conditions for a considerable period of time when racemization rates are reduced. Only when sediments are unfrozen does appreciable racemization take place. Or, less likely, *Salix* and *Populus* racemize at a much different rate than *Picea*. In addition, the relatively low variations of D/L ratios of aspartic acid in both units suggest that each unit was deposited under one sedimentary cycle with no major period of non-deposition or erosion.

## FRESHWATER MOLLUSCS AND BONE

### Introduction

Freshwater molluscs and bone are plentiful in certain stratigraphic horizons in sections throughout the Bell, Bluefish, and Old Crow Basins. So far, eight genera of molluscs have been identified from the Old Crow Basin. These consist of *Pisidium, Valvata, Bulimus, Amnicola, Sphaerium, Lymnaea, Gyraulus,* and *Anadonta.* With the exception of *Anadonta,* the specimens are only about 2 to 4 mm in size, commonly occurring in bits and pieces, making them difficult to identify. No matter how abundant, a single stratigraphic horizon has no more than one or two genera present. Fragments of bone varying in size from 2 to 15 cm are found widely scattered along horizons, sometimes occurring with molluscs. Identification of the animal by the type of bone is usually impossible. Although a great deal is known about the diagenesis of amino acids in bone (Oakley 1963; Hare 1969, 1974; Bada 1972; Bada et al., 1973; Bada and Heffman 1975; Dungworth 1978; Dungworth et al., 1973, 1976; Kvenvolden 1975; Schroeder and Bada 1976) very little is known about freshwater molluscs. Therefore, D/L ratios of molluscs were computed (along with bone) to determine their usefulness in relative age dating, correlating, confirming, or aiding in geological interpretation of stratigraphic units and to give an indication of the amount of racemization to be expected from molluscs and bone of various ages subjected to permafrost conditions.

## Old Crow Composite Section

Bluffs, on the order of 50 m high, expose Quaternary deposits in several locations along the Old Crow River and its tributaries in the Old Crow Basin (Fig. 1). These lie about 30 km north of the settlement of Old Crow. Correlation of equivalent units between bluffs has been difficult because key beds or other stratigraphic markers are not always present, deposits in various stratigraphic positions are similar, and lateral facies changes take place within deposits of the same age. Preliminary studies of the geology, biology, and archaeology of several sections has enabled construction of a composite section (Fig. 4). The composite section consists of a lower lacustrine clay unit that underlies unconformably reworked lacustrine sediments containing abundant wood and some molluscs. The overlying unit, reaching thicknesses of 15 m consists mostly of well-bedded, laminated sand and silt alluvium containing wood, peat, molluscs, and bone. The upper contact forms an unconformity that has been the source of abundant bones, some of which may be artifacts. Above this is a unit of well-bedded silt containing wood, molluscs, and bone and several horizons marked by disturbance from cryoturbation, ice wedge casts, and minor unconformities. This underlies a well-bedded lacustrine clay and silt unit commonly overlain by peat and forms the upper surface over much of the Old Crow Basin. In places, fluvial sands have been deposited in high level channels.

A series of $^{14}C$ dates and a fission track data from tephra have aided in dating the younger sediments. The tephra, lying just below the unconformity that separates the sand, silt unit from the silt unit has been estimated at between 30,000 and 80,000 years old (J.A. Westgate, U. of Toronto, personal communication, 1977). $^{14}C$ dates from the overlying silt unit have yielded dates of between about 32,000 and more than 53,000 years B.P. (GSC 2507, 2574, 2676, 2739) and a 12,460 year old date (I - 3574) from the uppermost sand unit. From these dates and other geological evidence it seems reasonable to place a Late Wisconsin age ($\pm$10,000–$\pm$25,000 years) on the upper lacustrine unit deposited when continental glaciers covered areas to the east. The underlying silt unit and at least the upper part of the sand, silt unit is Mid-Wisconsin in age ($\pm$25,000–$\pm$50,000 years). The Mid-Wisconsin was characterized by a climate probably colder than today's but not as severe as during a major glacial advance. Below the tephra, the sand, silt unit is probably Mid-

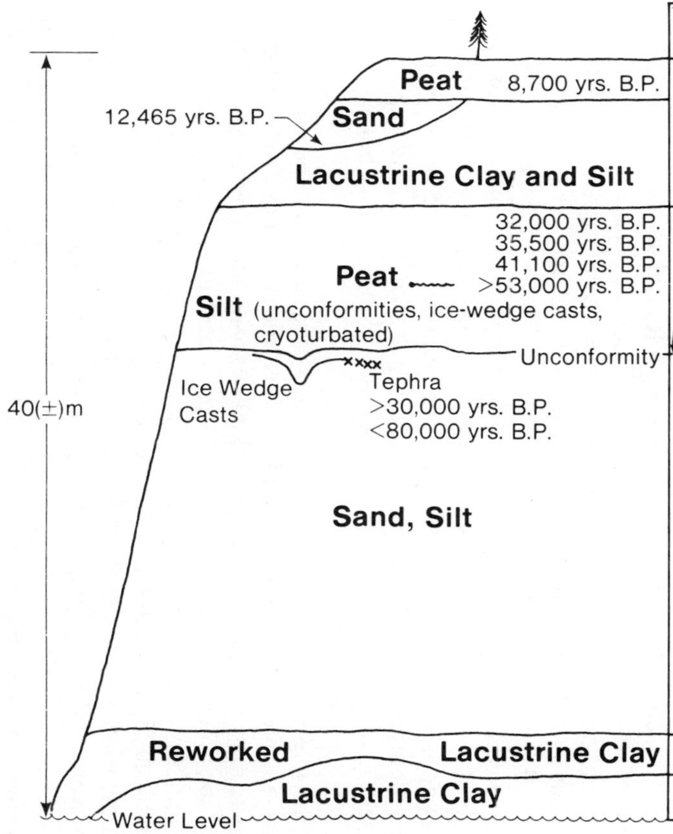

| Stratigraphy | Bone ASP | Shells (Undiff.) ASP | Pisidium ASP | Pisidium & Valvata ASP | Sphaerium ASP | Pisidium & Amnicola ASP | Valvata & Bulimus ASP | Valvata ASP |
|---|---|---|---|---|---|---|---|---|
| Peat 8,700 yrs. B.P. | | | | | | | | |
| Sand (12,465 yrs. B.P.) | | | | | | | | |
| Lacustrine Clay and Silt | | 0.22<br>0.28 | | | | | | |
| Silt (unconformities, ice-wedge casts, cryoturbated)<br>Peat: 32,000 yrs. B.P.; 35,500 yrs. B.P.; 41,100 yrs. B.P.; >53,000 yrs. B.P. | 0.11<br>0.38<br>0.35<br>0.24 | 0.31<br>0.29 | 0.24 | 0.34 | | | | 0.28<br>0.30 |
| Unconformity<br>Ice Wedge Casts<br>Tephra >30,000 yrs. B.P. <80,000 yrs. B.P.<br>Sand, Silt | 0.29<br>0.23 | 0.23<br>0.28<br>0.17<br>0.34<br>0.46<br>0.28<br>0.43<br>0.43<br>0.31 | 0.34<br>0.40 | 0.33 | 0.27 | 0.35 | 0.42 | |
| Reworked Lacustrine Clay | | | | | | 0.47 | | |
| Lacustrine Clay | | | | | | | | |
| Water Level | | | | | | | | |

40(±)m

**Figure 4.** Composite section from the Old Crow Basin showing D/L ratios of aspartic acid of bone and shell specimens at their approximate stratigraphic position.

Wisconsin with perhaps the lower sediments Early Wisconsin (> ± 50,000 – ± 100,000). The underlying reworked lacustrine clay appears to have resulted from the thawing and redeposition of ice-rich sediment derived from the unit below. This and the presence of abundant wood suggest a relatively warm period, perhaps the Sangamon Interglaciation (> ± 100,000 years) or a warming in the Early Wisconsin. The underlying lacustrine unit is believed to be associated with the melting of glaciers that were present to the east either in Early Wisconsin or Illinoian time.

The above discussion suggests that the units of the composite section have been under climates as severe or more severe than today's throughout most of their history, therefore, under continuous permafrost conditions. Thawing to unknown depths and duration took place when horizons such as those indicated by unconformities were exposed at the surface, and probably complete thawing of sediments below the water sediment interface during deposition of each unit.

## Results

Bone was sampled from various stratigraphic horizons from several sections and analyzed for D/L ratios. When the D/L ratios of aspartic acid of the various samples are placed in their stratigraphic position in the composite section, they do not support the geological interpretation. Commonly, the ratios are greater in younger sediments than in older ones (Fig. 4). The ratios vary between 0.11 and 0.38, with the higher ratios (0.35, 0.38) found on the unconformity below the silt unit and lower ratios (0.23, 0.29) in the lower part of the sand, silt unit. These inconsistencies can be explained geologically by the redeposition of older bones in younger sediments. An alternative explanation lies in the chemistry of diagenesis. The bone is fragmented, so taxa and frequently the type of bone could not be identified. The discrepancies are larger than one would expect for different taxa, bone type and/or variations within the same bone type of the same genus.

In the Old Crow region, therefore, D/L ratios of bone have been ruled out as an aid in stratigraphic studies. Identifiable bones from primary deposits are rare but not from reworked deposits that have concentrated many specimens in one place, such as on recent point bars. These offer potential for a variety of amino acid studies such as separation of various assemblages into relative age categories.

As mentioned in the Introduction, several varieties of freshwater molluscs were collected from stratigraphic positions from various sections in the Old Crow Basin. In the beginning of the study it was hoped that individual species or genera would not have to be separated. There are very few individuals in certain horizons, they are of small size, are difficult to identify and are often fragmented. Unfortunately, the results presented in Figure 4 attest to the necessity of single genus analysis.

In the column "Shells undifferentiated" (Fig. 4), which also includes a high proportion of fragmentary pieces, it can be seen that D/L ratios of aspartic acid give inconsistent results when considered with the geological interpretation of the stratigraphy. There is a trend for the older specimens to give higher D/L ratios, but the variations are such that little can be concluded. In the remainder of the columns, when genus differentiation is made, D/L ratios of aspartic acid become useful. *Pisidium* ratios increase with age, with 0.24 from shells collected on the uniformity below the silt unit and 0.34 and 0.40 from the lower part of the sand, silt unit. *Valvata* ratios are higher than *Pisidium* ratios in the same stratigraphic position, with 0.30 on the unconformity and 0.28 in materials dated about 32,000 years B.P. in the upper part of the silt unit. *Sphaerium* found just below the unconformity has a value of 0.27, suggesting that it racemizes more slowly than *Valvata* and perhaps *Pisidium*. Where two genera had to be analyzed together in order to have sufficient sample, the results, which obviously have to be used with caution, give ratios that are fairly compatible with those of a single genus. Higher ratios occur in older sediments as would be expected, although an anomaly exists with *Pisidium* and *Valvata* where the D/L ratio of aspartic acid is 0.34 from the unconformity, and 0.33 from the middle part of the sand, silt unit.

From these analyses, the following preliminary conclusions regarding D/L ratios of aspartic acid of freshwater molluscs with a similar history can be made:

1. Analysis of bulk samples cannot be used reliably for correlation and relative age dating.
2. Two genera analyzed together may offer some aid in correlation and relative age dating but must be used with caution.
3. Single genera appear to be reliable for correlation and relative age dating.
4. Different genera vary in their racemization rates.
5. Aspartic acid of freshwater molluscs subjected to a long period of permafrost conditions racemize at a rate such that D/L ratios of aspartic acid should be useful for correlation and relative age dating throughout the Pleistocene Period.

## CONCLUSIONS

D/L ratios of aspartic acid from specimens subjected to a long period of permafrost conditions are useful for correlating equivalent sedimentary units and determining relative age dates for material at least 100,000 years old and probably for the entire Quaternary. The wide fluctuation of temperatures that sediments have been subjected to since deposition makes it difficult to determine the average temperature history of a specimen and therefore precludes determination of accurate absolute dates.

Although desalting in the analytical procedure is time consuming, wood appears to be very useful for our purposes. Genus determination may be necessary for detailed work but as the results indicate, racemization of at least two genera (*Salix* and *Populus*) of about the same age gives similar D/L ratios. Freshwater molluscs also proved to be valuable, but D/L ratios must be determined and compared for the same genus, although racemization rates of various genera do not vary greatly. Mixing of two genera can give useful information if used with caution. The fragmented nature of bone specimens made identification of genus and type of bone impossible; therefore, the results were inconsistent and not useful for our purposes.

## *ACKNOWLEDGMENTS*

The authors wish to thank members of the Yukon Refugium Project: R.E. Morlan (Archeological Survey of Canada), C.E. Schweger (University of Alberta), and O.L. Hughes and J.V. Matthews (Geological Survey of Canada) for helpful discussions on the Quaternary stratigraphy of northern Yukon. Financial support from the National Museum of Man and the Natural Sciences and Engineering Research Council of Canada (NRC A3864) is gratefully acknowledged.

## *REFERENCES*

Bada, J. L., The dating of fossil bones using racemization of isoleucine, *Earth Planet. Sci. Lett.*, **15**, 223–231, 1972.

Bada, J. L. and P. M. Helfman, Amino acid racemization dating of fossil bones, *World Archaelogy*, **7**, 160–175, 1975.

Bada, J. L., K. A. Kvenvolden, and E. Peterson, Racemization of amino acids in bones, *Nature*, **245**, 308–310, 1973.

Brown, R. J. E., Permafrost in Canada, *Canada Geological Survey Map 1246A*, 1969.

Dungworth, G., Optical configuration and the racemization of amino acids in sediments and fossils — a review, *Chem. Geol.*, **17**, 135–153, 1976.

Dungworth, G., A. W. Schwartz, and L. van de Leemput, Compositions and racemization of amino acids in mammoth collagen determined by gas and liquid chromatography, *Comp. Biochem. Physiol.*, **53B**, 473–480, 1976.

Dungworth, G., N. J. Vinchen, and A. W. Schwartz, Racemization of aliphatic amino acids in fossil collagen of Pleistocene, Pliocene and Miocene ages, in *Advances in Organic Geochemistry,* B. Tissot and F. Bienner, eds., Editions Techniq., Paris, 659–700, 1973.

Engel, M. H., J. E. Zumberge, and B. Nagy, Kinetics of amino acid racemization in *Sequoiadendron giganteum* heartwood, *Anal. Biochem.,* **82,** 415–422, 1977.

Hare, P. E., Geochemistry of proteins, peptides and amino acids, in *Org. Chem.,* G. Eglinton and M. J. T. Murphy, eds., Springer-Verlag, New York, pp 438–463, 1969.

Hughes, O. L., Surficial geology of northern Yukon Territory and northwestern District of Mackenzie, Northwest Territories, *Geological Survey of Canada Paper 69-36,* 1972.

Irving, W. N., Pleistocene archeology in eastern Beringia, in *Early Man in America — From a Circum-Pacific Perspective,* A. L. Bryan, ed., Archaeological Researches International, Edmonton, pp. 96–101, 1978.

Kvenvolden, K. A., Advances in the geochemistry of amino acids, in *Annu. Rev. Earth Planet Sci.,* **3,** F. A. Donath, ed. Annual Reviews, Palo Alto, 183–212, 1975.

Lee, C., J. L. Bada, and E. Person, Amino acids in modern and fossil woods, *Nature,* **259,** 183–186, 1976.

Morlan, R. E., Early Man in northern Yukon Territory; perspectives as of 1977, in *Early Man in America — From a Circum-Pacific Perspective,* A. L. Bryan, ed., Archaeological Researches International, Edmonton, pp. 78–95, 1978.

Morlan, R. E. and J. V. Matthews, Jr., New dates for early man, *GEOS, Winter 1978,* 2–5, 1978.

Oakley, K. P., Analytical methods of dating bones, in *Science in Archaeology,* D. Brothwell and E. Higgs, eds., Thames and Hudson, London, 24–34, 1963.

Schroeder, R. A. and J. L. Bada, A view of the geochemical applications of the amino acid racemization reaction, *Earth Sci. Rev.,* **12,** 347–391, 1976.

# Amino Acid Bone Dating: A Feasibility Study, South San Francisco Bay Region, California

**KENNETH R. LAJOIE**
*U.S. Geological Survey, Menlo Park, California*

**ETTA PETERSON**
*NASA-Ames Research Center, Moffett Field, California*

**BERT A. GEROW**
*Stanford University, Stanford, California*

*ABSTRACT*

Fourteen fossil samples from seven geologically and radiometrically dated archeological and paleontological sites in the south San Francisco Bay region (Fig. 1) provide the basis for a feasibility study of amino acid racemization bone dating. Preliminary amino acid results and site and sample information are presented in tabular and graphical form to show general relationships.

The data do not reveal any criteria by which bone samples or amino acid results can be accepted or rejected for dating purposes. However, we think our data are more than comparable to that previously published for a single restricted area and that much more experimental work must be done before amino acid racemization in bone can be used independently to date geological deposits and archeological materials.

## INTRODUCTION

The samples analyzed are late Pleistocene to late Holocene in age. Radiocarbon dates (Table 1) for the sample localities are consistent with

**Table 1. Radiocarbon Dates**

| Site | Laboratory Number[a] | Material[b] | $^{14}C$ Age (years BP) |
|---|---|---|---|
| Castro (C) | I-7590 | shell (A) | 3,410 ± 100 |
| | | | −700[c] |
| | | | 2,710 |
| University Village (UV) | L-187A | charcoal | 2,950 ± 350[e] |
| | I-7591 | shell (O) | 3,050 ± 85[e] |
| | I-7592 | shell (O) | 3,265 ± 85[e] |
| | L-187B | charcoal | 3,400 ± 300[e] |
| | | average = | 3,166 |
| Stanford (S) | | | |
| S-III | I-7589 | charcoal | 2,270 ± 80 |
| | UCLA-1425B | bone (H) | 4,350 ± 125[e] |
| S-II | UCLA-1425A | bone (H) | 4,400 ± 270[e] |
| | | average = | 4,375 |
| S-I | UCLA-1861 | bone (H) | 5,130 ± 70[f] |
| BART (B) | W-2463 | organic debris | 4,900 ± 250[g] |
| Sunnyvale (SV)[h] | I-6977 | charcoal | 4,460 ± 95 |
| | I-8084 | shell (S) | 10,110 ± 260 |
| | I-6476 | shell (S) | 10,430 ± 150 |
| | | averagc = | 10,270 |
| Mountain View (MV)[i] | I-6298 | wood | 20,820 ± 380 |
| | I-6677 | wood | 21,600 ± 460 |
| | I-6625 | wood | 21,960 ± 520 |
| | I-6679 | wood | 22,340 ± 500 |
| | I-6477 | wood | 23,000 ± 500 |
| | I-6844 | wood | 23,150 ± 520 |
| | I-6625 | wood | 23,600 ± 520 |
| | | average = | 22,350 |

[a] *I* = Isotopes Incorporated; *UCLA* = University of California, Los Angeles; *W* = U.S. Geological Survey, Washington, D.C.; *L* = Lamont Geological Observatory.

[b] bone (*H*) = human (*Homo sapiens sapiens*); shell (*A*) = abalone (*Haliotis cracherodii* and *H. sp.*); shell (*S*) = snail (*Limnea sp.* and *Physa sp.*); shell (*o*) = oyster (*Ostrea lurida*).

[c] 700 year reservoir correction for open coast marine shell (S. Robinson 1978, written communication).

[d] Gerow 1974.

[e] Oakley et al., 1975.

[f] Henn et al., 1972.

[g] See Figure 6.

[h] All wood samples and bone samples (*MV-B, MV-T* and *MV-M*; Table 2) from this site occur 5–7 m below the ground surface in older fan deposits (Figure 3).

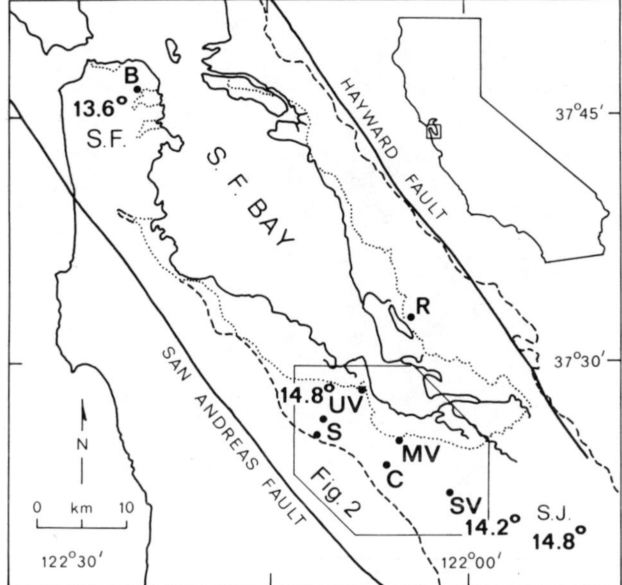


**Figure 1.** Location map, south San Francisco Bay region. Dotted line is the upper boundary of salt marsh around Bay. Dashed line is the upper boundary of the alluvial plain. 13.6°, 14.8°, 14.2°, and 14.8°C are mean annual air temperatures in San Francisco, Redwood City, Santa Clara, and San Jose, respectively (U.S. Department of Commerce, 1964). *B* = BART (Bay Area Rapid Transit District) site, *R* = Ryan site, *UV* = University Village site, *S* = Stanford sites (I, II, and III), *MV* = Mountain View site, *C* = Castro site, and *SV* = Sunnyvale site.

archeological and paleontological data, and agree with stratigraphic relationships among informal geologic units (Figs. 2 and 3) mapped throughout the San Francisco Bay region.

Sample data and a few environmental parameters—mean annual temperature, burial depth, ground-water conditions, soil pH, and residual nitrogen—are summarized in Table 2. The available data are inadequate to describe each sample locality, but this limited information does provide the background for initial evaluation of the amino acid data.

## RESULTS

Of the seven amino acids analyzed, aspartic acid has the highest D/L values (Table 3), as would be expected because of its relatively rapid rate of racemization. There is an inconsistent order among D/L values for the

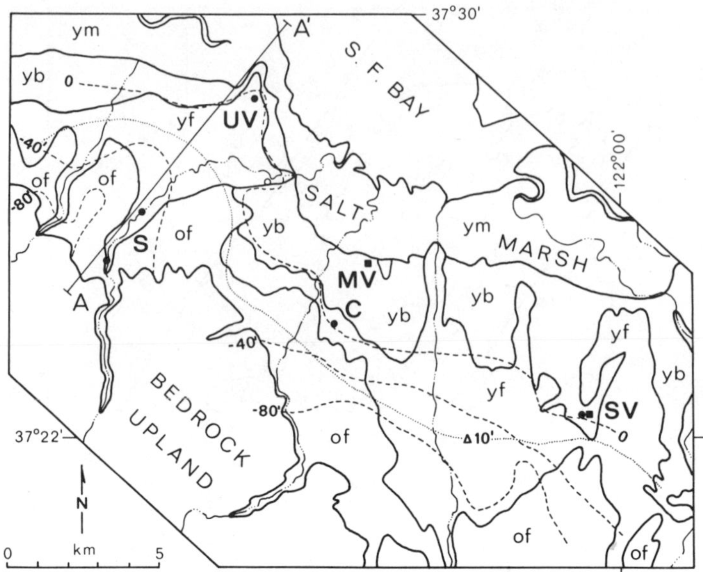


**Figure 2.** Geologic map of part of the south San Francisco Bay region (Fig. 1) modified from Helly and Brabb (1971), Lajoie et al. (1974), and Lajoie and Helly (1975). A-A′ is the cross section of Figure 3. Informal geologic units: *of* = older fan deposits, *yf* = younger fan deposits, *yb* = younger basin deposits, *ym* = younger estuarine mud deposits (all Holocene). Dashed lines = depth of water table (Cooper 1926). Dotted line = 10 foot fluctuation of water table (Clarke 1924). Solid circles are archeological sites and solid squares are paleontological sites.

other six amino acids. The ratios for these amino acids are all low, and the variations are close to the analytical uncertainty. The most significant aspects of the D/L results are the extremely high values for three samples from the Sunnyvale site (SV-A, SV-B, and SV-T) and the relatively low values for the three samples from the Mountain View site (MV-M, MV-T, and MV-B), the oldest samples analyzed (Table 2).

## DISCUSSION

Theoretically, D/L values should increase with age and effective temperature. However, there is virtually no correlation between D/L values for aspartic acid, the amino acid data most widely used for bone dating, and age in our samples (Fig. 4). Therefore, some factor other than age is expressed.

**Table 2. Sample Information**[a]

| Sample | Material[b] | $N_2\%$[h] | Age[c] | | MAT[d] (°C) | Depth (m) | Ground Water[e] | pH[f] | $CaCO_3$[g] |
|---|---|---|---|---|---|---|---|---|---|
| 1. S-III | human tibia + rib | 0.83 | 2,270 | $A_2$ | 14.8 | 2.9 | (+) | 6.9 | no |
| 2. C | human cranium | 0.52 | 2,710 | $A_3$ | 14.5 | 2.1 | (+,o) | 7.7 | yes |
| 3. UV-A | antler (elk) | 2.94 | 3,170 | $A_3$ | 14.8 | ≤2.3[i] | (+,o) | 7.7 | yes |
| 4. UV-B | human cranium | 1.41 | | | 14.8 | 2.3[i] | (+,o) | 7.7 | yes |
| 5. S-II | human cranium | 0.51 | 4,375 | $A_1$ | 14.8 | 5.0[j] | (+) | 6.4 | no |
| 6. SV-A | antler (elk) | 0.17 | 4,460 | $A_2$ | 14.2 | 2.7 | (o) | >7.0 | yes |
| 7. SV-B | human radius | 0.08 | 4,500 | $A_3$ | 14.2 | 2.1 | (o) | 7.4 | yes |
| 8. B | human tibia | 3.54 | 4,900 | $A_2$ | 13.6 | 13.7 | (−) | 6.3 | no |
| 9. S-I | human cranium | 1.37 | 5,130 | $A_1$ | 14.2 | 6.1[k] | (+) | 6.4 | no |
| 10. SV-T | nonhuman tibia | 0.16 | 4,500–10,300 | $A_3$ | 14.2 | 1.3 | (o) | >7.0 | yes |
| 11. MV-B | camel tibia | — | | | | | | | |
| 12. MV-T | camel l. incisor | — | 22,3500 | $A_3$ | 14.5 | 5.5–7.0 | (−) | <7.0 | no |
| 13. MV-M | camel mandible | — | | | | | | | |

[a] Underlined values are estimates. Samples listed in order of increasing age.
[b] Human = *Homo sapiens sapiens*; camel = *Camelops hesternus*.
[c] See Table 2. $A_1$ = direct date; $A_2$ = direct association; $A_3$ = indirect association.
[d] Mean annual air temperature interpolated from values in Figure 1.
[e] (+) = above water table; (−) = below water table; (o) = within fluctuation zone.
[f] pH on rewetted matrix material in laboratory.
[g] Secondary carbonate on sample or in environment.
[h] Dumas method.
[i] Gerow and Force, 1968.
[j] Gerow, 1974.
[k] Heizer and McCown, 1950.

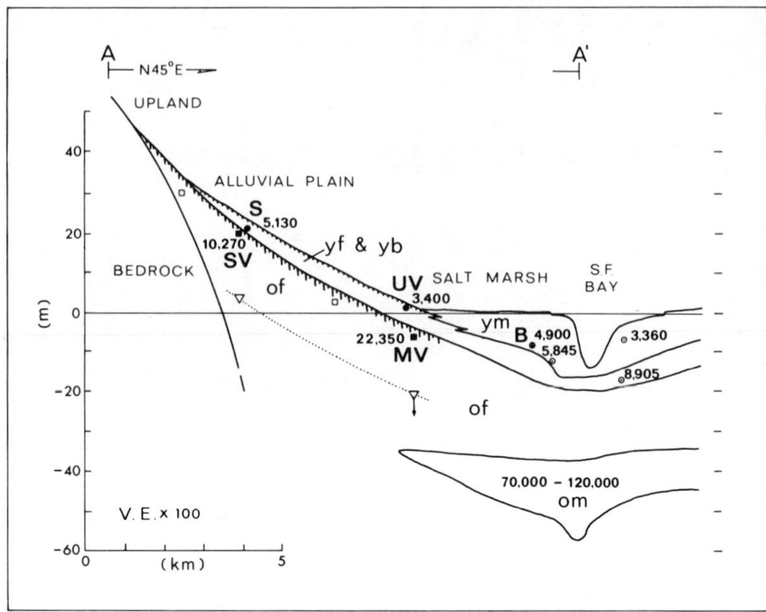


**Figure 3.** Geologic cross section A-A′ (Fig. 2) modified from Atwater et al. (1977) and Borcherdt et al. (1975). See Table 1 for $^{14}C$ date sample localities *S*, *SV*, *UV*, *MV* and *B*. $^{14}C$ dates beneath San Francisco Bay (circles with dots) from Atwater et al. (1977). Date on older estuarine mud (*om*) based on estimated Sangamon age of unit consistent with amino acid D/L values on oysters from cores (Atwater 1978, written communication). The open triangles and dotted line represent the depth of 70,000 year old deposits based upon sedimentation rates derived from the depths and ages of the samples from *SV* and *MV* localities. The sample localities *SV*, *MV*, and *B* are projected onto cross section along contour. Heavy hachures = moderately developed soil; and light hachures = weakly developed soil.

We have no site-specific data with which to evaluate diagenetic temperature, the factor thought to dominate racemization kinetics. However, we can approach the problem indirectly. Ground temperature changes with depth, so there should be some consistent relationship between D/L values and depth if temperature is a dominant factor. Our data show no systematic relationship between these two parameters (Fig. 5). However, the D/L values from the Sunnyvale (SV) and Stanford (S) sites, two different but internally uniform environments, appear to be inversely related to burial depth (dashed lines, Fig. 5).

We can evaluate the importance of temperature by calculating the aspartic acid age for each of our Holocene samples, using an equation that requires an independently dated calibration sample that has experienced the same temperature history as the samples to be dated (Bada and Protsch

Table 3. Amino Acid D/L data[a]

| Sample[b] | Age[c] | Asp[d] | Ala | Val | Leu | Pro | Phen | Glu |
|---|---|---|---|---|---|---|---|---|
| 1. S-III | 2,270 | 0.10 | — | 0.023 | 0.024 | 0.022 | 0.038 | 0.036 |
| | | 0.10 | 0.028 | 0.023 | 0.033 | 0.030 | 0.045 | 0.036 |
| 2. C | 2,710 | 0.13 | 0.061 | 0.025 | 0.022 | 0.025 | 0.026 | 0.040 |
| 3. UV-A | | 0.096 | 0.028 | 0.026 | 0.034 | 0.033 | 0.044 | 0.045 |
| 4. UV-B | 3,170 | 0.125 | 0.030 | 0.026 | 0.043 | 0.052 | 0.042 | 0.060 |
| 5. S-II | 4,375 | 0.092 | 0.022 | 0.016 | 0.019 | 0.024 | 0.029 | 0.025 |
| 6. SV-A | 4,460 | 0.49 | 0.21 | 0.074 | 0.059 | 0.081 | 0.081 | 0.12 |
| 7. SV-B | 4,500 | 0.42 | 0.27 | 0.54 | 0.089 | 0.092 | 0.092 | 0.16 |
| 8. B | 4,900 | 0.069 | 0.021 | 0.019 | 0.020 | 0.027 | 0.034 | 0.025 |
| 9. S-I | 5,130 | 0.089 | 0.023 | 0.013 | 0.017 | 0.026 | 0.030 | 0.026 |
| 10. SV-T | 4,500–10,300 | 0.517 | 0.433 | 0.104 | 0.245 | 0.126 | — | 0.179 |
| 11. MV-B | | 0.25 | 0.074 | 0.049 | 0.052 | 0.036 | 0.096 | 0.10 |
| 12. MV-T | 22,350 | 0.10 | 0.027 | 0.020 | 0.026 | 0.022 | 0.029 | 0.038 |
| 13. MV-M | | 0.078 | 0.022 | 0.020 | 0.027 | 0.025 | 0.034 | 0.031 |

[a] Determined by gas chromatography using optically active N-trifluoroacetyl 2+ butyl ester derivative on 150 foot by 0.02 inch columns coated with Carbowax 20M, OV-225 and Dexsil 400.
[b] Sample numbers used on Figures 4, 5, 7, and 8.
[c] See Table 1. Samples are listed in order of increasing age.
[d] Asp = aspartic acid, Ala = alanine, Val = valine, Leu = leucine, Pro = proline, Phen = phenylalanine, Glu = glutamic acid.

1973). If we assume all our Holocene samples have experienced similar temperature histories, an assumption supported by the rather uniform climate of the region (Fig. 1), and use one of the samples for calibration (S-I), the aspartic acid ages for the other samples should agree with their $^{14}C$ ages. The extremely large errors ($-61$ to $+1280\%$) in the aspartic acid ages (Table 4) clearly indicate that temperature, as estimated by mean annual air temperature, is not the primary factor controlling racemization in these samples.

The relatively high aspartic acid D/L value (0.42) and resultant age estimate (62,000 years) for the Sunnyvale skeleton (SV-B) deserve further comment (Table 4). Bada and Helfman (1975) published a 70,000 year age estimate for his skeleton based on their high aspartic acid D/L values (0.522 and 0.498) and a racemization rate constant ($K_{asp}$) derived from a poorly documented calibration sample in southern California. We believe geologic, archeological, and anthropometric relationships, and radiocarbon dates establish the age of the Sunnyvale skeleton at about 4500 years

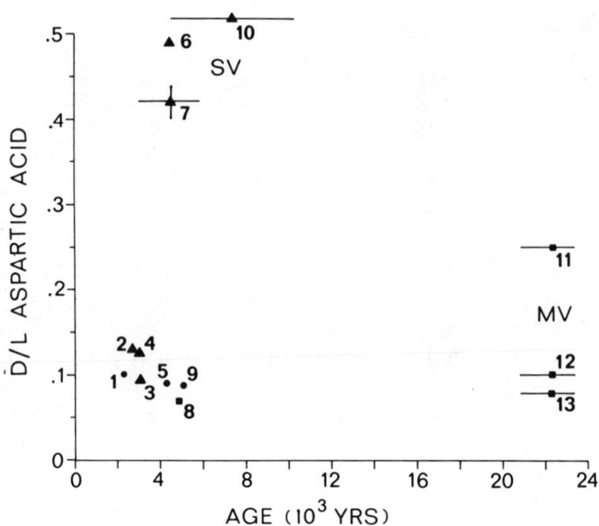


**Figure 4.** Aspartic acid D/L values versus sample age (Tables 2 and 3). Circles and squares represent samples above and below the water table, respectively; triangles represent samples within the fluctuating zone of the water table. Open symbols indicate estimates. Sample numbers are from Table 2. Horizontal bars are ranges in dates for *MV* site (Table 1) and uncertainty in ages for *SV* site (Fig. 6).

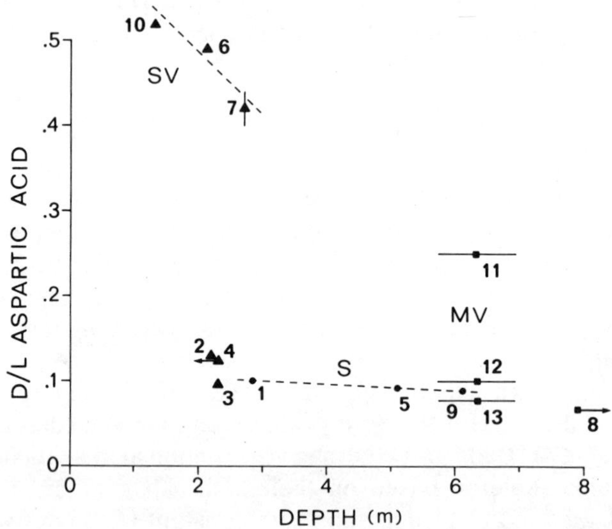


**Figure 5.** Aspartic acid D/L values versus burial depth of sample (Tables 2 and 3). Dashed lines show apparent inverse relationships for the Sunnyvale (*SV*) and Stanford (*S*) sites. See caption of Fig. 4 for explanation of symbols. For sample numbers see Table 2.

**Table 4. Comparison of $^{14}C$ and Aspartic Acid Ages of Five Archeological Samples**

| Sample | $^{14}C$ Age | $D/L_{asp}$ | Aspartic Acid Age[a] | % Error |
|---|---|---|---|---|
| S-III | 2270 | 0.100 | 6,880 | 200 |
| C | 2710 | 0.130 | 11,700 | 330 |
| UV-B | 3170 | 0.125 | 10,900 | 244 |
| S-II | 4375 | 0.092 | 5,600 | 28 |
| SV-B | 4500[b] | 0.42 | 62,000 | 1,280 |
| B | 4900 | 0.069 | 1,900 | −61 |

[a] Using S-I as the calibration sample; $D/L_{asp} = 0.089$, $K_{asp} = 6.284 \times 10^{-6}$ (Table 5).

$$\text{Age} = \frac{0.5}{K_{asp}}\left[\ln\left(\frac{1 + D/L_{asp}}{1 - D/L_{asp}}\right) - 0.114\right],$$

where $K_{asp}$ = calibration and $D/L_{asp}$ = sample. (Bada and Protsch 1973). 0.114 = laboratory constant for zero-age samples (NASA laboratory).

[b] Estimated from radiocarbon age for SV-A (Figure 6).

**Table 5. Aspartic Acid Rate Constants ($K_{asp}$) in Calibration Samples and Calculated Aspartic Acid Ages for Sunnyvale Human Skeleton (SV-B)**

| Calibration Sample | $^{14}C$ Age | $D/L_{asp}$ | $K_{asp}$[a] | Aspartic Acid Age SV-B[b] | % Error[c] |
|---|---|---|---|---|---|
| S-III | 2,270 | 0.100 | $1.909 \times 10^{-5}$ | 20,500 | 355 |
| C | 2,710 | 0.130 | $2.721 \times 10^{-5}$ | 14,300 | 220 |
| UV-B | 3,170 | 0.125 | $2.166 \times 10^{-5}$ | 18,000 | 300 |
| S-II | 4,375 | 0.092 | $8.060 \times 10^{-6}$ | 48,300 | 970 |
| B | 4,900 | 0.069 | $2.471 \times 10^{-6}$ | 158,000 | 3,400 |
| S-I | 5,130 | 0.089 | $6.284 \times 10^{-6}$ | 62,000 | 1,280 |
| MV-M | 22,350 | 0.078 | $9.467 \times 10^{-7}$ | 413,000 | 9,080 |
| MV-B | 22,350 | 0.250 | $8.878 \times 10^{-6}$ | 44,000 | 880 |

[a] $K_{asp} = \frac{0.5}{t}\left[ln\left(\frac{1 + D/L_{asp}}{1 - D/L_{asp}}\right) - 0.114\right]$, where $t$ = age of calibration sample (Bada and Protsch 1973).

[b] Calculated by solving the equation above for $t$ using $K_{asp}$ for each calibration sample and $D/L_{asp} = 0.42$ for SV-B (Table 4).

[c] Using 4500 years as the age for SV-B (Figure 6).

(Fig. 6). The skeleton is fully modern (*Homo sapiens sapiens*) in physical characteristics. Of 32 cranial and postcranial measurements and indices available for the Sunnyvale female, all fall within corresponding ranges for 48 females from the Ryan site (*R*, Fig. 1) dated at 430 ± 80 to 1650 ± 85 years BP by $^{14}C$; 24 differ from corresponding means by less than one sigma of standard deviation (Gerow, unpublished data). If we use our own Holocene human samples for calibration, the aspartic acid ages for the Sunnyvale skeleton (SV-B) range from 18,000 to 158,000 years BP (300% to 1280% error, Table 5). It is biologically inconceivable that a human population would remain morphologically unchanged for such periods of time.

If, for the sake of argument, we assume the Sunnyvale skeleton is late Pleistocene in age, we should use a local calibration sample of Pleistocene age to estimate its aspartic acid age. If we assume that there are only minor generic effects on racemization between human and camel bone, we can use our bone samples from the Mountain View site (MV-B and MV-M, Table 3) for calibration. The resultant ages for the Sunnyvale skeleton, 44,000 and 413,000 (Table 5), are even more extreme than the ages derived by using the Holocene calibration samples.

The broad range of aspartic acid ages (1900 to 413,000 years, Tables 4 and 5) for all our archeological samples ($^{14}C$ dated at 2270 to 5130 years BP) using local calibration samples indicates that the temperature his-

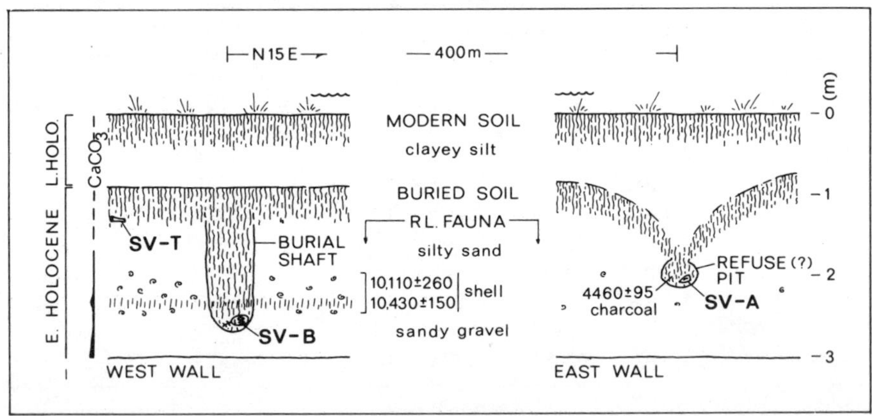


**Figure 6.** Sunnyvale (*SV*) site. Stratigraphic relationships in Sunnyvale East Drainage Channel between Central Expressway and Wolfe Road, Sunnyvale, California. *SV-A* = antler (elk?) wedge artifact, *SV-B* = human skeleton, *SV-T* = nonhuman tibia fragment. Geologic units: *yb* = younger basin deposits, and *of* = older fan deposits (Figs. 2 and 3). *RL* = Rancholabrean fauna (U.S. Geological Survey fossil locality USGS-1218). See Table 1 for $^{14}C$ data, Table 2 for site–specific data and Table 3 for amino–acid data.

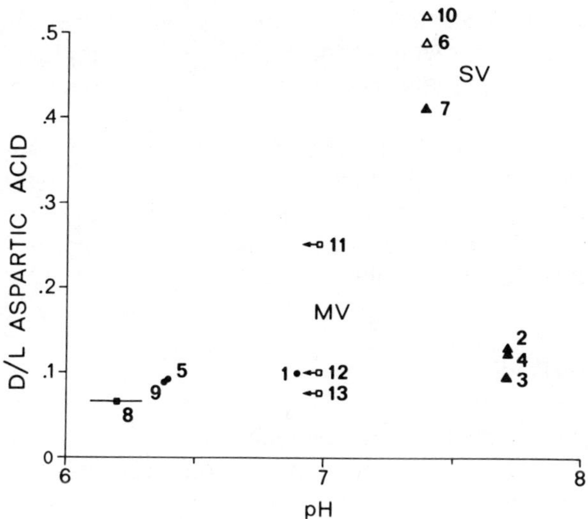


**Figure 7.** Aspartic acid D/L values versus pH of environment (Tables 2 and 3). See caption of Figure 4 for explanation of symbols. Open symbols are estimates. Sample numbers from Table 2.

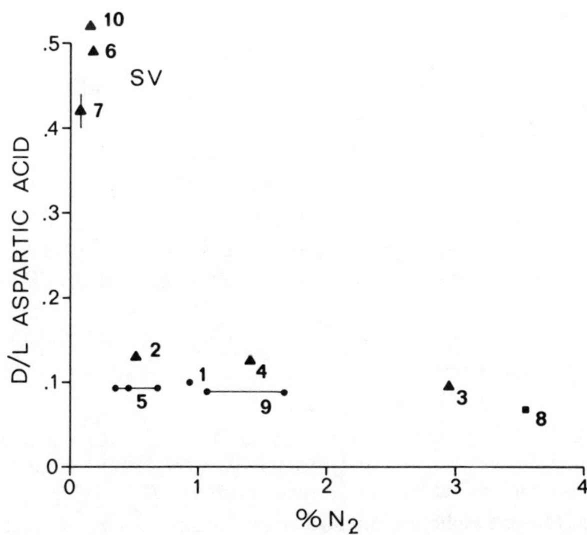


**Figure 8.** Aspartic acid D/L values versus nitrogen content of samples (Tables 2 and 3). See caption of Figure 4 for explanation of symbols. Sample numbers from Table 2.

tories for each site are extremely different or that temperature is not the sole factor controlling racemization in these bone samples. None of our samples or their immediate environments show evidence of heating by fire.

If temperature is not the primary factor controlling racemization in bone, then what is? Our data show no simple correlation between aspartic acid D/L values and environmental pH (Fig. 7). Nitrogen content, a possible indicator of sample preservation, also shows no consistent relationship with aspartic acid D/L values (Fig. 8).

## CONCLUSIONS

Amino acid D/L values from 14 independently dated fossil samples in the south San Francisco Bay region show no consistent relationship to age, burial depth, mean annual temperature, soil pH, or residual nitrogen. Apparently, racemization in bone is controlled by a complex, poorly understood combination of factors. The limited data we present indicate that much more experimental work must be done before D/L values in bone can be used independently to date geological deposits and archeological remains.

### *ACKNOWLEDGMENTS*

Paul Russell found and helped excavate the human skeletal remains (SV-B) in the Sunnyvale East Drainage Channel. Edward Helley, Charles Repenning and Polly Bickle assisted in the excavation of these remains. Marvin Washington found the antler wedge sample (SV-A) in the archeological refuse pit in the Sunnyvale Channel. Charles Repenning identified all the vertebrate paleontological samples. Edward Helley collected some of $^{14}C$ dated sampies from the Sunnyvale and Mountain View sites.

### *REFERENCES*

Atwater, B. F., C. W. Hedel, and E. J. Helley, Late Quaternary depositional history, Holocene sea-level changes, and vertical crustal movement, Southern San Francisco Bay, California, U.S. Geological Survey *Professional Paper 1014*, 1977.

Bada, J. L. and P. Masters Helfman, Amino acid racemization dating of fossil bones, *World Archeo.*, 7(2), 160–173, 1975.

Bada, J. L. and R. Protsch, Racemization reaction of aspartic acid and its use in dating fossil bones, *Proc. Nat. Acad. Sci. U.S.A.*, **70**, 1331–1334, 1973.

Borcherdt, R. D. et al., Predicted geologic effects of a postulated earthquake, in *Studies for Seismic Zonation of the San Francisco Bay Region,* R. D. Borcherdt, ed., U.S. Geological Survey *Professional Paper 941-A,* A88–A96, 1975.

Clarke, W. O., Ground water in Santa Clara Valley, California, U.S. Geological Survey *Water-Supply Paper 519,* 1924.

Cooper, W. S., Vegetational development upon alluvial fans in the vicinity of Palo Alto, California, *Ecology,* **7**(1), 1–30, 1926.

Gerow, B. A., Comments on Fredrickson's "Cultural Diversity," *J. Calif. Anthropol.,* **1,** 239–246, 1974.

Gerow, B. A. with R. W. Force, *An Analysis of the University Village Complex: With a Reappraisal of Central California Archaeology,* Stanford University Press, Stanford, CA, 1968.

Heizer, R. F. and T. D. McCown, The Stanford skull, a probable early man from Santa Clara County, California, *U. Calif. Archeol. Surv. Reports,* **6,** 1–18, Berkeley, 1950.

Helley, E. J. and E. E. Brabb, Geologic map of late Cenozoic deposits, Santa Clara County, California, U.S. Geological Survey *Miscellaneous Field Studies Map* MF-429, scale 1:62,500, 1971.

Henn, W., T. Jackson, and J. Schlocker, Buried human bones at the BART site San Francisco, *Calif. Geol.,* **25,**(7), 208–209, 1972.

Lajoie, K. R., E. J. Helley, D. R. Nichols, and D. R. Burke, Unconsolidated and semiconsolidated sedimentary deposits, San Mateo County, California, U.S. Geological Survey *Miscellaneous Field Studies Map* MF;575, scale 1:62,500, 1974.

Lajoie, K. R. and E. J. Helley, Differentiation of sedimentary deposits for purposes of seismic zonation, in *Studies for Seismic Zonation of the San Francisco Region,* R. D. Borcherdt, ed., U.S. Geological Survey *Professional Paper 941-A,* A39–A51, 1975.

Oakley, K. P., B. G. Campbell, and Theya Ivitsky Molleson, eds., *Catalogue of Fossil Hominoids,* London, 1975.

U.S. Department of Commerce, Decennial Census of United States Climate—Climatic Summary of the United States—supplement for 1951 through 1960, *Climatography of the United States,* No. 86-4, 1964.

# γ-Carboxyglutamic Acid in Fossil Bone

**KENNETH KING, JR.**
*Lamont-Doherty Geological Observatory of Columbia University*

## *ABSTRACT*

γ-Carboxyglutamic acid (Gla), a recently discovered amino acid in fresh bone matrix, has been found in fossil bones as old as 50,000 yr B.P. Laboratory and field data have demonstrated that the absolute concentration of preserved Gla is a useful quantitative index of *in situ* leaching. Gla was measured in 12 well-dated fossil samples for which racemization data were also available. No significant alteration by *in situ* leaching was detected. However, the data suggest that the general concordance observed between racemization ages and radiocarbon dates may have resulted from a close matching of the calibration and unknown samples in terms of leaching history, composition, and degree of racemization. It is therefore recommended that Gla content, amino acid concentration, and amino acid composition be routinely measured with the D/L ratio in both the calibration and unknown samples to minimize the possibility of introducing significant errors in the calibration step.

Gla was not detectable in the calcified skeletons of six invertebrate species from five phyla. These data raise the possibility that invertebrates as a group may lack the enzymatic capability for biosynthesizing Gla. In any case, the usefulness of Gla as an index of leaching history in fossils appears to be limited to vertebrate calcified tissues such as bone, teeth, and otoliths.

## INTRODUCTION

Stimulated by the discovery of a γ-carboxyglutamic acid (Gla)-containing protein in modern bone by Hauschka et al. (1975) three studies have been conducted to explore possible applications for this unusual amino acid in the biogeochemical study of fossils: (1) A laboratory and field evaluation of Gla as a possible leaching index for resolving the question of the effect of *in situ* leaching on racemization rates in fossil bone (King 1978a). (2) The application of Gla content as a leaching index to a series of well-dated fossil bones to determine the relationship between leaching history and amino acid racemization parameters (King and Bada 1979). (3) A survey of modern calcified tissues to determine the probable distribution of Gla in fossil skeletons other than bone (King 1978b).

## SAMPLE PREPARATION AND ANALYSIS

### Studies 1 and 2: Evaluation and Application of Gla as a Leaching Index

The samples of fresh bovine cortical bone used in the initial laboratory leaching experiment were taken from the central section of the tibia of slaughtered cows. After cleaning to remove connective tissue and marrow, the bone was carefully washed and then ground while frozen to a particle size less than 250 μm. A continuous leaching process was simulated by sealing approximately 100 mg of this modern bone powder in glass culture tubes under nitrogen with 2 ml of ultrapure water, maintaining the tubes at 160°C, and changing the water every 2 hours. Samples of the leached bone were withdrawn at 8 hour intervals over the total 96 hour period.

After evaporating the leached bone samples to dryness in a vacuum centrifuge, 2 mg aliquots were hydrolyzed directly in 4 *M* NaOH at 110°C for 22 hours under nitrogen (Hugli and Moore 1972). Alkaline hydrolysis is required to release the peptide-bound Gla for amino acid analysis. Gla is acid-labile and readily converts by decarboxylation to glutamic acid during standard acid hydrolysis. To each of the hydrolysates, a measured amount of norleucine was added as an internal standard. Excess cold (0°C) 12 *M* HCl was then added to neutralize the hydrolysates and precipitate $Na^+$ as NaCl. After centrifugation, the supernatants were freeze-dried and stored at −20°C until analyzed.

In the fossil bone samples in Studies 1 and 2 (Fig. 1 and 2), the dense cortical bone was dissected for analysis from the more porous, cancellous tissue. The outer several millimeters of all exposed surfaces were removed and discarded prior to an ultrasonic cleaning of the bone fragments in a series of dilute (1%) sodium hexametaphosphate washes and ultrapure water rinses. Preliminary experiments (King, unpublished information)

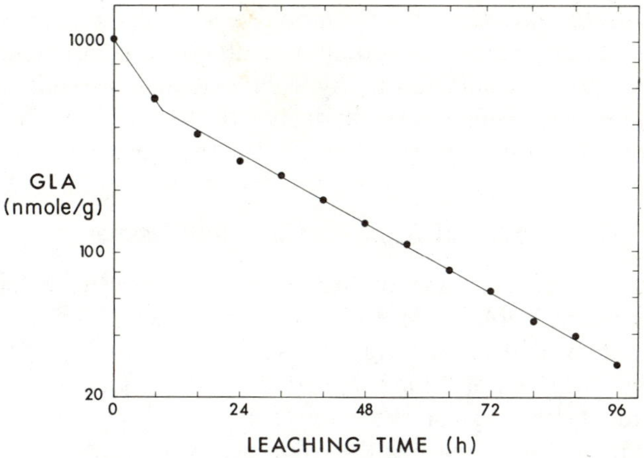


**Figure 1.** γ-Carboxyglutamic acid (Gla) concentration in modern bovine bone as a function of leaching time. (From King 1978a).

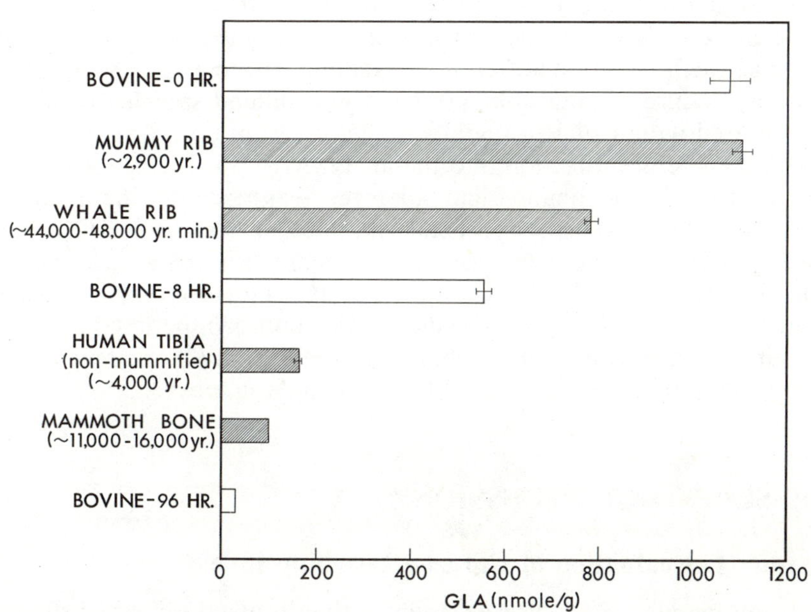


**Figure 2.** γ-Carboxyglutamic acid (Gla) concentration in four fossil bone samples. Shaded bars represent the means of duplicate analyses of single fossil samples. Reference values from three artificially leached bovine samples (Fig. 1) are included as open bars. Standard deviations are shown by the error bars. (From King 1978a).

indicate that this cleaning procedure with sodium hexametaphosphate yields more accurate and reproducible results for Gla concentration than obtained by partial dissolution by hydrochloric acid. Alkaline hydrolysis and subsequent processing of the hydrolysates for the fossil samples (Fig. 2) was identical to that described above for the fresh bone material.

### Study 3: Gla Survey of Modern Calcified Tissues

Following a careful scraping of the outer surfaces, the calcified tissue samples shown in Table 1 were cleaned in a series of 1% sodium hexametaphosphate washes and ultrapure water rinses and air-dried to constant weight. Aliquots of 2 mg were used for the organic-rich vertebrate samples and of 100 mg for the invertebrate skeletons. The procedure described above for preparing the alkaline hydrolysates of fossil bone samples was modified as follows: (1) samples were dissolved in cold (0°C) 6 *N* HCl and freeze-dried prior to hydrolysis; and (2) the alkaline hydrolysates of the 100 mg invertebrate samples were desalted by precipitating the excess $Ca^{2+}$ with concentrated (48%) HF, centrifuging and freeze-drying the recovered supernatant (Hare 1969).

The freeze-dried hydrolysates prepared in all three studies were dissolved in the first elution buffer (0.2 *M* sodium citrate, pH 2.68) according to the procedure of Hauschka (1977). These diluted samples were then each filtered with a Flath-Lundin syringe (0.22 μm polycarbonate membrane filter) before injection into the analyzer.

A Durrum D-500 amino acid analyzer was used at the picomole ($\sim 10^{-10}$g) level of sensitivity. Total Gla content (nanomoles Gla/g calcified tissue) was calculated using: (1) a ninhydrin color yield for Gla equivalent to 48% of that for glutamic acid (Glu) (determined by the conversion of Gla to Glu by acid hydrolysis of both synthetic Gla and the Gla-containing bovine prothrombin fragment 1, and (2) the recovery of norleucine to measure the loss of Gla in sample processing.

## RESULTS AND DISCUSSION

### Study 1: Evaluation of Gla as a Leaching Index

The semi-log plot of Gla versus leaching time shown in Figure 1 suggests that the absolute concentration of Gla in bone varies inversely with leaching intensity. During the course of the experiment, Gla concentration was reduced 97% from an initial value of 1100 nmol/g of bone to 30 nmol/g after 96 hours exposure to continuous leaching with ultra-pure water at

**Table 1. Distribution of γ-Carboxyglutamic Acid (Gla) in Calcified Tissues**

| | Major Anion | Major Mineral Phase | Organic Matrix[a] | Gla[b] (nmol/g tissue) |
|---|---|---|---|---|
| Vertebrate tissues | | | | |
| Shark tooth (*Prionace glauca*) | $PO_4^{3-}$ | Fluorapatite | C | 494.6(7.0) |
| Fish otolith (*Morone saxatilis*) | $CO_3^{2-}$ | Aragonite | NC | 40.1(1.2) |
| Hen egg shell (*Gallus domesticus*) | $CO_3^{2-}$ | Calcite | NC | <0.1 |
| Bovine cortical bone (*Bos taurus*) | $PO_4^{3-}$ | Hydroxyapatite | C | 1076.4(43.6) |
| Invertebrate tissues | | | | |
| Foraminiferal tests (*Globorotalia menardii*) | $CO_3^{2-}$ | Calcite | NC | <0.1 |
| Bivalve mollusc shell (*Mercenaria mercenaria*) | $CO_3^{2-}$ | Aragonite | NC | <0.1 |
| Sea urchin spine (*Heterocentrotus mammillatus*) | $CO_3^{2-}$ | Calcite | NC | <0.1 |
| Blue crab carapace (*Callinectes sapidus*) | $CO_3^{2-}$ | Calcite | NC | <0.1 |
| Brachiopod (articulate) shell (*Terebratulia transversa*) | $CO_3^{2-}$ | Calcite | NC | <0.1 |
| Brachiopod (inarticulate) shell (*Lingula anatina*) | $PO_4^{3-}$ | Hydroxyapatite | NC | <0.1 |

[a] C = collagenous; NC = noncollagenous.
[b] Figures are mean values and standard deviations (in parentheses) for duplicate analyses of single samples. (From King 1978b).

160°C. The leaching curve generated by the 13 samples is composed of two linear segments. The longer segment extending from 8 to 96 hours, reflects a nearly constant rate of Gla disappearance. The initial 8-hour period of more rapid Gla loss coincides with the period in which most of the collagen is removed by leaching (King, unpublished information).

Encouraged by the promising inverse correlation between Gla concentration and leaching intensity in modern bone under controlled laboratory conditions (Fig. 1), four natural fossil bones having a range of leaching histories were analyzed (Fig. 2). Gla was found in measurable amounts in all fossil samples and its concentration varied by an order of magnitude among the four samples, (i.e., with a range of 1100–100 nanomoles/g).

The highest fossil Gla content was found in an Egyptian mummy rib radiocarbon dated at approximately 2900 yr. B.P. This sample represents a case of minimal leaching because it was dissected from a mummy sufficiently well-preserved to allow paleopathological diagnosis of a skin disease (Zimmerman and Clark 1976). The Gla value of 1100 nano moles/g in this fossil is identical to that of unleached fresh bovine bone.

The next highest concentration of Gla (780 nanomoles/g) occurred in a fossil whale rib sample with an estimated age of approximately 50,000 years B.P. (G.H. Miller, personal communication). This sample was collected from the permafrost layer of a raised marine delta in a fjord at Baffin Island, Canada. Geologic and paleoclimatic considerations suggest that this sample had been totally protected from leaching in permafrost for about 80% of the time since burial. Alternating conditions of freezing and melting probably occurred seasonally for some unknown portion of the past 8000–9000 years. This would have resulted in some leaching. The observed Gla content of this sample is 28% lower than the concentration in unleached, modern bovine bone.

The extremely low levels of Gla in the other two fossil samples were not unexpected. The 12th Dynasty Egyptian nonmummified human tibia (165 nanomoles/g) and the Colorado Dutton site mammoth bone (101 nanomoles/g) were both recovered unprotected from depositional environments conducive to leaching.

The field data represented by these four fossil samples (Fig. 2), when compared with the Gla contents of laboratory-leached bovine bone samples (Fig. 1), strongly suggest that Gla concentration of fossil bones and *in situ* leaching intensity are inversely related in the natural environment.

Because Gla content appears to be a valid quantitative index of *in situ* leaching in fossil bones, Study 2 was undertaken to apply this new parameter in an investigation of the possible role of *in situ* leaching in the alteration of racemization parameters used for amino acid dating.

## Study 2: Application of Gla as a Leaching Index to Evaluate Role of Leaching in the Alteration of Racemization Parameters

Despite the clear potential of the amino acid racemization reactions for estimating the absolute ages of fossil bones (Bada and Schroeder 1975; Schroeder and Bada 1976), an important question remains unanswered: What is the effect of *in situ* leaching on apparent racemization rates (Hare 1974; Hare et al., 1974)? Using γ-carboxyglutamic acid (Gla) as a quantitative index of leaching, the relationship between leaching history and the corresponding D/L aspartic acid ratio for 12 well-dated fossil samples was examined in an effort to resolve this problem.

The fossil bone samples studied consisted of six calibration/unknown sample pairs which encompassed a wide range of ages, environmental temperatures, and leaching histories. Detailed data on each sample are published elsewhere (King and Bada 1979), and the thermal calibration technique involving calibration/unknown sample pairs has been previously described (Bada and Protsch 1973). Key findings for each pair that are relevant to this study will be discussed below and are summarized in Figure 3. Although not indicated in Figure 3, the correct pairing of the calibration sample with the corresponding unknown sample will be made in the following discussion.

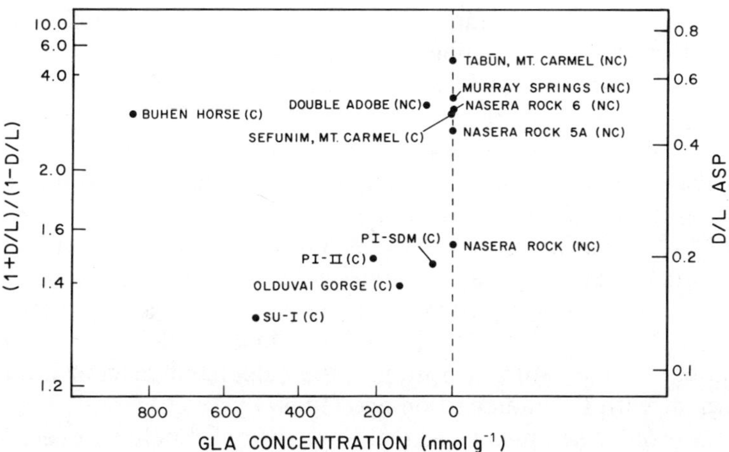


**Figure 3.** Relationship between aspartic acid racemization parameters and degree of *in situ* leaching. Intensity of leaching increases with decreasing values of Gla content. The parameter $\ln[(1 + D/L) / (1 - D/L)]$ is the key variable in the equation to calculate racemization ages (Bada and Protsch 1973). C = collagenous matrix; NC = non-collagenous matrix.

The pair of samples exhibiting the lowest degree of leaching (Buhen Horse and Mindanao SU-1) came from localities having nearly identical air temperatures: 26°C and 27°C, respectively. However, their environments were quite different in terms of annual rainfall: the Buhen Horse sample was found in the arid Sudan (0.4 cm/yr precipitation), and the Mindanao SU-1 sample was collected from an estuarine swamp in the Philippines (232 cm/yr precipitation). This environmental difference is reflected by their Gla values of 845 and 516 nmol $g^{-1}$, respectively. Expressed as percentages of the Gla concentration measured in fresh bovine bone (1076 nmol $g^{-1}$ bone (King 1978a) ), the Gla content of the Buhen Horse sample is 79% and the Mindanao sample, 48%. Both samples have the collagenous organic matrices expected in slightly to moderately leached specimens (Hare 1974). Despite a measurable difference in leaching histories reflected by the Gla contents, the two samples exhibit very similar racemization rate constants ($k_{asp}$) when the environmental temperatures are taken into account.

Both of the La Jolla Paleo-Indian samples have experienced substantial leaching as evidenced by their low Gla values. Although sample SDM 16709 has a Gla content equivalent to only 5% of fresh bovine bone, its organic matrix is still collagenous in composition. When the D/L ratio of SDM 16709 is used to calibrate the racemization reaction, the resulting racemization age for W-12/76 is in good agreement with the radiocarbon date (8300 yr B.P.).

The sites at which the Olduvai Gorge and Nasera Rock Shelter samples were collected are only about 30 km apart. Both have been heavily leached, although the less intensely leached Olduvai Gorge sample (13% Gla content relative to fresh bovine bone) still displays a collagenlike pattern. The calculated racemization age of the Nasera Rock sample (using the Olduvai Gorge specimen for calibration) is within 8% of its radiocarbon date (2200 yr B.P.).

The Murray Springs and Double Adobe samples, very similar in age, are also intensively leached as shown by the Gla contents and the absence of collagen in both organic matrices. As a result of good agreement between the D/L aspartic acid ratios of the Double Adobe sample and the calibrating sample (Murray Springs), the calculated racemization age is concordant with the radiocarbon age (10,400 yr B.P.).

The two older samples from the Nasera Rock Shelter (levels 5A and 6) have similar radiocarbon ages, 21,600 and 22,900, respectively. Both have experienced severe leaching and do not show detectable Gla or a collagenous organic matrix. The racemization age of level 6, based on a calibration with level 5A, is about 13% higher than the radiocarbon age.

The oldest pair of samples is represented by two from Mt. Carmel in

Israel. Both are leached to an extreme degree as shown by the absence of Gla in the Tabūn sample (noncollagenous) and the trace of Gla in the Sefunim specimen (collagenous). Although there is an age difference of approximately 14,000 yr between the Sefunim calibration sample (~27,000 yr B.P.) and the Tabūn sample (41,900 yr B.P.), the calculated racemization age for the Tabūn specimen is within 8% of the radiocarbon age.

These data indicate concordance of aspartic acid racemization age or rate constant ($k_{asp}$) with radiocarbon data for each of the six unknown samples despite measurable differences in leaching history between the unknown and calibration samples. Thus, *in situ* leaching appears to have a negligible effect on the racemization parameters calculated from these six pairs of fossil bone samples.

How can this finding be reconciled with the original laboratory results showing significant alteration of racemization rates with intensive leaching (Hare 1974; Hare et al., 1974)? An examination of the leaching data, as plotted in Figure 3, suggests an answer to this question and also aids in defining the conditions under which leaching could adversely affect racemization calculations.

The 12 samples segregate into two major clusters (Fig. 3): a low aspartic acid D/L ratio group ($\bar{X}$ = 0.18) and a high aspartic acid D/L ratio group ($\bar{X}$ = 0.51). The lower cluster of five samples can be characterized as young in age ($\bar{X}$ = 4100 yr), moderately leached (only one sample has <0.1 nmol $g^{-1}$ Gla), and predominantly collagenous in composition (only one sample contains a noncollagenous organic matrix). In contrast, the other group of seven samples with higher D/L values has a mean age of 12,400 yr, generally exhibits intense leaching (five samples have undetectable or trace amounts of Gla), and contains predominantly noncollagenous matrices (only two samples display a collagen pattern).

It is probably significant that, with only one exception (the slightly leached Buhen Horse specimen), the sample used to calibrate the racemization reaction plots in the same group as the unknown sample to be dated. This observation suggests the possibility that each of the calibration samples has, in effect, calibrated the racemization reaction in the corresponding unknown for degree of leaching as well as environmental temperature.

The need for intragroup calibration probably stems from key compositional differences between the two groups produced, in part at least, by *in situ* leaching. Lacking detailed information on the key properties (e.g., molecular weight, amino acid composition, and degree of racemization) of each matrix component, comparative total amino acid compositions provide evidence for the requirement of compositional similarity between calibration and unknown samples. For example, four of the six concordant racemization calculations involved sample pairs having organic matrices of similar composition (King and Bada 1979).

### Study 3: Survey of Gla in Calcified Tissues

The most striking result of this study was that no Gla (<0.1 nanomoles/g tissue) was detected in any of the invertebrate calcified tissues which were selected from five phyla (Table 1).

Three of the four vertebrate calcified tissues contained detectable levels of Gla: the apatitic shark tooth (495 nanomoles/g), the aragonitic fish otolith (40 nanomoles/g), and the previously reported (King 1978a) apatitic bovine bone (1076 nanomoles/g). No Gla was found in the calcitic hen egg shell.

These data suggest that the presence of Gla in calcifiable matrices may be restricted to vertebrate tissues. Further, it is interesting to note that coagulating hemolymphs (the invertebrate analogs of Gla-containing vertebrate plasmas) also lack Gla. This observation, in conjunction with the absence of Gla in the invertebrate calcified tissues, raises the possibility that the invertebrates as a group may lack the necessary enzymes to biosynthesize Gla for any physiological purpose.

Although the functions of the Gla-containing proteins in the mineralization process have not yet been elucidated, it is clear from these data (Table 1) that their presence in the organic matrix is not obligatory for the initiation or regulation of invertebrate calcification. It is particularly noteworthy that protein-bound Gla is not required by the inarticulate brachiopod *Lingula anatina* for the formation of a phosphatic shell containing hydroxyapatite closely resembling bone mineral.

## CONCLUSIONS

$\gamma$-Carboxyglutamic acid (Gla) content is a valuable parameter for quantitatively assessing the integrated *in situ* leaching history of fossil bones. Because Gla has a very limited distribution in nature and has not been detected in microorganisms (Zytkovicz and Nelsestuen 1976; Hauschka 1977), the indigenous Gla concentration is unlikely to be altered by contaminating organic matter from the environment.

Although the laboratory-observed perturbation of racemization rates by leaching may be a potential problem in the natural environment, the effect of *in situ* leaching apparently can be minimized and virtually eliminated by careful compositional matching of the calibration and unknown samples. It is therefore recommended that Gla content, amino acid composition, and amino acid concentration be measured with the D/L ratio in both the calibration and unknown samples to minimize the possibility of introducing significant errors in the calibration step. In addition to improving the reliability of the racemization dating method, the routine measurement and reporting of these parameters would, over time, generate a valuable data base for profiling the leaching-racemization relationship at ambient temperatures.

## *ACKNOWLEDGMENTS*

The author thanks P. E. Hare and P. V. Hauschka for helpful discussions during these studies. This research was supported by National Science Foundation Grant OCE75-18136. Contribution no. 2945 of Lamont–Doherty Geological Observatory of Columbia University.

## *REFERENCES*

Bada, J. L. and R. Protsch, Racemization reaction of aspartic acid and its use in dating fossil bones, *Proc. Nat. Acad. Sci. U.S.A.*, **70**, 1331–1334, 1973.

Bada, J. L. and R. T. Schroeder, Amino acid racemization reactions and their geochemical implications, *Naturwissenschaften*, **62**, 71–79, 1975.

Hare, P. E., Geochemistry of proteins, peptides, and amino acids, in *Organic Geochemistry*, G. Eglinton and M. T. J. Murphy, eds., Springer-Verlag, New York, pp. 438–463, 1969.

Hare, P. E., Amino acid dating of bone—the influence of water, *Carnegie Inst. Wash. Year Book 73*, 576–581, 1974.

Hare, P. E., D. J. Ortner, and D. W. Von Endt, Amino acid dating of bone and teeth, *Geol. Soc. Amer. Annu. Mtng. Abstr.*, **6**, 778, 1974.

Hauschka, P. V., Quantitative determination of γ-carboxyglutamic acid in proteins, *Anal. Biochem.*, **80**, 212–223, 1977.

Hauschka, P. V., J. B. Lian, and P. M. Gallop, Direct identification of the calcium-binding amino acid, γ-carboxyglutamate, in mineralized tissue, *Proc. Nat. Acad. Sci. U.S.A.*, **72**, 3925–3929, 1975.

Hugli, T. E. and S. Moore, Determination of the tryptophan content of proteins by ion exchange chromatography of alkaline hydrolysates, *J. Biol. Chem.*, **247**, 2828–2834, 1972.

King, K., Jr., γ-Carboxyglutamic acid in fossil bones and its significance for amino acid dating, *Nature*, **273**, 41–43, 1978a.

King, K., Jr., Distribution of γ-carboxyglutamic acid in calcified tissues, *Biochim. Biophys. Acta*, **542**, 542–546, 1978b.

King, K., Jr. and J. L. Bada, Effect of in situ leaching on amino acid racemization rates in fossil bone, *Nature*, **281**, 135–137, 1979.

Schroeder, R. A. and J. L. Bada, A review of the geochemical applications of the amino acid racemization reaction, *Earth Sci. Rev.*, **12**, 347–391, 1976.

Zimmerman, M. R. and W. H. Clark, A possible case of sub-corneal pustular dermatosis in an Egyptian mummy, *Arch. Dermat.*, **112**, 204–205, 1976.

Zytkovicz, T. H. and G. L. Nelsestuen, γ-Carboxyglutamic acid distribution, *Biochim. Biophys. Acta*, **444**, 344–348, 1976.

# Amino Acids in Bristlecone Pine: An Evaluation of Factors Affecting Racemization Rates and Paleothermometry

JOHN E. ZUMBERGE
MICHAEL H. ENGEL
BARTHOLOMEW NAGY
*Laboratory of Organic Geochemistry, Department of Geosciences, University of Arizona*

*ABSTRACT*

The extent of amino acid racemization, moisture content, and the abundances of terpenes, sugars and melanoidin-type substances have been determined in two bristlecone pine logs of the same age, 3210 to 3240 B.C. The samples were interior portions of heartwood from two logs collected at elevations of 2804 m and 3453 m. At the lower elevation site the mean annual temperature is +6°C; the log rested on soil (pH ≃ 6.3) containing more moisture than the ground at the higher elevation. The annual mean temperature at the higher elevation is 0°C and the log rested on practically dry ground (pH ≃ 8.0). The aspartic acid D/L ratio and the soluble sugar content were lower, and the moisture, terpene, and melanoidin-type substances were more abundant in the sample from the lower elevation site than in the sample from the higher elevation. Prevailing conditions in paleoenvironments affect aspartic acid racemization rates. Paleothermometry must therefore consider the paleoenvironmental history of bristlecone pine remnants.

## INTRODUCTION

Racemization rates of amino acids are dependent on both time and temperature. If the age of a sample, for example, wood, is accurately known, it is possible to derive its temperature history, provided other influential factors, such as the chemical, physical and microbiological conditions prevailing in the preserving environment, are also known. Enzymatic activity is restricted mainly to the cambium of a living tree; heartwood cells in living trees are essentially "dead." No proteins are being synthesized in these cells; only their remnants are present (Grozdits and Ifju 1973).

Bristlecone pines provide the oldest dendrochronologically dated wood samples. The exact ages ($\pm$<1 year) of many of these specimens from the White Mountains of California have been determined by the dendrochronological technique of cross-correlating (crossdating) annual tree rings from living trees and long-dead remnants from the present to more than 8000 years BP (Ferguson 1968, 1970; LaMarche 1969, 1974; LaMarche and Harlan 1973). The heartwood of the bristlecone pines is at present the only candidate among plants for yielding a continuous paleotemperature record based on amino acid racemization of more than 8,000 years. However, the older bristlecone pine samples are not parts of living trees. They are logs and fragments scattered about in various preserving environments which may affect amino acid racemization rates. The purpose of this study was to determine some of the factors that affect the amino acid racemization rates in two bristlecone pine logs of identical age but preserved at different elevations in different environments.

Dendrochronology can not only serve as an accurate dating method but can also reveal certain paleoclimatic indicators, for example, the response of annual tree-ring widths to climatic variations such as prevailing moisture (Fritts 1966; LaMarche 1974). Thus, certain species of the order Coniferales, families Abietaceae and Taxodiaceae, such as bristlecone pine (*Pinus longaeva*) and sequoia (*Sequoiadendron giganteum*) are of special significance to more than one aspect of paleoclimatic research. According to Bailey (1970), the bristlecone pines (family Abietaceae, genus *Pinus*) consist of two species, *P. aristata* Engelm. and *P. longaeva* D. K. Bailey. The latter species occurs mainly in California, Nevada, and Utah, and the former in Colorado, New Mexico, and Arizona. This classification is supported by monoterpene chemotaxonomy (Zavarin and Snajberk 1973). A related species, *P. balfouriana*, occurs west and north of the bristlecone pine *P. longaeva* in California. These three species do not seem to be closely related to other members of the genus *Pinus*.

The study of amino acid racemization in the heartwood of a sequoia tree (Engel et al., 1977) illustrated the utility of amino acid racemization for determining paleotemperatures in one plant species. The sequoia samples were extracted from an individual tree which was living when cut down in 1915 and was subsequently stored. Activation energies and Arrhenius frequency factors of the amino acid racemization reactions were determined from the results of elevated temperature, laboratory simulation experiments on young (1683–1781 A.D.) sequoia heartwood samples. The first-order reversible rate constant was determined from the extent of aspartic acid racemization in older (210 B.C., 375 A.D., and 1000 A.D.) dendrochronologically dated, unheated heartwood samples obtained from the same section of the sequoia tree as those used for the elevated temperature experiments. The activation energies, frequency factors, and rate constant permitted the calculation of the average temperature experienced by aspartic acid during the past 2185 years. This calculated temperature (6°C) was in remarkably close agreement with modern mean annual temperatures (7°C) recorded by weather stations near the location of the sequoia tree and also in agreement with inferred paleotemperatures. A similar study of bristlecone pine that covers a considerably longer period of time than sequoia is reported here.

## BRISTLECONE PINE SAMPLES

Interior unweathered heartwood portions of two bristlecone pine specimens were obtained from V. C. LaMarche (Sample No. CAM 15C) and C. W. Ferguson (Sample No. TRL 71-52) of the Laboratory of Tree-Ring Research, The University of Arizona. The two samples are of the same age (from 3210 to 3240 B.C.) determined by dendrochronological techniques. Neither sample has been part of a living tree for approximately 3000 years; they have both been preserved as broken logs for this period of time. The two logs were collected at different elevations and from dissimilar environments in the White Mountains of California. Sample CAM was collected from an exposed sandstone slope (Figs. 1 and 2) at 3453 m elevation just above the timberline on Campito Mountain (latitude: 37° 30′ N; longitude: 118° 12′ W). The TRL sample was found partly buried under talus boulders (Figs. 3 and 4) at a lower elevation, 2804 m in Wyman Canyon (latitude: 37° 27′ N; longitude 118° 10′ W). The TRL log was lying in dolomitic soil between and partially beneath dolomitic talus blocks and showed some decay on its side in contact with the soil. The CAM log was lying on sandstone rocks with little contact with soil and was well preserved throughout. According to Wright and Mooney (1965), the dolomitic soil has a higher moisture content than the sandstone

**Figure 1.** Bristlecone pine sample CAM 15C. The log is lying on a north facing sandstone slope on Campito Mountain in the White Mountains, California; elevation: 3,453 m. The log is ~0.3 m in diameter. (Photograph by D. Campbell.)

soil. Also, the sandstone soil is slightly acidic (pH = 6.3), and the dolomitic soil is slightly alkaline (pH = 8.0). Present mean annual temperatures are approximately 0°C at 3453 m and +6°C at 2804 m in the White Mountains, based on available U.S. Weather Bureau records (LaMarche 1973).

## EXPERIMENTAL

All glassware employed in the following experiments was cleaned in hot concentrated $H_2SO_4/HNO_3$ (85/15, v:v). All solvents were freshly redistilled from spectral or pesticide grade reagents. All procedural steps requiring nitrogen gas used electronic grade $N_2$ that had been passed through a molecular sieve purifier.

The bristlecone pine samples obtained for analysis were sliced into

**Figure 2.** View toward the south of Campito Mountain showing the exposed site. Sample CAM 15C was collected above the timber line indicated by the arrow. (Photograph by V. C. LaMarche, Jr.)

approximately 0.1–0.5 mm thick shavings with a solvent-cleaned scalpel. Unlike some other sample preparation processes, this procedure ensured that the wood was not subjected to heating that might affect its amino acid components. Shavings from interior portions from two previously untreated bristlecone pine samples were analyzed for D and L amino acids using the following procedures. The shavings were first thoroughly washed in a fritted glass filter with 1.5 *N* HCl (double distillated) followed by triple glass distilled water to extract soluble amino acids, which were subsequently analyzed. The extracted samples were then hydrolyzed in 6 *N* HCl (double distilled) at 100°C for 24 hours under nitrogen in sealed Pyrex test tubes to yield bound amino acid fractions. Next, the filtrates of the hydrolyzates were evaporated under nitrogen, redissolved in 2 ml double distilled 0.06 *N* HCl, centrifuged to remove any insoluble residue, and placed on Bio Rad AG 50-X8, 50-100 mesh cation exchange columns. The columns had been prewashed consecutively with 2 *N* NaOH, $H_2O$, 1.5 *N* HCl, 6 *N* HCl, and $H_2O$. Twenty-five ml of triple distilled $H_2O$ containing 50 μl of a dilute phenolphthalein solution in methanol was then passed through the columns. The amino acids were subsequently eluted with 1 *N* $NH_4OH$. After evaporating the eluates to dryness, anhydrous

**Figure 3.** Bristlecone sample TRL 71-52. The remnant was partially covered by dolomite talus boulders; some of them had been removed before the photograph was taken. Wyman Canyon, White Mountains, California; elevation: 2804 m. The remnant is pie-shaped in cross section witha diameter of ~1 m. (Photograph by C. W. Ferguson.)

2–4 *N* (+)-2-butanol/HCl was added to the residues, and the reaction mixtures were heated at 100°C for 3 hours. Next, the excess butanol was evaporated and $CH_2Cl_2$ and pentafluoropropionic anhydride were added to each residue. After a minimum of 2 hours the samples were evaporated to dryness, redissolved in $CH_2Cl_2$, and the resulting volatile diastereomeric derivatives of the amino acid hydrolysis products were separated on Carbowax-20M coated, 46 m × 0.5 mm i.d. stainless steel (S.C.O.T.) and/or 61 m × 0.5 mm i.d. nickel (W.C.O.T.) capillary columns. The D/L ratios of the amino acids were determined by an Infotronics CRS-208 digital integrator coupled to the Perkin-Elmer Model F-11 gas chromatograph.

The two bristlecone pine samples were also analyzed for other constituents. First, the volatile contents were determined. Duplicate samples (thin shavings) of both the CAM and TRL heartwoods were heated at 78°C in a vacuum oven for 13 hours in the presence of strong dessicants. Second, to determine the nature of the volatile substances, the two different wood samples were heated at 79°C under vacuum in a quartz pyr-

**Figure 4.** View toward the west of the south slope of Wyman Canyon showing the talus; sample TRL 71-52 was collected close to the center right of the photograph. (Photograph by C. W. Ferguson.)

olysis vessel directly connected to the gas inlet system of an organic mass spectrometer (Hitachi RMU-6E). The evolving volatiles were trapped during heating at liquid nitrogen temperatures and were subsequently introduced into the mass spectrometer. Terpenes from both samples were further characterized by infrared spectroscopy and combined gas chromatography/mass spectrometry. Bristlecone pine shavings from both samples were extracted with $CHCl_3$, using ultrasonication. Concentrated terpene fractions were analyzed between NaCl discs in a Perkin-Elmer 137B Infracord Spectrophotometer. Aliquots of the $CHCl_3$ extracts were injected onto a 30.5 m × 0.5 mm Apiezon-L coated capillary column in a Perkin-Elmer 226 gas chromatograph which was directly connected via a molecular separator to a Hitachi RMU-6E mass spectrometer.

In another set of experiments, the two bristlecone pine samples were extracted with hot triple distilled water, and the resulting extracts were analyzed for carbohydrates by the classical phenol-sulfuric acid method (Dubois et al., 1956). The water extracts were also analyzed by gas chromatography after the solutions were evaporated to dryness and the residues were silylated with Tri Sil reagent (Pierce Chemical Co.) to obtain the trimethylsilyl (TMS) derivatives of carbohydrates. The TMS derivates were separated on a SE-30 capillary column (46 m × 0.5 mm) after the excess Tri Sil reagent was evaporated under $N_2$ and after the residue was dissolved in hexane. In one series of gas chromatographic experiments co-injections of derivatized sugar standards together with both extracts helped characterize the eluates. The water extracts were also analyzed by ultraviolet spectroscopy, using a Beckman DK-2A Ratio Recording Spectrophotometer for the presence of possible melanoidin-type polymers.

Elevated temperature kinetic experiments of amino acid racemization in wood must be performed for paleotemperature studies in order to obtain the necessary kinetic data, that is , the activation energies and Arrhenius frequency factors. The thin shavings of the CAM and TRL samples were heated at various temperatures (148°, 130°, 120.5°, 109.5°C) and for different lengths of time under nitrogen in sealed Pyrex tubes with only their current intrinsic moisture contents. The experimental variables are listed in Table 1. No attempts were made in this set of experiments to reconstitute the water content of living bristlecone pines because, as mentioned above, the samples have not been part of living tress for approximately 3000 years. After being heated, the samples were washed with 1.5 *N* HCl to separte the free amino acid fractions from the bound amino acids. Both free and bound fractions were hydrolyzed in 6 *N* HCl prior to derivatization and determination of D/L ratios. Next, the derivatized amino acids were quantitatively analyzed in a manner identical to the analysis of the D/L ratios described above. Both the amino acid contents of the samples and the D/L ratios of their free and bound amino acid fractions were determined. In this context, the free fraction is considered to consist of those amino acids and/or small peptides which can be removed from the wood with 1.5 *N* HCl.

In another set of experiments, further attempts were made to determine whether the racemization of amino acids in bristlecone pine follows first order reversible kinetics. Three other TRL 71-52 samples from the same log were studied. The ages of these samples were 3555 B.C., 3855 B.C., and 4205 B.C. The D/L ratios of the amino acids were determined without heating the samples (except during hydrolysis and derivatization).

The accuracy of D/L amino acid ratios may be suspect if the samples were contaminated by microorganisms because certain microbiota contain D-amino acid enantiomers *in vivo* besides the usual L-form. Consequently, other TRL samples were analyzed in order to evaluate potential contaminations from microbial amino acids. A visibly pitted, weathered, and rotten series of tree rings (from 3575 to 3675 B.C.; sample No. TRL 71-52) were analyzed for amino acid abundances and D/L ratios; both the free and the bound amino acid fractions were examined. A closely ad-

**Table 1. D/L Ratios and Amino Acid Abundances Derived from Elevated Temperature Experiments on Bristlecone Pine**

| Sample | Temperature (°C) | Time (hours) | Fraction | D/L | | | | | ng/mg wood | | |
|---|---|---|---|---|---|---|---|---|---|---|---|
| | | | | Asp | Glu | Pro | Phe | Leu | Asp | Glu | Pro |
| TRL | unheated | — | Total[a] | 0.088 | 0.051 | 0.046 | 0.050 | 0.042 | 123 | 132 | 122 |
| CAM | unheated | — | Total | 0.102 | 0.041 | 0.055 | 0.019 | 0.044 | 44 | 42 | 43 |
| TRL | 109.5 | 601.5 | Bound | 0.466 | 0.182 | 0.120 | 0.105 | 0.095 | 53 | 68 | 69 |
| CAM | 109.5 | 601.5 | Bound | 0.471 | 0.151 | 0.054 | 0.061 | 0.044 | 32 | 37 | 58 |
| TRL | 109.5 | 601.5 | Free | 0.574 | 0.260 | 0.405 | 0.220 | 0.178 | 0.5 | 1 | 1 |
| CAM | 109.5 | 601.5 | Free | 0.595 | 0.226 | 0.306 | 0.239 | n.d.[b] | 1 | 4 | 2 |
| TRL | 120.5 | 72.0 | Bound | 0.196 | 0.069 | 0.039 | 0.039 | 0.036 | 86 | 86 | 88 |
| CAM | 120.5 | 72.0 | Bound | 0.134 | 0.060 | 0.030 | 0.038 | 0.026 | 85 | 88 | 99 |
| TRL | 120.5 | 72.0 | Free | 0.359 | 0.157 | 0.168 | 0.222 | 0.065 | 3 | 4 | 2 |
| CAM | 120.5 | 72.0 | Free | 0.301 | 0.200 | 0.141 | 0.259 | 0.057 | 4 | 6 | 4 |
| TRL | 130.0 | 96.0 | Bound | 0.359 | 0.099 | 0.055 | 0.072 | 0.053 | 43 | 53 | 48 |
| CAM | 130.0 | 96.0 | Bound | 0.380 | 0.128 | 0.042 | 0.058 | 0.043 | 22 | 23 | 29 |
| TRL | 130.0 | 96.0 | Free | 0.539 | 0.225 | 0.350 | 0.352 | 0.112 | 6 | 11 | 6 |
| CAM | 130.0 | 96.0 | Free | n.d. | n.d. | n.d. | n.d. | n.d. | n.d. | n.d. | n.d. |
| TRL | 148.0 | 96.0 | Bound | 0.738 | 0.365 | 0.189 | 0.231 | 0.207 | 26 | 50 | 49 |
| CAM | 148.0 | 96.0 | Bound | 0.754 | 0.361 | 0.096 | 0.125 | 0.152 | 14 | 33 | 55 |
| TRL | 148.0 | 96.0 | Free | 0.652 | 0.449 | 0.589 | 0.406 | 0.419 | 4 | 14 | 7 |
| CAM | 148.0 | 96.0 | Free | >0.425 | <0.470 | 0.244 | 0.282 | 0.171 | 1 | 2 | 2 |

[a] Total = Bound fraction, since no free amino acids were detected.
[b] n.d. = Not determined.

jacent set of unweathered and not rotten rings in the same sample (from 3687 to 3707 B.C.) was also analyzed for the sake of comparison.

## RESULTS AND DISCUSSION

The results of the experiments designed to determine the D/L amino acid ratios from multiple subsamples of the CAM 15C and TRL 71-52 bristlecone pine logs were, for example, for aspartic acid: CAM, D/L = 0.102 ± .008; TRL, D/L = 0.088 ± .007. D/L ratios of other amino acids are listed in Table 1. These D/L results were obtained from the bound amino acid fraction, which essentially corresponds to the total amino acid fraction, since relatively few free amino acids were detected in the two logs. These two bristlecone pine samples are the same age but they have experienced different temperature histories. It was expected that sample CAM collected at the higher elevation would exhibit less aspartic acid racemization because it was exposed to lower temperatures than sample TRL, provided that all of its other physical and chemical parameters were comparable with the TRL sample. The results listed above do not agree with a simple theory of amino acid racemization; however, they could open up interesting possibilities such as the effects of the preserving environment.

In an attempt to understand the seemingly anomalous D/L ratios, the $H_2O$, terpene and sugar contents of the CAM and TRL samples were determined. The purpose of these analyses was to detect any differences in the preserving environments from chemical data. Sample CAM (collected at the high elevation) lost 9.8% of its weight, and sample TRL (from the lower elevation site) lost 17.8% after heating in a vacuum oven. Analyzing the volatiles in the mass spectrometer gave the following results. The relative abundances of various ions from $H_2O$ ($m/e$ = 18, 17) and $C_{10}$ terpens ($m/e$ = 77, 92, 93, 121, 136) showed that the composition of the CAM volatiles were: 87.5% $H_2O$ and 12.5% $C_{10}$ terpenes. The composition of the volatiles released from the TRL sample was: 75.4% $H_2O$ and 24.6% $C_{10}$ terpenes. It was calculated that the CAM sample contained 8.6% $H_2O$ and 1.2% $C_{10}$ terpenes by weight and that sample TRL contained 13.4% $H_2O$ and almost four times as much $C_{10}$ terpenes as CAM, that is, 4.4% by weight. Thus, the TRL sample contains considerably more water than the CAM sample.

Infrared spectra of the chloroform extracts (obtained by ultrasonication) of the two bristlecone pines indicated C-H stretching and bending vibrations at 2910 $cm^{-1}$ and 1450 $cm^{-1}$, respectively, together with the carbonyl (C=O) stretching vibration at 1695 $cm^{-1}$. An absorption band at 1590 $cm^{-1}$ was apparently caused by a C=C stretching vibration.

Gas chromatograms of the chloroform extracts of samples CAM and TRL and the mass spectral identifications of the numbered peaks are shown in Figure 5. The identifications were based on the comparison of the unknown mass spectra with published spectra and/or with those of standards obtained from the same mass spectrometer under identical operating conditions as were used for the unknowns. Figures 6 and 7 show mass spectra corresponding to two gas chromatographic peaks of the chloroform extracts from samples CAM and TRL and mass spectra of corresponding terpene standards. Gas chromatographic retention times helped to confirm the mass spectral identifications of the terpenes. Carbonyl containing compounds were not detected by the gas chromatographic/mass spectrometric analyses; the reason for this could be that the polar carbonyl compounds were not properly eluted from the nonpolar Apiezon-L phase in the column. The gas chromatograms (Fig. 5) show significant differences in the abundances and distribution of the terpenes between the two bristlecone pine samples. The major component of sample TRL is the $C_{10}$ monoterpene, 3-carene; sample CAM completely lacks this compound. The major component of CAM is another $C_{10}$ monoterpene, α-pinene. TRL also contains α-pinene with a 3-carene/α-pinene ratio of approximately 2.5. Both samples contain limonene and camphene along with some sesquiterpenes. These results show that the analyzed resin components of bristlecone pine samples CAM and TRL differ both quantitatively and qualitatively. This could be caused by *post mortem* differential weathering or by inter- or intra-species variations. Differences in monoterpene composition between the bristlecone pine species *P. aristata* and *P. longaeva* were observed by Zavarin and Snajberk (1973).

The carbohydrate contents of the CAM and TRL samples were also different. Sample CAM contained 3.0% soluble sugars, and the TRL sample contained 1.3% sugars by weight as determined by the phenol-sulfuric acid analysis of the $H_2O$ extract. Arabinose was the only sugar which has yet been identified as a major carbohydrate component in both samples by gas chromatography of TMS derivatives; CAM contained about three times as much arabinose as TRL. It is generally known that L-arabinose exists in the free state in many coniferous trees. The following sugars were not present as major constituents in the water extracts of the heartwood samples: ribose, fucose, xylose, rhamnose, glucose, fructose, galactose, and sucrose. However, other major $H_2O$ soluble unknown components, having different retention times than these sugars, were present in varying quantities in the CAM and TRL samples.

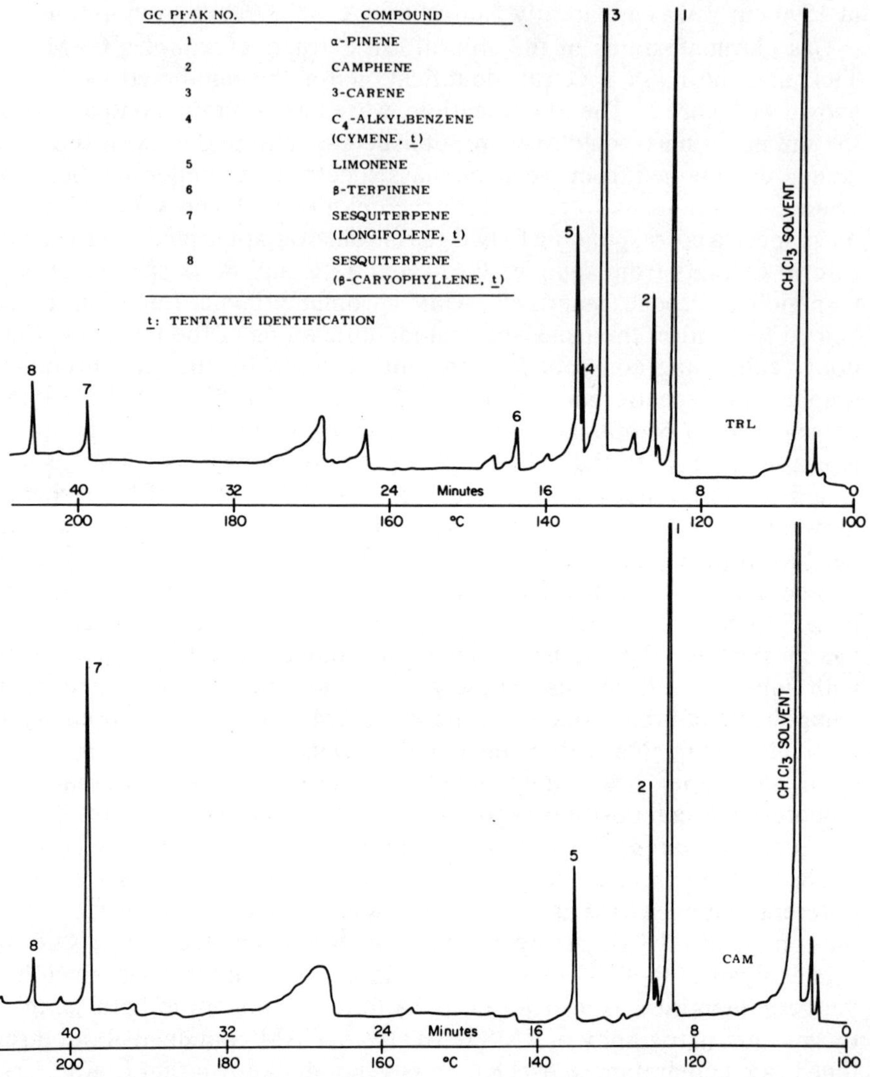


**Figure 5.** Gas chromatograms showing terpene components in chloroform extracts of bristle cone pine samples TRL and CAM. Mass spectral identifications of the numbered peaks are listed on the diagram. Apiezon-L (S.C.O.T.) 30.5 m × 0.5 mm, stainless steel capillary column.

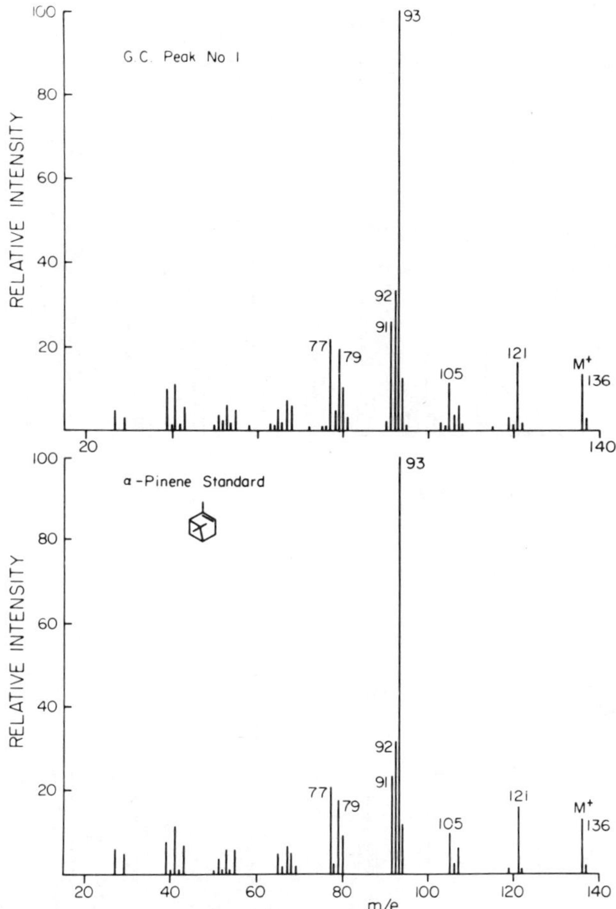


**Figure 6.** Comparison of the mass spectra of peak No. 1 (on the gas chromatograms shown in Fig. 5) with that of a α-pinene standard.

Ultraviolet spectroscopy of the aqueous extracts of the two bristlecone pine samples revealed an absorption band at 280 nm; this band was more intense in the TRL sample than in CAM. This absorption effect may indicate a higher concentration of a soluble melanoidin-type polymer and/or furaldehydes in the TRL sample. The melanoidin-type polymers may form from soluble amino acids and carbohydrates in the presence of water (e.g., Hedges 1976). The higher water and lower carbohydrate contents of the TRL sample, as compared with the CAM sample, supports the possibility that carbohydrates may have polymerized with amino acids to form melanoidin-type substances in the TRL log.

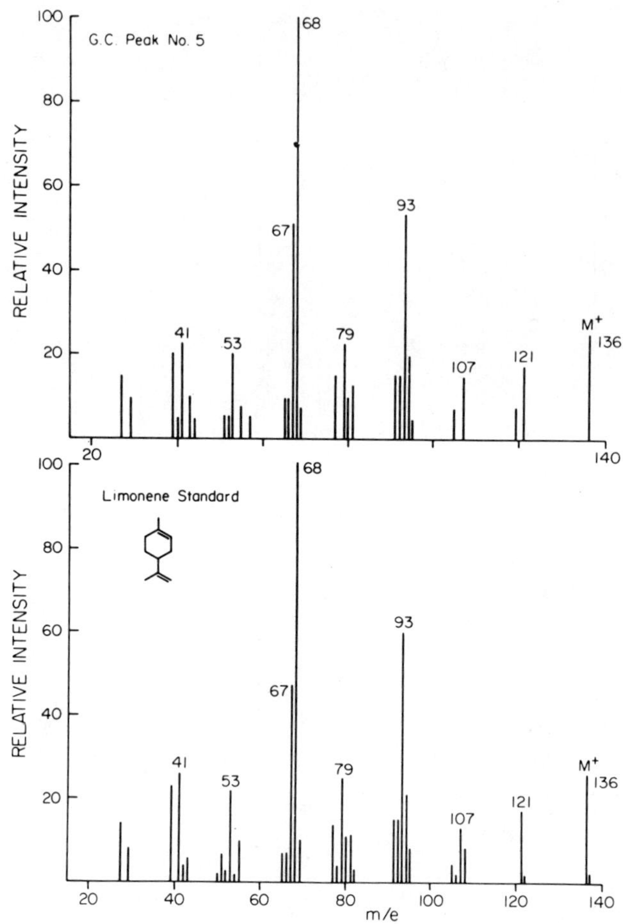


**Figure 7.** Comparison of the mass spectra of peak No. 5 (on the gas chromatograms shown in Fig. 5) with that of a limonene standard.

The results of the amino acid kinetic experiments revealed both similarities and dissimilarities between the two bristlecone pine logs. The amino acid D/L ratios and abundances from the elevated temperature kinetic experiments are given in Table 1. Free amino acids and possibly small peptides were obtained in larger quantities from the heated than from the unheated wood samples, although the free fractions still comprised only a small percentage of the total amino acid content. In almost every case, the free fraction exhibited more racemization than the bound fraction. It is possible that the free amino acids obtained by heating are

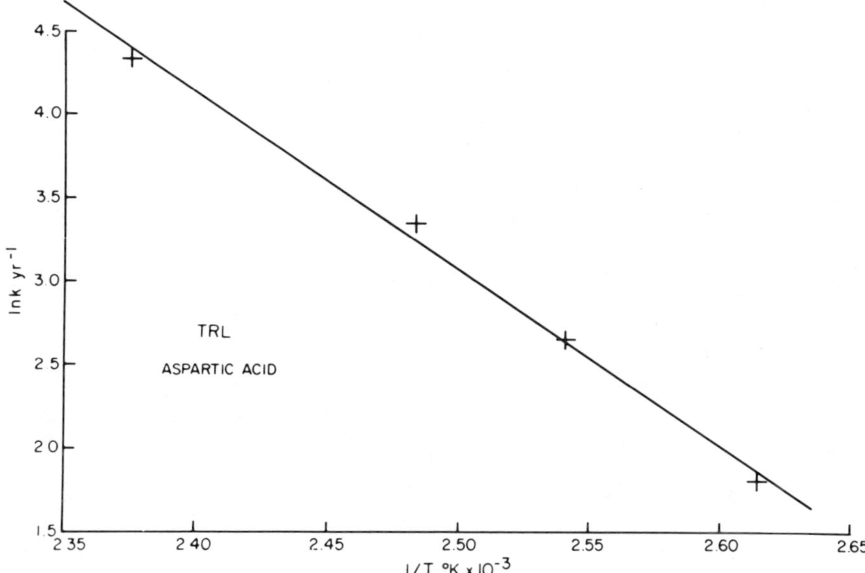


**Figure 8.** Arrhenius plot of the racemization reaction calculated from the total aspartic acid fraction of sample TRL.

the products of hydrolysis of polypeptides and of the concomitant release of many terminal amino acids. Kriausakul and Mitterer (1978) have shown that terminal amino acids racemize faster than the interior amino acids in peptides during elevated temperature experiments. Also, the racemization of the released free amino acids may have been catalyzed by sugars during early stages of melanoidin formation (Zumberge 1979).

First order reversible rate constants ($k_1$) were calculated for the free and bound amino acid fractions of the heated TRL and CAM samples using the following equation:

$$ln\left[\frac{1 + D/L}{1 - D/L}\right]_t - ln\left[\frac{1 + D/L}{1 - D/L}\right]_{t=0} = 2k_1t.$$

The logarithmic term at $t = 0$ was evaluated using the D/L ratios obtained from the bristlecone pine heartwoods prior to heating; this term also includes the extent of racemization by hydrolysis and by (–)-2-butanol impurities.

Plots of ln $k_1$ against 1/T°K yield $E_a$ (activation energies) and $ln\ A$ (frequency factors) based on the Arrhenius equation ($ln\ k_1 = ln\ A - E_a/RT$). Arrhenius plots for aspartic and glutamic acids of the TRL samples are shown in Figures 8 and 9, respectively. The $ln\ k_1$ ($yr^{-1}$) values used

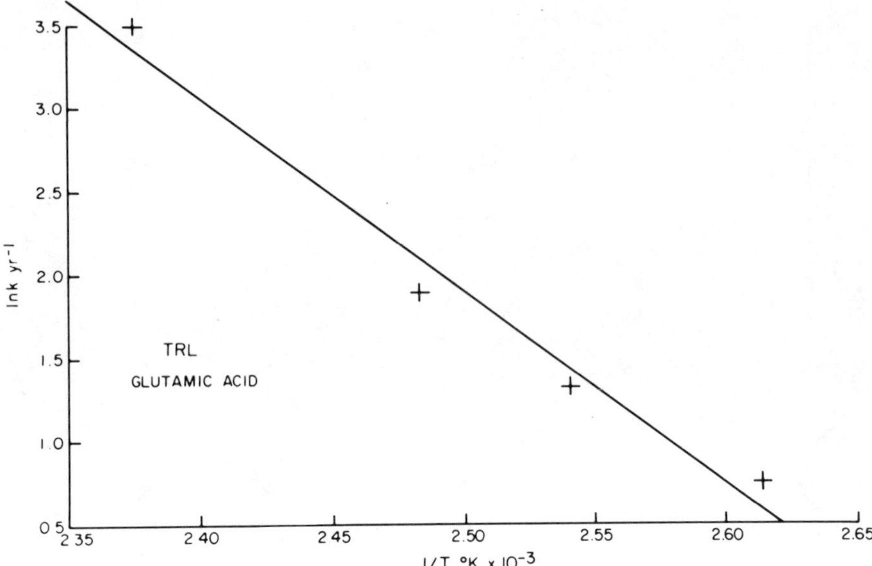


**Figure 9.** Arrhenius plot of the racemization reaction calculated from the total glutamic acid fraction of sample TRL.

for plotting Figures 8 and 9 were determined from the data listed in Table 1. These rate constants were partially optimized for the constraints which will be discussed below and were corrected for the amount of racemization in the unheated wood. The calculation of selected $ln\ k_1$ values were based on the D/L ratios of the total amino acid fractions, that is, the weighted averages of the D/L ratios of both the bound and free fractions. In the majority of cases, the total amino acid D/L ratios were only slightly higher than the bound D/L ratios because the abundance of the more racemized free fraction was relatively small. ln $k_1$, values of the bound fractions were also calculated. Table 2 lists the $E_a$ and $ln\ A$ values for aspartic and glutamic acids in both the total and bound fractions for TRL and CAM samples, together with the error ranges and correlation coefficients, $R^2$. $E_a$ and $ln\ A$ were calculated using a least squares, best fit computer program which included weighting the experimental error of the D/L ratios and resultant $ln\ k_1$ values. The errors in slope $(-E_a/R)$ and $y$-axis intercept ($ln\ A$ yr$^{-1}$) were estimated by measuring the error range in the $y$-intercept at $1/T \neq 0$ because data points were available only in the temperature region studied, and consequently the full range of scatter (which determines the $y$-intercept error at $1/T = 0$) could not be established closer to $1/T = 0$. The value of $ln\ A$, but not its error range, was determined at

**Table 2. Energies of Activation, Frequency Factors, and Correlation Coefficients from Bristlecone Pine Samples**

| | $E_a$ (kcal/mole) | ln A ($yr^{-1}$) | $R^{2a}$ |
|---|---|---|---|
| TRL Sample | | | |
| Asp Total | 21.01 ± 0.05 | 29.51 ± 0.06 | 0.9950 |
| Asp Bound | 21.28 ± 0.05 | 29.83 ± 0.06 | 0.9991 |
| Glu Total | 23.07 ± 0.19 | 30.92 ± 0.24 | 0.9753 |
| Glu Bound | 22.50 ± 0.23 | 30.13 ± 0.29 | 0.9515 |
| CAM Sample | | | |
| Asp Total[b] | 24.37 ± 1.20 | 33.45 ± 1.51 | 0.8635 |
| Asp Bound | 24.78 ± 1.20 | 33.93 ± 1.51 | 0.8529 |
| Glu Total[b] | 25.03 ± 0.19 | 33.33 ± 0.24 | 0.9747 |
| Glu Bound | 25.65 ± 0.21 | 34.04 ± 0.27 | 0.9673 |

[a] Perfect correlation coefficient: $R^2$ = 1.0000.
[b] The D/L ratio for the total fraction from the CAM 130°C experiment (see Table 1) was estimated from the extent of racemization in other free fractions because the 130°C free fraction was not determined.

*1/T* = 0. Slope error was calculated by holding the variables constant ($ln\ k_1$ and *1/T* in the Arrhenius equation) and varying the *y*-intercept (*ln A*) by the maximum permissible error defined by the scatter of the experimental data points. It should be noted that although the best fit lines of most of the Arrhenius plots have relatively high correlation coefficients, the calculated slopes, *y*-intercepts and their errors were deduced from a line derived from only four points. The error range is greater for aspartic acid in the CAM sample, as is shown in Table 2.

It is important that the wood samples be heated at appropriate temperatures for the optimum (and not excessive) lengths of time. This minimizes the effects of secondary reactions (i.e., nonracemization reactions); one example may be the formation of melanoidins. Such reactions will affect the extent of racemization. It was noted with both sequoia and bristlecone pine that extensive degradation of amino acids may occur if samples are heated too long above approximately 110°C. Under such conditions the resulting rate constants will have erroneous values and the kinetic data will yield inaccurate paleotemperatures. For example, Figure 10 shows the relationship between the quantities (ng/mg wood) of bound aspartic acid, glutamic acid, and proline with increasing heating time of the TRL sample at 120°C. Apparently the amino acids are either decomposing (e. g., decarboxylation or deamination) with increasing time of heating or they are reacting with other components (e.g., sugars) to form

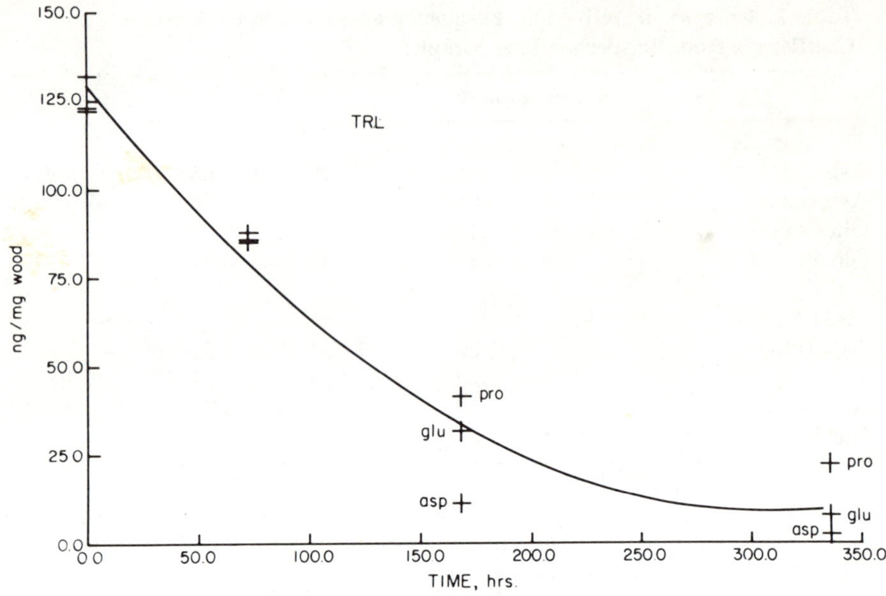


**Figure 10.** Decrease in the concentration of bound proline, aspartic and glutamic acids from sample TRL with increasing time at 120°C.

polymer-like material. Small quantities of free amino acids are produced during heat treatment but not in sufficient quantities to account for the substantial decrease in the abundances of bound amino acids.

Figure 11 shows the scatter and limitations of rate constants of bristle cone pine derived from elevated temperature kinetic experiments. This illustration emphasizes the need for establishing and applying all appropriate constraints including the optimum temperature when performing kinetic experiments, as was noted above. One of the constraints is that the rate constants and other kinetic data be derived from elevated temperature experiments when amino acid loss is minimal during heating. Evaluation of heating different bristlecone pine samples for various lengths of time at specific temperatures has been reported by Lee (1975).

Additional experiments to better define rate constants yielded the following results. These experiments were conducted on three additional TRL 71-52 samples, the ages of which were: 3555 B.C., 3855 B.C., and 4205 B.C. Table 3 shows that racemization in the unheated TRL samples generally follows first order reversible kinetics with the exception of the racemization of aspartic acid in the 4205 B.C. sample. The same integrated equation was used to calculate $k_1$ as was used in the elevated temperature experiments except that the D/L ratios at time zero were determined to

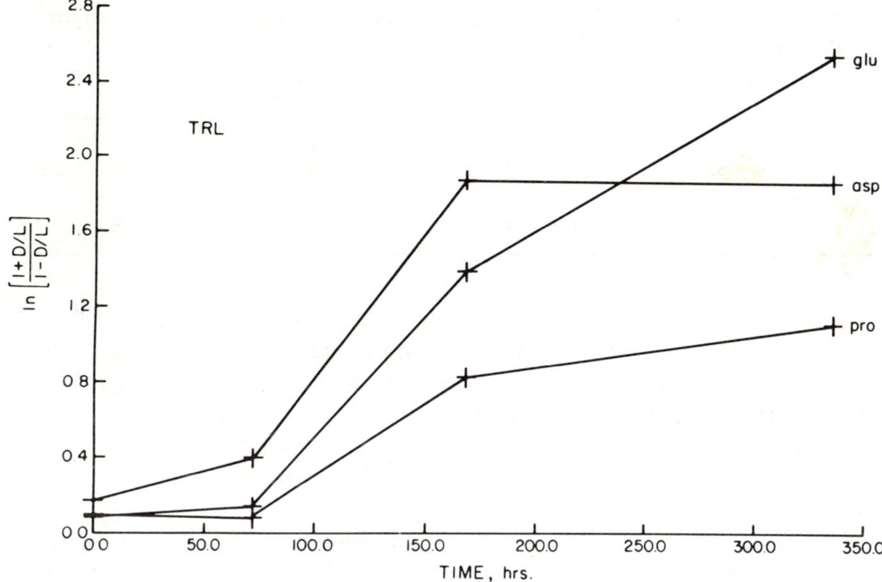


**Figure 11.** Nonlinear increase of the racemization of bound proline, aspartic, and glutamic acids with increasing time at 120°C from sample TRL.

be 0.021 and 0.0175 for aspartic and glutamic acids, respectively. These initial ratios were the results of (−)-2-butanol impurities and slight racemization during hydrolysis. The average rate constants ($k_1$) for aspartic acid and glutamic acid from the TRL samples are $1.27 \pm .05 \times 10^{-5}\ yr^{-1}$ and $0.69 \pm .04 \times 10^{-5}\ yr^{-1}$, respectively. The CAM sample (high elevation) yielded $k_1$ values for aspartic acid and glutamic acid of $1.56 \times 10^{-5}\ yr^{-1}$ and $0.45 \times 10^{-5}\ yr^{-1}$, respectively. CAM samples of different ages were not available for analysis.

**Table 3. D/L Ratios and $k_1$ Values for Aspartic and Glutamic Acids from Unheated TRL Samples**

| Age, years | D/L asp | D/L glu | $\ln\left[\frac{1+D/L}{1-D/L}\right]_{asp} - 0.042$ | $\ln\left[\frac{1+D/L}{1-D/L}\right]_{glu} - 0.035$ | $k_{1asp} \times 10^{-5} yr^{-1}$ | $k_{1glu} \times 10^{-5} yr^{-1}$ |
|---|---|---|---|---|---|---|
| 5203 | 0.088 | 0.051 | 0.134 | 0.067 | 1.29 | 0.65 |
| 5533 | 0.092 | 0.055 | 0.143 | 0.075 | 1.29 | 0.68 |
| 5833 | 0.096 | 0.060 | 0.151 | 0.085 | 1.29 | 0.73 |
| 6183 | 0.094 | 0.062 | 0.147 | 0.089 | 1.19 | 0.72 |

**Table 4. Calculated Effective Temperatures from Bristlecone Pine Racemization Kinetics**

| TRL Sample | °C | CAM Sample | °C |
|---|---|---|---|
| asp total | −13.9 ± 0.8 | asp total | 2.3 ± 16.5 |
| asp bound | −12.5 ± 0.8 | asp bound | 4.0 ± 16.3 |
| glu total | − 1.8 ± 2.7 | glu total | 2.8 ± 2.6 |
| glu bound | − 3.6 ± 3.4 | glu bound | 5.4 ± 2.9 |

Inserting the above values and corresponding errors for $k_1$ into the Arrhenius equation along with activation energies and frequency factors determined from the elevated temperature kinetic experiments (Table 2), permits the calculation of average effective temperatures during the past 5000 to 6000 years. This corresponds to the temperature experienced by the amino acids in the bristlecone pines. The calculated temperatures are listed in Table 4.

The results of the microbiological contamination control experiments on a TRL 71-52 sample showed differences in the amino acid D/L ratios between the rotten and nonrotten portions of the log. The concentrations and D/L ratios of aspartic acid, glutamic acid, proline, and alanine for both the rotten and adjacent nonrotten samples are listed in Table 5. These results show that the abundance of bound amino acids in the rotten wood is more than ten times higher than in the adjacent nonrotten wood. The bound D/L aspartic acid ratios are essentially the same between the rotten and nonrotten portions. However, alanine showed an unusually high D/L ratio in the free amino acid fraction of the rotten wood. Presumably, the high concentration of amino acids in the rotten sample could be the result of reasonably well preserved remnants of microorganisms; the high alanine D/L ratio may be attributed to certain microbial polypeptides which can be enriched *in vivo* with the D-enantiomers.

**Table 5. Amino Acid Abundances and D/L Ratios of Rotten and Adjacent Nonrotten Bristlecone Pine (TRL 71-52)**

| | | Abundances (ng/mg) | | | | D/L Ratios | | | |
|---|---|---|---|---|---|---|---|---|---|
| Sample | Fraction | asp | glu | pro | ala | asp | glu | pro | ala |
| Rotten | Bound | 1592 | 1994 | 1102 | 1452 | 0.089 | 0.096 | 0.046 | 0.080 |
| | Free | 26 | 93 | 6 | 7 | 0.068 | 0.052 | 0.042 | 0.233 |
| Nonrotten | Bound | 125 | 160 | 108 | 60 | 0.085 | 0.043 | 0.043 | 0.133 |
| | Free | 9 | 18 | —[a] | 2 | — | 0.034 | — | — |

[a] Below quantitative detection limit.

## CONCLUSIONS

Comparison of the effective temperatures calculated from aspartic and glutamic acid racemization in the TRL and CAM samples with present mean annual temperatures in the White Mountains shows reasonable agreement for one of the samples (CAM). Furthermore, the experimental precision ($<1°C$) for TRL is closely approaching the error range needed for determining paleotemperature variations during the past ~8000 years. The results are noteworthy in view of the many factors that affect bristle cone pine logs preserved in different environments for the past 3000 years.

These factors include the microenvironment (chemical, physical, and biological) in the wood which evidently affects the racemization of amino acids. For example, the intrinsic moisture content and resin (e.g., $C_{10}$ monoterpene) composition and distribution between the two bristlecone pine samples are markedly different. Differences between these two components could well have an effect on the rate of amino acid racemization within the woody tissues and may also serve as paleoenvironment indicators. The difference in $H_2O$ content between the two bristlecone pine samples may be related to variations in the preserving environments. The TRL log was lying on dolomitic soil which had a higher moisture content than the sandstone soil (site of the CAM log); the TRL log contains more $H_2O$ than the CAM log. In addition, there may be some surface water runoff at the TRL site during the spring melt period.

Variation in arabinose content (CAM > TRL) and presumable related polymerization reactions could also have an effect on amino acid racemization. The differences in sugar content may explain the greater abundance of D-aspartic acid in the CAM samples because at least one sugar is known to increase the racemization rate of aspartic acid (Zumberge 1979). Differences in the pH values of the two soils may also affect the rate of racemization if pH variations are reflected within the wood remnants. As was shown above, microbial contamination of the heartwood probably did not affect the apparent racemization rates of the two samples studied in an attempt to define some of the factors affecting bristlecone pine paleothermometry. A better understanding of the factors affecting racemization should improve the accuracy in the amino acid paleothermometry of bristlecone pine logs.

Calculated effective heartwood temperatures need to be accurately correlated with actual atmospheric paleotemperatures. This includes the determination of the insulation effect of wood in trees. The reason for this is that amino acid racemization rates in older wood will be affected by the heat insulation of younger wood. Thermocouples inserted at various depths within a living tree indicate that outside air temperature fluctuations are reflected inside a tree in a regular manner but, as expected, in decreasing amplitudes. Also, as was shown by more than 1000 hours of temperature recording over a six month period, the mean temperature inside the tree (13.7 cm depth) was less than 1°C warmer than the mean outside (air) temperature (Nagy et al., 1980).

The results described in this report are interpreted to suggest that determination of accurate and specific paleotemperatures within acceptable temperature error ranges appears feasible after comprehensive information on the effect of preserving environments, kinetic experimental constraints, matrix and wood insulation effects become available for bristlecone pine. It is also possible that amino acid racemization and ancillary studies may be helpful for defining the nature of the preserving paleoenvironments of bristlecone pine.

## *ACKNOWLEDGMENTS*

The authors wish to thank Drs. B. Bannister, C. W. Ferguson, and V. C. LaMarche, Jr., of the Laboratory of Tree-Ring Research. The University of Arizona, for critically reading this manuscript and offering comments. Dr. C. W. Ferguson (supported by NSF Grant No. EAR 78-04436) provided the dated TRL bristlecone pine samples, and Dr. V. C. LaMarche, Jr. provided the dated CAM bristlecone pine samples. Mr. V. A. S. McCord offered helpful advice, and Dr. D. S. Sklarew determined the total carbohydrate contents. The research described in this report was supported by the Climate Dynamics Program, Climate Dynamics Research Section, National Science Foundation, Grant No. ATM 76-23118.

## *REFERENCES*

Bailey, D. K, Phytogeography and taxonomy of *Pinus* subsection *Balfourianae, Ann. Missouri Bot. Gard.*, **57,** 210–249, 1970.

Dubois, M., F. A. Gilles, J. K. Hamilton, P. A. Rebers, and F. Smith, Colorimetric method for determination of sugars and related substances, *Anal. Chem.*, **28,** 350–356, 1956.

Engel, M. H., J. E. Zumberge, and B. Nagy, Kinetics of amino acid racemization in *Sequoiadendron giganteum* heartwood, *Anal. Biochem.*, **82,** 415–422, 1977.

Ferguson, C. W., Bristlecone pine: Sciences and esthetics, *Science,* **159,** 839–846, 1968.

Ferguson, C. W., Dendrochronology of bristlecone pine, *Pinus aristata*: Establishment of a 7484-year chronology in the White Mountains of eastern central California, U.S.A., in *Radiocarbon Variations and Absolute Chronology,* Nobel Symposium 12, I. U. Olsson, ed., Almquist and Wiksell, Stockholm, pp. 237–259, 1970.

Fritts, H. C., Growth-rings of trees: Their correlation with climate, *Science,* **154,** 973–979, 1966.

Grozdits, G. A. and G. Ifju, Nitrogen distribution in eastern hemlock and its relation to wood formation, *Wood Science,* **6,** 1–8, 1973.

Hedges, J., The formation and clay mineral reactions of melanoidin, *Carnegie Inst. Wash. Year Book 75,* 792–800, 1976.

Kriausakul, N. and R. M. Mitterer, Isoleucine epimerization in peptides and proteins: Kinetic factors and application to fossil proteins, *Science,* **201,** 1011–1014, 1978.

LaMarche, V. C., Jr., Environment in relation to age of bristlecone pines, *Ecology,* **50,** 53–59, 1969.

LaMarche, V. C., Jr., Holocene climatic variations inferred from treeline fluctuations in the White Mountains, California, *Quat. Res.,* **3,** 632–660, 1973.

LaMarche, V. C., Jr., Paleoclimatic inferences from long tree-ring records, *Science,* **183,** 1043–1048, 1974.

LaMarche, V. C., Jr. and T. P. Harlan, Accuracy of tree-ring dating of bristlecone pine for calibration of the radiocarbon time scale, *J. Biophys. Res.,* **78,** 8849–8858, 1973.

Lee, C. L., Biological and geochemical implications of amino acids in seawater, wood and charcoal, Ph.D. dissertation, Scripps Inst. Oceanogr., Univ. of Calif., 178 pp., 1975.

Nagy, B., J. E. Zumberge, and M. H. Engel, Thermal regime within a living palo verde tree (*Cercidium floridum*) in response to air temperature fluctuations, in preparation, 1980.

Wright, R. D. and H. A. Mooney, Substrate-oriented distribution of bristlecone pine in the White Mountains of California, *The American Midland Naturalist,* **73,** 257–284, 1965.

Zavarin, E. and K. Snajberk, Variability of the wood monoterpenoids from *Pinus aristata, Biochemical Systematics,* **1,** 39–44, 1973.

Zumberge, J. E., Effect of glucose on aspartic acid racemization, *Geochim. Cosmochim. Acta,* **43,** 1443–1448, 1979.

# Necessity of Reporting Amino Acid Compositions of Fossil Bones Where Racemization Analyses Are Used for Geochronological Applications; Inhomogeneities of D/L Amino Acids in Fossil Bones

**HENK J. KESSELS**
*Laboratory of Organic Geochemistry, Department of Exobiology, University of Nijmegen, Holland*

**GRAHAM DUNGWORTH**
*Geochemistry Branch, Exploration and Production Division, BP Research Center, Sunbury-on-Thames, Middlesex*

## INTRODUCTION

Two years ago a collaborative program between the organic geochemical laboratories at Scripps Institution of Oceanography and the University of Nijmegen was begun. Subsequently, Dr. Keith Kvenvolden of the United States Geological Survey, Menlo Park, joined this program, which formed the basis of an inter-laboratory comparison of racemization analyses of fossil bones. The aim of this program was to compare D/L ratios of aspartic acid, glutamic acid, and alanine from a comprehensive suite of fossil bones (>20 separate specimens) which were to be cleaned, hydrolyzed, and their total amino acid content isolated by a standard procedure developed at Scripps. Recent reviews of the analytical procedures used to measure D/L enantiomers (Schroeder and Bada 1976; Kvenvolden 1975; Dungworth 1976; Williams and Smith 1977) have described three separation methods in current use (not including liquid column separation of isoleucine diastereoisomers). The three methods are: the ion-exchange chromatography of synthesized diastereomeric dipeptides by the method of Manning and Moore (1968), used at Scripps; the resolution of N-TFA(−)-2 butyl diastereoisomers on gas chromatographic capillary columns (Charles et al., 1963), at USGS, Menlo Park; and the resolution of N-TFA-methyl ester enantiomers on optically active stationary phases (Gil-Av et al., 1966), at Nijmegen. This collaborative program represented a first attempt to obtain unbiased concordance or discordance between D/L ratios determined in separate laboratories. The results were collated and shown to reveal very good agreement for D/L aspartic acid, reasonable agreement for D/L glutamic acid, and good agreement for D/L alanine, thereby vindicating the accuracy of these three analytical tools.

During the study of the suite of fossil bones at Nijmegen we expanded our techniques to provide D/L ratios for eight of the amino acid constituents of fossil bones. Also, results of work we had previously published stressed the necessity of obtaining amino acid compositions (mole %) of fossil bones which were to be used for "age-dating." The reason for this cautionary note was that the application of amino acid racemization was in a preliminary stage, and the kinetics of racemization of amino acids in contemporary collagens were being applied in a highly simplistic manner to reactions occurring over geologically extended periods of time, in fossil collagens. With the exception of earlier work by Wyckoff's group (Wyckoff 1972), actual compositions of these "fossil collagens" were rarely published. More often than not, radical differences of amino acid composition in fossil bones are recognized when compared with data obtained from fresh bone analyses. Furthermore, the amino acid content of fossil bones is often only 1% or less of fresh bone analyses, and it has been assumed tacitly that (a) the Arrhenius parameters determined for the kinetics of racemization of amino acids in fresh bone are valid for the kinetics operating within the depleted amino acid residues in fossil bones or (b) rate constants are independent of protein composition and mineral matrix effects.

For these reasons we decided to determine the bound amino acid distribution and content of the suite of fossil bones at our disposal. We adopted another analytical technique, gas chromatography of (+)-2-butyl diastereoisomers, to enhance the former D/L measurements and to obtain D/L ratios of proline, which is incapable of resolution on our optically active dipeptide phase. Finally, since we had published data on the inhomogeneities of D/L ratios in deer antler of Pleistocene age and had revealed the presence of two discrete amino acid compositions posterior and anterior to the boney burr (Dungworth et al., 1975), we considered it essential to establish whether inhomogeneities existed in the larger bone specimens. The results of this study are presented here, since several of these samples have been used for "age-dating" purposes.

## DISCUSSION OF RESULTS

In Table 1 the results of amino acid distributions (mole %) are presented for fossil bones from various localities. In Table 1 the amino acid distributions are essentially indistinguishable from analyses of fresh collagen, with only one exception. Fossil mammoth collagen from Santa Rosa Island displays a decreased abundance of glycine with more pronounced depletions of hydroxyproline and serine. Acidic amino acids have doubled. The overall trends are shown in Table 1, which are similar to previous reports by Wyckoff's group. Of greater interest is the loss of protein from these fossils, Santa Rosa Mammoth retaining about 0.3% of the protein present in fresh bone.

The amino acid distributions in Table 2 have been collated because they are atypical distributions, seldom recognizable as former collagenous distributions. Acidic amino acids often predominate. Vertessozollos and Scripps Horse scapula contain more than 40 mole % of aspartic and glutamic acids and are similar in composition to those reported by Hare for extensively leached modern bones. All of these samples contain drastically reduced amino acid contents, Vertessozollos, Scripps Horse Scapula and Lothagam retaining approximately 0.1%, 0.05%, and 0.01% of the content expected for fresh bone. Palaeomagnetic dates of a bone with a slightly lower D/L aspartic acid ratio, from Czechoslovakia, suggest that the Vertessozollos bone is of great antiquity, *ca*. 500,000–700,000 years BP (J. Bada, personal communication). However, such samples with atypical distributions and highly depleted amino acid contents should be regarded as unsuitable for geochronological applications for the following reasons.

First, rate constants are undoubtedly related to protein composition. Second, it has not been established that the Arrhenius parameters are invariant during the period of burial. This may be answered by studies of the kinetics of racemization of younger and reliably dated fossil bones which display correspondingly greater depletions of their original amino acid content. This is a long-term study, and it may temporarily suffice to date only specimens that retain amino acid distributions which are collagenous in composition.

As part of the collaborative program we measured the D/L enantiomeric ratios of aspartic acid, glutamic acid, and alanine for this suite of bones and the D/L ratios of five other amino acids (see Table 3). For each sample locality the top set of figures refers to gas chromatographic analyses of NTFA amino acid methyl esters on capillary columns coated with the optically active stationary phase NTFA-L-phenylalanyl-L-leucyl cyclohexyl ester. The bottom set of figures refers to gas chromatographic analyses of NTFA amino acid-(+)-2-butyl esters on capillary columns coated

**Table 1. Amino Acid Compositions (mole %) of Fossil Bones**

| | | Buhen | SDM 16709 No. 1 | SDM 16709 No. 2 | Santa Rosa | CHA BM13 | CHA BM13 | CHA BM2 | Marin (bound) | Marin (free) | Palegawra 11224 |
|---|---|---|---|---|---|---|---|---|---|---|---|
| Basic | Histidine | 0 | 0 | 0 | 0 | 0 | 0.4 | 0 | 0.5 | 0.4 | 0.1 |
| | Lysine | 2.8 | 3.3 | 2.6 | 3.4 | 2.9 | 2.5 | 2.6 | 2.8 | 2.5 | 2.3 |
| | Hydroxylysine | 0.4 | 0.3 | 0 | 0 | 0.3 | 0 | 0 | 0.3 | 0.2 | 0.4 |
| | Ornithine | 0.2 | 0.6 | 0.6 | 0.6 | 0.4 | 0.4 | 0.4 | 0.2 | 0.2 | 0.7 |
| Acidic | Aspartic acid | 6.0 | 5.9 | 6.4 | 12.4 | 7.2 | 8.8 | 15.0 | 5.4 | 7.2 | 6.0 |
| | Glutamic acid | 9.1 | 8.0 | 8.7 | 14.2 | 8.7 | 9.6 | 9.2 | 7.9 | 9.4 | 8.5 |
| Neutral | Hydroxyproline | 9.3 | 9.7 | 9.9 | 4.9 | 9.3 | 8.8 | 9.0 | 9.5 | 9.4 | 10.2 |
| | Threonine | 1.6 | 1.7 | 1.6 | 1.9 | 2.1 | 2.1 | 1.9 | 1.9 | 1.9 | 2.0 |
| | Serine | 2.9 | 2.7 | 2.5 | 1.6 | 3.1 | 2.9 | 2.8 | 3.2 | 3.3 | 2.6 |
| | Proline | 12.6 | 12.7 | 12.4 | 9.6 | 12.4 | 11.6 | 11.1 | 13.2 | 12.6 | 12.9 |
| | Glycine | 34.8 | 32.6 | 32.2 | 26.7 | 31.9 | 31.7 | 27.5 | 33.8 | 32.2 | 33.0 |
| | Alanine | 11.9 | 13.4 | 13.5 | 11.7 | 12.5 | 12.6 | 11.8 | 12.3 | 12.1 | 12.9 |
| | Valine | 3.2 | 3.1 | 3.8 | 5.1 | 3.5 | 3.9 | 3.8 | 3.1 | 2.7 | 3.0 |

| | | | | | | | | | | | |
|---|---|---|---|---|---|---|---|---|---|---|---|
| | Leucine | 2.7 | 2.6 | 2.7 | 4.0 | 2.9 | 2.5 | 2.4 | 3.0 | 2.9 | 2.4 |
| | Allo-iso-leucine | 0 | 0 | 0 | 0 | 0 | 0 | 0 | 0 | 0 | 0 |
| | Isoleucine | 1.2 | 1.2 | 1.4 | 1.8 | 1.3 | 1.4 | 1.2 | 1.2 | 1.0 | 1.3 |
| | γ-aminobutyric acid | 0.3 | 0.9 | 0.2 | 0.4 | 0.1 | 0.2 | 0.2 | 0.1 | 0.2 | 0.1 |
| Aromatic | Phenylalanine | 1.0 | 1.1 | 1.0 | 1.4 | 0.9 | 0.8 | 0.8 | 1.4 | 1.3 | 0.9 |
| | Tyrosine | 0.2 | 0 | 0 | 0 | 0 | 0 | 0 | 0.1 | 0 | 0 |
| Sulphur | Methionine | 0 | 0.4 | 0.6 | 0 | 0.6 | 0 | 0.5 | 0.2 | 0.6 | 0.7 |
| Total amino acids | (μmol/g. dry bone) | 950 | 180 | 87 | 5 | 65 | 51 | 12 | 1000 | 130 | 75 |
| Amino sugars (mol % of total amino acids) | Glucosamine | | | | | 0.1 | 0.3 | 0.8 | | 0.2 | |
| | Galactosamine | | | | | 1.4 | 1.9 | 2.6 | | 1.0 | |

Table 2. Amino Acid and Amino Sugar (mole %) Compositions of Fossil Bones

| | | Fresh bone | Tarkhan 15560 | Ventes-sozollos | Double Adobe | Florida spring | CHA BM7 | Riverside | Scripps femur. | Scripps scap. | Lothagam |
|---|---|---|---|---|---|---|---|---|---|---|---|
| Basic | Histidine | 0.6 | 0.6 | | 0.9 | 0.8 | 0 | 0.7 | 0 | 0 | 0 |
| | Lysine | 3.0 | 4.1 | 3.7 | 2.7 | 3.7 | 2.8 | 2.6 | 2.5 | 4.0 | 2 |
| | Hydroxy-lysine | 0.6 | 0.3 | 0.6 | 0.2 | 0.2 | 0.2 | 0 | 0 | 0 | 0 |
| | Ornithine | 0 | 1.2 | 1.8 | 0.9 | 1.1 | 1.0 | 0.9 | 1.5 | 1.3 | 0 |
| Acidic | Aspartic acid | 5.6 | 3.1 | 23.2 | 21.9 | 11.2 | 20.7 | 16.7 | 22.5 | 21.8 | 13 |
| | Glutamic acid | 8.4 | 7.3 | 18.0 | 17.5 | 12.9 | 15.3 | 18.0 | 17.0 | 23.7 | 8 |
| Neutral | Hydroxy-proline | 9.2 | 4.6 | | 0.7 | 4.1 | 2.6 | 1.0 | 0 | 0 | 0 |
| | Threonine | 2.1 | 3.1 | 2.1 | 2.1 | 2.1 | 2.7 | 2.3 | 1.6 | 0 | 0.4 |
| | Serine | 3.6 | 2.5 | 2.3 | 2.1 | 1.9 | 2.5 | 1.8 | 1.7 | | 0.8 |
| | Proline | 12.5 | 14.4 | 3.0 | 4.3 | 9.4 | 6.4 | 4.8 | 4.6 | | 18 |
| | Glycine | 31.9 | 15.6 | 12.1 | 14.9 | 21.8 | 19.0 | 18.3 | 12.7 | 15.0 | 26 |
| | Alanine | 12.1 | 16.7 | 5.4 | 9.5 | 9.8 | 10.8 | 13.6 | 8.7 | 8.8 | 11 |
| | Valine | 3.0 | 8.3 | 10.4 | 7.0 | 6.0 | 7.3 | 7.1 | 8.6 | 8.5 | 9 |

| | | | | | | | | | | | |
|---|---|---|---|---|---|---|---|---|---|---|---|
| | Leucine | 3.2 | 9.9 | 8.9 | 7.9 | 6.2 | 3.5 | 5.6 | 10.1 | 8.8 | 6 |
| | Allo-iso-leucine | 0 | 0.2 | 1.0 | 0.4 | 0.2 | 0.1 | | 0 | 0.2 | 0.6 |
| | Isoleucine | 1.3 | 3.7 | 4.7 | 3.7 | 3.2 | 2.7 | 3.3 | 6.2 | 5.5 | 3 |
| | γ-amino butyric acid | 0 | 1.2 | 1.0 | 1.0 | 0.7 | 0.7 | 1.3 | 0.5 | 0.7 | 3 |
| Aromatic | Phenyl-alanine | 1.7 | 3.2 | 1.7 | 2.4 | 2.3 | 1.2 | 1.1 | 1.9 | 1.7 | 0 |
| | Tyrosine | 0.6 | 0 | 0 | 0.1 | 1.5 | 0.3 | 0.8 | 0 | 0 | 0 |
| Sulphur | Methionine | 0.5 | 0 | 0 | 0 | 1.1 | 0.4 | 0 | 0 | 0 | 0 |
| Total amino acids | (μmol/g. dry bone) | 1700 | 10 | 2 | 3 | 10 | 3 | 4 | 1.5 | 0.8 | 0.2 |
| Amino sugars (mol % of total amino acids) | Gluco-samine | 0 | 5.2 | 4.3 | | 1.8 | 2.9 | 1.0 | 4.4 | 1.5 | 0 |
| | Galacto-samine | 0 | 5.8 | | 3.2 | 3.0 | 8.2 | 2.4 | 7.3 | 2.6 | 7 |

**Table 3. Racemization Analyses of Fossil Bones**

| Sample Locality | D/L Ratio | | | | | | | |
|---|---|---|---|---|---|---|---|---|
| | Ala | Val | Ile | Leu | Pro | Asp | Phe | Glu |
| Tarkhan, Egypt F 15560[b] | 0.233 | 0.035 | 0.051 | 0.044 | —[a] | 0.436 | 0.087 | 0.227 |
| | 0.237 | 0.039 | 0.041 | 0.037 | 0.038 | 0.395 | 0.087 | 0.122 |
| Vertessozollos, Hungary | 0.419 | 0.120 | 0.138 | 0.497 | — | 0.557 | — | 0.23 |
| | 0.426 | 0.117 | 0.106 | 0.425 | 0.186 | 0.508 | — | 0.245 |
| Double Adobe, Arizona | 0.234 | 0.053 | 0.067 | — | — | 0.459 | — | 0.161 |
| Warm Spring, Florida | 0.115 | 0.052 | 0.081 | 0.064 | — | 0.527 | 0.189 | 0.160 |
| | 0.114 | 0.058 | 0.071 | 0.064 | 0.051 | 0.493 | 0.151 | 0.160 |
| Gua Cha BM 2 | | | | | | | | |
| | 0.046 | 0.020 | 0.024 | 0.036 | 0.041 | 0.364 | 0.050 | 0.091 |
| Gua Cha BM 7 | | | | | | | | |
| | 0.151 | 0.057 | 0.061 | — | 0.080 | 0.395 | 0.135 | 0.157 |
| Gua Cha BM 13 | 0.039 | 0.013 | 0.014 | — | — | 0.221 | 0.040 | 0.068 |
| | 0.033 | 0.008 | 0.020 | 0.023 | 0.025 | 0.184 | 0.040 | 0.060 |
| Riverside Palaeo-Indian | 0.270 | 0.054 | 0.059 | 0.070 | — | 0.448 | — | 0.33 |
| | 0.269 | 0.067 | — | 0.071 | 0.019 | 0.380 | 0.165 | 0.32 |
| Scripps Horse, femur | 0.39 | 0.04 | 0.04 | — | — | 0.45 | — | 0.17 |
| | 0.37 | 0.05 | 0.04 | — | 0.14 | 0.45 | — | 0.21 |
| Scripps Horse, scapula | 0.34 | 0.06 | 0.035 | — | | 0.58 | — | 0.25 |
| | 0.34 | 0.07 | — | — | 0.165 | 0.53 | — | 0.19 |
| Lothagam, E. Africa | 0.474 | 0.190 | 0.182 | — | 0.373 | 0.575 | — | 0.307 |

| | | | | | | | | |
|---|---|---|---|---|---|---|---|---|
| Buhen Horse, Sudan | 0.064 | 0.028 | 0.033 | — | — | 0.457 | 0.05 | 0.098 |
| | 0.058 | 0.034 | 0.045 | 0.04 | 0.05 | 0.44 | 0.06 | 0.11 |
| SDM 16709 No. 1 | 0.030 | 0.014 | 0.014 | — | — | 0.203 | 0.05 | 0.06 |
| | 0.020 | 0.015 | 0.015 | 0.016 | 0.028 | 0.185 | 0.036 | 0.065 |
| No. 2 | | | | | | | | |
| | 0.043 | 0.018 | 0.015 | 0.028 | 0.042 | 0.235 | — | 0.080 |
| Santa Rose Isl. Mammoth | 0.11 | 0.028 | 0.031 | — | — | 0.37 | 0.13 | 0.10 |
| | 0.11 | 0.030 | | 0.035 | 0.05 | 0.35 | | 0.12 |
| Marin, Calif. 152[c] | 0.011 | 0.005 | — | — | — | 0.091 | — | — |
| Palegawra, Iran 11224 | 0.039 | 0.020 | — | 0.039 | — | 0.233 | — | 0.085 |
| | 0.036 | 0.023 | 0.029 | 0.045 | 0.044 | 0.201 | 0.061 | 0.075 |
| Asych, U.S.S.R. | 0.474 | 0.11 | — | 0.232 | — | 0.621 | — | 0.227 |
| | 0.49 | 0.12 | 0.18 | — | — | 0.69 | — | 0.36 |
| Tarkhan, Egypt F 1691[b] | 0.118 | 0.030 | 0.040 | 0.045 | | 0.477 | 0.150 | 0.131 |
| | 0.119 | 0.027 | 0.038 | 0.035 | 0.037 | 0.438 | 0.059 | 0.135 |
| Kûlna Cave, Czech. 11[c] | 0.010 | 0.006 | — | — | — | 0.10 | 0.052 | 0.031 |
| Palegawra, Iran 12121 | 0.260 | 0.053 | 0.073 | 0.077 | — | 0.393 | 0.125 | 0.204 |
| | 0.25 | 0.056 | 0.057 | 0.087 | 0.079 | 0.35 | 0.084 | 0.22 |
| Tura, Egypt[b] | 0.114 | 0.025 | 0.031 | 0.042 | — | 0.433 | 0.126 | 0.165 |
| | 0.113 | 0.027 | 0.037 | 0.045 | 0.062 | 0.384 | 0.072 | 0.172 |
| Marin, Calif. 152 (bound)[c] | 0.008 | 0.005 | — | 0.009 | — | 0.064 | — | 0.004 |
| (free)[c] | 0.026 | 0.005 | — | 0.016 | — | 0.219 | — | — |

[a] Not determined due to the presence of interfering peaks. For each sample locality the top set of figures refer to analyses of NTFA methyl esters and the bottom set of figures to analyses of NTFA (+) 2 butyl esters.
[b] Hydrolysis in 6*M* HCL for 6 hours.
[c] Hydrolysis in 6*M* HCL for 4 hours.

**Table 4. Interlaboratory Comparison of Racemization Analyses of Fossil Bones**

| Sample Locality and Identification Number | D/L Aspartic Acid | | | D/L glutamic Acid | | | D/L Alanine | | |
|---|---|---|---|---|---|---|---|---|---|
| | Scripps | Nijmegen | USGS | Scripps | Nijmegen | USGS | Scripps | Nijmegen | USGS |
| *EXTRACTS* | | | | | | | | | |
| 1. Tarkhan, Egypt, F1691 | 0.475 | 0.477 | 0.48 | 0.18 | 0.131 | 0.16 | 0.22 | 0.118 | 0.12 |
| 2. Tura, Egypt, Middle Kingdom | 0.416 | 0.433 | 0.44 | 0.21 | 0.165 | 0.16 | 0.15 | 0.114 | 0.1 |
| 3. Palegawra, Iran, 12121 | 0.470 | 0.393 | — | — | 0.204 | — | — | 0.260 | — |
| 4. Külna Cave, Czech, 11 | 0.092 | 0.099 | — | — | 0.031 | — | — | 0.0096 | — |
| 5. Warm Mineral Springs, Florida, human femur, 85019 | 0.55 | 0.527 | 0.54 | 0.19 | 0.160 | 0.20 | — | 0.115 | 0.1 |
| 6. Buhen Horse, Sudan | 0.474 | 0.457 | 0.48 | 0.16 | 0.098 | 0.16 | 0.14 | 0.064 | 0.0 |
| 7. Gua Cha, Malaysia, BM13 | 0.244 | 0.264 | — | 0.10 | 0.113 | — | 0.083 | 0.077 | — |

| | | | | | | | | | |
|---|---|---|---|---|---|---|---|---|---|
| *BONES* | | | | | | | | | |
| 1. Tarkhan, Egypt, F1556 | 0.453 | 0.436 | 0.44 | 0.31 | 0.227 | 0.14 | 0.18 | 0.223 | 0.2 |
| 2. Double Adobe, Arizona | 0.500 | 0.459 | — | 0.21 | 0.161 | — | — | 0.234 | — |
| 3. Marin, Calif., 152 | 0.080 | 0.091 | 0.06 | — | — | 0.03 | — | 0.011 | 0.0 |
| 4. Asych, USSR | 0.73 | 0.621 | 0.67 | ~0.3 | 0.227 | 0.33 | ~0.5 | 0.474 | 0.4 |
| 5. Palegawra, Iran, 11224 | 0.244 | 0.233 | — | — | 0.085 | — | — | 0.039 | — |
| 6. Vertessozollos, Hungary | 0.504 | 0.557 | 0.48 | ~0.2 | 0.230 | 0.24 | ~0.5 | 0.419 | 0.4 |
| 7. Riverside, California Paleo-indian skeleton | 0.437 | 0.41 | — | — | 0.33 | — | — | 0.27 | — |
| 8. Scripps Horse (a) Leg | 0.503 | 0.45 | — | — | 0.17 | — | — | 0.39 | — |
| (b) Scapula | 0.556 | 0.58 | — | 0.23 | 0.25 | — | 0.41 | 0.34 | — |
| 9. Gua Cha, Malaysia, BM13 | 0.244 | 0.221 | — | 0.10 | 0.068 | — | 0.083 | 0.039 | — |
| 10. Santa Rose Island, Calif. Mammoth | 0.28 | 0.372 | — | — | 0.10 | — | — | 0.11 | — |
| 11. La Jolla, Calif., SDM16709 | 0.189 | 0.208 | 0.12 | — | 0.060 | 0.04 | — | 0.030 | 0.05 |
| 12. Yuha Burial, California | 0.56 | — | 0.52 | 0.33 | — | 0.33 | 0.47 | — | 0.28 |

with FFAP. In addition, samples containing amino acid distributions atypical of collagen are underlined. Generally, good agreement is evident for most D/L values. There are several exceptions resulting from co-elution of nonprotein amino acids with protein amino-acid enantiomers. A notable case is D and L leucine where L-serine often co-elutes with D-leucine on the optically active dipeptide phase.

Table 4 reveals the concordance of D/L aspartic acid, glutamic acid, and alanine measurements at Scripps, Nijmegen, and the U.S. Geological Survey. These interlaboratory comparative measurements are discussed in another presentation.

Bacterial activity in sediments, soils, and fossils have been reported; and their pertinence to contamination of fossil bones and molluscs has been noted by several authors (Wyckoff 1972; Gil-Av 1976; Dungworth et al., 1975; Schroeder and Bada 1976). Anomalously high D/L values for alanine and glutamic acid have been related to bacterial activity in soils (Gil-Av 1975; Pollock et al., 1977) and lake sediments (Dungworth et al., 1977). These D-amino acids are associated with the murein structure of bacteria.

In this study it is apparent that all samples with amino acid distributions atypical of collagen compositions yield anomalously high D/L ratios for alanine and glutamic acid, with the exception of the Santa Rosa Mammoth. Furthermore, bones from Palagawra, Iran (12121), and Asych, U.S.S.R., are also suspect. The second Palagawra sample does not display this anomaly. Of course, if bacterial activity is contemporaneous with burial, this may not affect the results when applied to geochronological determinations. At present there does not appear to be evidence to support a bacterial contribution of D-aspartic acid in soils and sediments. For the remaining samples the order of decreasing rates of racemization is similar to the order determined in laboratory kinetic experiments and to the natural order found in many fossil bones, although the interpretative value of this order has been critically questioned by Williams and Smith (1977).

The data in Table 3 for Palagawra 11224 and 12121 have been used for geochronological calculations. An age was derived for Palagawra 11224 (5574 yr BP) by use of Bada et al. (1974) radiocarbon data for Palagawra UCLA 1714 D sample (collagen radiocarbon age 13600 ± 460 yr BP; *in situ* rate constant $k$ asp = 2.34 $10^{-5}$ $yr^{-1}$). An age for Palagawra 12121 was derived using this *in situ* rate constant for aspartic acid racemization. The $t$ = 0 constants in the rate expression were 0.18 for aspartic acid, 0.05 for alanine, 0.02 for valine, 0.03 for isoleucine, 0.04 for leucine, 0.05 for proline, 0.06 for phenylalanine, and 0.10 for glutamic acid. The derived ages for the five other amino acids are shown in Table 5. The exceptionally

**Table 5. Amino Acid Racemization Ages**[a]

| D/L ratio of amino acid in | | Rate constant $yr^{-1}$ | Derived amino acid age yr |
|---|---|---|---|
| Palagawra 11224 | | | |
| Aspartic acid | 0.217 | $2.34 \times 10^{-5}$ | 5574 |
| Palagawra 11224 and 12121 | | | |
| Alanine | 0.0375 | $2.25 \times 10^{-6}$ | 105,000 ± 17,000 |
| | 0.255 | | |
| Valine | 0.0215 | $2.06 \times 10^{-6}$ | 21,600 ± 4,000 |
| | 0.0545 | | |
| Isoleucine | 0.029 | $2.51 \times 10^{-6}$ | |
| | 0.065 | | 19,900 ± 3,200 |
| Leucine | 0.0425 | $4.04 \times 10^{-6}$ | |
| | 0.083 | | 15,600 ± 4,000 |
| Proline | 0.044 | $3.41 \times 10^{-6}$ | |
| | 0.079 | | 15,900 |
| Aspartic acid | | $2.34 \times 10^{-5}$ | |
| | 0.372 | | 12,850 ± 1,080 |
| Phenylalanine | 0.061 | $5.57 \times 10^{-6}$ | |
| | 0.105 | | 9,700 to 17,200 |
| Glutamic acid | 0.080 | $5.41 \times 10^{-6}$ | |
| | 0.212 | | 30,500 ± 7,000 |

[a] For each amino acid the top-set and bottom-set of D/L ratios refer to samples 11224 and 12121, respectively.

high ages for alanine and glutamic acid are expected, since they are believed to be due to bacterial activity. The remaining amino acid dates are quite consistent, but they may be fortuitous owing to the large error inherent in determining $t = 0$ values. An advantage may be gained by calculating individual amino acid dates rather than relying solely upon aspartic acid or isoleucine, although with samples as young as Palagawra 11224, accurate measurement of low D/L values limits the use of other amino acids.

During preparation of these bone samples we analyzed for differences of D/L ratios between surface scrapings and the interstices of bone. The magnitude of these differences is evident in Table 6 for four of the samples. Interestingly, contamination by the naturally occurring L-amino acids would be expected to produce lower D/L ratios on the surface. Although this is generally true for the results presented here, the rigorous cleaning

**Table 6. Inhomogeneity of D/L Amino Acid Enantiomers in Fossil Bones**

| | D/L Ratio[a] | | | | | | | |
|---|---|---|---|---|---|---|---|---|
| Sample Locality | Ala | Val | Ile | Leu | Pro | Asp | Phe | Glu |
| Gua Cha, BM2 | | | | | | | | |
| bone interstices | 0.046 | 0.020 | 0.024 | — | 0.041 | 0.364 | 0.050 | 0.091 |
| surface scrapings | 0.035 | 0.016 | 0.016 | — | 0.033 | 0.226 | 0.038 | 0.071 |
| Gua Cha, BM 13 | | | | | | | | |
| bone interstices | 0.036 | 0.010 | 0.017 | 0.023 | 0.025 | 0.202 | 0.040 | 0.064 |
| surface scrapings | 0.043 | 0.016 | 0.021 | | 0.039 | 0.226 | 0.050 | 0.075 |
| Scripps Horse scapula | | | | | | | | |
| bone interstices | 0.340 | 0.065 | 0.035 | — | 0.165 | 0.55 | — | 0.22 |
| surface scrapings | 0.270 | 0.064 | 0.075 | — | — | 0.47 | — | 0.18 |
| Scripps Horse femur | | | | | | | | |
| bone interstices | 0.38 | 0.045 | 0.040 | — | 0.14 | 0.45 | — | 0.19 |
| surface scrapings | 0.27 | 0.070 | — | — | — | 0.43 | — | — |
| Gua Cha BM 7 | | | | | | | | |
| bone interstices | 0.151 | 0.057 | 0.061 | — | 0.080 | 0.395 | 0.135 | 0.157 |
| pigmented zone | 0.232 | 0.069 | 0.059 | — | 0.115 | 0.488 | 0.184 | 0.187 |
| sand infill | 0.128 | 0.040 | 0.039 | — | 0.063 | 0.372 | 0.101 | 0.200 |

[a] Analyzed by capillary gas chromatography of NTFA amino acid (+)2-butyl esters; FFAP stationary phase.

procedure removes free amino acids. These D/L ratios represent values determined for hydrolysis of the "bound" fraction.

The last sample in Table 6 was an exception. This specimen, Gua Cha BM7, contained a sand infill. At the interface between sand and bone, a zone extended into the bone matrix by about 4 to 5 mm and was colored purple by unidentified pigmentation. Material was removed from these three zones and analyzed for D/L ratios individually. Except for the D-alloisoleucine/L-isoleucine ratio, consistently higher D/L amino acid ratios were found in the pigmented zone than in the unpigmented bone. But, with the exception of D/L glutamic acid, consistently low D/L amino acid ratios were found in the sand infill compared to nonpigmented bone. It is difficult to reconcile these results if one accepts that they are not the result of the analytical errors involved in measurement. Transition elements in the pigmented zone may have enhanced racemization, and the apparent differences between the sand infill and bone may be due to experimental error. Alternatively, these differences may be real and may represent catalysis by metal cations or by the greater fluctuations of pH expected in the sand infill. Such data are still tentative, and where sample size permits, bones should be sectioned off and tested for the homogeneity of D/L amino acids at discrete intervals prior to geochronological or palaeotemperature determinations.

To summarize, our results have emphasized: (1) The need to determine amino acid distribution and content of all samples used for geochronologic determinations and to select only those with collagen distributions. (2) The need to determine Arrhenius parameters for amino acid racemization in reliably dated fossil bones in order to ascertain whether or not *in situ* derived rate constants are time invariant. (3) The need to undertake routine checks of the homogeneity of D/L ratios throughout fossil bones.

## *REFERENCES*

Bada, J. L., R. A. Schroeder, R. Protsch, and R. Berger, Concordance of collagen-based radiocarbon and aspartic acid racemization ages, *Proc. Nat. Acad. Sci. U.S.A.*, **71,** 914–917, 1974.

Charles, R., G. Fischer, and E. Gil-Av, Resolution of (±) 2-n-alkanols by gas-liquid partition chromatography, *Isr. J. Chem.*, 1, 234–235, 1963.

Dungworth, G., Optical configuration and racemization of amino acids in sediments and fossils. A review, *Chem. Geol.*, **17,** 135–153, 1976.

Dungworth, G., M. Thijssen, J. Zuurveld, W. Van derVelden, and A. W. Schwartz, Distribution of amino acids, amino sugars, purines, and pyrinidines in a Lake Ontario sediment core, *Chem. Geol.*, **19,** 295–308, 1977.

Dungworth, G., J. A. Th. Vrenker, and A. W. Schwartz, Amino Acid compositions of Pleistocene collagens, *Comp. Biochem. Physiol.*, **51B,** 331–335, 1975.

Gil-Av, E., Present status of enantiomeric analysis by gas chromatography, *J. Mol. Evol.*, **6,** 131–144, 1975.

Gil-Av, E., B. Feibush, and R. Charles-Sigler, Separation of enantiomers by gas-liquid chromatography with an optically active phase, *Tetrahedron Lett.*, **10,** 1009–1015, 1966.

Kvenvolden, K. A., Advances in the geochemistry of amino acids, *Annu. Rev. Earth Planet. Sci.*, **3,** 183–212, 1975.

Manning, J. M. and S. Moore, Determination of D- and L-amino acids by ion exchange chromatography as L-D and L-L dipeptides, *J. Biol. Chem.*, **243,** 5591–5597, 1968.

Pollock, G. E., C-N. Cheng, and S. E. Cronin, Determination of the D and L isomers of some protein amino acids present in soils, *Anal. Chem.*, **49,** 2–7, 1977.

Schroeder, R. A. and J. L. Bada, A Review of the geochemical applications of the amino acid racemization reaction, *Earth Sci. Rev.*, **12,** 347–391, 1976.

Williams, K. M. and G. G. Smith, A critical evaluation of the application of amino acid racemization to geochronology and geothermometry, *Origins of Life*, **8,** 91–144, 1977.

Wyckoff, R. W. G., *The Biochemistry of Animal Fossils,* Scientechnica, Bristol, 1972.

# Author Index

# Subject Index